SpringerWienNewYork

Gernot Beer
Ian Smith
Christian Duenser

The Boundary Element Method with Programming

For engineers and scientists

SpringerWienNewYork

Univ.-Prof. DI Dr. Gernot Beer
Institute for Structural Analysis,
Graz University of Technology, Austria

Prof. Ian M. Smith
University of Manchester, United Kingdom

DI Dr. Christian Duenser
Institute for Structural Analysis,
Graz University of Technology, Austria

Printed in Germany

SpringerWienNewYork is part of Springer Science Business Media
springer.at

Typesetting: Camera ready by the authors
Printing: Strauss GmbH, 69509 Mörlenbach, Germany

Printed on acid-free and chlorine-free bleached paper

With 254 figures and 17 tables
SPIN: 11957096

Library of Congress Control Number: 2008922295

ISBN 978-3-211-71574-1 SpringerWienNewYork

Contents

Preface

This is a sequel to the book "Programming the Boundary Element Method" by G. Beer published by Wiley in 2001. The scope of this book is different however and this is reflected in the title. Whereas the previous book concentrated on explaining the implementation of a limited range of problems into computer code and the emphasis was on programming, in the current book the problems covered are extended, the emphasis is on explaining the theory and computer code is not presented for all topics. The new topics covered range from dynamics to piezo-electricity. However, the main idea, to provide an explanation of the Boundary Element Method (BEM), that is easy for engineers and scientists to follow, is retained. This is achieved by explaining some aspects of the method in an engineering rather than mathematical way.

Another new feature of the book is that it deals with the implementation of the method on parallel processing hardware. I. M. Smith, who has been involved in programming the finite element method for decades, illustrates that the BEM is "embarrassingly parallelisable". It is shown that the conversion of the BEM programs to run efficiently on parallel processing hardware is not too difficult and the results are very impressive, such as solving a 20 000 element problem during a "coffee break".

Due to the fact that, compared to the Finite Element Method, a significantly smaller group of people are working in this field the development of the method is lagging considerably behind. The often quoted comparison that the method is a "Cinderella", dominated by her "big sister", the Finite Element Method, and whose beauty is hidden away, is still true and we hope that the reader will see the beauty of the method in the examples on industrial applications and the advanced topics presented at the end.

The book includes some innovative development work carried out by the small but very active group at the Institute for Structural Analysis, Graz University of Technology, Austria under the leadership of G. Beer. The main scope of their research is to further develop the method, so that it can be applied to a much wider range of practical problems in engineering, one particular application of interest being in the field of geotechnical engineering, especially underground excavation.

COMPUTER PROGRAMS

All software (libraries and programs) can be downloaded free of charge from the website
http://www.ifb.tugraz.at/BEM

Acknowledgements

This book would not have been possible without the research effort by the small but very active group of scientists working on boundary element methods at the Institute for Structural Analysis, Graz University of Technology (Katherina Riederer, Andre Maues Brabo Pereira, Klaus Thöni, Plinio Glauber Carvalho de Prazeres, Thomas Rüberg and Jürgen Zechner). The Austrian Science Fund (FWF) and the European Commission (under its framework program for research and technical development) contributed significantly to the funding of the research effort.The complete set of fundamental solutions presented in the Appendix has been supplied by Tatiana Souza Antunes Ribeiro a former PhD student at the institute. Katherina Riederer supplied the two examples for Chapter 18 (Advanced topics) on heterogeneous domains and linear inclusions. Andre Periera made significant contributions to Chapter 14 (Dynamic Analysis).

The authors are grateful to Sylvia Beer for proofreading the manuscript and for her valuable suggestions. Thanks are also due to the companies that gave the opportunity to apply the method to the real engineering problems reported in Chapter 17: Lahmeyer International (Bernhard Stabel), Geoconsult and Schoeller Bleckmann Austria. The cooperation with Kuwait University (Abdullah Ebrahim) led to the application in reservoir engineering. Last but not least our thanks go to our families for their support.

1

Preliminaries

A journey of a thousand miles
begins with a single step
Lao-tzu, *Chinese philosopher*

1.1 INTRODUCTION

Nearly all physical phenomena occurring in nature can be described by differential equations and boundary conditions. In the solution of these **boundary value problems** we aim to determine a response to given boundary conditions. For example we may be interested in determining the response of the rock mass due to the excavation of a tunnel, or the response of a structure to dynamic excitations of its foundations (caused by an earthquake). Analytical solutions of boundary value problems, i.e. solutions that satisfy both the differential equations (DE) and the boundary conditions (BCs), can only be obtained for few problems with very simple boundary conditions. For example, analytical solutions exist for the excavation of a circular tunnel in a homogeneous rock mass, not really a realistic scenario for practical tunnelling. To be able to solve real life problems, the engineer must revert to approximate solutions. Two approaches can be taken: instead of satisfying both the DE and the BCs, one can attempt to satisfy only one of the two and minimise the error in satisfying the other one. In the first approach (based on the original idea of Ritz[1]) solutions are proposed that satisfy the boundary conditions exactly. The error in satisfying the differential equation is then minimised. This is the well known Finite Element Method. In the alternative (proposed by Trefftz[2]), the assumed functions satisfy the DE exactly and the error in the satisfaction of the boundary conditions is minimised.

Most readers of this book will be familiar with the finite element method. In the most common version of this method we subdivide the domain into elements and approximate the response to a specified loading with functions which are defined at element level, i.e., are piecewise continuous. This subdivision is necessary because in practice it is impossible to determine functions that cover the whole domain and at the same time satisfy the boundary conditions (as originally proposed by Ritz). The parameters of these functions, which are the values of the unknowns at the nodes where elements are connected to each other, are determined by minimising the error in satisfying the DE. This can be done using *residual methods,* where the integral of the error is minimised and this involves a domain integral. A violation of the DE may occur at any point in the domain, but the variation of the unknown is chosen in such a way that the error in the satisfaction of the DE over the whole domain is a minimum. In continuum mechanics, for example, this means that the chosen functions will usually not satisfy exactly the equilibrium conditions at specified points.

Figure 1.1 shows an example of a finite element mesh for the three-dimensional analysis of sequential excavation and construction of a tunnel. A plane of symmetry is applied, so that only half of the tunnel is discretised. Note that to model the rock mass through which the tunnel is driven, which for all practical purposes can be assumed to be infinite, we must make a 'box' of solid elements. At the outer boundaries of this box, unless we use infinite elements, we either set displacements to zero or apply stress boundary conditions, which represent the *in situ* stress. The mesh shown here has approximately 100 000 degrees of freedom and a solution took several hours on a PC. Note that small jumps occur in the contours of maximum compressive stress, between elements indicating a lack of satisfaction of equilibrium locally.

The second approach to solving this problem (based on the original idea of Trefftz) does not require the subdivision of the domain into elements because the functions used for approximating the solution inside the domain are chosen to be those which exactly satisfy the governing differential equations. In a similar way as with the FEM the error in satisfying the boundary conditions is minimised and this now involves a boundary integral. Numerically, this integral can be evaluated by subdividing the boundary into elements over which the values (for example, tractions or displacements in the case of continuum mechanics) are interpolated, much in the same way as with the FEM. The advantage of the method is obvious: the dimensionality of the problem is reduced by one order, i.e. only a surface instead of a volume integral is required. This means that the number of unknowns is reduced dramatically, especially for three-dimensional problems, because unknowns occur only on the problem boundary. Other advantages are that the DE is satisfied exactly everywhere in the domain and that infinite domain problems are easy to deal with.

As an example, Figure 1.2 shows the boundary element mesh for the same tunnel as analysed by the FEM in Figure 1.1. This mesh has approx. 1000 degrees of freedom and took 3 minutes to solve on a PC. The stress contours computed and drawn on the excavation surface and a user defined plane inside the rock mass show no jumps as they are seen in FEM results. Since functions must be found which exactly satisfy the governing differential equation (DE) the BEM requires a solution of the DE. This solution must be as simple as possible because, as will be seen in the chapter on implementation, this is crucial for efficiency. Unfortunately, the simplest solutions which we can find (fundamental solutions) are due to concentrated loads or sources and are singular, i.e., have infinite values at certain points. This property has to be taken into account when integrating these functions over boundary elements. This will make the numerical integration procedure more complicated than is the case with finite elements.

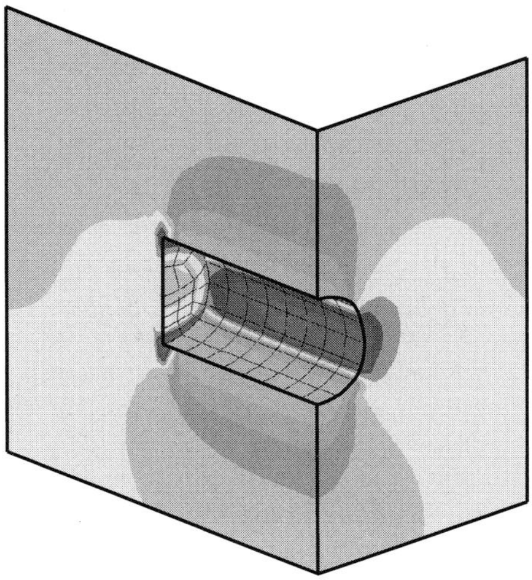

Figure 1.1 Finite element mesh for the analysis of tunnel excavation. Left side: mesh with contours of z-displacement; right side: detail with contours of maximum compressive stress

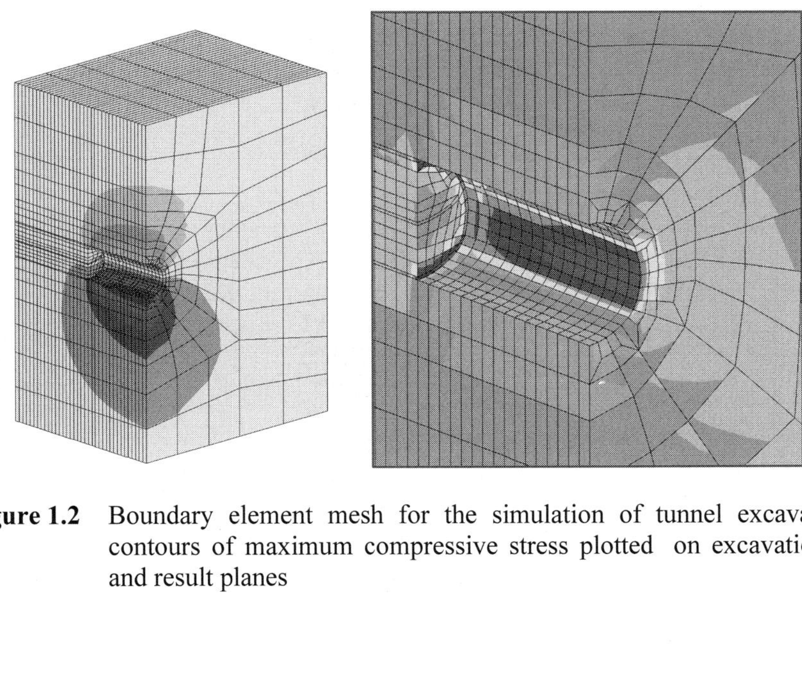

Figure 1.2 Boundary element mesh for the simulation of tunnel excavation with contours of maximum compressive stress plotted on excavation surface and result planes

There has been a general misconception that because a fundamental solution of the problem must exist for the BEM to work, the method can only be applied to linear problems with homogeneous material. As will be shown in this book, non-linear problems can almost as easily be solved as with the FEM, by the repeated solution of linear problems and special methods may be employed to solve problems with heterogeneous material properties.

1.2 OVERVIEW OF BOOK

This book is designed to be used as basis for a course on the BEM or for self study. It is recommended that chapters be read consecutively as later chapters build on material discussed earlier. Throughout the book, the reader will build a suite of subprograms, which perform the various tasks needed for the numerical implementation of the BEM. Various exercises are included which allow the reader to test the programs written and experience how the method works.

We start with an introduction to the FORTRAN 95 programming language. FORTRAN, which stands for FORMula TRANslation is still the most widely used language for programming engineering applications and is easier to learn and more efficient than other high level languages such as C++. However, there is no reason why the procedures outlined in some detail in this book could not be implemented in another language.

The next chapter deals with the way in which we can describe the geometrical boundary of a problem and boundary conditions in a numerical way. This is done by subdividing the surface into small elements and by interpolating between nodal values. This is essential for the later treatment of integral equations. With the aid of the examples we can not only test the subroutines developed but also get an understanding of the error introduced by the approximations used to describe boundaries.

Another fundamental building block is the description of the material response. In Chapter 4 we introduce basic concepts of elasticity and potential flow and develop fundamental solutions, that is, simple solutions which satisfy the governing differential equations. These will be central to our subsequent deliberations.

Next we introduce the concepts of boundary element methods using the method originally proposed by Trefftz. Although this very simple method cannot be used for general purpose programs, it serves very well to explain the fundamental ideas of the method. A small computer program can be developed to solve some simple problems. Again, this will serve as a tool for learning by experience.

The direct boundary element method used in the majority of BEM software is introduced next. Here we will use the reciprocal theorem by Betti, which is well known to engineers to obtain an integral equation. The major task in the implementation however, is to solve the integral equations numerically.

The next chapter on numerical implementation therefore deals with the evaluation of integrals using numerical integration. Those familiar with isoparametric finite elements will recognise the Guass Quadrature method used. However, in contrast to its use in the

FEM, one must be very careful to select the number of integration points, as they are dependent on how close the singularity is to the integration region. This is the most difficult and crucial part in the implementation of the BEM. The integration over the boundary surface is carried out over a boundary element and the contributions of all elements which describe a boundary are then added. We will see that this is very similar to the assembly procedure in the FEM.

After the numerical treatment of the integral equations we end up with a system of equations. In contrast to the FEM, the coefficient matrix is fully populated and unsymmetrical. Standard Gauss elimination can be used but, for large systems, the storage requirement and the computation times may be reduced considerably by iterative solvers, such as conjugate gradient methods. Such special solution techniques are introduced in the next chapter. Here we also find that the method is "embarrassingly parallelisable" i.e. that excellent speed up rates can be achieved with special hardware.

The primary results obtained from the analysis are values of displacement or traction at the boundary depending on the boundary condition specified. In contrast to the FEM, primary results do not include values in the interior of the domain but these are computed by post-processing. In Chapter 9 it is explained how the stresses at the boundary and in the interior can be obtained from boundary displacements and tractions. This is indeed an advantage of the method, because the user has free choice of the locations where results are obtained.

We now have all the building blocks together and are able to compile a computer program that is able to solve two and three-dimensional problems in elasticity and potential flow, depending on which fundamental solution is used. In Chapter 10 we apply the program developed to test examples and find out what level of accuracy can be obtained in comparison with the FEM.

For inhomogeneous domains, where we can not obtain a fundamental solution, we introduce the concept of multiple regions, where the domain is subdivided into sub-regions, similar to the FEM. There is an additional advantage in this concept, because sparseness is introduced in the system of equations. We will also find out in a later chapter that the multi-region method allows contact and excavation problems to be solved in an elegant way.

In the next chapter we deal with problems that involve corners and geometry which changes with time, as is the application to sequential excavation/construction of a tunnel.

Because elements only exist on the boundary the BEM has difficulty dealing with problems where forces are applied inside the domain. These forces can be loosely termed "body forces". It will be shown that an additional volume integral has to be considered. For body forces that are constant the volume integral can be transformed into a surface integral. However, if the body forces are not constant throughout the domain the volume integral needs to be evaluated numerically. This can be done by using internal cells, which look like finite elements, but do not involve any additional degrees of freedom, as they are only used for integration. The implementation of this procedure, discussed in chapter 13 also allows the solution of problems in elasto- and visco-plasticity. Body forces of a different kind (mass forces) occur in the case of dynamics, but their treatment with the BEM is quite different to the FEM and this is discussed in Chapter 14.

In Chapter 15 we show that the solution of non-linear problems follows similar procedures as in the FEM and that the general solution algorithm is similar. Here two types of non-linear problems are discussed in more detail: plasticity and contact problems.

It is possible to couple the BEM with the FEM thus getting the 'best of both worlds'. In Chapter 16, methods of coupling are presented. Basically, a stiffness matrix of the BE region is obtained and assembled with the FEM stiffness matrices. Since many general purpose programs allow the input of a user defined element stiffness matrix this may be used to extend the capabilities of a reader's FEM code.

To demonstrate that the method also works for large scale industrial problems, Chapter 17 shows some applications of the boundary and coupled method in engineering. The purpose of this chapter is twofold: firstly it shows how complex problems, as they invariably occur in real life, can be simplified and how a suitable boundary element mesh is obtained. Secondly it shows the advantage of the BEM and the coupled BEM/FEM in terms of user friendliness and computing time.

The last chapter deals with topics which were still subject to research at the writing of the book. The first deals with the efficient treatment of heterogeneous ground conditions the other with the consideration of linear inclusions such as reinforcement and rock bolts. The application in piezo-electricity shows the flexibility of the method to deal with any problem whose fundamental solution is known.

By the end of this book the reader should have an understanding of how the method works, of its potential and how it can be implemented into a computer program.

1.3 MATHEMATICAL PRELIMINARIES

A good consistent notation is essential to any textbook. For the development and explanation of numerical methods two notations are used by engineers: matrix and tensor notation. Traditionally, textbooks on the BEM have use tensor notation, whereas those about the FEM have used matrices, although this is rapidly changing. The main notation chosen for this book is the matrix notation.

There are two reasons for this: firstly, the book which is probably still the most widely read on numerical modelling, "The Finite Element Method", by O.C. Zienkiewicz and R.L Taylor[3], uses matrix notation throughout. Since we hope to attract more engineers to the BEM, this was one motivation. The other reason is that books on the BEM that use tensor notation have to revert to matrix notation at some stage, for example when discussing the assembly of the system of equations. Thus the book attempts to avoid two different notations.

However, when discussing fundamental solutions and their derivatives it transpires that tensor notation is much easier to use. Therefore in this book we have made a compromise in that for this case only we revert to a simplified version of the tensor notation.

In the following we discuss some basic mathematics which will be used in this book and also attempt a comparison of matrix and tensor notation.

1.3.1 Vector algebra

Vectors are used to represent a displacement/force or to define the position of a point relative to a set of Cartesian axes. We define the position of a point in 3-D space with respect to Cartesian axes x, y, z (Figure 1.3) as

$$\mathbf{x} = \begin{Bmatrix} x \\ y \\ z \end{Bmatrix} \tag{2.1}$$

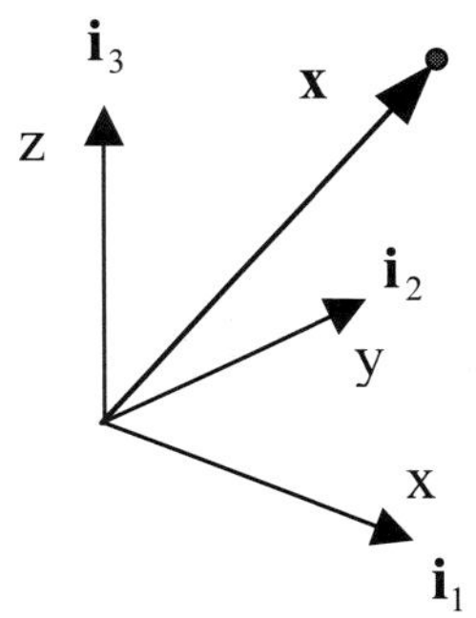

Figure 1.3 Position vector **x** defining a point in space

Alternatively, we may represent the point in terms of Cartesian coordinates x_i, where i=1,2,3 (the last number is also referred to as *range*).

The components are specified with respect to a set of orthogonal coordinate axes, which are defined by base vectors of unit length, $\mathbf{i}_i$ and which have the property:

$$\mathbf{i}_i \bullet \mathbf{i}_j = \delta_{ij} \quad \begin{Bmatrix} 1 \text{ for } i=j \\ 0 \text{ for } i \neq j \end{Bmatrix} \tag{2.2}$$

where $(\ )\bullet(\ )$ denotes the scalar (dot) product

$$\mathbf{i}_i \bullet \mathbf{i}_j = i_{1x} \cdot i_{2x} + i_{1y} \cdot i_{2y} + i_{1z} \cdot i_{2z} \tag{2.3}$$

and δ_{ij} is known as the Kronecker delta.

Vector **x** may then be represented in indicial notation as

$$\mathbf{x} = \sum_{i=1}^{3} x_i \mathbf{i}_i = x_i \mathbf{i}_i \tag{2.4}$$

where the Einstein summation convention has been used for the last term. This convention specifies that for any index which is repeated and which does not appear on the left hand side, a summation of all terms within the *range* is implied.

Another vector quantity is the displacement which can be written either as

$$\mathbf{u} = \begin{Bmatrix} u_x \\ u_y \\ u_z \end{Bmatrix} \tag{2.5}$$

in matrix notation or $\mathbf{u} = u_i \mathbf{i}_i$ in indicial notation

Coordinate transformation

If we want to express the location of a point, $\mathbf{x}$ in another orthogonal coordinate system ($\overline{\mathbf{x}}$) the directions of which are given by unit vectors $\mathbf{v}_1$, $\mathbf{v}_2$, $\mathbf{v}_3$ then in matrix notation we write

$$\overline{\mathbf{x}} = \mathbf{T}_g \mathbf{x} \tag{2.6}$$

where the transformation matrix is defined as

$$\mathbf{T}_g = \begin{bmatrix} \mathbf{v}_1 & \mathbf{v}_2 & \mathbf{v}_3 \end{bmatrix} \tag{2.7}$$

Alternatively, we may write in indicial notation

$$\overline{x}_i = \Lambda_{ij} x_j \tag{2.8}$$

where

$$\Lambda_{ij} = \mathbf{v}_i \bullet \mathbf{i}_j \tag{2.9}$$

Projection of one vector onto another

If we want to compute the projection of a vector onto a direction specified by a unit vector $\mathbf{v}$, then it is very convenient to use the dot product. For example, the component, u' of the displacement $\mathbf{u}$ in the direction specified by $\mathbf{v}$ is given by

$$u' = \mathbf{u} \bullet \mathbf{v} \tag{2.10}$$

The angle θ between the two vectors is computed by

$$cos\theta = \frac{1}{|\mathbf{u}|} \mathbf{u} \bullet \mathbf{v} \tag{2.11}$$

where the length of vector **u** is given by

$$|\mathbf{u}| = \sqrt{u_x^2 + u_y^2 + u_z^2} \tag{2.12}$$

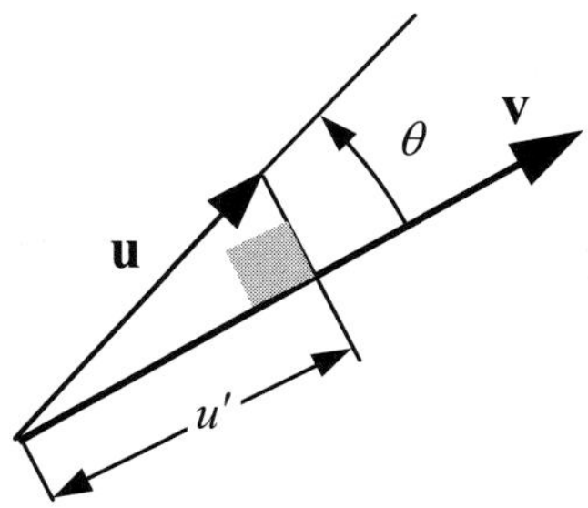

Figure 1.4 Projection of vector

Derivatives of vectors

The derivatives of the displacement vector may be written as

$$\frac{\partial \mathbf{u}}{\partial x} = \mathbf{u}_{,x} = \begin{Bmatrix} \dfrac{\partial u_x}{\partial x} \\ \dfrac{\partial u_y}{\partial x} \\ \dfrac{\partial u_z}{\partial x} \end{Bmatrix} \;;\; \frac{\partial \mathbf{u}}{\partial y} = \mathbf{u}_{,y} = \begin{Bmatrix} \dfrac{\partial u_x}{\partial y} \\ \dfrac{\partial u_y}{\partial y} \\ \dfrac{\partial u_z}{\partial y} \end{Bmatrix} \;;\; \frac{\partial \mathbf{u}}{\partial z} = \mathbf{u}_{,z} = \begin{Bmatrix} \dfrac{\partial u_x}{\partial z} \\ \dfrac{\partial u_y}{\partial z} \\ \dfrac{\partial u_z}{\partial z} \end{Bmatrix} \tag{2.13}$$

In indicial notation we simply write

$$\frac{\partial u_i}{\partial x_j} = u_{i,j} \tag{2.14}$$

1.3.2 Stress and strain

Stresses and strains are tensorial quantities. In the indicial notation the strain tensor is defined by

$$\varepsilon_{ij} = \frac{1}{2}\left(u_{i,j} + u_{j,i}\right) \tag{2.15}$$

In this book, however, we use a notation originally proposed by Timoshenko[4]. We define a *pseudo-vector* of strain, i.e., a matrix with one column:

$$\boldsymbol{\varepsilon} = \begin{Bmatrix} \varepsilon_x \\ \varepsilon_y \\ \varepsilon_z \\ \gamma_{xy} \\ \gamma_{yz} \\ \gamma_{xz} \end{Bmatrix} = \begin{Bmatrix} \dfrac{\partial u_x}{\partial x} \\ \dfrac{\partial u_y}{\partial y} \\ \dfrac{\partial u_z}{\partial z} \\ \dfrac{\partial u_x}{\partial y} + \dfrac{\partial u_y}{\partial x} \\ \dfrac{\partial u_y}{\partial z} + \dfrac{\partial u_z}{\partial y} \\ \dfrac{\partial u_x}{\partial z} + \dfrac{\partial u_z}{\partial x} \end{Bmatrix} \tag{2.16}$$

Note that in the *pseudo-vector* notation we only have 6 strain components, whereas the symmetric strain tensor has 9. Also note that the ½ term is missing for the shear strains in order to achieve consistency between the tensor and matrix operations. The index number of the location of the strain or stress components for matrix notation and tensor notation is given in Table 1.1

Table 1.1 Index numbering for strain and stress

Notation	Index number					
Matrix	1	2	3	4	5	6
Tensor	11	22	33	12&21	23&32	31&13
	xx	yy	zz	xy&yx	yz&zy	zx&xz

Similarly, the stress tensor σ_{ij} can be written as a *pseudo-vector*

$$\boldsymbol{\sigma} = \begin{bmatrix} \sigma_x & \sigma_y & \sigma_z & \tau_{xy} & \tau_{yz} & \tau_{xz} \end{bmatrix}^{\mathrm{T}} \tag{2.17}$$

1.4 CONCLUSIONS

At the beginning of this chapter we have shown on an example in geomechanics that substantial gains can be made with the BEM, in terms of mesh generation and solution times. These gains are most pronounced for problems involving infinite or semi-infinite domains. Other examples where the BEM seems to be superior to the FEM is for problems where boundary stresses are important, e.g. in Mechanical Engineering. Examples of this will be shown later.

The main purpose of this book is to encourage the use of the method. The simple computer programs included contain all the necessary building blocks for building more advanced and more specific computer programs for research or industrial applications.

In conclusion the reader should see this book as an advanced introduction to the BEM, with some basic building blocks for computer programming.

1.5 REFERENCES

1. Ritz, W. (1909) Über eine Methode zur Lösung gewisser Variations-Probleme der mathematischen Physik. *Journal für reine und angewandte Mathematik,* **135**:1-61.
2. Trefftz, E. (1926) *Ein Gegenstück zum Ritz'schen Verfahren.* Proc. 2nd int. Congress in Applied Mechanics, Zurich, pp 131.
3. Zienkiewicz O.C. and Taylor R.L. (2000) The Finite Element Method - Fifth Edition. Butterworth-Heinemann, UK.
4. Timoshenko, S.P. and Goodier, J.N. (1970) Theory of Elasticity. McGraw-Hill, London.

2
Programming

Art is only pleasing if it
has the character of lightness

J.W. von Goethe

2.1 STRATEGIES

Although the first idea which provided the background for the boundary element method dates back to the early 1900s, the method only emerged when digital computers became available. This is because, except for the simplest problems, the number of computations required is too large for 'hand calculation'.

The implementation into a computer application basically consists of giving the processor a series of instructions, or tasks, to perform. In the early days these instructions had to be given in complicated machine code and writing them was mainly the domain of specialised programmers. Very soon higher level languages were developed which made the programming task easier and this had the additional advantage that code developed could run on any hardware. One of these languages, especially developed for scientists and engineers, was FORTRAN. In the past decades, the language has undergone tremendous development. Whereas with FORTRAN IV the writing of programs was rather lengthy and tedious and the code difficult to follow, the new facilities of FORTRAN 90/95/2000 (F90) make it suitable for writing short, readable code. This has mainly to do with features that do away with the need to use statement numbers and the availability of powerful array and matrix manipulation tools. Today, any engineer should be able to write a program in a rather short time.

When developing a relatively large program, such as will be attempted in this book, it is important to use the concept of modular programming. This means that the task has to be divided into many subtasks. Therefore, we will develop a library of procedures to perform certain tasks, for example, computing the value or the derivative of a function at a certain point. The sub-procedures or functions can be called as needed from the main program or from other procedures.

In the following, we will give a short introduction to some features of F90 that will be used in this book. Here, it will be assumed that readers already have some knowledge of FORTRAN. A more detailed description of F90 is given by Smith[1].

We will also introduce in this chapter the notation used in this book, especially with respect to vectors and matrices. A short introduction to matrix algebra and vectors will also be given.

2.2 FORTAN 90/95/2000 FEATURES

2.2.1 Representation of numbers

Numbers are stored in the computer in binary form. Real numbers are stored in two parts: one consists of the digits that make up the number, the other of the exponent. The exact way in which a number is stored depends on the hardware used. For real numbers, either 4 or 8 bytes could be allocated for storage in earlier Fortrans, for example by declaring the variable REAL*4 ("single precision") or REAL*8 ("double precision"). However, since the storage is machine dependent, we do not know exactly how many digits can be stored in either mode. Fortran now provides a facility for specifying the precision in number of digits with a "KIND *parameter*".

First, one must interrogate the processor as to which value of KIND provides a certain number of digits. For example

```
IWP= SELECTED_REAL_KIND(16)
```

would assign to parameter IWP, the KIND number which would give 16 significant digits (if the processor can achieve it). The declaration of the **REAL** variable A would then be

```
REAL(KIND=IWP) :: A
```

or

```
REAL(IWP) :: A
```

which replaces the former type of declaration, for example

```
REAL*8 A
```

The precision in which the numbers are stored is significant because numerical round-off may occur if two numbers, which differ greatly in magnitude, are subtracted from each other. If a great number of such operations are carried out, the error produced by the round-off may become significant. For example, in the solution of a large system of equations by Gauss elimination, there are many subtractions and therefore a high precision storage is necessary, whereas other parts of the program may not be so sensitive. In the programs developed in this book except in the chapter on parallel processing we use **REAL**(KIND=8) instead of **REAL**(iwp).

2.2.2 Arrays

F90 has powerful features to handle arrays. An array can be of rank 1,2,3 etc. A rank 1 array is a vector, a rank 2 array a matrix. In this book we will distinguish between real vectors (identified by a lowercase bold letter) and matrices with one column, or pseudo-vectors (identified by lowercase bold Greek letters).

The *shape* of an array indicates the number of elements in each dimension. For example the vector

$$\mathbf{v} = \begin{Bmatrix} v_x \\ v_y \\ v_z \end{Bmatrix} \tag{2.1}$$

would be of *shape* (3,1) , whereas the matrix

$$\mathbf{A} = \begin{bmatrix} a_{11} & a_{12} \\ a_{21} & a_{22} \\ a_{31} & a_{32} \end{bmatrix} \tag{2.2}$$

would be of *shape* (3,2).

The declaration of the two arrays in F90 would be

```
REAL :: V(3),A(3,2)
```

One of the most important features of F90, however, is that arrays may be declared dynamically, that is, the programmer does not need to know the dimensions of an array when writing the program, but these can be read, or calculated, at run time. Since array dimensions in the BEM will depend on number of nodes and/or number of degrees of freedom, this is a particularly useful feature. To declare an array A, whose dimensions are known at run-time, we write:

```
REAL, ALLOCATABLE :: A(:,:)
```

In the program, we can then allocate the dimensions of the array with computed values of dimensions I,J by:

```
ALLOCATE(A(I,J))
```

When the array is no longer used, then the space in memory can be freed by

```
DEALLOCATE(A)
```

An array may be assigned initial values by the statement

```
REAL :: V(3)=(/ 1.0, 0.0, 0.0 /)
```

2.2.3 Array operations

FORTRAN 90 has features for array and vector operations, which simplifies the manipulation of arrays from the programmer's point of view. The operations include matrix/vector additions and subtractions, multiplication and vector product. They also include operations on part of the arrays, examining arrays, determining max/min values, gathering of submatrices etc.

Matrix Addition: If all the coefficients of A are to be added then one can simply write

```
A= A+B
```

Multiplication by a scalar: If all coefficients of A are to be multiplied by a scalar (say 3.0), then this would translate into

```
A= A*3.0
```

or

```
A= A*3.0 _IWP
```

Although A and the scalar 3.0 clearly have different shapes, 3.0 is said to be "broadcast" to all the coefficients of A.

Operation on selected coefficients of a matrix: For example, to add the first 3 coefficients of the second column of array A to array B and store it in A one needs a single statement instead of a loop

```
A(1:3,2)= A(1:3,2) + B(1:3,2)
```

Transpose of a matrix: The transpose of a matrix means exchanging rows and columns. For example, the transpose of vector **v** defined above is given by:

$$\mathbf{v}^{\mathrm{T}} = \lfloor v_x \quad v_y \quad v_z \rfloor \tag{2.3}$$

this translates into

```
VT= TRANSPOSE(V)
```

the resulting vector $\mathbf{v}^T$ would be of shape (1,3).

Matrix multiplication: The multiplication of two matrices of shapes (1,3) and (3,2) gives a result of shape (1,2). For example

$$\mathbf{v}^T\mathbf{A} = \begin{bmatrix} v_x & v_y & v_z \end{bmatrix} \begin{bmatrix} a_{11} & a_{12} \\ a_{21} & a_{22} \\ a_{31} & a_{32} \end{bmatrix} = \begin{bmatrix} v_x a_{11} + v_y a_{21} + v_z a_{31} & , v_x a_{12} + v_x a_{22} + v_x a_{32} \end{bmatrix} \quad (2.4)$$

translates into

B= **MATMUL**(VT,A)

It is obvious that for matrix multiplication to be possible, the shapes of the matrices to be multiplied have to obey certain rules.

Vector dot product: The vector "dot" product of two vectors

$$\mathbf{v}_1 = \begin{Bmatrix} v_{1x} \\ v_{1y} \\ v_{1z} \end{Bmatrix} \quad ; \quad \mathbf{v}_2 = \begin{Bmatrix} v_{2x} \\ v_{2y} \\ v_{2z} \end{Bmatrix} \quad (2.5)$$

is a scalar and is defined as:

$$x = \mathbf{v}_1 \bullet \mathbf{v}_2 = v_{1x}v_{2x} + v_{1y}v_{2y} + v_{1z}v_{2z} \quad (2.6)$$

this translates into

X= **DOT_PRODUCT**(V1,V2)

Maximum value in an array: To find the element of array **A** in (2.2) which has the maximum value one writes

AMAX= **MAXVAL**(A)

Location of maximum value: To find the location of the maximum element of array **A**, NMAX, execute the statement

NMAX= **MAXLOC**(A)

Upper bound of an array: Sometimes it is useful for the program to find out what shape an array had when it was assigned. This will be used extensively in SUBROUTINES in order to reduce parameter lists. The statement

N= **UBOUND**(A,1)

will return the number of the last row of **A,** which is 3 whereas

M= **UBOUND**(A,2)

will return the number of the last column of **A,** which is 2.

Check on array elements: Another useful function is one which checks if all elements of an array fulfil a certain condition. For example

ALL(A >0)

or

ALL(A >0.0_IWP)

will return a logical .TRUE. if all elements of **A** are greater than "zero".

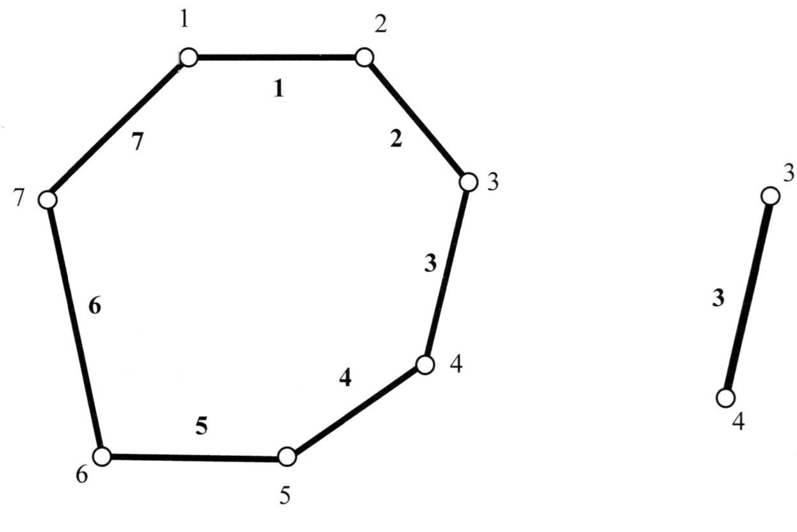

Figure 2.1 Two-dimensional boundary element mesh and connectivity of element 3

Sum of array elements: Instead of summing the coefficients of an array requiring at least one loop, with the intrinsic function **SUM** we may calculate the sum of all coefficients of an array by simply writing

C= **SUM**(A)

Masking can be used to sum only coefficients which fulfil certain conditions. For example

```
C= SUM(A, MASK=A>0.0)
```

would sum only coefficients of A which are greater than “zero”.

Gathering and scattering: A feature in F90 makes the ‘gathering’ and ‘scattering’ of values, which we will need later, very simple. To explain these operations consider a two-dimensional mesh of boundary elements in Figure 2.1.

The nodes of the mesh where elements are connected with each other can be numbered in two different ways: locally and globally. When referring, for example, to the unknown u (e.g. temperature in the case of heat conduction problems) one has two vectors, a global one

$$\{u\} = \begin{Bmatrix} u_1 \\ u_2 \\ u_3 \\ u_4 \\ u_5 \\ u_6 \\ u_7 \end{Bmatrix} \tag{2.7}$$

and a local one defined at element level, for example, for the two nodes of element 3:

$$\{u\}^3 = \begin{Bmatrix} u_1 \\ u_2 \end{Bmatrix}^3 \tag{2.8}$$

For element 3 we may define a ‘connectivity vector’ of dimension 2, which contains the global node numbers of the two nodes of the element

```
CONNECTIVITY=(/3,4/)
```

The ‘scatter’ operation is where the locally defined unknowns are put into the global vector

```
U_GLOBAL(CONNECTIVITY)= U_LOCAL
```

This statement would put u_1 and u_2 of element 3 into locations 3 and 4 of the global vector. The ‘gather’ operation would do the opposite, i.e.

```
U_LOCAL= U_GLOBAL(CONNECTIVITY)
```

would put the global values of u_1 and u_2 into the local positions 1 and 2.

2.2.4 Control

Various features which can be used to control the flow of the program have been improved and new ones added. With these new features it should no longer be necessary to have **GOTO** statements and statement numbers, features which sometimes made programs very difficult to read. A new feature is the **SELECT CASE** which replaces the computed **GOTO**. This feature allows us to control which parts of the code are executed under certain conditions.

For example the coding

```
SELECT CASE(NUMBER_OF_FREEDOMS)
        CASE(1)
                Coding for one degree of freedom
        CASE(2)
                Coding for two degrees of freedom
        CASE DEFAULT
                Error message
END SELECT
```

would execute two different types of instructions, depending on the degrees of freedom per node (i.e., potential vs. elasticity problems) and would issue an error message if another value is encountered.

The IF statement is mainly used for controlling execution. It has been improved in that symbols which are familiar to engineers can be used.

For example the operators :

.NE.	can be written as	/=
.EQ.		==
.GT.		>
.GE.		>=
.LT.		<
.LE.		<=

The DO loop has also been improved. It is possible to give each DO loop a name which enhances readability. Also, there is an easier possibility of exiting a loop when a certain condition is reached. For example, in an iteration loop the condition for exiting may be that a convergence has been achieved. The code

```
Iteration_loop:        &
DO ITER=1,NITERS
    Statements
    IF(CONVERGED) EXIT
    Statements
END DO &
Iteration_loop
```

would exit the loop completely if the value of CONVERGED is .TRUE.

Another nice feature is CYCLE. For example the coding

```
Element_loop:      &
DO NEL=1,NELEM
    Statements 1
    IF(NEL >= NELB) CYCLE
    Statements 2
END DO &
Element_loop
```

would skip Statements 2 if NEL becomes greater or equal to NELB and would continue with the next value of NEL.

2.2.5 Subroutines and functions

Subroutines and functions perform frequently used tasks and split a complex problem into smaller ones. For example, to normalise a vector we may define a Subroutine Vector_norm as

```
SUBROUTINE VECTOR_NORM(V,VLEN)
 !----------------------------------------
 ! Normalise vector
 !----------------------------------------
 REAL, INTENT(INOUT) :: V(:)      !  Vector to be normalised
 REAL, INTENT(OUT)   :: VLEN      !  Length of vector
 VLEN= SQRT( SUM(V*V))
 IF(ABS(VLEN)<1.E-10) RETURN
 V= V/VLEN
 RETURN
END SUBROUTINE VECTOR_NORM
```

Two things are of note here. Firstly, in the declaration of variables in the parameter list, we may specify if a parameter is to be used for input (IN), output (OUT) or input and output (INOUT). This not only helps to clarify the readability of the code, but also protects variables from being changed by accident in the subprogram. Secondly, we do not need to specify the dimension of vector V, since this will be determined in the program calling the Subroutine. For example, the calling program will have

```
REAL :: V(3)
.
.
CALL VECTOR_NORM(V,VLEN)
```

Another very useful feature which we will use in the book is that a function can also return an array. For example, we may write a function for determining the vector ex-product of two vectors as:

```
FUNCTION VECTOR_EX(V1,V2)
 !---------------------------------------
 !  Returns vector x-product v1xv2
 !  where v1 and v2 are dimension 3
 !---------------------------------------
  REAL, INTENT(IN) :: V1(3),V2(3)          !   Input
  REAL             :: VECTOR_EX(3)         !   Result
  VECTOR_EX(1)=V1(2)*V2(3)-V2(2)*V1(3)
  VECTOR_EX(2)=V1(3)*V2(1)-V1(1)*V2(3)
  VECTOR_EX(3)=V1(1)*V2(2)-V1(2)*V2(1)
 RETURN
 END FUNCTION VECTOR_EX
```

In the calling program we use this function in this way

```
REAL  :: V1(3),V2(3),V3(3)
.
.
V3= VECTOR_EX(V1,V2)
```

2.2.6 Subprogram libraries and common variables

As indicated previously, for developing large programs, it is convenient to subdivide the big task into smaller ones. This means that a library of subroutines will be developed. There are basically two ways in which these subroutines were able to communicate with each other in earlier Fortrans: via parameter lists or via COMMON blocks. F90 has replaced the somewhat tedious COMMON block structure by the MODULE and USE statements. A MODULE is simply a set of declarations and/or subroutines. If a program or subprogram wants to use the declarations and subroutines, it simply has a USE statement at the beginning. For example, to define some variables which are used by subprograms we specify

```
MODULE Common_Variables
   REAL(IWP) :: A, B
   REAL(IWP), ALLOCATABLE :: C(:)
END MODULE Common_Variables
```

This replaces the COMMON statements. Any program or sub-program that uses the common declarations has a **USE** statement such as:

```
PROGRAM TEST
  USE Common_Variables
  -
  -
  -
END PROGRAM TEST
```

To help with the management of large programs it is convenient to group subroutines into different files. The MODULE facility can be used for this purpose. For example, we may group all subroutines which have to do with the geometrical description of boundary elements into a module Geometry_lib

```
MODULE Geometry_lib
  REAL :: Pi= 3.149
  CONTAINS
    SUBROUTINE Shape ...
    END SUBROUTINE Shape
    ...
END MODULE Geometry_lib
```

2.3 CHARTS AND PSEUDO CODE

Even though the features of F90 have made programs readable, there is still a need to show the general layout of the program in a simple way. Flow charts, as used in the early days of programming, are not useful because they do not illustrate the essential features of a program's structure.

Instead, structure charts and pseudo code are used, i.e., a FORTRAN-like code which gives but a general description of what to do. Since nested DO LOOPS are complicated to read in a FORTRAN code they can be explained better in a structure chart. For instance, the chart for the example of two-dimensional numerical integration discussed in the next chapter is shown in Figure 2.2.

The advantage of the structure chart is that the structure of the nested do loop can be clearly seen. Another feature where structure charts may be useful is in IF statements, especially when they are complicated. For example, if we wish to check all diagonal elements of the coefficient matrix and take appropriate action if they are negative, zero or positive, the structure chart in Figure 2.3 can be used.

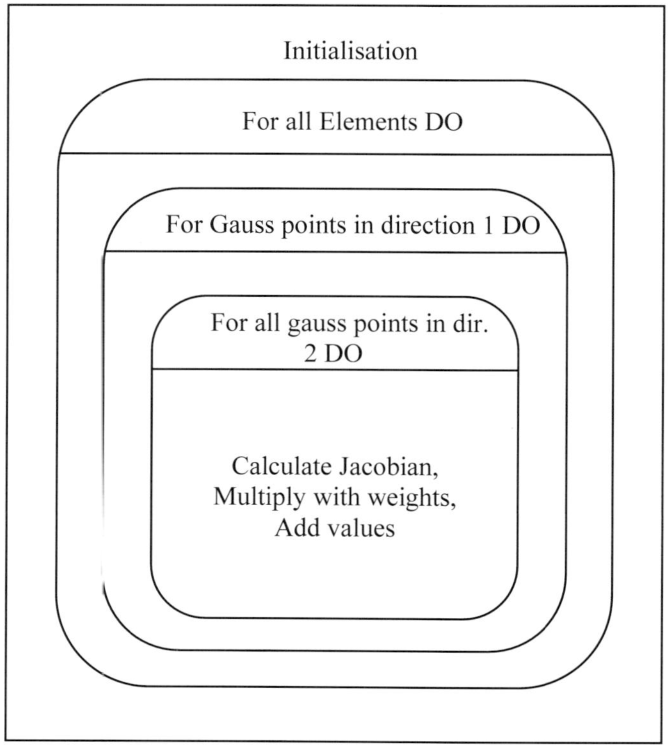

Figure 2.2 Example of a structure chart, nested DO-loop

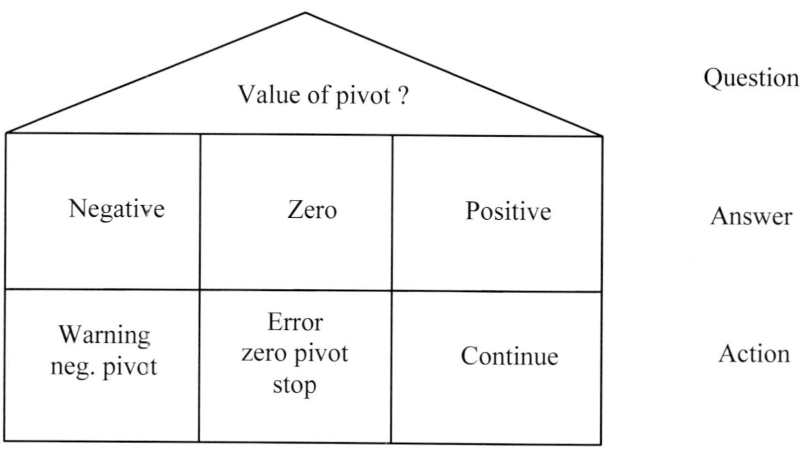

Figure 2.3 Example of a structure chart, IF statement

2.4 PARALLEL PROGRAMMING

As problem sizes grow, or analyses types become more ambitious, computer processing time can inhibit the design process[2,3]. For example, if a nonlinear analysis (Chapter 15) takes hours to complete, the designer loses interest because interaction between computation and design is hampered. Alternatively, design of many systems depends on a statistical evaluation of their responses and so a single analysis is not sufficient. Stochastically it may be necessary to complete several hundred analyses before a statistically significant result can be reported.

For these reasons it is important to minimise computer analysis time, and the presently available means of so doing is called "parallel processing". The idea is to carry out the computation on NPES processors, usually of similar type, connected in parallel. If the computational work can be shared equally amongst the processors, under perfect conditions computation time is reduced by a factor NPES. At the time of writing, systems with 1000 processors are quite common, meaning that an analysis taking 1 hour on a single processor could be completed in 4 seconds in parallel. More affordably, 10 standard PCs coupled together in parallel would reduce the hour of computation to a 6 minute "coffee break".

A second advantage of working in parallel is that data can be distributed across the processors and so much larger problems can be analysed.

2.4.1 Message Passing using MPI

While there are several ways of organising programs to run in parallel, we concentrate here on "message passing" using a portable system called MPI – "Message Passing Interface"[4]. When computations are subdivided and assigned to the various processors, a time will come when information has to be shared, or exchanged between processors. The job of MPI is to handle these exchanges – a process called "communication". Since computation is fast and communication slow, an aim will be to optimise the ratio between them.

By "portable" in the above description, we mean that execution of a program in parallel should appear to the user to be independent of the processing hardware being employed and MPI satisfies this requirement. It is a de-facto standard and consists of subroutines, callable from FORTRAN by means of the usual CALL statement. For example

```
CALL MPI_ALLREDUCE(F1,F,Ndofs,MPI_REAL8,MPI_SUM,MPI_COMMWORLD,ier)
```

collects the sum of the distributed arrays F1, of length Ndofs, from all processors and returns the result to F. The KIND of the arrays would have to match the Fortran **REAL*8** precision. The significance of the other parameters need not concern the applications programmer.

2.4.2 Using MPI on a "Supercomputer"

Current "supercomputers" all have parallel architectures of some kind and so it has been essential to make the use of MPI simple. Therefore at compile time a command like "f90" to compile a serial program is simply replaced by, for example, "mpif90" for a parallel program using MPI. At runtime, a command like "mpirun" will have a parameter specifying the number of parallel processors requested.

2.4.3 Using MPI on PC "Clusters"

A low cost entry to parallel computing can be achieved by linking PCs together using for example an Ethernet for communication between machines. In the 1990s such groupings were often termed "Beowulf" clusters and there are publications[5] describing how to set these up. The basic steps might be as follows:

1. Choose N identical PCs. (In practice the PCs need not be identical but for beginners this is a simplifying step).
2. Install the same (e.g. Linux) operating system on each PC.
3. Connect the PCs together using for example an Ethernet.
4. Give each PC a distinct IP address ....X where X =1, N.
5. Install a version of MPI on each PC.
6. Compile the same program (linking to an MPI Library – see below) on each PC.
7. Configure MPI (put a list of IP addresses in a configuration file) to "talk" to all PCs.

An obvious pitfall is that all machines must be running identical software and it is easy to forget to update modifications.

Two MPI libraries which can be freely downloaded are MPICH[6] and LAM MPI[7]. The parallel program is launched by typing a command from one of the PCs (implementation-specific). The program and input data must be in the same location (directory) on each PC (or compute node). When the program is launched from the "master" PC, each "slave" PC will run the same program, in parallel, but will work on its own data. Most Beowulf-type software is Unix or Linux, based but Microsoft have recently put forward a "Microsoft Compute Cluster" enabling parallel processing from a Microsoft Windows environment.

In Chapter 8, programs are listed which enable Boundary Element computations to be processed in parallel by any system capable of supporting MPI, ranging from clusters of PCs to "supercomputers".

2.5 BLAS LIBRARIES

Since computations by the Boundary Element Method make extensive use of array manipulations, it is sensible to make use of software which facilitates this, if available. BLAS, "Basic Linear Algebra Subroutine" libraries permit three levels of array processing: vector-vector, matrix-vector and matrix-matrix[8]. In Chapter 8 we shall use BLAS subroutine DGEMV to carry out matrix by vector multiplications. On some processors this can lead to significant speed-up in comparison with the Fortran MATMUL.

2.6 PRE- AND POST-PROCESSING

For large problems it is very tedious, or impossible, to produce a file using a text editor which contains all input data the program needs for analysis. For example, the specification of the coordinates of all nodes and the connectivity of all elements may involve thousands of lines. It is common practice, therefore, to use preprocessors with a graphical interface which allow the user to specify the problem geometry and loading and which automatically generate the necessary information. Unfortunately, even though FORTRAN has developed very sophisticated features for computation, there are no built in tools available for graphical display, as there are, for example, with C++. If one wants to develop graphic capabilities and user interfaces, one must use special libraries, such as that supplied, for example, by Interactive Software Services (INTERACTER[9]).

Results obtained from boundary element programs can be displayed in a variety of ways. The simplest is to print out values of displacements at nodal points and surface stresses inside boundary elements. In addition, the values at interior points can be printed out. Printed numbers are appropriate for the small examples used in this book, but for larger problems, one cannot do without graphical display. Indeed this will be what will 'sell' any numerical method to the engineering community. A few examples of graphical display will be shown in Chapter 17 (Applications).

General purpose graphical pre- and postprocessors are freely available and sometimes quite inexpensive (for example, GID by UPC[10]). Therefore, the topic of pre- and post-processing will not be discussed in this book. Since the small test examples used here only require a few lines of input, we can do without preprocessors. However, the reader is encouraged to enhance the software by providing a suitable interface for the programs to be developed in this book with existing preprocessing packages.

2.7 CONCLUSIONS

In this chapter we have given a short overview of some of the features of F90 the latest dialect of FORTRAN which we are going to use. There are as many programming styles as there are programmers and each programmer will no doubt claim that his/hers is best.

The aim in good programming should be to produce efficient, readable and easy to check code. The last is a very stringent requirement in quality control. Easy to read programs sometimes also tend to be efficient; however a small gain in efficiency should not be made, if clarity is sacrificed. For example sometimes it is clearer and also more efficient not to use a DO loop if less than 4 cycles will occur. If permutations of indices have to be made such as in the fundamental solutions shown later, it is often better to generate all of the coefficients using the editor's copy and paste facility, since the code can be checked much faster visually.

In the past, sub-programs had either many COMMON blocks or long parameter lists. These were needed to pass variables between SUBROUTINES and the main program. Fortunately F90 has done away with COMMON blocks and the number of parameters for SUBROUTINES can be further reduced by the dynamic array allocation, the use of UBOUND and the USE statement. However one must very carefully consider which variables should be declared in the Common Module as explained previously and which should be declared in each subroutine.

Regarding the programs presented in this book, we claim neither that they are very efficient nor that this is the only way that the procedures outlined may be implemented. Indeed, we encourage the reader to think of different ways in which the theory can be converted efficiently and elegantly into code. In the programs that we present here we have placed our emphasis on readability. Otherwise there would be no point in including the code in the text. In many cases we have sacrificed efficiency and have limited ourselves to solving small problems, because we do not use direct access files for storing values, but assume instead that all data required fit into RAM. With the dramatic increase in the amount of RAM available on standard PCs this, however, is not likely to become a main issue throughout the lifetime of this book especially if parallel processing is used. With regard to efficiency some rearranging of DO loops, would be necessary so that computations which only need to be carried out once are not unnecessarily carried out many times. This occurs especially in the subroutines for the integration of Kernel-shape function products. However, such rearrangement would have made the programs more difficult to follow and therefore was not implemented.

The programs were developed on a Visual Fortan[11] compiler. However, since only standard F90 features have been used, the source code should be able to be compiled with any FORTRAN90, FORTRAN95 or FORTRAN 2000 compiler.

2.8 EXERCISES

Exercise 2.1.

Given are two integer arrays of rank one named Inci1, Inci2 with element node numbers. Write a **LOGICAL FUNCTION** Match(Inci1,Inci2) which returns .TRUE. if all the numbers of Inci2 match all the numbers in Inci1. Note that the sequence of the numbers in Inci1 and Inci2 will in general not be the same. The dimension of both arrays will be declared in the calling program.

Exercise 2.2.
Write a **REAL FUNCTION DETERMINANT**(A) which computes the determinant of the matrix **A** which can be of shape (2,2) or (3,3).

Exercise 2.3.
Given is a sub-matrix **A** of shape (2,2) and a matrix **B** whose shape is declared in the calling program. Write a **SUBROUTINE ASSEMBLE**(A,B,I,J) which assembles the sub-matrix **A** into the matrix **B** at location i,j (see Figure below)

$$\begin{array}{c} \\ j \\ \\ \\ \end{array}\begin{array}{c} \quad\quad\quad\quad i \\ \begin{bmatrix} \cdots & \cdots & \cdots & \cdots \\ \cdots & A_{11} & A_{12} & \\ \cdots & A_{21} & A_{22} & \\ \cdots & & & \end{bmatrix}\end{array}$$

2.9 REFERENCES

1. Smith I.M. (1995), Programming in FORTRAN90, J.Wiley.
2. Smith I. M. and Griffiths D. V., (2004) Programming the Finite Element Method, 4th ed, J.Wiley.
3. ParaFEM Web Reference (2004), http://www.parafem.org.uk.
4. Message Passing Interface Forum (1994) MPI: A message Passing Interface Standard, *International Journal of Supercomputer Applications* **8**:3-4.
5. Sterling, T.L., Salmon, J., Becker, D.J. and Savarese, D.F. (1999) How to Build a Beowulf, The MIT Press.
6. MPICH Web Reference (2007), http://www-unix.mcs.anl.gov/mpi/mpich/
7. LAM MPI Web Reference (2007), http://www.lam-mpi.org/
8. Dongarra, J.J. and Walker, D.W. (1995) Software Libraries for Linear Algebra Computations on High Performance Computers,SIAM Rev. **37**(2):151-180
9. Interactive Software Services (1999), INTERACTER Subroutine Reference.
10. GID web reference: http://gid.cimne.upc.es
11. Digital Equipment Corp. (1997) Digital FORTRAN Language Reference Manual

3

Discretisation and Interpolation

Nature is indifferent
towards the difficulties it
causes to a mathematician

Fourier

3.1 INTRODUCTION

One of the fundamental requirements for numerical modelling is a description of the problem, its boundaries, boundary conditions and material properties, in a mathematical way. The exact definition of the shape of a complicated boundary would require the specification of the location (relative to the origin of a set of axes) of a large number of points on the surface (indeed an exact definition will take an infinite number). In order to be able to model such problems with a reasonable amount of input data, only a limited number of points may be defined and the shape between the points approximated by functions. This is known as *solid modelling*[1]. Solid modelling is being used, for example, to describe the shape of car bodies in mechanical engineering and ore bodies in mining, for the purpose of generating displays on computer graphics terminals. Thus, a new form of car body can be visualised, in perspective, from various angles, even before a scale model is built and the location and grade of ore bodies can be displayed for optimising excavation strategies in mine planning.

In the following we will discuss one and two-dimensional boundary elements as defined by the number of intrinsic (element) coordinate directions. One-dimensional elements exist in two-dimensional Cartesian space and two-dimensional elements in three-dimensional space. Thus, in this chapter, we consider *discretisation* methods used in the boundary method and start building the library of subroutines needed later. For the treatment of non-linear problems discussed in Chapter 15 we will also need two- and three-dimensional cells, which are also discussed here.

3.2 ONE-DIMENSIONAL BOUNDARY ELEMENTS

One-dimensional elements are used for the description of a boundary in the x-y plane. The first step in the description of the boundary is to specify a discrete number of points on the boundary (Figure 3.1). Next we specify an interpolation between these points. In the simplest case we have linear segments (or boundary elements) which connect two nodes i and j, the positions of which are defined by Cartesian coordinates. For each element it is convenient to define a local (*intrinsic*) coordinate ξ which follows the direction of the element, equals zero at the centre and has the value of $\pm$ 1 at the ends (Figure 3.2).

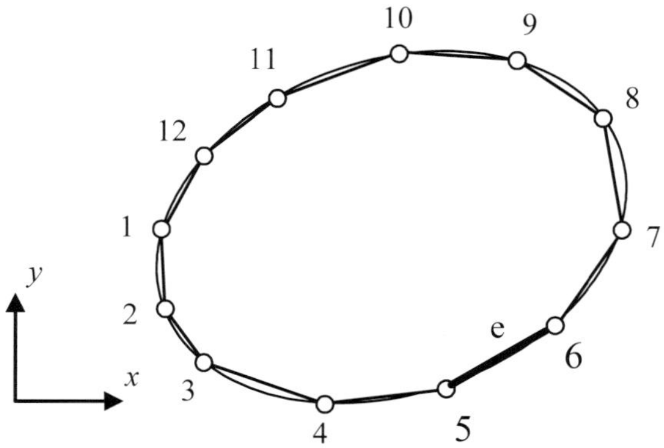

Figure 3.1 Plane domain, boundary approximated by linear elements

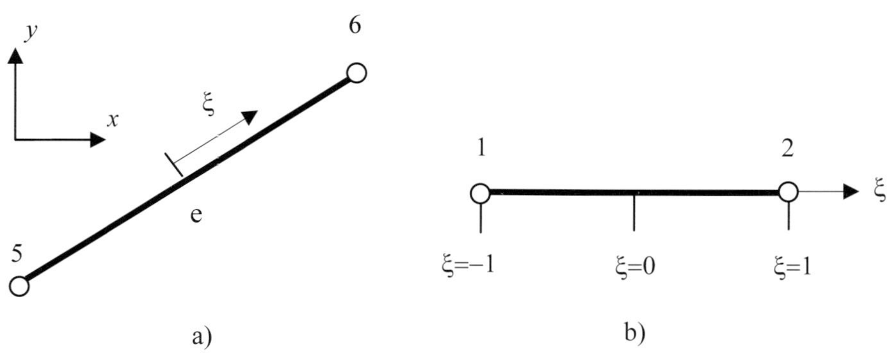

Figure 3.2 Boundary element shown in a) global and b) local (intrinsic) coordinate space

It can easily be verified that the Cartesian coordinates of a point on element e with the intrinsic coordinate ξ are given by

$$x(\xi)=\frac{x_5+x_6}{2}+\frac{x_6-x_5}{2}\xi$$
$$y(\xi)=\frac{y_5+y_6}{2}+\frac{y_6-y_5}{2}\xi \qquad (3.1)$$

This equation can be checked by substituting $\xi = -1$ and $\xi = +1$ to obtain the coordinates of nodes 5 and 6.

It is now convenient to substitute for the global coordinates:

$$\left.\begin{aligned} x_5 &= x_1^{\,e} \\ y_6 &= y_1^{\,e} \end{aligned}\right\} \text{ first node of element e}$$

$$\left.\begin{aligned} x_5 &= x_2^{\,e} \\ y_6 &= y_2^{\,e} \end{aligned}\right\} \text{ second node of element e}$$

In this way we establish a link between local and global numbering of nodes.

The global numbers of nodes which belong to the element are referred to as '*element incidences*' or '*element connectivity*'. In the example in Fig. 3.1, the connectivity of element e is *5,6*. The sequence in which the element node numbers are entered will be significant later, as it will affect the direction of the *outward normal*. From now on we will work with the local numbering system and use the element incidences to 'gather' coordinates from the global values, as explained in the previous chapter.

We can rewrite equation (3.1) as

$$x(\xi)=\frac{1}{2}(1-\xi)x_1^e+\frac{1}{2}(1+\xi)x_2^e$$
$$y(\xi)=\frac{1}{2}(1-\xi)y_1^e+\frac{1}{2}(1+\xi)y_2^e \qquad (3.2)$$

or in abbreviated form

$$\begin{Bmatrix} x \\ y \end{Bmatrix}=\sum_{n=1}^{L} N_n(\xi)\begin{Bmatrix} x_n^{\,e} \\ y_n^{\,e} \end{Bmatrix} \qquad (3.3)$$

where L is the number of element nodes and N_n are element 'shape' functions. Equation (3.3) can be written in matrix notation

$$\mathbf{x}=\sum N_n(\xi)\,\mathbf{x}_n^e \qquad (3.4)$$

where **x** is a vector containing coordinates of a point on element e and $\mathbf{x}_n^e$ is a vector of coordinates of the nth node of element e.

For the two-node element just derived, the shape functions are (Figure 3.3)

$$N_1 = \frac{1}{2}(1-\xi)$$
$$N_2 = \frac{1}{2}(1+\xi) \tag{3.5}$$

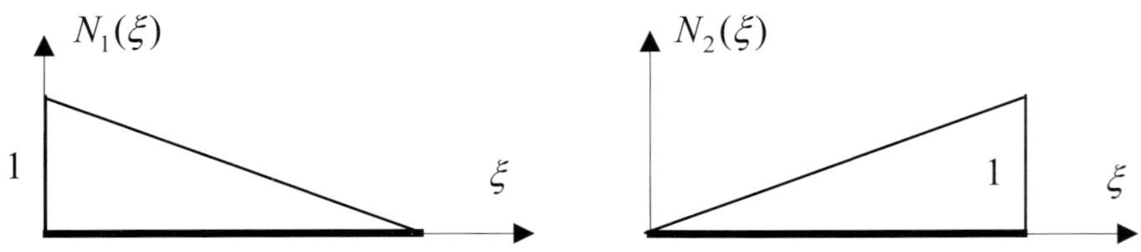

Figure 3.3 Linear shape functions

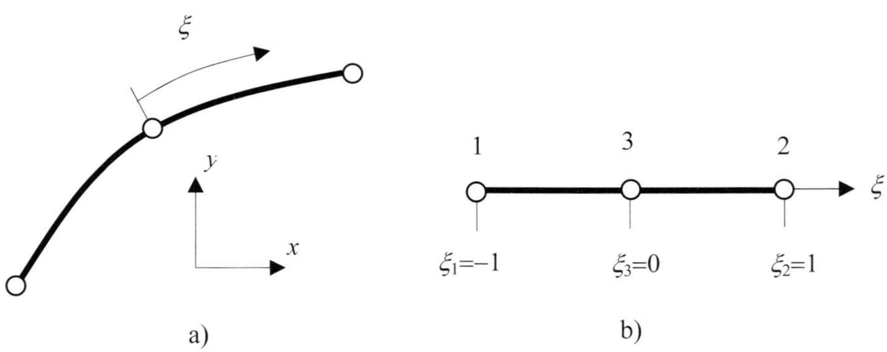

Figure 3.4 Quadratic element shown in a) global and b) local coordinate space

The shape functions may be also expressed by

$$N_n = \frac{1}{2}(1+\xi_n \xi) \tag{3.6}$$

where the local coordinates of the 2 nodes are given by

$$\xi_1 = -1$$
$$\xi_2 = 1 \tag{3.7}$$

Complicated shapes can be more accurately described by a smaller number of elements with three nodes and quadratic shape functions (Figure 3.4). The coordinate ξ now follows the element shape, i.e., is curvilinear and the third node is placed at ξ = 0. The shape function associated with the mid-side node is a parabola, which has unit value at the third node and zero value at the other nodes, that is,

$$N_3(\xi) = 1 - \xi^2 \tag{3.8}$$

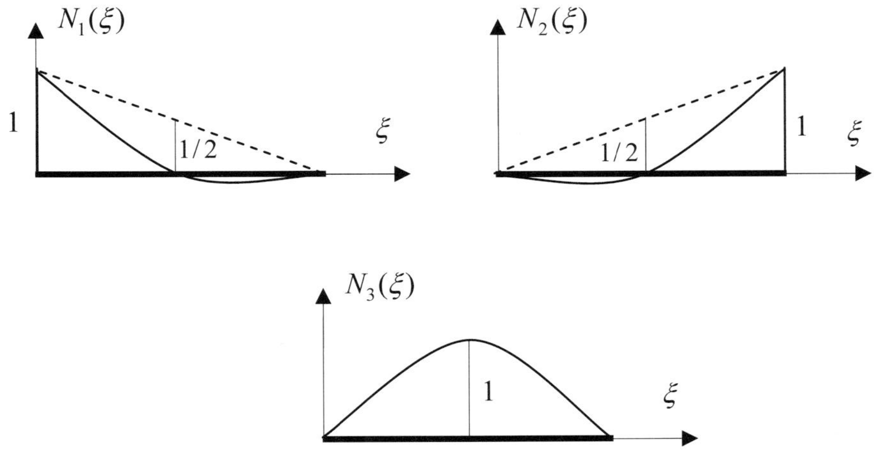

Figure 3.5 Quadratic shape functions

The corner node shape functions can be obtained by subtracting half of the centre node function from each of the linear shape functions (Figure 3.5)

$$N_n = \frac{1}{2}(1 + \xi_n \xi) - \frac{1}{2} N_3 \qquad n = 1,2 \tag{3.9}$$

The shape functions presented here have so far not been derived mathematically but written down intuitively. Shape functions derived this way have been called *Serendipity* functions[2]. It can be seen that the shape functions derived so far have the following properties

$$\begin{aligned} N_n(\xi_n) &= 1 \\ N_n(\xi_i) &= 0 \quad for \quad i \neq n \\ \sum N_n(\xi) &= 1 \end{aligned} \tag{3.10}$$

The mathematical derivation of functions which satisfy conditions (3.10) is possible using Lagrange polynomials[3]. For the parabolic elements the Lagrange shape functions are defined as

$$L_i(\xi) = A_{i1} A_{i2} A_{i3} \tag{3.11}$$

where

$$\begin{aligned} A_{in} &= \frac{\xi - \xi_n}{\xi_i - \xi_n} \quad \textit{for} \quad i \neq n \\ A_{in} &= 1 \quad\quad\quad \textit{for} \quad i = n \end{aligned} \tag{3.12}$$

The reader can verify that the Lagrange and Serendipity shape functions are identical for one-dimensional elements. However, Lagrange polynomials will be used to construct shape functions for two-dimensional elements, which differ from the Serendipity shape functions.

3.3 TWO-DIMENSIONAL ELEMENTS

For the description of the boundary of three-dimensional problems two-dimensional boundary elements are used. The elements are also used for defining cells for the evaluation of volume integrals for plane problems. Their derivation is analogous to that of the one-dimensional elements described previously, except that two intrinsic coordinates (ξ, η) are used, as shown in Figure 3.6.

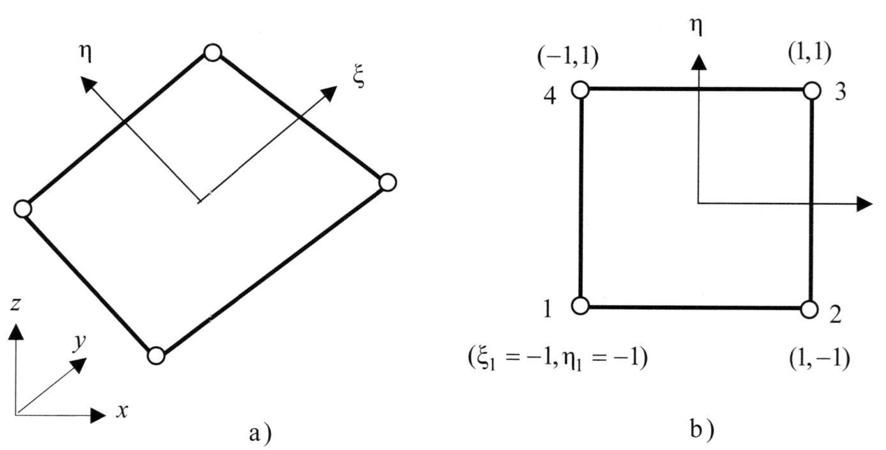

Figure 3.6 Quadrilateral boundary element in a) global and b) local coordinate system

The Cartesian coordinates of a point with intrinsic coordinates (ξ, η) are obtained by

$$\mathbf{x} = \sum_{n=1}^{L} N_n(\xi,\eta)\,\mathbf{x}_n^e \tag{3.13}$$

where for boundary elements

$$\mathbf{x} = \begin{pmatrix} x \\ y \\ z \end{pmatrix} \tag{3.14}$$

For cells we have

$$\mathbf{x} = \begin{pmatrix} x \\ y \end{pmatrix} \tag{3.15}$$

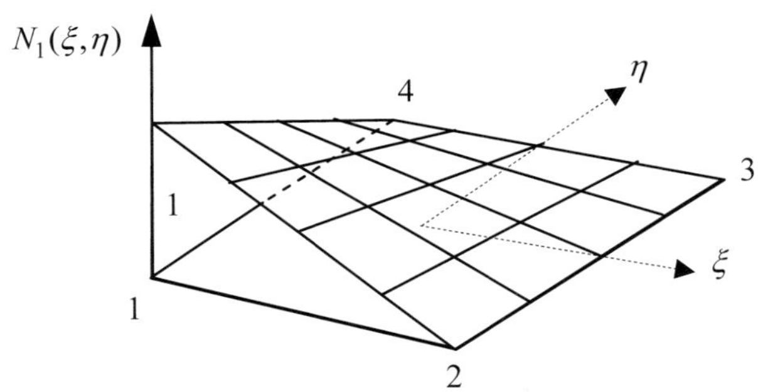

Figure 3.7 Bilinear shape function N_1

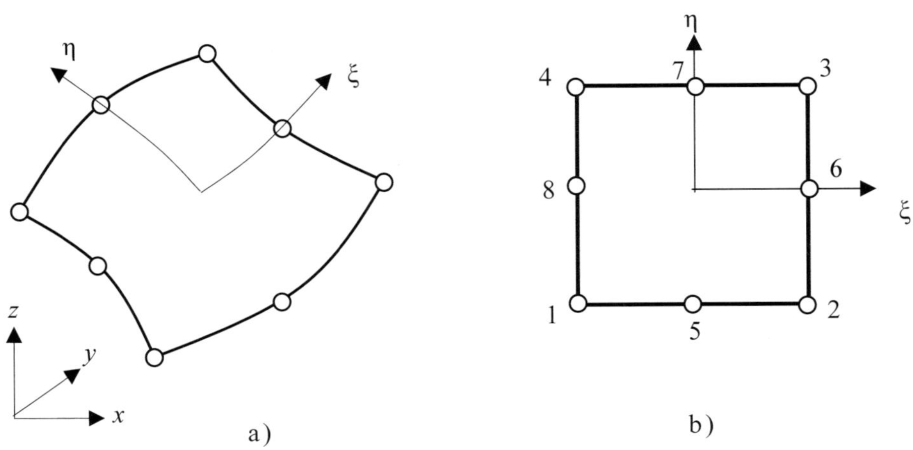

Figure 3.8 Quadratic Serendipity element

Bilinear shape functions are used:

$$N_n = \frac{1}{2}\left(1+\xi_n\xi\right)\frac{1}{2}\left(1+\eta_n\eta\right) \tag{3.16}$$

The shape function N_1 is shown in Figure 3.7. It describes a curved surface consisting of straight lines in the ξ, η directions. The surface, also called a hyper-surface, has been a widely used shape for concrete shells, because the formwork is simple to construct.

Table 3.1 Intrinsic coordinates of nodes

n	ξ_n	η_n
1	-1.0	-1.0
2	1.0	-1.0
3	1.0	1.0
4	-1.0	1.0
5	0.0	-1.0
6	1.0	0.0
7	0.0	1.0
8	-1.0	0.0

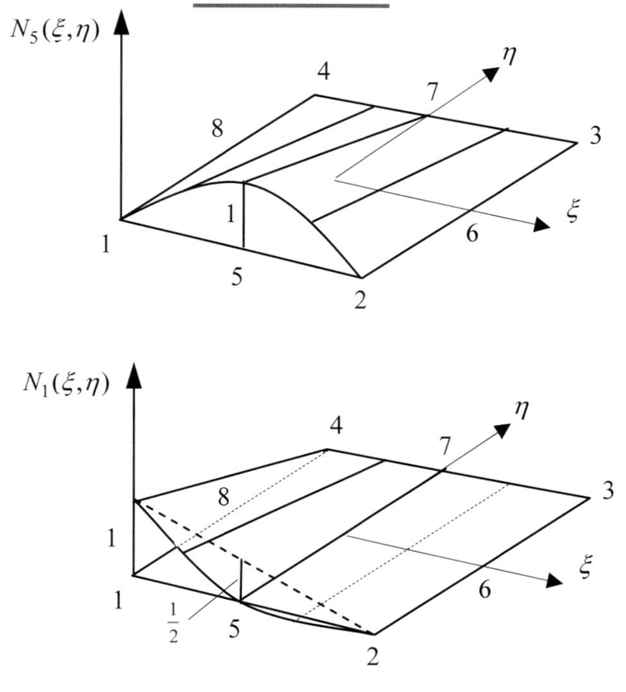

Figure 3.9 Shape functions for mid-side and corner nodes

Again we can derive a higher order element by adding mid-side nodes on the element sides. A quadratic element is shown in Figure 3.8 and the local coordinates of nodes are shown in Table 3.1. The shape functions for the mid-side nodes are given by

$$
\begin{aligned}
N_n &= \frac{1}{2}\left(1-\xi^2\right)\left(1+\eta_n\eta\right) \quad \textit{for } n=5,7 \\
N_n &= \frac{1}{2}\left(1-\eta^2\right)\left(1+\xi_n\xi\right) \quad \textit{for } n=6,8
\end{aligned}
\tag{3.17}
$$

The corner node functions are constructed in a similar way as for the one-dimensional element (Figure 3.9)

$$
\begin{aligned}
N_1 &= \frac{1}{4}(1-\xi)(1-\eta)-\frac{1}{2}N_5-\frac{1}{2}N_8 \\
N_2 &= \frac{1}{4}(1+\xi)(1-\eta)-\frac{1}{2}N_5-\frac{1}{2}N_6 \\
N_3 &= \frac{1}{4}(1+\xi)(1+\eta)-\frac{1}{2}N_6-\frac{1}{2}N_7 \\
N_4 &= \frac{1}{4}(1-\xi)(1+\eta)-\frac{1}{2}N_7-\frac{1}{2}N_8
\end{aligned}
\tag{3.18}
$$

By writing down the shape functions in this manner, it is possible to derive elements with variable numbers of nodes by deleting appropriate terms. For example, for an element with no midside node 5, a linear function is assumed between nodes 1 and 2 and the shape functions are obtained by simply setting $N_5 = 0$.

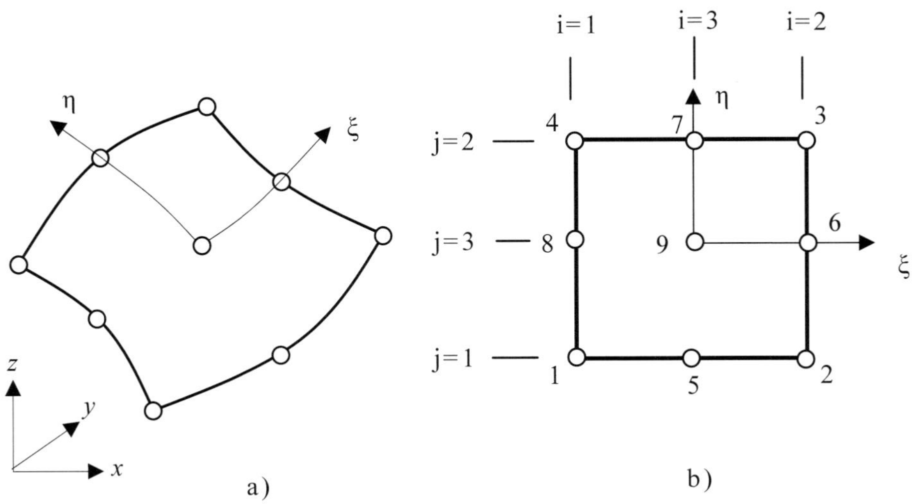

Figure 3.10 Quadratic Lagrange element in a) global and b) local coordinate system

If the element shape functions for the quadratic element are derived from Lagrange polynomials, then there is an additional node at the centre of the element (Figure 3.10). The shape functions are given by

$$L_{n(i,j)} = A_{i1} A_{i2} A_{i3} B_{j1} B_{j2} B_{j3} \tag{3.19}$$

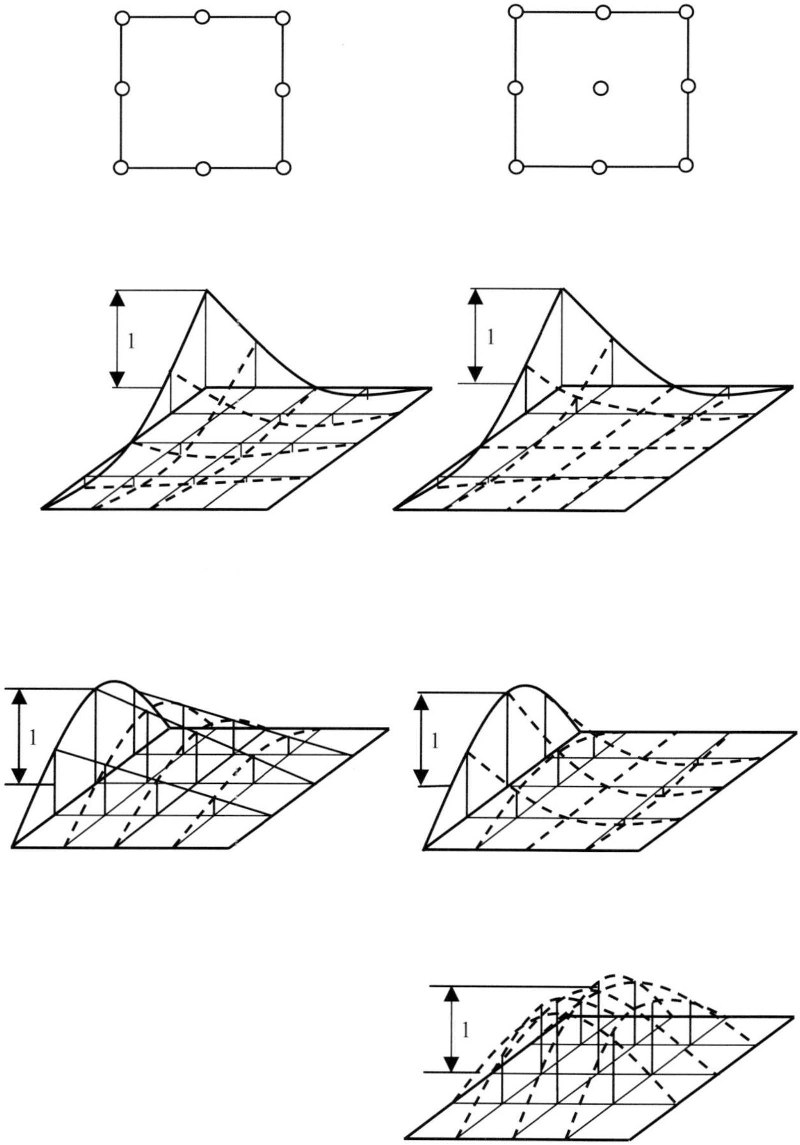

Figure 3.11 Serendipity and Lagrange shape functions

$A_{i,l}$ is defined in equation (3.12) and

$$B_{jm} = \frac{\eta - \eta_m}{\eta_j - \eta_m} \quad if\ j \neq m$$
$$B_{jm} = 1 \qquad if\ j = m \tag{3.20}$$

where i and j are the column and row numbers of the nodes. This numbering is defined in Figure 3.10. The nodes are given by

$$\begin{array}{lll} n(1,1) = 1 & n(2,1) = 2 & n(3,1) = 5 \\ n(1,2) = 4 & n(2,2) = 3 & n(3,2) = 7 \\ n(1,3) = 8 & n(2,3) = 6 & n(3,3) = 9 \end{array}$$

The Serendipity and Lagrange shape functions are compared in Figure 3.11
The Lagrange element has an additional 'bubble mode' and is, therefore, able to describe complicated shapes more accurately. Triangular elements can be formed from quadrilateral elements, by assigning the same global node number to two or three corner nodes. Such degenerate elements are shown in Figure 3.12.

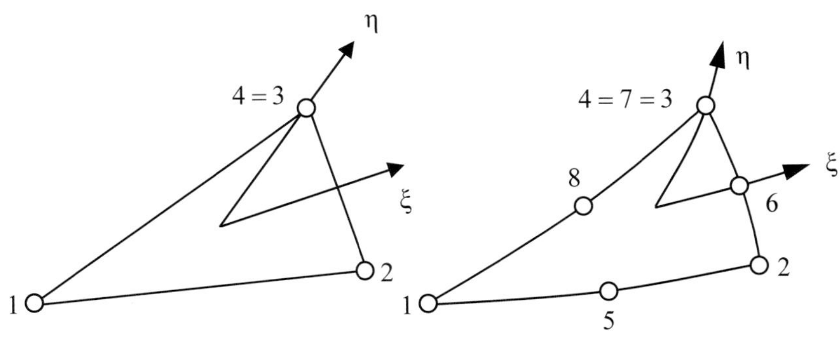

Figure 3.12 Linear and quadratic degenerate elements

Alternatively triangular elements may be defined using the iso-parametric concept. In Figure 3.13 we show a triangular element in the global and local coordinate system. The shape functions for the transformation are defined as[4]

$$N_1(\xi,\eta) = 1 - \xi - \eta$$
$$N_2(\xi,\eta) = \xi \tag{3.21}$$
$$N_3(\xi,\eta) = \eta$$

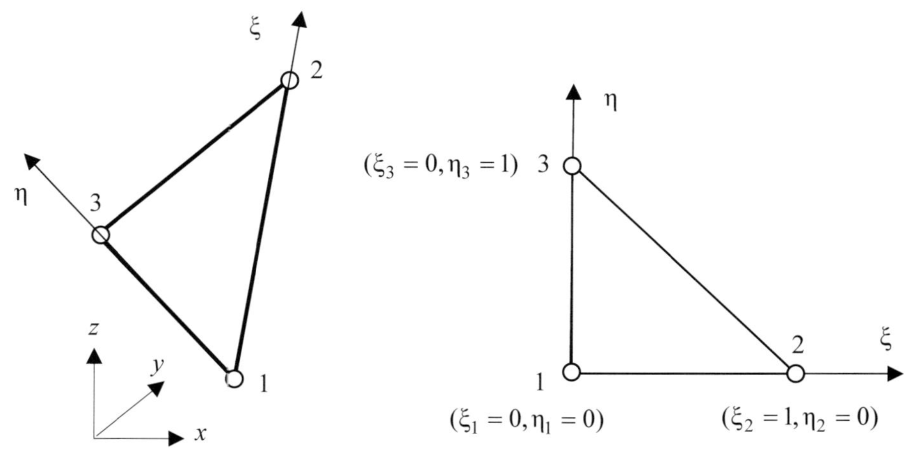

Figure 3.13 Triangular linear element in global and local coordinate system

As can be seen in Figure 3.14 the shape functions are represented by planes.

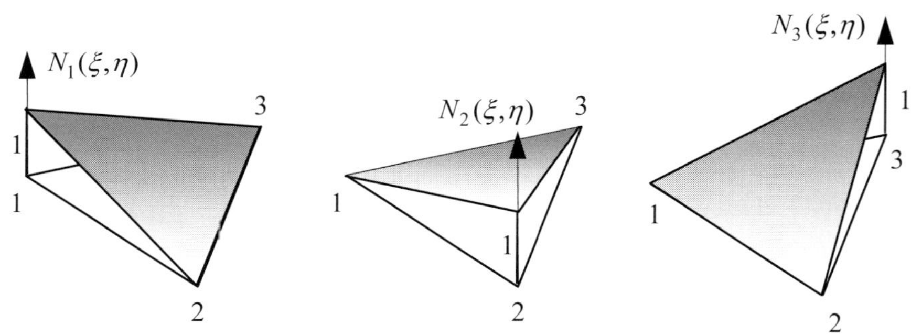

Figure 3.14 Shape functions of linear triangular boundary element

It is also possible to define a triangular element with a quadratic shape function. The shape functions for the mid-side nodes are given by

$$\begin{aligned} N_4 &= 4\xi(1-\xi-\eta) \\ N_5 &= 4\xi\eta \\ N_6 &= 4\eta(1-\xi-\eta) \end{aligned} \tag{3.22}$$

The corner node functions are constructed in a similar way as for the previous elements

$$\begin{aligned} N_1 &= (1-\xi-\eta) - \frac{1}{2}N_4 - \frac{1}{2}N_6 \\ N_2 &= \xi - \frac{1}{2}N_4 - \frac{1}{2}N_5 \\ N_3 &= \eta - \frac{1}{2}N_5 - \frac{1}{2}N_6 \end{aligned} \tag{3.23}$$

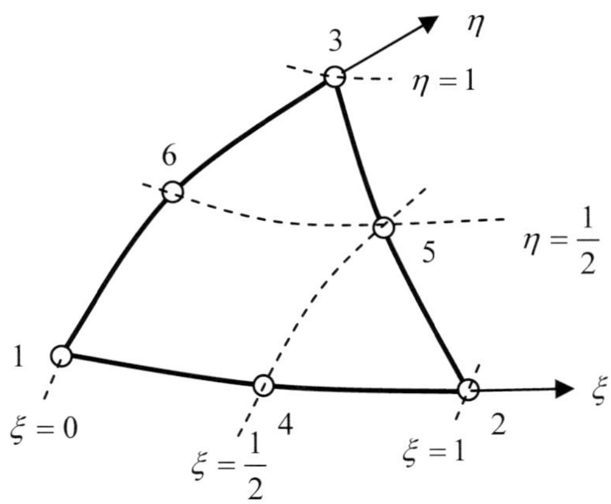

Figure 3.15 Triangular quadratic element

3.4 THREE-DIMENSIONAL CELLS

For the description of cells for 3-D problems three-dimensional elements are used. Their derivation is analogous to that of the two-dimensional elements described previously, except that now three intrinsic coordinates (ξ, η, ζ) are used, as shown in Figure 3.16. The Cartesian coordinates of a point with intrinsic coordinates (ξ, η, ζ) are obtained by

$$\mathbf{x} = \sum_{n=1}^{8} N_n(\xi,\eta,\zeta)\mathbf{x}_n^e \tag{3.24}$$

Bilinear shape functions are used for the quadrilateral element in Figure 3.16

$$N_n = \frac{1}{8}(1+\xi_n\xi)(1+\eta_n\eta)(1+\zeta_n\zeta) \tag{12.1}$$

where local coordinates of the nodes are defined in Table 3.2. For the description of cells with a quadratic shape function, see for example [4].

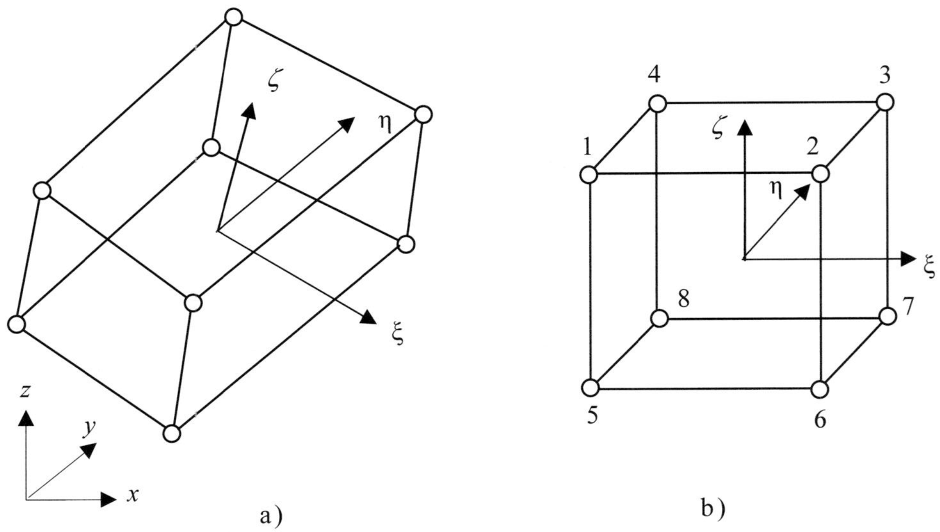

Figure 3.16 3-D cell element in a) global and b) local coordinate system

Table 3.2 Local coordinates of nodes for 3-D cells

n	ξ_n	η_n	ζ_n	n	ξ_n	η_n	ζ_n
1	-1.0	-1.0	1.0	5	-1.0	-1.0	-1.0
2	1.0	-1.0	1.0	6	1.0	-1.0	-1.0
3	1.0	1.0	1.0	7	1.0	1.0	-1.0
4	-1.0	1.0	1.0	8	-1.0	1.0	-1.0

3.5 ELEMENTS OF INFINITE EXTENT

It is sometimes necessary to describe surfaces of infinite extent. Examples are found in geomechanics, where either the surface of the ground extends to infinity or a tunnel can be assumed to be infinitely long. To describe the geometry of an element of infinite extent in one intrinsic coordinate direction, we may use special shape functions[5] which tend to infinity, as the intrinsic coordinate tends to +1. For the one-dimensional element shown in Figure 3.17 the coordinate transformation

$$\mathbf{x}(\xi) = \sum_{n=1}^{3} N_n^{\infty}(\xi)\mathbf{x}_n^e \tag{3.25}$$

results in infinite Cartesian coordinates at $\xi = 1$ if the shape functions are taken to vary as follows:

$$N_1^{\infty} = -2\xi/(1-\xi) \quad \text{and} \quad N_2^{\infty} = (\xi+1)/(1-\xi) \tag{3.26}$$

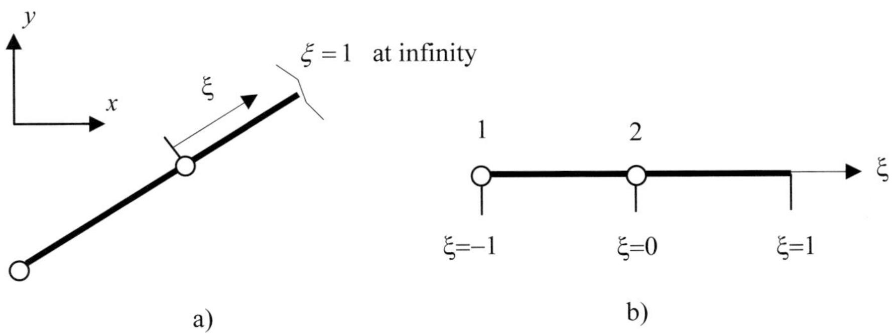

Figure 3.17 One-dimensional infinite element in a) global and b) local coordinate space

Note that the element is finite in the local coordinate space and therefore can be treated the same way as a finite boundary element for the integration.

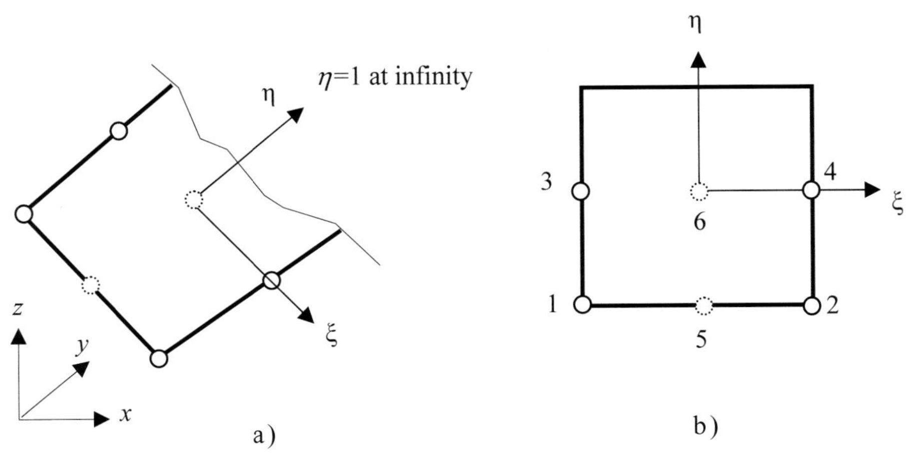

Figure 3.18 Two-dimensional infinite element in a) global and b) local coordinate space.

The concept can be extended to two-dimensions. The geometry of the two-dimensional element shown in Figure 3.18, for example, is described by

$$\mathbf{x} = \sum_{n=1}^{4(6)} N_{m(n)}(\xi) N_{k(n)}^{\infty}(\eta) \mathbf{x}_n^e \tag{3.27}$$

where $N_{m(n)}(\xi)$ are linear or quadratic Serendipity shape functions as presented for the one-dimensional finite boundary elements, $N_{k(n)}^{\infty}(\eta)$ are the same infinite shape functions as for the one-dimensional element, with η substituted for ξ and the values for *m(n)* and *k(n)* are given in Table 3.3

Table 3.3 Values for *m* and *k* in Equation (3.27)

n	*m*	*k*
1	1	1
2	2	1
3	2	2
4	1	2
5	3	1
6	3	2

3.6 SUBROUTINES FOR SHAPE FUNCTIONS

Here we start building our library of Subroutines for future use. We create routines for the calculation of Serendipity, infinite and Lagrange shape functions. Only the listing for the first one is shown here.

As explained in Chapter 3, some variables will be defined as global, that is, as accessible to all the subroutines in a MODULE and all programs which use them via the USE statement. The dimensions for the array Ni, which contains the shape functions, depend on the type of element and will be set by the main program.

```
SUBROUTINE Serendip_func(Ni,xsi,eta,ldim,nodes,inci)
!---------------------------------
! Computes Serendipity shape functions Ni(xsi,eta)
! for one and two-dimensional (linear/parabolic) finite
! boundary elements
!---------------------------------
REAL,INTENT(OUT)    :: Ni(:)     ! Array with shape function
REAL,INTENT(IN)     :: xsi,eta! intrinsic coordinates
INTEGER,INTENT(IN):: ldim      ! element dimension
INTEGER,INTENT(IN):: nodes     ! number of nodes
INTEGER,INTENT(IN):: inci(:)! element incidences
REAL:: mxs,pxs,met,pet         ! temporary variables
SELECT CASE (ldim)
CASE(1)! one-dimensional element
```

```
    Ni(1)= 0.5*(1.0 - xsi);  Ni(2)= 0.5*(1.0 + xsi)
    IF(nodes == 2) RETURN!  linear element finished
    Ni(3)=  1.0 - xsi*xsi
    Ni(1)= Ni(1) - 0.5*Ni(3); Ni(2)= Ni(2) 0.5*Ni(3)
 CASE(2)! two-dimensional element
    mxs=1.0-xsi; pxs=1.0+xsi; met=1.0-eta; pet=1.0+eta
    Ni(1)= 0.25*mxs*met ; Ni(2)= 0.25*pxs*met
    Ni(3)= 0.25*pxs*pet ; Ni(4)= 0.25*mxs*pet
    IF(nodes == 4) RETURN!  linear element finished
    IF(Inci(5) > 0) THEN   !zero node = node missing
     Ni(5)= 0.5*(1.0 -xsi*xsi)*metNi(1)= Ni(1) - 0.5*Ni(5) ;
     Ni(2)= Ni(2)0.5*Ni(5)
    END IF
    IF(Inci(6) > 0) THEN
     Ni(6)= 0.5*(1.0 -eta*eta)*pxs
     Ni(2)= Ni(2) - 0.5*Ni(6) ; Ni(3)= Ni(3) - 0.5*Ni(6)
    END IF
    IF(Inci(7) > 0) THEN
     Ni(7)= 0.5*(1.0 -xsi*xsi)*pet
     Ni(3)= Ni(3) - 0.5*Ni(7) ; Ni(4)= Ni(4)- 0.5*Ni(7)
    END IF
    IF(Inci(8) > 0) THEN
     Ni(8)= 0.5*(1.0 -eta*eta)*mxs
     Ni(4)= Ni(4) - 0.5*Ni(8) ; Ni(1)= Ni(1) - 0.5*Ni(8)
    END IF
 CASE DEFAULT !   error message
 CALL Error_message('Element dimension not 1 or 2')
 END SELECT
 RETURN
 END SUBROUTINE Serendip_func
```

3.7 INTERPOLATION

In addition to defining the shape of the solid to be modelled, we will also need to specify the variation of physical quantities (displacement, temperature, traction, etc.) in an element. These can be interpolated from the values at the nodal points.

3.7.1 Isoparametric elements

The value of a quantity q at a point inside an element e can be written as

$$q = \sum \bar{N}_n q_n^e \tag{3.28}$$

where q_n^e is the value of the quantity at the nth node of element e and $\bar{N}_n$ are interpolation functions (Figure 3.19).

If for a particular element the same functions are used for the element shape and for the interpolations of physical quantities inside the element, then the element is called 'isoparametric' (i.e., same number of parameters).

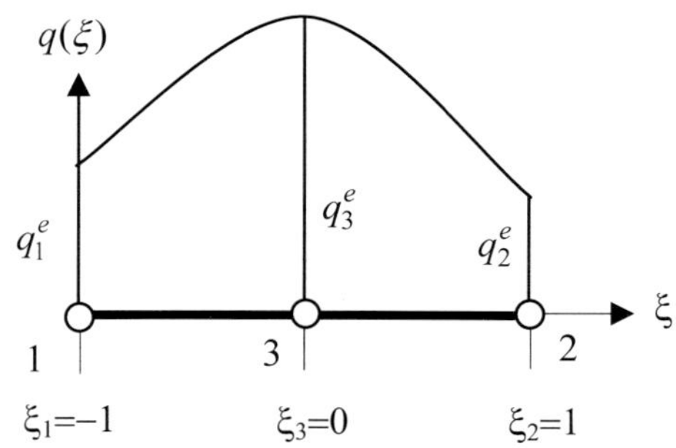

Figure 3.19 Variation of q along a quadratic 1-D boundary element (in local coordinate system)

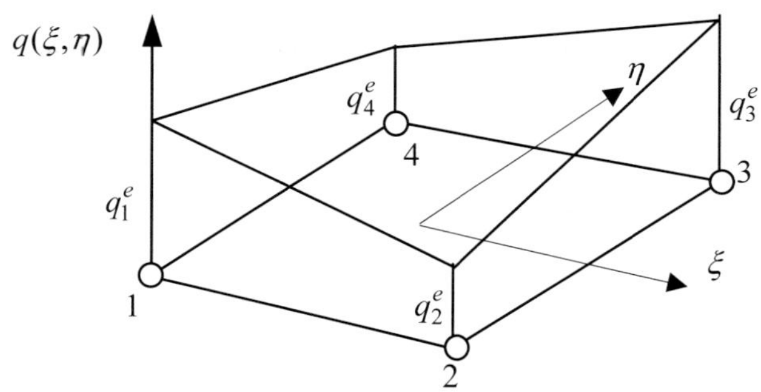

Figure 3.20 Interpolation of q over a linear 2-D element

The variation of physical quantities on the surface of two-dimensional elements or inside plane elements can be described (Figure 3.20)

$$q(\xi,\eta)=\sum \bar{N}_n(\xi,\eta)q_n^e \tag{3.29}$$

Note than q may be a scalar or a vector (i.e. may refer to tractions **t** or displacements **u**). The physical quantities are defined for each element separately, so they can be discontinuous at nodes shared by two elements as shown in Figure 3.21. If Serendipity or Lagrange shape functions are used only C^0 continuity can be enforced between elements by specifying the same function value for each element at a shared node.

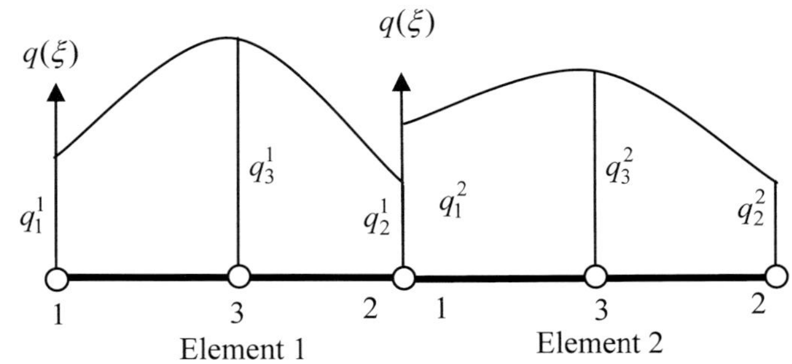

Figure 3.21 Variation of q with discontinuous variation at common element nodes

3.7.2 Infinite elements

For the one-dimensional infinite element we can assume that the displacements and tractions decay from node 1 to infinity with *o(1/r)* and *o(1/r²)* respectively, or that they remain constant. The former corresponds to a surface that extends to infinity, but the loading is finite, the latter corresponds to a the case where both the surface and the loading extends to infinity (this corresponds to *plane strain* conditions). For the one-dimensional "decay" infinite element we have

$$\mathbf{u} = N_{u1}^{\infty} \cdot \mathbf{u}_1 \quad ; \quad \mathbf{t} = N_{t1}^{\infty} \cdot \mathbf{t}_1 \tag{3.30}$$

where

$$N_{u1}^{\infty} = \frac{1}{2}(1-\xi) \quad ; \quad N_{t1}^{\infty} = \frac{1}{4}(1-\xi)^2 \tag{3.31}$$

For the "plane strain" infinite element the variation is given simply by

$$\mathbf{u} = \mathbf{u}_1 \quad ; \quad \mathbf{t} = \mathbf{t}_1 \tag{3.32}$$

For the two-dimensional "decay" infinite element we have

$$\mathbf{u} = \sum_{n=1}^{2(3)} N_n(\xi) N_{u1}^{\infty}(\eta) \mathbf{u}_n^e \quad ; \quad \mathbf{t} = \sum_{n=1}^{2(3)} N_n(\xi) N_{t1}^{\infty}(\eta) \mathbf{t}_n^e \tag{3.33}$$

Where $N_n(\xi)$ are linear or quadratic Serendipity shape functions as presented for the one-dimensional finite boundary elements and $N_{t1}^{\infty}(\eta)$ and $N_{u1}^{\infty}(\eta)$ are the same infinite shape functions as for the one-dimensional element with η substituted for ξ.
For the two-dimensional "plane strain" infinite element we have

$$\mathbf{u} = \sum_{n=1}^{2(3)} N_n(\xi)\mathbf{u}_n^e \quad ; \quad \mathbf{t} = \sum_{n=1}^{2(3)} N_n(\xi)\mathbf{t}_n^e \tag{3.34}$$

3.7.3 Discontinuous elements

Later we will see that in some cases it is convenient to interpolate q not from the nodes that define the geometry but from other (interpolation) nodes that are moved inside the element.

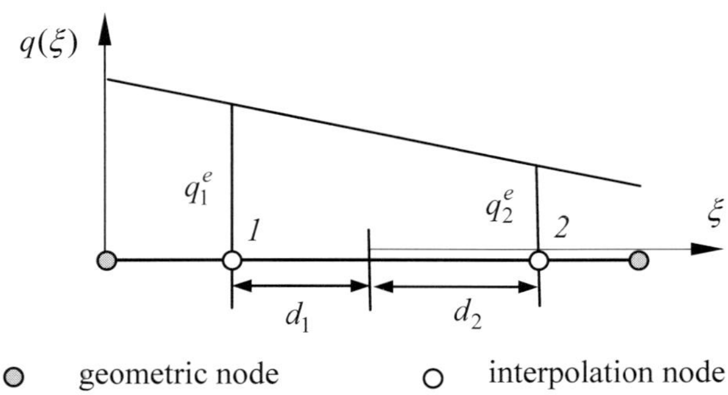

Figure 3.22 One dimensional linear discontinuous element

This type of element will be used in the Chapter on corners and edges to avoid a multiple definition of the traction vector. For the one-dimensional linear element in Figure 3.22 we have

$$q(\xi) = \sum \bar{N}_n(\xi)\bar{q}_n^e \tag{3.35}$$

Where $\bar{q}_n^e$ are the values of q at the interpolation nodes and the interpolation functions are

$$\bar{N}_1(\xi) = \frac{1}{(d_1 + d_2)}(d_1 - \xi) \quad ; \quad \bar{N}_2(\xi) = \frac{1}{(d_1 + d_2)}(d_2 - \xi) \tag{3.36}$$

Here d_1, d_2 are absolute values of the intrinsic coordinate of the interpolation nodes.

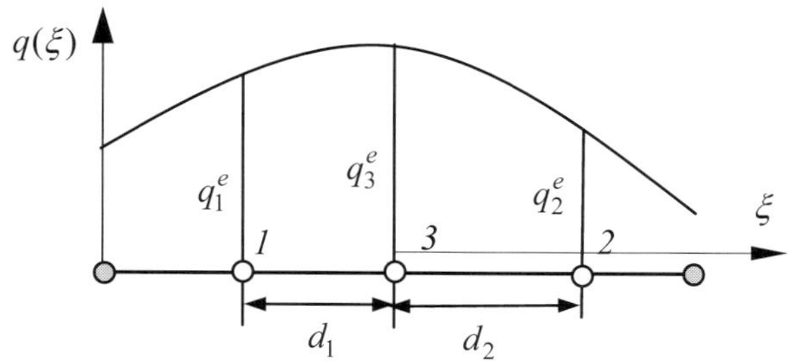

Figure 3.23 One dimensional quadratic discontinuous element

It can be easily verified that for d_1=d_2=1 the shape functions for the continuous element are obtained. For a quadratic element we have

$$\bar{N}_1(\xi) = \frac{1}{(d_1+d_2)}(d_1-\xi)(-1-\frac{\xi}{d_2}) \; ; \; \bar{N}_2(\xi) = \frac{1}{(d_1+d_2)}(d_2-\xi)(-1-\frac{\xi}{d_1})$$
$$\bar{N}_3(\xi) = \frac{1}{d_1 d_2}(d_1-\xi)(d_2+\xi) \tag{3.37}$$

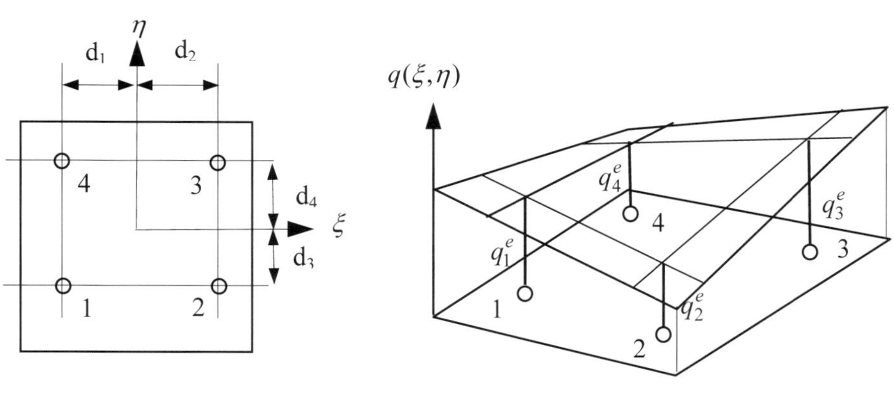

Figure 3.24 Two-dimensional linear discontinuous element

For the two-dimensional linear element shown in Figure 3.24 the shape functions are given for the corner nodes by

$$\bar{N}_1(\xi,\eta) = \frac{1}{c}(d_1-\xi)(d_3-\eta) \; ; \; \bar{N}_2(\xi,\eta) = \frac{1}{c}(d_2-\xi)(d_3-\eta)$$
$$\bar{N}_3(\xi,\eta) = \frac{1}{c}(d_2-\xi)(d_4-\eta) \; ; \; \bar{N}_4(\xi,\eta) = \frac{1}{c}(d_1-\xi)(d_4-\eta) \tag{3.38}$$
$$c = (d_1+d_2)(d_3+d_4)$$

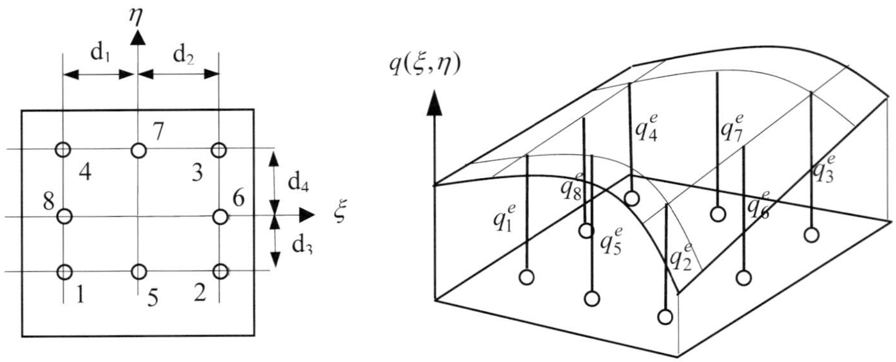

Figure 3.25 Two-dimensional quadratic discontinuous element

For the quadratic element in Figure 3.25 we have for the corner nodes

$$
\begin{aligned}
N_1(\xi,\eta) &= \frac{1}{(d_1+d_2)(d_3+d_4)}(d_1-\xi)(d_3-\eta)(-1-\frac{\xi}{d_2}-\frac{\eta}{d_4}) \\
N_2(\xi,\eta) &= \frac{1}{(d_1+d_2)(d_3+d_4)}(d_2-\xi)(d_3-\eta)(-1-\frac{\xi}{d_1}-\frac{\eta}{d_4}) \\
N_3(\xi,\eta) &= \frac{1}{(d_1+d_2)(d_3+d_4)}(d_2-\xi)(d_4-\eta)(-1-\frac{\xi}{d_1}-\frac{\eta}{d_3}) \\
N_4(\xi,\eta) &= \frac{1}{(d_1+d_2)(d_3+d_4)}(d_1-\xi)(d_4-\eta)(-1-\frac{\xi}{d_2}-\frac{\eta}{d_3})
\end{aligned}
\tag{3.39}
$$

and for the mid side nodes:

$$
\begin{aligned}
N_5(\xi,\eta) &= \frac{1}{d_1 d_2(d_3+d_4)}(d_1-\xi)(d_2+\xi)(d_3-\eta) \\
N_6(\xi,\eta) &= \frac{1}{d_3 d_4(d_1+d_2)}(d_2-\xi)(d_3+\xi)(d_4-\eta) \\
N_7(\xi,\eta) &= \frac{1}{d_1 d_2(d_3+d_4)}(d_1-\xi)(d_2+\xi)(d_4-\eta) \\
N_8(\xi,\eta) &= \frac{1}{d_3 d_4(d_1+d_2)}(d_1-\xi)(d_3+\xi)(d_4-\eta)
\end{aligned}
\tag{3.40}
$$

3.8 COORDINATE TRANSFORMATION

Sometimes it might be convenient to define the coordinates of a node in a local Cartesian coordinate system. A local coordinate system is defined by the location of its origin, $\mathbf{x}_0$ and the direction of the axes. In two dimensions we define the direction with two vectors as shown in Figure 3.26a. The global coordinates of a point specified in a local coordinates system $\overline{\mathbf{x}}$ are given by

$$\mathbf{x} = \mathbf{x}_0 + \mathbf{T}_g \overline{\mathbf{x}} \tag{3.41}$$

where $\mathbf{x}_0$ is a vector describing the position of the origin of the local axes. For two-dimensional problems the geometric transformation matrix is given by

$$\mathbf{T}_g = \begin{bmatrix} v_{1x} & v_{2x} \\ v_{1y} & v_{2y} \end{bmatrix} \tag{3.42}$$

where $\mathbf{v}_1, \mathbf{v}_2$ are orthogonal unit vectors specifying the directions of $\overline{x}, \overline{y}$.

For three-dimensional problems a local (orthogonal) coordinate system is defined by unit vectors $\mathbf{v}_1, \mathbf{v}_2, \mathbf{v}_3$ as shown in Figure 3.26b. The transformation matrix for a 3-D coordinate system is given by

$$\mathbf{T}_g = \begin{bmatrix} v_{1x} & v_{2x} & v_{3x} \\ v_{1y} & v_{2y} & v_{3y} \\ v_{1z} & v_{2z} & v_{3z} \end{bmatrix} \tag{3.43}$$

The inverse relationship between local and global coordinates is given by

$$\overline{\mathbf{x}} = \mathbf{x}_0 + \mathbf{T}_g^T \mathbf{x} \tag{3.44}$$

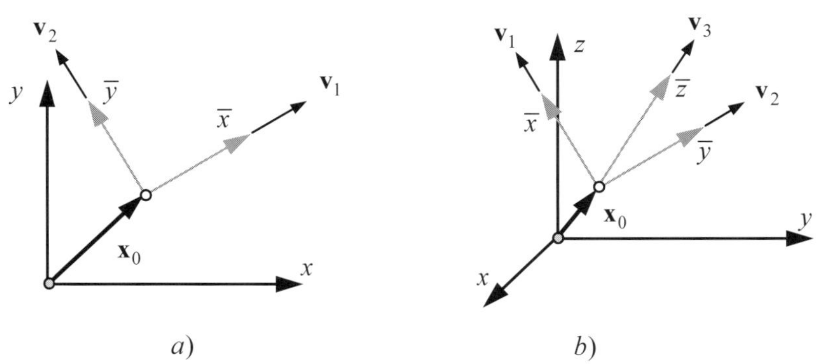

Figure 3.26 Local coordinate systems a) 2-D and b) 3-D

3.9 DIFFERENTIAL GEOMETRY

In the boundary element method it will be necessary to work out the direction normal to a line or surface element.

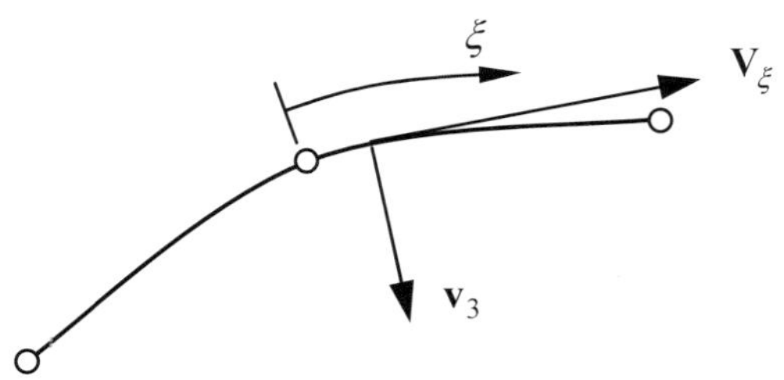

Figure 3.27 Vectors normal and tangential to a one-dimensional element

The best way to determine these directions is by using vector algebra. Consider a one-dimensional quadratic boundary element (Figure 3.27). A vector in the direction of ξ can be obtained by

$$\mathbf{V}_\xi = \frac{\partial}{\partial \xi}\mathbf{x} \tag{3.45}$$

By the differentiation of equation (3.4) we get

$$\mathbf{V}_\xi = \frac{\partial}{\partial \xi}\mathbf{x} = \sum_{n=1}^{3} \frac{\partial N_n}{\partial \xi}\mathbf{x}_n^e \tag{3.46}$$

A vector normal to the line element, $\mathbf{V}_3$, may then be computed by taking the cross-product of $\mathbf{V}_\xi$ with a unit vector in the z-direction ($\mathbf{v}_z$):

$$\mathbf{V}_3 = \mathbf{V}_\xi \times \mathbf{v}_z \tag{3.47}$$

This vector product can be written as:

$$\mathbf{V}_3 = \begin{Bmatrix} V_{3x} \\ V_{3y} \\ V_{3z} \end{Bmatrix} = \begin{Bmatrix} \dfrac{dx}{d\xi} \\ \dfrac{dy}{d\xi} \\ \dfrac{dz}{d\xi} \end{Bmatrix} \times \begin{Bmatrix} 0 \\ 0 \\ 1 \end{Bmatrix} = \begin{Bmatrix} \dfrac{dy}{d\xi} \\ -\dfrac{dx}{d\xi} \\ 0 \end{Bmatrix} \tag{3.48}$$

The length of the vector $\mathbf{V}_3$ is equal to

$$V_3 = |\mathbf{V}_3| = \sqrt{\frac{dy}{d\xi}^2 + \frac{dx}{d\xi}^2} = J \tag{3.49}$$

and therefore the unit vector in the direction normal to a line element is given by

$$\mathbf{v}_3 = \mathbf{V}_3 \frac{1}{V_3} \tag{3.50}$$

It can be shown that the length of $\mathbf{V}_3$ represents also the real length of a unit segment ($\Delta\xi = 1$) in local coordinate space (this is also known as the *Jacobian* J of the transformation from local to global coordinate space).

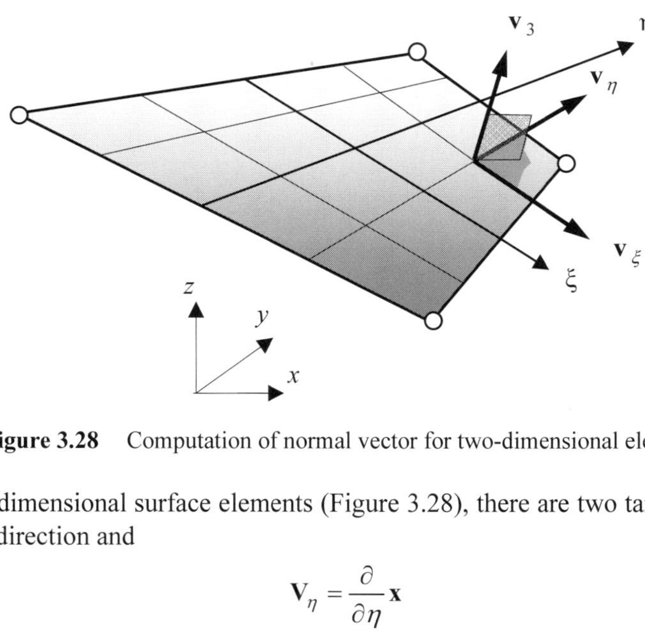

Figure 3.28 Computation of normal vector for two-dimensional elements

For two-dimensional surface elements (Figure 3.28), there are two tangential vectors, $\mathbf{V}_\xi$ in the ξ-direction and

$$\mathbf{V}_\eta = \frac{\partial}{\partial \eta}\mathbf{x} \tag{3.51}$$

in the η-direction, where

$$\frac{\partial}{\partial \eta}\mathbf{x} = \sum \frac{\partial N_n}{\partial \eta}\mathbf{x}_n^e \tag{3.52}$$

The vector normal to the surface may be computed by taking the cross-product of $\mathbf{V}_\xi$ and $\mathbf{V}_\eta$:

$$\mathbf{V}_3 = \mathbf{V}_\xi \times \mathbf{V}_\eta \tag{3.53}$$

that is

$$\mathbf{V}_3 = \begin{Bmatrix} \dfrac{\partial x}{\partial \xi} \\ \dfrac{\partial y}{\partial \xi} \\ \dfrac{\partial z}{\partial \xi} \end{Bmatrix} \times \begin{Bmatrix} \dfrac{\partial x}{\partial \eta} \\ \dfrac{\partial y}{\partial \eta} \\ \dfrac{\partial z}{\partial \eta} \end{Bmatrix} = \begin{Bmatrix} \dfrac{\partial y}{\partial \xi}\dfrac{\partial z}{\partial \eta} - \dfrac{\partial y}{\partial \eta}\dfrac{\partial z}{\partial \xi} \\ \dfrac{\partial z}{\partial \xi}\dfrac{\partial x}{\partial \eta} - \dfrac{\partial z}{\partial \eta}\dfrac{\partial x}{\partial \xi} \\ \dfrac{\partial x}{\partial \xi}\dfrac{\partial y}{\partial \eta} - \dfrac{\partial x}{\partial \eta}\dfrac{\partial y}{\partial \xi} \end{Bmatrix} \tag{3.54}$$

As indicated previously the unit normal vector $\mathbf{v}_3$ is obtained by first computing the length of the vector:

$$V_3 = \sqrt{V_{3x}^2 + V_{3y}^2 + V_{3z}^2} = J \tag{3.55}$$

This is also the real area of a segment of size 1x1 in the local coordinate system, or the *Jacobian* of the transformation. The normalised vector in the direction perpendicular to the surface of the element is given by

$$\mathbf{v}_3 = \mathbf{V}_3 \frac{1}{V_3} \tag{3.56}$$

It should be noted here that $\mathbf{v}_\xi, \mathbf{v}_\eta$ are not orthogonal to each other. An orthogonal system of axes is required for the definition of strains and stresses needed later. Here we assume that the first axis defined by vector $\mathbf{v}_1$ is in the direction of $\mathbf{v}_\xi$. The second axis is defined by:

$$\mathbf{v}_2 = \mathbf{v}_1 \times \mathbf{v}_3 \tag{3.57}$$

The computation of the normal vector requires the derivatives of the shape functions. These are computed by SUBROUTINE Serendip_deriv shown below.

```
SUBROUTINE Serendip_deriv(DNi,xsi,eta,ldim,nodes,inci)
!---------------------------------
! Computes Derivatives of Serendipity shape functions
! for one and two-dimensional (linear/parabolic)
! finite boundary elements
!---------------------------------
REAL,INTENT(OUT)    :: DNi(:,:)  ! Derivatives of Ni
REAL, INTENT(IN)    :: xsi,eta   ! intrinsic coordinates
INTEGER,INTENT(IN)::  ldim       ! element dimension
INTEGER,INTENT(IN)::  nodes      ! number of nodes
INTEGER,INTENT(IN)::  inci(:)    ! element incidences
```

```
REAL:: mxs,pxs,met,pet          ! temporary variables
SELECT CASE (ldim)
   CASE(1)                      ! one-dimensional element
      DNi(1,1)= -0.5
      DNi(2,1)= 0.5
      IF(nodes == 2)RETURN      !  linear element finished
      DNi(3,1)=  -2.0*xsi
      DNi(1,1)= DNi(1,1) - 0.5*DNi(3,1)
      DNi(2,1)= DNi(2,1) - 0.5*DNi(3,1)
   CASE(2)                      ! two-dimensional element
      mxs= 1.0-xsi
      pxs= 1.0+xsi
      met= 1.0-eta
      pet= 1.0+eta
      DNi(1,1)= -0.25*met
      DNi(1,2)= -0.25*mxs
      DNi(2,1)=  0.25*met
      DNi(2,2)= -0.25*pxs
      DNi(3,1)=  0.25*pet
      DNi(3,2)=  0.25*pxs
      DNi(4,1)= -0.25*pet
      DNi(4,2)=  0.25*mxs
      IF(nodes == 4) RETURN    ! linear element finshed
      IF(Inci(5) > 0) THEN    ! zero node = node  missing
       DNi(5,1)= -xsi*met
       DNi(5,2)= -0.5*(1.0 -xsi*xsi)
       DNi(1,1)= DNi(1,1) - 0.5*DNi(5,1)
       DNi(1,2)= DNi(1,2) - 0.5*DNi(5,2)
       DNi(2,1)= DNi(2,1) - 0.5*DNi(5,1)
       DNi(2,2)= DNi(2,2) - 0.5*DNi(5,2)
      END IF
      IF(Inci(6) > 0) THEN
       DNi(6,1)= 0.5*(1.0 -eta*eta)
       DNi(6,2)= -eta*pxs
       DNi(2,1)= DNi(2,1) - 0.5*DNi(6,1)
       DNi(2,2)= DNi(2,2) - 0.5*DNi(6,2)
       DNi(3,1)= DNi(3,1) - 0.5*DNi(6,1)
       DNi(3,2)= DNi(3,2) - 0.5*DNi(6,2)
      END IF
      IF(Inci(7) > 0) THEN
       DNi(7,1)= -xsi*pet
       DNi(7,2)= 0.5*(1.0 -xsi*xsi)
       DNi(3,1)= DNi(3,1) - 0.5*DNi(7,1)
       DNi(3,2)= DNi(3,2) - 0.5*DNi(7,2)
       DNi(4,1)= DNi(4,1) - 0.5*DNi(7,1)
       DNi(4,2)= DNi(4,2) - 0.5*DNi(7,2)
      END IF
      IF(Inci(8) > 0) THEN
       DNi(8,1)= -0.5*(1.0-eta*eta)
       DNi(8,2)= -eta*mxs
       DNi(4,1)= DNi(4,1) - 0.5*DNi(8,1)
```

```
        DNi(4,2)= DNi(4,2) - 0.5*DNi(8,2)
        DNi(1,1)= DNi(1,1) - 0.5*DNi(8,1)
        DNi(1,2)= DNi(1,2) - 0.5*DNi(8,2)
       END IF
      CASE DEFAULT            !   error message
 CALL Error_message('Element dimension not 1 or 2' )
END SELECT
RETURN
END SUBROUTINE Serendip_deriv
```

The computation of the vector normal to the surface and the Jacobian is combined in one SUBROUTINE Normal_Jac.

```
SUBROUTINE Normal_Jac(v3,Jac,xsi,eta,ldim,nodes,inci,coords)
!-------------------------------------------------------
! Computes normal vector and Jacobian
!-------------------------------------------------------
REAL,INTENT(OUT)    :: v3(:)      ! Vector normal to point
REAL,INTENT(OUT)    :: Jac        ! Jacobian
REAL, INTENT(IN)    :: xsi,eta    ! intrinsic coords of point
INTEGER,INTENT(IN)  :: ldim       ! element dimension
INTEGER,INTENT(IN)  :: nodes      ! number of nodes
INTEGER,INTENT(IN)  :: inci(:)    ! element incidences
REAL, INTENT(IN)    :: coords(:,:)! node coordinates
REAL,ALLOCATABLE    :: DNi(:,:)   ! Derivatives of Ni
REAL,ALLOCATABLE    :: v1(:),v2(:)! Vectors in xsi,eta dir
INTEGER :: Cdim    !   Cartesian dimension
Cdim= ldim+1
!Allocate temporary arrays
ALLOCATE (DNi(nodes,Cdim),V1(Cdim),V2(Cdim))
!Compute derivatives of shape function
Call Serendip_deriv(DNi,xsi,eta,ldim,nodes,inci)
! Compute vectors in xsi (eta) direction(s)
DO I=1,Cdim
  V1(I)= DOT_PRODUCT(DNi(:,1),COORDS(I,:))
  IF(ldim == 2) THEN
    V2(I)= DOT_PRODUCT(DNi(:,2),COORDS(I,:))
  END IF
END DO
!Compute normal vector
IF(ldim == 1) THEN
  v3(1)= V1(2)
  v3(2)= -v1(1)
ELSE
  V3= Vector_ex(v1,v2)
END IF
!Normalise
CAll Vector_norm(V3,Jac)
DEALLOCATE (DNi,V1,V2)
RETURN
END SUBROUTINE Normal_Jac
```

3.10 INTEGRATION OVER ELEMENTS

The functions to be integrated over elements will be quite complex so they require numerical treatment. Therefore the main reason for selecting a range of +1 to –1 for the intrinsic coordinates is to enable the use of numerical integration over the elements.

3.10.1 Integration over boundary elements

To compute the real length S^e of an element using local integration variables we have

$$S^e = \int_{-1}^{1} J d\xi \tag{3.58}$$

where the *Jacobian* J is given by equation (3.49).

Similarly, the area of a two-dimensional boundary element A^e is computed by

$$A^e = \int_{-1}^{+1}\int_{-1}^{+1} J\, d\xi d\eta \tag{3.59}$$

where the J is given by equation (3.55). For a one-dimensional infinite element the *Jacobian* is given by

$$J = \frac{\partial x}{\partial \xi} = \frac{\partial N_1^{\infty}}{\partial \xi} x_1^e + \frac{\partial N_2^{\infty}}{\partial \xi} x_2^e = \frac{2}{(1-\xi^2)}(x_2^e - x_1^e) \tag{3.60}$$

3.10.2 Integration over cells

The integration over 2-D cells is identical to the 2-D boundary elements. For 3-D cells the volume is computed by

$$A^e = \int_{-1}^{+1}\int_{-1}^{+1}\int_{-1}^{+1} J\, d\xi d\eta d\varsigma \tag{3.61}$$

where the *Jacobian* is given by

$$J = \mathrm{Det}\begin{pmatrix} \dfrac{\partial x}{\partial \xi} & \dfrac{\partial y}{\partial \xi} & \dfrac{\partial z}{\partial \xi} \\ \dfrac{\partial x}{\partial \eta} & \dfrac{\partial y}{\partial \eta} & \dfrac{\partial z}{\partial \eta} \\ \dfrac{\partial x}{\partial \zeta} & \dfrac{\partial y}{\partial \zeta} & \dfrac{\partial z}{\partial \zeta} \end{pmatrix} \tag{3.62}$$

3.10.3 Numerical integration

In numerical integration schemes, the integral is approximated by a sum of values of the integrand evaluated at certain points, times a weighting function. For the integration of function $f(\xi)$, for example we can write

$$I = \int_{-1}^{+1} f(\xi)d\xi \approx \sum_{i=1}^{I} W_i f(\xi_i) \tag{3.63}$$

In the above, W_i are weights and ξ_i are the intrinsic coordinates of the integration (*sampling*) points. If the well known trapezoidal rule is used, for example, then I=2, the weights are 1 and the *sampling* points are at +1 and –1. That is

$$I = \int_{-1}^{+1} f(\xi)d\xi \approx f(-1)+f(1) \tag{3.64}$$

However, the trapezoidal rule is much too inaccurate for the functions that we are attempting to integrate. The Gauss Quadrature with a variable number of integration points can be used to integrate more accurately. In this method it is assumed that the function to be integrated can be replaced by a polynomial of the form

$$f(\xi) = a_0 + a_1\xi + a_2\xi^2 + \ldots + a_p\xi^p \tag{3.65}$$

where the coefficients are adjusted in such a way as to give the best fit to $f(\xi)$. We determine the number and location of the sampling points, or *Gauss* points, and the weights by the condition that the given polynomial is integrated exactly.

Table 3.4 Gauss point and degree of polynomial

No. of Gauss points, I	Degree of polynomial p
1	1 (linear)
2	3 (cubic)
3	5 (quintic)

Table 3.5 Gauss point coordinates and weights

I	ξ_i	W_i
1	0.0	2.0
2	0.57735 , -0.57735	1.0,1.0
3	0.77459, 0.0 , -0.77459	0.55555, 0.88888, 0.55555

We find that with increasing degree of polynomial p, we need an increasing number of Gauss points. Table 3.4 gives an overview of the number of Gauss N points needed to integrate a polynomial of degree p up to degree 5. The computed location of the sampling points and the weights are given in Table 3.5 for one to three Gauss points (data for up to 8 Gauss points are given in the program listing). It should be noted here that in the application of numerical integration later in this book the integrands can not be replaced by polynomials. However, it can be assumed that as the rate of variation of the functions is increased more integration points will be required.

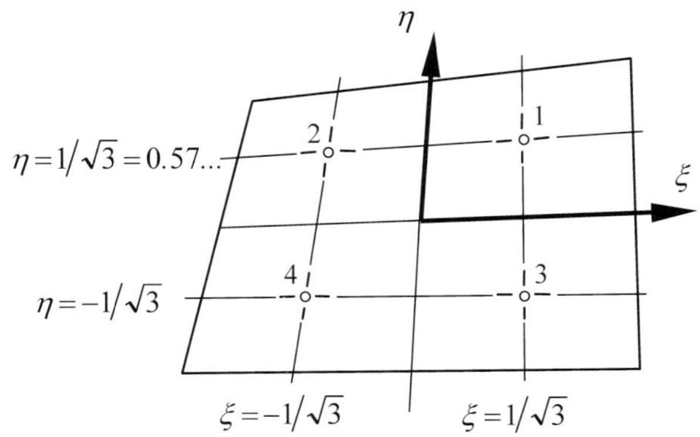

Figure 3.29 Gauss integration points for a two-dimensional element

If we apply the numerical integration to two-dimensional elements or cells then a double sum has to be specified

$$I = \int_{-1}^{+1}\int_{-1}^{+1} f(\xi,\eta)\,d\xi d\eta \approx \sum_{i=1}^{I}\sum_{j=1}^{J} W_i W_j f(\xi_i,\eta_j) \tag{3.66}$$

The Gauss integration points for a two-dimensional element and a 2x2 integration are shown in Figure 3.29. For the integration over 3-D cells we have:

$$I = \int_{-1}^{+1}\int_{-1}^{+1}\int_{-1}^{+1} f(\xi,\eta)\,d\xi d\eta \approx \sum_{i=1}^{I}\sum_{j=1}^{J}\sum_{k=1}^{K} W_i W_j W_k f(\xi_k,\eta_i,\zeta_k) \tag{3.67}$$

A subroutine can be written which returns the Gauss point coordinates and weights depending on the number of Gauss points for an integration order of up to 8.

```
SUBROUTINE Gauss_coor(Cor,Wi,Intord)
!-------------------------------------
! Returns Gauss coords and Weights for up to 8 Gauss points
!-------------------------------------
REAL, INTENT(OUT)   :: Cor(8)    !   Gauss point coordinates
REAL, INTENT(OUT)   :: Wi(8)     !   weigths
INTEGER,INTENT(IN)  :: Intord    !   integration order
SELECT CASE (Intord)
CASE(1)
  Cor(1)= 0.
  Wi(1) = 2.0
CASE(2)
  Cor(1)= .577350269  ; Cor(2)= -Cor(1)
  Wi(1) = 1.0 ;  Wi(2) = Wi(1)
CASE(3)
  Cor(1)= .774596669  ; Cor(2)= 0.0 ; Cor(3)= -Cor(1)
  Wi(1) = .555555555  ; Wi(2) = .888888888 ; Wi(3) = Wi(1)
CASE(4)
  Cor(1)= .861136311 ; Cor(2)= .339981043 ; Cor(3)= -Cor(2)
  Cor(4)= -Cor(1)
  Wi(1) = .347854845 ; Wi(2) = .652145154 ; Wi(3) = Wi(2)
  Wi(4) = Wi(1)
CASE(5)
  Cor(1)= .906179845 ; Cor(2)= .538469310 ; Cor(3)= .0
  Cor(4)= -Cor(2) ; Cor(5)= -Cor(1)
  Wi(1)= .236926885 ; Wi(2)= .478628670 ; Wi(3)= .568888888
  Wi(4) = Wi(2) ; Wi(5) = Wi(1)
CASE(6)
  Cor(1)=.932469514 ; Cor(2)=.661209386 ; Cor(3)=.238619186
  Cor(4)= -Cor(3) ;  Cor(5)= -Cor(2) ; Cor(6)= -Cor(1)
  Wi(1)= .171324492 ; Wi(2)= .360761573 ; Wi(3)= .467913934
  Wi(4) = Wi(3) ; Wi(5) = Wi(2) ; Wi(6) = Wi(1)
CASE(7)
  Cor(1)=.949107912 ; Cor(2)=.741531185 ; Cor(3)=.405845151
  Cor(4)= 0.
  Cor(5)= -Cor(3) ;Cor(6)= -Cor(2) ;Cor(7)= -Cor(1)
  Wi(1)= .129484966 ; Wi(2)= .279705391 ; Wi(3)= .381830050
  Wi(4) = .417959183
  Wi(5) = Wi(3) ; Wi(6) = Wi(2) ; Wi(7) = Wi(1)
CASE(8)
  Cor(1)=.960289856 ; Cor(2)=.796666477 ; Cor(3)=.525532409
  Cor(4)= .183434642
  Cor(5)= -Cor(4) ; Cor(6)= -Cor(3) ; Cor(7)= -Cor(2)
  Cor(8)= -Cor(1)
  Wi(1)= .101228536 ; Wi(2)= .222381034 ; Wi(3)= .313706645
  Wi(4) = .362683783
  Wi(5)= Wi(4) ; Wi(6)= Wi(3) ; Wi(7)= Wi(2) ; Wi(8)= Wi(1)
CASE DEFAULT
  CALL Error_Message('Gauss points not in range 1-8')
END SELECT
END SUBROUTINE Gauss_coor
```

3.11 PROGRAM 3.1: CALCULATION OF SURFACE AREA

We now have developed sufficient library subroutines for writing our first program. The program is intended to calculate the length or surface area of a boundary described by boundary elements. First we define the libraries of subroutines to be used. The names after the **USE** statement refer to the **MODULE** names in the source code which can be downloaded from the web. There are three types of libraries:

- The Geometry_lib, which contains all the shape functions, derivative of the shape functions and the routines to compute the Jacobian and the outward normal.
- The Utility_lib, which contains utility subroutines for computing, for example, vector ex-products, normalising vectors and printing error messages.
- The Integration_lib, which contains Gauss point coordinates and weights.

```
PROGRAM Compute_Area
!---------------------------------------------
!   Program to compute the length/surface area
!   of a line/surface modelled by boundary elements
!---------------------------------
USE Geometry_lib ; USE Utility_lib ; USE Integration_lib
IMPLICIT NONE
INTEGER :: ldim,noelem,nelem,lnodes,maxnod,node,Cdim
INTEGER,ALLOCATABLE :: inciG(:,:)! Incidences
INTEGER,ALLOCATABLE :: inci(:)! Incidences one element
REAL,ALLOCATABLE :: corG(:,:) ! Coordinates (all nodes)
REAL,ALLOCATABLE :: cor(:,:)  ! Coordinates one element
REAL,ALLOCATABLE :: v3(:)     ! Normal vector
REAL :: Gcor(8),Wi(8) ! Gauss point coords and weights
REAL :: Jac, xsi, eta, Area
OPEN(UNIT=10,FILE='INPUT.DAT',STATUS='OLD')
OPEN(UNIT=11,FILE='OUTPUT.DAT',STATUS='UNKNOWN')
READ(10,*) ldim,lnodes,noelem,intord
WRITE(11,*) ' Element dimension=',ldim
WRITE(11,*) ' No. of elem.nodes=',lnodes
WRITE(11,*) ' Number of elements=',noelem
WRITE(11,*) ' Integration order =',intord
Cdim= ldim+1    !Cartesian dimension
ALLOCATE(v3(Cdim))
ALLOCATE(inciG(8,noelem))! Allocate global incid. Array
DO nelem=1,noelem
  READ(10,*) (inciG(n,nelem),n=1,lnodes)
END DO
maxnod= MAXVAL(inciG)
ALLOCATE(corG(Cdim,0:maxnod)!Allocate array for coords
corG(:,0)= 0.0!Node No 0 means node is missing
DO node=1,maxnod
  READ(10,*) (corG(i,node),i=1,Cdim)
```

```
END DO
ALLOCATE(inci(lnodes),cor(Cdim,lnodes))
CALL Gauss_coor(Gcor,Wi,Intord)! Gauss coordinates and weigths
Area= 0.0 !   Start sum for area/length
Element_loop: &
DO nelem=1,noelem
  inci=  inciG(:,nelem)!   Store incidences locally
  cor= corG(:,inci)!   gather element coordinates
  SELECT CASE (ldim)
   CASE (1)!   One-dim. problem determine length
    Gauss_loop:&
    DO I=1,INTORD
      xsi= Gcor(i)
      CALL Normal_Jac(v3,Jac,xsi,eta,ldim,lnodes,inci,cor)
      Area= Area + Jac*Wi(i)
    END DO &
    Gauss_loop
   CASE (2)!   Two-dim. problem determine area
    Gauss_loop1:&
    DO I=1,INTORD
      DO j=1,INTORD
        xsi= Gcor(i)
        eta= Gcor(j)
        CALL Normal_Jac(v3,Jac,xsi,eta,ldim,lnodes,inci,cor)
        Area= Area + Jac*Wi(i)*Wi(j)
      END DO
    END DO &
    Gauss_loop1
   CASE DEFAULT
  END SELECT
END DO &
Element_loop
IF(ldim == 1) THEN
 WRITE(11,*) ' Length =',Area
ELSE
 WRITE(11,*) ' Area =',Area
END IF
END PROGRAM Compute_Area
```

We define allocable arrays for storing the incidences of all elements, the incidences of one element, the coordinates of all node points, the coordinates of all nodes of one element and the vector normal to the surface. The dimensions of these arrays depend on the element dimension (one-dimensional, two-dimensional), the number of element nodes (linear/parabolic shape function) and the number of elements and nodes. The dimension of these arrays will be allocated once this information is known.

The first executable statements read the information necessary to allocate the dynamic arrays and the integration order to be used for the example. Here we use two files INPUT.DAT and OUTPUT.DAT for input and output. The input file has to be created by the user before the program can be run. The FORMAT of inputting data is free-field, that is, numbers can be separated by blanks. After reading the general information the incidences (connectivity) are read for all elements and stored in array InciG. While reading this information, we find the largest node number, information which we need for allocating the dimension for the array containing node coordinates and which we do next before reading the node coordinates. We make use of the new feature in F90, that allows the subscripts of an array to start with zero, because a transition element that has the midside node missing will have a node number of 0 in the incidences. We assign zero coordinates to node number 0.

We loop over all elements describing the boundary. For each element we get the Gauss point coordinates and weightings, by a call to Gaus_coor, which correspond to the integration order Specified by Intord. We then add all the Gauss point contributions, i.e. the Jacobians computed (by a call to **Normal_jac**) for each Gauss point multiplied by the weighting. Note that there are two cases to be considered: for a one-dimensional case, that is, if we work out a length of a curve, only one DO LOOP is required. For two-dimensional cases, that is, when we work out surface areas, two nested DO LOOPS are required (see equation 3.66).

3.12 CONCLUDING REMARKS

In this chapter we have dealt with methods for describing the geometry of a problem and have concentrated on describing problem boundaries. The method consists of subdividing the boundary into small elements and is commonly known as *discretisation.* The concept of isoparametric elements was introduced, where we use interpolation functions to describe the boundary surface in terms of nodal values and the variation of known or unknown values. We have laid here the foundation for Chapter 6 (Boundary Element Methods), where we will use the concepts described. We find that, once we use this advanced discretisation method in the BEM, the analytical integration is no longer possible. Therefore, we have also introduced the Gauss Quadrature method of numerical integration, most commonly used in numerical work. For general purpose programs using the isoparametric concept, the accuracy of the numerical integration will be crucial. We have started here our process of building a Subroutine library which will be needed later. A small program has been written which we can use to test the subroutines and to do numerical experiments.

3.13 EXERCISES

Exercise 3.1

Using program Compute_area calculate the length of a quarter circle using:

(a) one linear element
(b) two linear elements
(c) one quadratic element

Determine the discretisation error. Use 2x2 integration.

Exercise 3.2
Using program Compute_area, calculate the area of a quarter circle using:

(a) discretisation into one quadrilateral element, as shown in Figure 3.30 (a)
(b) discretisation into three quadrilateral elements, as shown in Figure 3.30 (b)

Plot the variation of the *Jacobian* over the element using the Gauss point values. Determine the discretisation error.

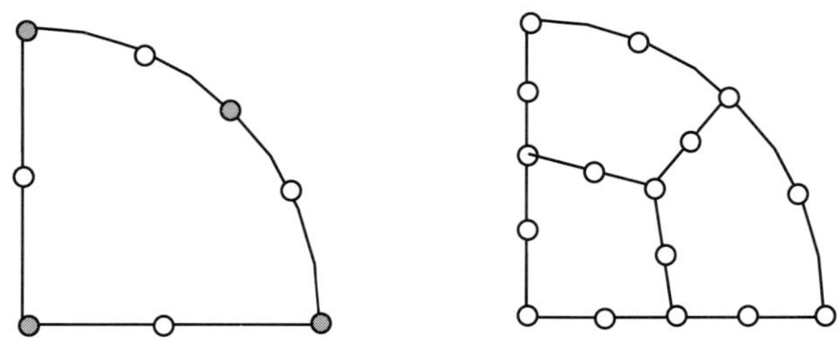

(a) (b)

Figure 3.30 Discretisations for determining the area of a quarter circle

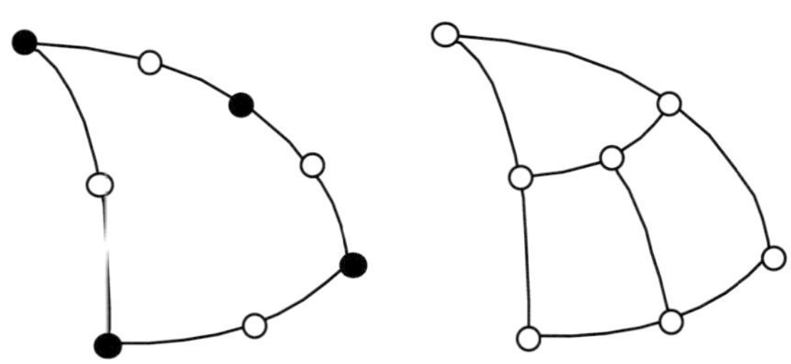

(a) (b)

Figure 3.31 Discretisations for determining the surface area of 1/8 sphere

Exercise 3.3.
Using program Compute_area, calculate the area of 1/8 of a sphere using

(a) discretisation into one quadrilateral element, as shown in Figure 3.31 (a)
(b) discretisation into three quadrilateral elements as shown in Figure 3.31 (b)

Plot the variation of the *Jacobian* over the element using the Gauss point values. Determine the discretisation error.

3.14 REFERENCES

1. Encarnacao, J., and Schlechtendahl, J. (1983) Computer Aided Design. Springer Verlag, Berlin.
2. Irons, B. M. (1966) Engineering applications of numerical integration in stiffness method, *J.A.I.A.A.*, **14**, 2035-7.
3. Zienkiewicz, O. C., Irons, B. M., Ergatoudis, J. G., Ahmad, S., and Scott, F. C. (1969) Iso-parametric and associate element families for two- and three-dimensional analysis. *Finite Element Methods in Stress Analysis* (eds. I. Holland and K. Bell), Tapir Press, Norway, Ch. 14.
4. Bathe K-J (1995), Finite Element Procedures. Prentice Hall.
5. Moser W., Duenser Ch., Beer G. (2004) Mapped infinite elements for three-dimensional multi-region boundary element analysis. *Int. J. Numer. Meth. Engng,* **61**: 317 - 328
6. Fraejis de Veubeke, B. (1965) Displacement and equilibrium models in the finite element method. *Stress Analysis*, (eds. O.C. Zienkiewicz and G.S. Holister), John Wiley, Chichester, Ch. 9.

4

Material Modelling and Fundamental Solutions

If you can measure what you are speaking about, and express it in numbers, you know something about it.

Lord Kelvin

4.1. INTRODUCTION

In addition to specifying the geometry of the problem, it is necessary to describe the physical response of the material in a mathematical way. This is done by defining the response characteristics of an infinitesimally small portion of the solid. The *constitutive* law establishes a relationship between heat flow and the temperature gradient or between strain and stress. The constants in such relationships are characteristic values or properties of the material. We distinguish between material properties which are direction independent (*isotropic* material), and those which are dependent on direction (*anisotropic* material). Furthermore, there are problems where the same properties apply everywhere (*homogeneous* problems) and where properties change from location to location (*non-homogeneous* problems).

In the material response we distinguish between linear and non-linear behaviour. For linear materials we can establish a unique (linear) relationship between stress/strain, temperature/heat flow or potential/fluid flow. For non-linear material behaviour, this relationship depends on the current state and can therefore only be written in incremental form. These problems are therefore dependent on the deformation (thermal) history.

As outlined previously, for the boundary element method a solution of the governing equation has to be available. In nearly all cases, the solution is obtained for very simple loading conditions (point load or source) and for infinite or semi-infinite domains. In the literature, these solutions are referred to as *fundamental solutions*, *Green's functions* or *Kernels*. Obviously, these solutions can only be found for linear material behaviour and for a homogeneous domain.

The fundamental solutions have to satisfy three conditions:

- Constitutive law
- Equilibrium or conservation of energy
- Compatibility or continuity

The last condition will be automatically satisfied for solutions which are continuous in the domain. In the following, we will first derive the governing differential equations and then present fundamental solutions for potential problems (heat flow and seepage) and for elasticity problems in two and three dimensions.

4.2. STEADY STATE POTENTIAL PROBLEMS

Heat conduction in solids and flow in porous media (seepage) are diffusion problems and can be treated concurrently, because they are governed by the same differential equation (Laplace).

Steady state heat flux or fluid flow $\mathbf{q}$ per unit area is related to temperature or potential u by

$$\mathbf{q} = -\mathbf{D}\nabla u \tag{4.1}$$

where the negative sign is due to the fact that the flow is always from higher to lower temperature/potential. The flow vector is defined as:

$$\mathbf{q} = \begin{Bmatrix} q_x \\ q_y \\ q_z \end{Bmatrix} \tag{4.2}$$

The conductivity/pemeabilty tensor $\mathbf{D}$ is given by

$$\mathbf{D} = \begin{bmatrix} k_{xx} & k_{xy} & k_{xz} \\ k_{yx} & k_{yy} & k_{yz} \\ k_{zx} & k_{zy} & k_{zz} \end{bmatrix} \tag{4.3}$$

where k_{xx} etc, are conductivities measured in [W/°K-m] in the case of thermal problems and permeabilities measured in [m/sec], in the case of seepage problems. The coefficients in $\mathbf{D}$ represent flow values for unit values of temperature gradient or potential gradient. It can be shown that $\mathbf{D}$ has to be symmetric and positive definite.

The differential operator ∇ for three-dimensional problems is defined as

$$\nabla = \begin{Bmatrix} \frac{\partial}{\partial x} \\ \frac{\partial}{\partial y} \\ \frac{\partial}{\partial z} \end{Bmatrix} \tag{4.4}$$

and for two-dimensional problems

$$\nabla = \begin{Bmatrix} \frac{\partial}{\partial x} \\ \frac{\partial}{\partial y} \end{Bmatrix} \tag{4.5}$$

The conservation of energy condition states that the outflow must be equal to the inflow, plus any flow per unit volume, $\hat{Q}$, generated by a source.

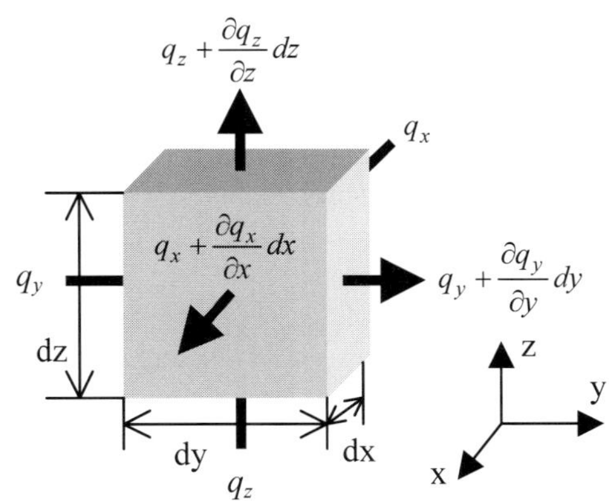

Figure 4.1 Heat flow in an infinitesimal cube

For the infinitesimal cube in Figure 4.1 this gives the following

$$\left(q_x + \frac{\partial q_x}{\partial x} dx\right) dydz + \left(q_y + \frac{\partial q_y}{\partial y} dy\right) dxdz + \left(q_z + \frac{\partial q_z}{\partial z} dz\right) dxdy = \tag{4.6}$$
$$q_x dydz + q_y dxdz + q_z dxdy + \hat{Q} dxdydz$$

After cancelling terms we obtain

$$\frac{\partial q_x}{\partial x}+\frac{\partial q_y}{\partial y}+\frac{\partial q_z}{\partial z}-\hat{Q}=0 \tag{4.7}$$

Substituting the Fourier law for isotropic material (i.e., $k_{xx}= k_{yy}= k_{zz}= k$ and $k_{xy}= k_{xz}= k_{yz}=0$) we obtain the governing differential equation for which we seek a fundamental solution.

$$k\left(\frac{\partial^2 u}{\partial x^2}+\frac{\partial^2 u}{\partial y^2}+\frac{\partial^2 u}{\partial z^2}\right)-\hat{Q}=0 \tag{4.8}$$

The simplest solution we can find is that of a concentrated source at point *P (source point)* of magnitude one in an infinite homogeneous domain. This means that internal heat or flow generation only occurs at one point *(P)* in the domain and is zero elsewhere. The function describing this variation is also referred to a *Dirac Delta* function which is defined as

$$\begin{aligned}&\delta(P\text{-}Q)=0 \qquad \text{when } P\neq Q\\ &\int_\Omega \delta(P\text{-}Q)\,d\Omega=1\end{aligned} \tag{4.9}$$

where Q is a point in the domain Ω. Due to a unit point source at P the temperature or potential at point *Q (field point)* can be written for the three-dimensional case as

$$U(P,Q)=\frac{1}{4\pi\, rk} \tag{4.10}$$

where $r=\sqrt{(x_Q-x_P)^2+(y_Q-y_P)^2+(z_Q-z_P)^2}$ is the distance between source point and field point (Figure 4.2).

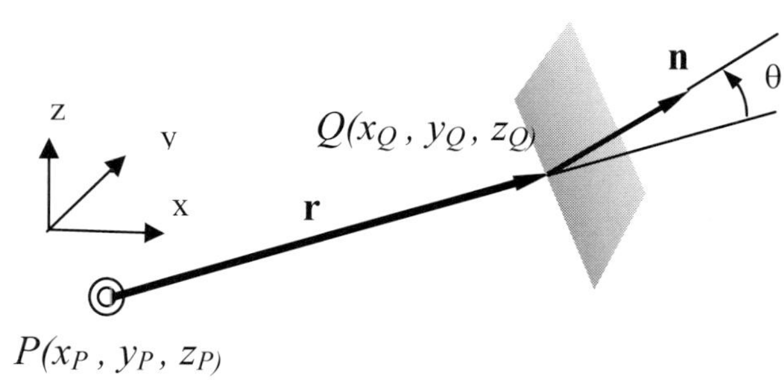

Figure 4.2 Notation for fundamental solution (three-dimensional potential problems)

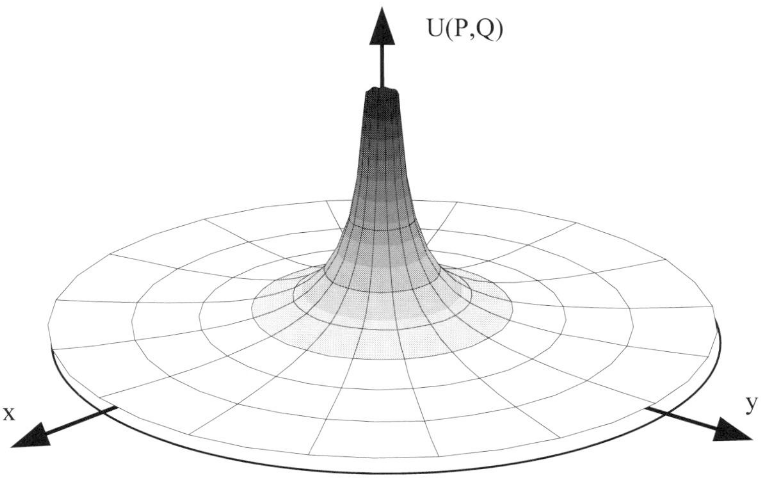

Figure 4.3 Variation of fundamental solution *U* (potential/temperature) in the x-y plane for 3-D potential problems (source at origin of coordinate system)

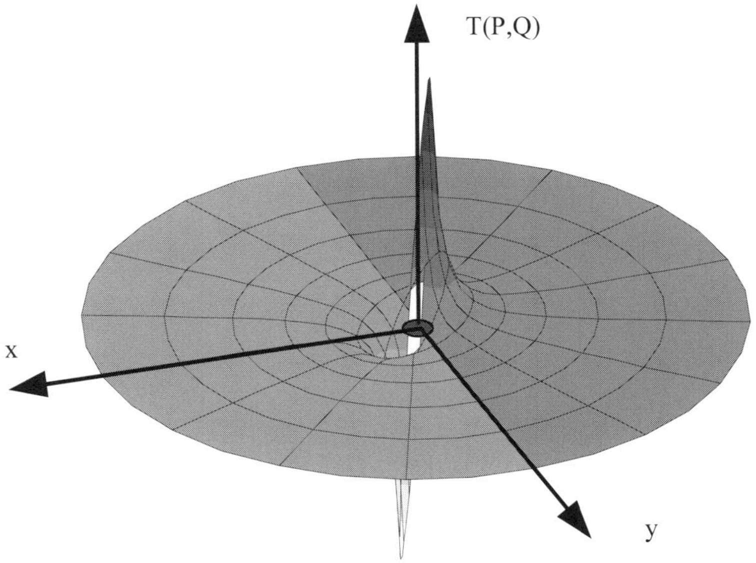

Figure 4.4 Variation of fundamental solution T for **n** = {1,0,0} (flow in x-direction) in x-y plane for 3-D potential problems

As we will see later, the flow in a direction normal to a boundary defined by a vector **n** is also required. For three-dimensional isotropic problems, the flow is computed by

$$T(P,Q) = -k\frac{\partial U}{\partial \mathbf{n}} = -k\left(n_x\frac{\partial U}{\partial x} + n_y\frac{\partial U}{\partial y} + n_z\frac{\partial U}{\partial z}\right) \tag{4.11}$$

The derivatives of U in the global directions are

$$\frac{\partial U}{\partial x} = \frac{-r_{,x}}{4\pi r^2 k} \quad ; \quad \frac{\partial U}{\partial y} = \frac{-r_{,y}}{4\pi r^2 k} \quad ; \quad \frac{\partial U}{\partial z} = \frac{-r_{,z}}{4\pi r^2 k} \tag{4.12}$$

where

$$r_{,x} = \frac{x_Q - x_P}{r} \quad ; \quad r_{,y} = \frac{y_Q - y_P}{r} \quad ; \quad r_{,z} = \frac{z_Q - z_P}{r} \tag{4.13}$$

Equation (4.11) can be rewritten as

$$T(P,Q) = \frac{\cos\theta}{4\pi r^2} \tag{4.14}$$

where θ is defined as the angle between the normal vector **n** and the distance vector **r**, i.e.

$$\cos\theta = \frac{1}{r}\mathbf{n} \bullet \mathbf{r} \quad with \quad \mathbf{n} = \left\{ n_x, n_y, n_z \right\}^{\mathrm{T}} \quad and \quad \mathbf{r} = \left\{ r_x, r_y, r_z \right\}^{\mathrm{T}} \tag{4.15}$$

The variation of kernels U and T is plotted in Figures 4.3 and 4.4. It can be seen that both solutions decay very rapidly from the value of infinity at the source. Whereas the fundamental solution for U is symmetric with respect to polar coordinates, the solution for T with the vector **n** pointing in x-direction (thus meaning flow in x-direction) is antisymmetric.

For a two-dimensional problem, the source is assumed to be distributed along a line of infinite length from $z = -\infty$ to $z = +\infty$ and the fundamental solutions are given by

$$U(P,Q) = \frac{1}{2\pi k} ln\left(\frac{1}{r}\right) \tag{4.16}$$

and

$$T(P,Q) = -k\frac{\partial U}{\partial \mathbf{n}} = -k\left(n_x\frac{\partial U}{\partial x} + n_y\frac{\partial U}{\partial y}\right) \tag{4.17}$$

where

$$\frac{\partial U}{\partial x} = \frac{-r_{,x}}{2\pi rk} \quad ; \quad \frac{\partial U}{\partial y} = \frac{-r_{,y}}{2\pi rk} \tag{4.18}$$

Equation (4.16) can also be rewritten as

$$T(P,Q) = \frac{\cos\theta}{2\pi r} \tag{4.19}$$

where

$$\cos\theta = \frac{1}{r}\mathbf{n} \bullet \mathbf{r} \quad with \quad \mathbf{n} = \{n_x, n_y\} \quad and \quad \mathbf{r} = \{r_x, r_y\} \tag{4.20}$$

Subroutines for the isotropic solutions are presented below.

```
MODULE Laplace_lib
REAL :: PI=3.14159265359
CONTAINS
  REAL FUNCTION U(r,k,Cdim)
  !   Fundamental solution for Potential problems
  !   Temperature/Potential isotropic material
  REAL,INTENT(IN)      :: r    !  Distance source and field point
  REAL,INTENT(IN)      :: k    !   Conducivity
  INTEGER,INTENT(IN)   :: Cdim !   Cartesian dimension (2-D,3-D)
  SELECT CASE (CDIM)
     CASE (2)             !  Two-dimensional solution
        U= 1.0/(2.0*Pi*k)*LOG(1/r)
     CASE (3)             !  Three-dimensional solution
        U= 1.0/(4.0*Pi*r*k)
     CASE DEFAULT
        U=0.0
        WRITE (11,*)'Cdim not equal 2 or 3 in Function U(...)'
  END SELECT
  END FUNCTION U
  REAL FUNCTION T(r,dxr,Vnorm,Cdim)
 !   Fundamental solution for Potential problems
 !   Flow, isotropic material
 REAL,INTENT(IN)::          r    !   Distance source and field point
 REAL,INTENT(IN)::    dxr(:)    !   r,x,r,y,r,z
 REAL,INTENT(IN):: Vnorm(:)     !   Normal vector
 INTEGER,INTENT(IN) ::  Cdim !   Cartesian dimension
 SELECT CASE (Cdim)
    CASE (2)              !  Two-dimensional solution
      T= DOT_PRODUCT (Vnorm,dxr)/(2.0*Pi*r)
    CASE (3)              !  Three-dimensional solution
      T= DOT_PRODUCT (Vnorm,dxr)/(4.0*Pi*r*r)
    CASE DEFAULT
      T=0.0
      WRITE (11,*)'Cdim not equal 2 or 3 in Function T(...)'
 END SELECT
 END FUNCTION T
END MODULE Laplace_lib
```

For anisotropic problems the fundamental solutions have been presented by Bonnet[1]. For example, the solution for temperature/potential is given by

$$U(P,Q)=\frac{1}{2\pi \bar{k}} ln\frac{1}{\bar{r}} \tag{4.21}$$

for two-dimensional problems and

$$U(P,Q)=\frac{1}{4\pi \bar{r}\bar{k}} \tag{4.22}$$

for three-dimensional problems, where

$$\bar{k} = \sqrt{det\,\mathbf{D}} \quad and \quad \bar{r} = \mathbf{r}^T\mathbf{D}^{-1}\mathbf{r} \tag{4.23}$$

For general anisotropy in three dimensions, **D** has 9 material parameters but, because of the property of symmetry, only 6 components need to be input. A special case of anisotropy exists where the material parameters are different in three orthogonal coordinate directions. This is known as *orthotropic* material. If these conductivities are defined in the direction of global coordinates, then all off-diagonal elements of **D** are zero. If we denote the conductivities in x,y and z-directions as k_1, k_2, k_3 then

$$\mathbf{D} = \begin{bmatrix} k_1 & 0 & 0 \\ 0 & k_2 & 0 \\ 0 & 0 & k_3 \end{bmatrix} \tag{4.24}$$

For this case the values in equation (4.23) are given by:

$$\bar{k} = \sqrt{k_1 k_2 k_3} \quad and \quad \bar{r} = \sqrt{r_x^2\frac{1}{k_1} + r_y^2\frac{1}{k_2} + r_z^2\frac{1}{k_3}} \tag{4.25}$$

4.3. STATIC ELASTICITY PROBLEMS

In solid mechanics applications, a relationship between stress and strain must be established. Stresses are forces per unit area inside a solid. They can be visualised by cutting the solid on planes parallel to the axes and by showing the traction vectors acting on these planes (in Figure 4.5).

The traction vectors acting on the three planes are defined as:

$$\mathbf{t}_1 = \begin{pmatrix} \sigma_x \\ \tau_{xy} \\ \tau_{xz} \end{pmatrix} \; ; \; \mathbf{t}_2 = \begin{pmatrix} \tau_{yx} \\ \sigma_y \\ \tau_{yz} \end{pmatrix} \quad \mathbf{t}_3 = \begin{pmatrix} \tau_{zx} \\ \tau_{zy} \\ \sigma_z \end{pmatrix} \tag{4.26}$$

The components of the traction vectors are also known as *stress* components.

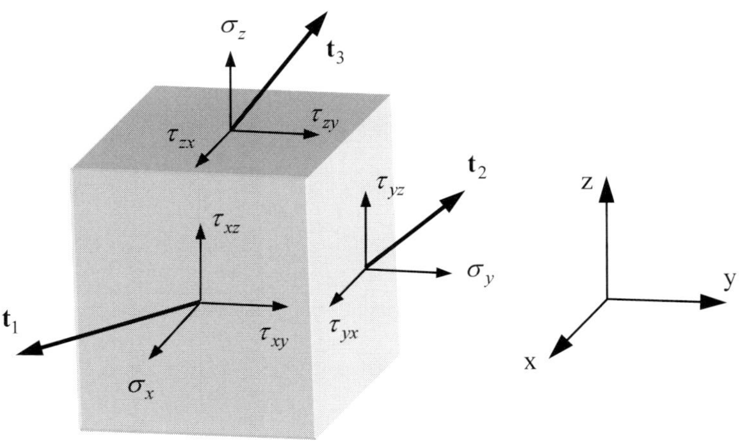

Figure 4.5 Tractions acting on the faces of an infinitesimal cube and stress components

Using the condition for rotational equilibrium ($\tau_{xy} = \tau_{yx}$; $\tau_{xz} = \tau_{zx}$; $\tau_{yz} = \tau_{zy}$) only 6 unique traction components remain and may be put into a *pseudo stress-vector*

$$\boldsymbol{\sigma} = \begin{Bmatrix} \sigma_x \\ \sigma_y \\ \sigma_z \\ \tau_{xy} \\ \tau_{yz} \\ \tau_{xz} \end{Bmatrix} \tag{4.27}$$

In *plane stress* problems, such as in thin plates subject to in-plane loading, all stresses associated with the *z* direction are assumed to be zero, i.e., $\sigma_z = \tau_{xz} = \tau_{yz} = 0$.

The components of traction vector **t** acting on a general plane defined by a normal vector **n** $\{n_x, n_y, n_z\}$ that is not parallel to one of the axis planes can be expressed in terms of stress components by (Figure 4.6)

$$
\begin{aligned}
t_x &= n_x\sigma_x + n_y\tau_{xy} + n_z\tau_{xz} \\
t_y &= n_y\sigma_y + n_x\tau_{xy} + n_z\tau_{yz} \\
t_z &= n_z\sigma_z + n_x\tau_{zx} + n_y\tau_{zy}
\end{aligned}
\qquad (4.28)
$$

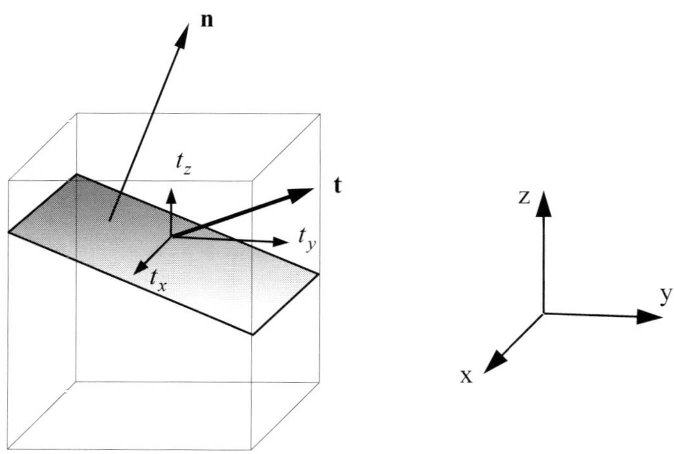

Figure 4.6 Definition of traction vector acting on a general plane

Infinitesimal strains are defined in terms of displacement components in the *x, y, z* directions (u_x, u_y, u_z) by

$$
\begin{aligned}
\varepsilon_x &= \frac{\partial u_x}{\partial x} \\
\varepsilon_y &= \frac{\partial u_y}{\partial y} \\
\varepsilon_z &= \frac{\partial u_z}{\partial z} \\
\gamma_{xy} &= \frac{\partial u_x}{\partial y} + \frac{\partial u_y}{\partial x} \\
\gamma_{yz} &= \frac{\partial u_y}{\partial z} + \frac{\partial u_z}{\partial y} \\
\gamma_{zx} &= \frac{\partial u_z}{\partial x} + \frac{\partial u_x}{\partial z}
\end{aligned}
\qquad (4.29)
$$

These can be put into a pseudo-vector

$$\boldsymbol{\varepsilon} = \begin{Bmatrix} \varepsilon_x \\ \varepsilon_y \\ \varepsilon_z \\ \gamma_{xy} \\ \gamma_{yz} \\ \gamma_{zx} \end{Bmatrix} \tag{4.30}$$

In matrix form this can be written as

$$\boldsymbol{\varepsilon} = \mathbf{B}\,\mathbf{u} \tag{4.31}$$

where **u** is a vector of displacements

$$\mathbf{u} = \begin{Bmatrix} u_x \\ u_y \\ u_z \end{Bmatrix} \tag{4.32}$$

and B is a differential operator matrix

$$\mathbf{B} = \begin{bmatrix} \frac{\partial}{\partial x} & 0 & 0 \\ 0 & \frac{\partial}{\partial y} & 0 \\ 0 & 0 & \frac{\partial}{\partial z} \\ \frac{\partial}{\partial y} & \frac{\partial}{\partial x} & 0 \\ 0 & \frac{\partial}{\partial z} & \frac{\partial}{\partial y} \\ \frac{\partial}{\partial z} & 0 & \frac{\partial}{\partial x} \end{bmatrix} \tag{4.33}$$

In some circumstances, simplifications can be made and certain strain components taken to be zero. A state of *plane strain* can be assumed, if the solid extends a long distance in the z-direction, the loading is uniform in this direction and $u_z = 0$ everywhere. We then have $\varepsilon_z = \gamma_{xz} = \gamma_{yz} = 0$. Another special case is a state of complete plane strain, in which derivatives in the z direction of all displacements are taken to be zero, but u_z may be non-zero.

This gives

$$
\begin{aligned}
\varepsilon_{x} &= \frac{\partial u_{x}}{\partial x} \\
\varepsilon_{y} &= \frac{\partial u_{y}}{\partial y} \\
\varepsilon_{z} &= 0 \\
\gamma_{xy} &= \frac{\partial u_{x}}{\partial y} + \frac{\partial u_{y}}{\partial x} \quad , \quad \gamma_{yz} = \frac{\partial u_{z}}{\partial y} \quad , \quad \gamma_{xz} = \frac{\partial u_{z}}{\partial x}
\end{aligned}
\tag{4.34}
$$

Complete plane strain can be split into the plane strain case already discussed and an antiplane strain or St Venant torsion component for which $\varepsilon_x = \varepsilon_y = \varepsilon_z = \gamma_{xy} = 0$ and

$$
\begin{aligned}
\gamma_{zy} &= \frac{\partial u_{z}}{\partial y} \\
\gamma_{zx} &= \frac{\partial u_{z}}{\partial x}
\end{aligned}
\tag{4.35}
$$

In complete plane strain it is possible to have shear strains and stresses acting in the *z*-direction.

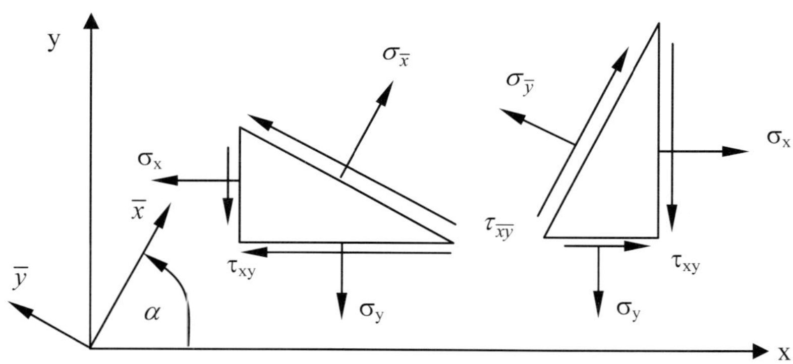

Figure 4.7 Transformation of stresses in two dimensions

Sometimes it is necessary to compute the magnitudes of stress or strain in directions which do not coincide with the global axes. In this case a *transformation* of stress or strain is necessary. The transformation of local stresses $\bar{\boldsymbol{\sigma}}$ acting on planes in the material parallel with the $\bar{x}, \bar{y}, \bar{z}$ axes to global stresses $\boldsymbol{\sigma}$ acting on cuts parallel with the *x, y, z* axes can be written as

$$
\bar{\boldsymbol{\sigma}} = \mathbf{T}_{\sigma} \boldsymbol{\sigma} \tag{4.36}
$$

For the two-dimensional case, in which the local axes are defined by a rotation about the z-axis, $\mathbf{T}_\sigma$ is obtained by the two transformations shown in Figure 4.7

$$\mathbf{T}_\sigma = \begin{bmatrix} \cos^2\alpha & \sin^2\alpha & -2\cos\alpha\sin\alpha \\ \sin^2\alpha & \cos^2\alpha & 2\sin\alpha\cos\alpha \\ \cos\alpha\sin\alpha & -\cos\alpha\sin\alpha & \cos^2\alpha - \sin^2\alpha \end{bmatrix} \tag{4.37}$$

For the stress transformation in three-dimensional space it is convenient to refer to the components of unit vectors in the directions of the local axes (Figure 4.8).

For example, we denote by

$$\mathbf{v}_1 = \begin{Bmatrix} v_{1x} \\ v_{1y} \\ v_{1z} \end{Bmatrix} \tag{4.38}$$

the unit vector in the direction of the $\bar{x}$ - axis.

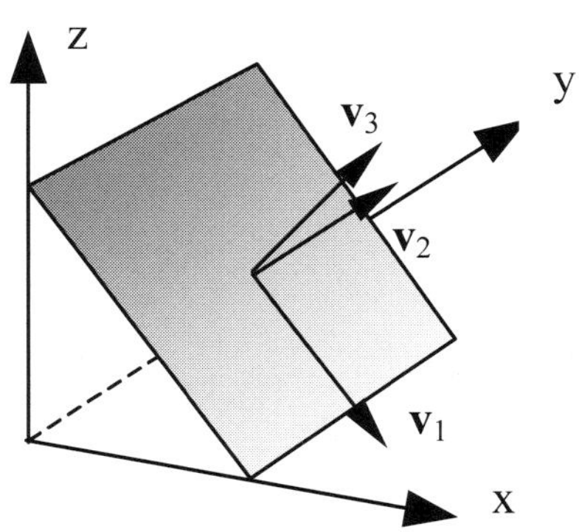

Figure 4.8 Definition of unit vectors for the transformation of stresses in 3-D

Similarly, $\mathbf{v}_2$ and $\mathbf{v}_3$ are unit vectors along the $\bar{y}$ - and $\bar{z}$ - axes. In terms of these vector components, the matrix $\mathbf{T}_\sigma$ is written as

$$\mathbf{T}_\sigma = \begin{bmatrix} \mathbf{T}_{\sigma 11} & \mathbf{T}_{\sigma 12} \\ \mathbf{T}_{\sigma 21} & \mathbf{T}_{\sigma 22} \end{bmatrix} \tag{4.39}$$

where

$$\mathbf{T}_{\sigma 11} = \begin{bmatrix} v_{1x}^2 & v_{2x}^2 & v_{3x}^2 \\ v_{1y}^2 & v_{2y}^2 & v_{3y}^2 \\ v_{1z}^2 & v_{2z}^2 & v_{3z}^2 \end{bmatrix}$$

$$\mathbf{T}_{\sigma 12} = \begin{bmatrix} 2v_{1x}v_{2x} & 2v_{2x}v_{3x} & 2v_{1x}v_{3x} \\ 2v_{1y}v_{2y} & 2v_{2y}v_{3y} & 2v_{1y}v_{3y} \\ 2v_{1z}v_{2z} & 2v_{2z}v_{3z} & 2v_{1z}v_{3z} \end{bmatrix}$$

$$\mathbf{T}_{\sigma 21} = \begin{bmatrix} v_{1x}v_{1y} & v_{2x}v_{2y} & v_{3y}v_{3x} \\ v_{1y}v_{1z} & v_{2y}v_{2z} & v_{3y}v_{3z} \\ v_{1x}v_{1z} & v_{2x}v_{2z} & v_{3x}v_{3z} \end{bmatrix} \tag{4.40}$$

$$\mathbf{T}_{\sigma 22} = \begin{bmatrix} v_{1x}v_{2y}+v_{1y}v_{2x} & v_{2x}v_{3y}+v_{2y}v_{3x} & v_{1x}v_{3y}+v_{1y}v_{3x} \\ v_{1y}v_{2z}+v_{1z}v_{2y} & v_{2y}v_{3z}+v_{2z}v_{3y} & v_{1y}v_{3z}+v_{1z}v_{3y} \\ v_{1x}v_{2z}+v_{1z}v_{2x} & v_{2x}v_{3z}+v_{2z}v_{3x} & v_{1x}v_{3z}+v_{1z}v_{3x} \end{bmatrix}$$

4.3.1 Constitutive equations

The elastic material response is governed by Hooke's law. For an isotropic material, this is in three dimensions

$$\begin{aligned} \varepsilon_x &= \frac{1}{E}\left[\sigma_x - \nu\left(\sigma_y + \sigma_z\right)\right] \\ \varepsilon_y &= \frac{1}{E}\left[\sigma_y - \nu\left(\sigma_x + \sigma_z\right)\right] \\ \varepsilon_z &= \frac{1}{E}\left[\sigma_z - \nu\left(\sigma_x + \sigma_y\right)\right] \\ \gamma_{xy} &= \frac{1}{G}\tau_{xy}, \quad \gamma_{yz} = \frac{1}{G}\tau_{yz}, \quad \gamma_{zx} = \frac{1}{G}\tau_{zx}, \end{aligned} \tag{4.41}$$

where E is the modulus of elasticity, ν the Poisson's ratio and G the shear modulus, given by

$$G = \frac{E}{2(1+\nu)} \tag{4.42}$$

Equation (4.41) can be conveniently written in matrix form

$$\boldsymbol{\varepsilon} = \mathbf{C}\boldsymbol{\sigma} \tag{4.43}$$

where matrix $\mathbf{C}$ is defined as

$$\mathbf{C} = \frac{1}{E}\begin{bmatrix} 1 & -\nu & -\nu & 0 & 0 & 0 \\ -\nu & 1 & -\nu & 0 & 0 & 0 \\ -\nu & -\nu & 1 & 0 & 0 & 0 \\ 0 & 0 & 0 & \frac{E}{G} & 0 & 0 \\ 0 & 0 & 0 & 0 & \frac{E}{G} & 0 \\ 0 & 0 & 0 & 0 & 0 & \frac{E}{G} \end{bmatrix} \tag{4.44}$$

The inverse relationship can be defined by

$$\sigma = \mathbf{D}\varepsilon \tag{4.45}$$

where

$$\mathbf{D} = \mathbf{C}^{-1} = C_1\begin{bmatrix} 1 & C_2 & C_2 & 0 & 0 & 0 \\ C_2 & 1 & C_2 & 0 & 0 & 0 \\ C_2 & C_2 & 1 & 0 & 0 & 0 \\ 0 & 0 & 0 & \frac{G}{C_1} & 0 & 0 \\ 0 & 0 & 0 & 0 & \frac{G}{C_1} & 0 \\ 0 & 0 & 0 & 0 & 0 & \frac{G}{C_1} \end{bmatrix} \tag{4.46}$$

with

$$C_1 = E\frac{1-\nu}{(1+\nu)(1-2\nu)} \quad ; \quad C_2 = \frac{\nu}{(1-\nu)} \tag{4.47}$$

A subroutine to compute the isotropic D-matrix is given below.

```
SUBROUTINE D_mat(E,ny,D,Cdim)
!-----------------------------------
!   Computes isotropic D-matrix
!   Plane-strain (Cdim= 2)
!   or 3-D       (Cdim= 3)
!-----------------------------------
REAL, INTENT(IN)    :: E        !  Young's modulus
REAL, INTENT(IN)    :: ny       !  Poisson's ratio
INTEGER,INTENT(IN)  :: Cdim     !  Cartesian Dimension
REAL, INTENT(OUT)   :: D(:,:)   !  D-matrix
REAL                :: c1,c2,G
c1= E*(1.0-ny)/( (1.0+ny)*(1.0-2.0*ny) )
c2= ny/(1.0-ny)
G = E/(2.0*(1.0+ny))
D = 0.0
SELECT CASE (Cdim)
CASE (2)
  D(1,1)= 1.0  ; D(2,2)= 1.0
  D(2,1)= c2   ; D(1,2)= c2
  D(3,3)= G/c1
CASE (3)             !    3-D
  D(1,1)= 1.0  ;  D(2,2)= 1.0  ;  D(3,3)= 1.0
  D(2,1)= c2   ;  D(1,3)= c2   ;  D(2,3)= c2
  D(1,2)= c2   ;  D(3,1)= c2   ;  D(3,2)= c2
  D(4,4)= G/c1 ;  D(5,5)= G/c1 ;  D(6,6)= G/c1
CASE DEFAULT
END SELECT
D= c1*D
RETURN
END SUBROUTINE D_mat
```

For a general anisotropic material, 21 material properties are required but it is usually not possible to determine these. However, special types of anisotropy may exist for the case where the material properties are different in orthogonal directions. We may have a *laminate* or *stratified* material where the material property is the same in two orthogonal directions but different in the third orthogonal direction (see Figure 4.9).

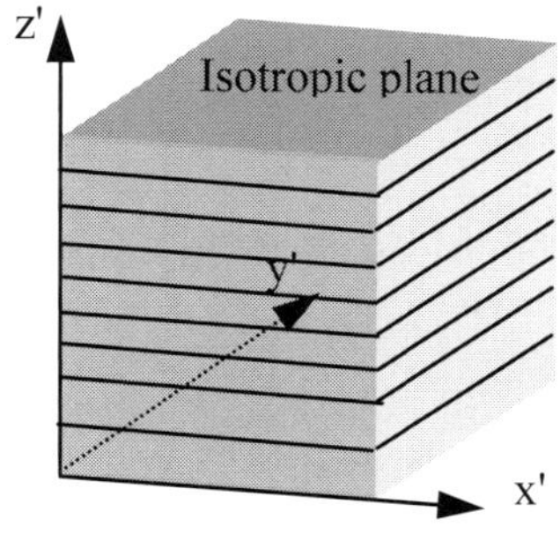

Figure 4.9 Definition of stratified material

Examples of this are a stratified rock mass, or fibre reinforced plastics. For a stratified material the elastic properties are defined as:

$$E_1 = \frac{\sigma_{x'}}{\varepsilon_{x'}} = \frac{\sigma_{y'}}{\varepsilon_{y'}} \quad ; \quad E_2 = \frac{\sigma_{z'}}{\varepsilon_{z'}}$$

$$G_1 = \frac{\tau_{x'y'}}{\gamma_{x'y'}} \quad ; \quad G_2 = \frac{\tau_{z'y'}}{\gamma_{z'y'}} = \frac{\tau_{z'x'}}{\gamma_{z'x'}} \tag{4.48}$$

$$\nu_1 = -\frac{\varepsilon_{y'}}{\varepsilon_{x'}} \quad ; \quad \nu_2 = -\frac{\varepsilon_{x'}}{\varepsilon_{z'}} = -\frac{\varepsilon_{y'}}{\varepsilon_{z'}}$$

For this case the D-matrix is

$$\mathbf{D}' = \begin{bmatrix} C_1' & C_2' & C_3' & 0 & 0 & 0 \\ C_2' & C_1' & C_3' & 0 & 0 & 0 \\ C_3' & C_3' & C_4' & 0 & 0 & 0 \\ 0 & 0 & 0 & G_1 & 0 & 0 \\ 0 & 0 & 0 & 0 & G_2 & 0 \\ 0 & 0 & 0 & 0 & 0 & G_2 \end{bmatrix} \tag{4.49}$$

with

$$\begin{aligned} &C_1' = n\left(1-\nu_2^{\,2}\right)C \quad ; \quad C_2' = n\left(\nu_1 + n\nu_2^{\,2}\right)C \quad ; \quad C_3' = n\nu_2(1+\nu_1)C \\ &C_4' = \left(1+\nu_1^{\,2}\right)C \quad ; \quad C = \frac{E_2}{(1-\nu_1)\left(1-\nu_1-2n\nu_2^{\,2}\right)} \quad ; \quad n = \frac{E_1}{E_2} \quad and \quad G_1 = \frac{E_1}{2(1+\nu_1)} \end{aligned} \tag{4.50}$$

If the orthogonal directions are not in the directions of the Cartesian coordinates, then the following transformation of the D-matrix has to be made

$$\mathbf{D} = \mathbf{T}_\sigma^T \mathbf{D}'\mathbf{T}_\sigma \tag{4.51}$$

where **D'** is defined in local coordinates.

4.3.2 Fundamental solutions

The governing differential equations are obtained from the condition of equilibrium. For two-dimensional problems these are

$$\begin{aligned} \frac{\partial \sigma_x}{\partial x} + \frac{\partial \tau_{xy}}{\partial y} + b_x = 0 \\ \frac{\partial \tau_{xy}}{\partial x} + \frac{\partial \sigma_y}{\partial y} + b_y = 0 \end{aligned} \tag{4.52}$$

where b_x and b_y are components of body force in x and y directions.

Substitution of the equations for strain (4.29) and the Hooke's law for plane strain conditions gives

$$G(\frac{\partial^2 u_x}{\partial x^2}+\frac{\partial^2 u_x}{\partial y^2})+(\lambda+G)(\frac{\partial^2 u_y}{\partial x\partial y}+\frac{\partial^2 u_y}{\partial x\partial y})+b_x=0$$

$$(\lambda+G)(\frac{\partial^2 u_x}{\partial x\partial y}+\frac{\partial^2 u_x}{\partial x\partial y})+G(\frac{\partial^2 u_y}{\partial x^2}+\frac{\partial^2 u_y}{\partial y^2})+b_y=0 \qquad (4.53)$$

where

$$\lambda=\frac{\nu E}{(1+\nu)(1-2\nu)} \qquad (4.54)$$

For the plane strain problem, the fundamental solution is obtained for point unit loads in x and y directions of magnitude 1, which are distributed to infinity in the +z and–z directions. The solution was first worked out by Lord Kelvin[2].

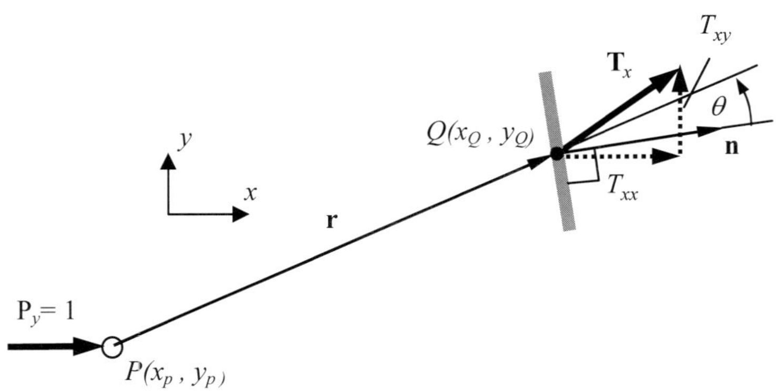

Figure 4.10 Notation for two-dimensional Kelvin solution (unit load in x-direction)

The solutions for the displacements in x and y directions due to a unit load in x-direction can be written as (Figure 4.10)

$$U_{xx}(P,Q)=C\left(C_1\ln\left(\frac{1}{r}\right)+r_x{}^2\right)$$

$$U_{xy}(P,Q)=C\,r_x\,r_y \qquad (4.55)$$

with

$$C=1/\left(8\pi G\left(1-\nu\right)\right) \quad , \quad C_1=3-4\nu$$

Note that the first subscript of U refers to the direction of the unit load whereas the second relates to the direction of the displacement.

We note that as the distance between source point P and field point Q approaches infinity the solution tends to negative infinity. This is due to the fact that the source is distributed along an infinite line and its resultant is infinite. As we will see later this does not present any difficulties because scaling is introduced for the coordinates which limit the maximum scaled distance to unity. The fundamental solution has a positive singularity when points P and Q coincide.

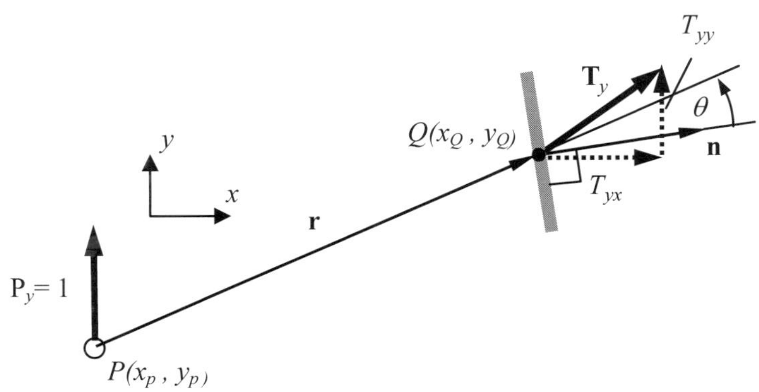

Figure 4.11 Notation for two-dimensional Kelvin solution (unit load in y-direction)

For a unit load in the y-direction we have

$$U_{yy}(P,Q) = C\left(C_1 \ln\left(\frac{1}{r}\right) + r_{,y}^{\ 2}\right)$$
$$U_{yx}(P,Q) = U_{xy}(P,Q) \tag{4.56}$$

the second equation indicating the symmetry of the solution.

Equations (4.55) und (4.56) can be written as a single equation as follows:

$$U_{ij}(P,Q) = C\left(C_1 \ln\frac{1}{r}\delta_{ij} + r_{,i}r_{,j}\right) \tag{4.57}$$

where x,y is substituted for i,j and

$$\delta_{ij} = 1 \quad \text{if} \quad i = j$$
$$\delta_{ij} = 0 \quad \text{if} \quad i \neq j \tag{4.58}$$

is the *Kronecker Delta*.

For the boundary element method we also need the solutions for the boundary stresses (tractions) acting on a surface with an outward normal direction of **n** (see figure 4.10).

The fundamental solutions for the *tractions* are obtained by first computing the fundamental solutions for the strains and then applying Hooke's law. The fundamental solutions for strains are obtained by taking the derivative of the displacement solution. The tractions at point Q due to a unit load at P in x-direction are given by

$$
\begin{aligned}
T_{xx}(P,Q) &= \frac{C_2}{r}\left(C_3 + 2r_{,x}^{\ 2}\right)\cos\theta \\
T_{xy}(P,Q) &= \frac{C_2}{r}\left[2r_{,x}r_{,y}\cos\theta + C_3\left[n_{,y}r_{,y} - n_{,x}r_{,x}\right]\right] \\
C_2 &= 1/\left(4\pi\left(1-\nu\right)\right) \quad , \quad C_3 = 1-2\nu \quad , \quad \cos\theta = \frac{1}{r}\mathbf{r}\bullet\mathbf{n}
\end{aligned}
\tag{4.59}
$$

where θ is defined in Figure 4.10.

For a unit load in the y-direction we have

$$
\begin{aligned}
T_{yy}(P,Q) &= \frac{C_2}{r}\left(C_3 + 2r_{,y}^{\ 2}\right)\cos\theta \\
T_{yx}(P,Q) &= \frac{C_2}{r}\left[2r_{,x}r_{,y}\cos\theta + C_3\left[n_{,y}r_{,x} - n_{,x}r_{,y}\right]\right]
\end{aligned}
\tag{4.60}
$$

The combined expression is

$$
T_{ij}(P,Q) = \frac{C_2}{r^2}\left[C_3\delta_{ij} + 2r_{,i}r_{,j}\cos\theta - C_3(1-\delta_{ij})(n_j r_{,i} - n_i r_{,j})\right] \tag{4.61}
$$

We note that the first part of the solution is symmetrical (i.e., the first part of T_{xy} equals T_{xy}) but the second part is not.

For the three-dimensional problem, the fundamental solution is obtained for point loads in x ,y and z directions.

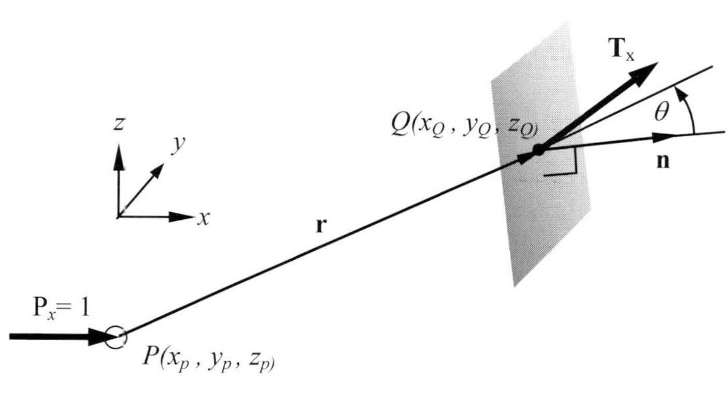

Figure 4.12 Notation for three-dimensional Kelvin solution (point load in x direction)

The solutions for the displacements in x,y and z directions due to a unit load in x-direction can be written as

$$
\begin{aligned}
&U_{xx}(P,Q)=\frac{C}{r}\left(C_1+r_{,x}{}^2\right)\\
&U_{xy}(P,Q)=C\frac{1}{r}r_{,x}r_{,y}\\
&U_{xz}(P,Q)=C\frac{1}{r}r_{,x}r_{,z}\\
&\textit{with}\\
&C=1/\left(16\pi G(1-\nu)\right)\quad,\quad C_1=3-4\nu
\end{aligned}
\tag{4.62}
$$

Now the solution approaches zero, as the distance between source point P and field point Q tends to infinity. However, this solution also approaches an infinite value, as r tends to zero. This fact will pose some problems with integrating the fundamental solutions which we will address later. The solution for load in x,y,z directions can be written as a combined Equation

$$U_{i,j}(P,Q)=C(C_1\delta_{ij}+r_{,i}r_{,j}) \tag{4.63}$$

The solutions for stresses acting on a boundary surface with an outward normal direction of **n** (see figure 4.8) are presented next.

The fundamental solutions for the *tractions* due to a unit load at P, in x-direction, are

$$
\begin{aligned}
&T_{xx}(P,Q)=\frac{C_2}{r^2}\left(C_3+3r_{,x}{}^2\right)\cos\theta\\
&T_{xy}(P,Q)=\frac{C_2}{r^2}\left[3r_{,x}\,r_{,y}\cos\theta+C_3\left[n_{,x}r_{,y}-n_{,y}r_{,x}\right]\right]\\
&T_{xz}(P,Q)=\frac{C_2}{r^2}\left[3r_{,x}\,r_{,z}\cos\theta+C_3\left[n_{,x}r_{,z}-n_{,z}r_{,x}\right]\right]
\end{aligned}
\tag{4.64}
$$

with

$$C_2=1/\left(8\pi(1-\nu)\right)\quad,\quad C_3=1-2\nu \tag{4.65}$$

The general solution for the tractions can be written as

$$T_{ij}(P,Q)=\frac{C_2}{r^2}\left[C_3\delta_{ij}+3r_{,i}r_{,j}\cos\theta-C_3(1-\delta_{ij})(n_{,j}r_{,i}-n_{,i}r_{,j})\right] \tag{4.66}$$

The Kelvin solutions for displacements are plotted in Figures 4.13 and 4.14. A small circle of exclusion is used to avoid plotting very high values near the singularity. The variation of the displacement in x-direction shows symmetry about the x- and y-axes. The variation of the displacements in y-direction shows anti-symmetry about both axes. The influence of the Poisson's ratio on the displacements perpendicular to the load axis can be clearly seen in Figure 4.14. Note that the finite element method has difficulty dealing with a Poisson's ratio of 0.5 (incompressible material) because of the definition of C_1 in equation (4.47) which would give an infinite value for $\nu = 0.5$.

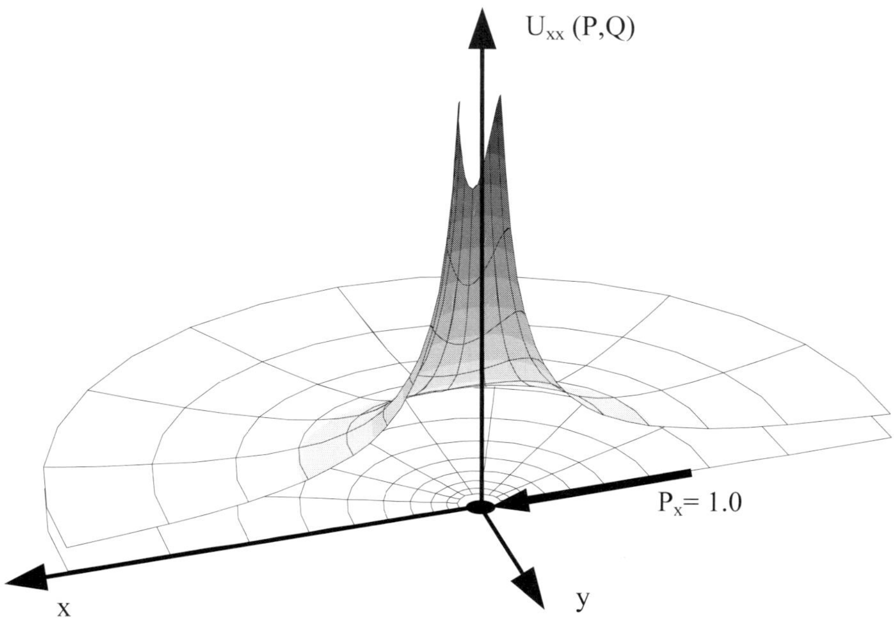

Figure 4.13 3-D Kelvin solution: variation of displacements in x-direction due to $P_x = 1.0$

Figure 4.15 shows the variation of the fundamental solution for the boundary traction in x-direction assuming that the vector normal to the boundary, **n** , points in the x-direction (this means that the computed traction is equivalent to the stress in the x-direction). We can see clearly that the fundamental solution is anti-symmetric about the y–axis and decays very rapidly from the singularity.

To implement the above equations in F90 we define functions UK and TK which return rank two arrays of dimension 2 or 3. The function only provides solutions for plane strain and 3-D problems. To obtain the solutions for plane stress problems simply substitute an effective Poisson´s ration of $\overline{\nu} = \nu/(1+\nu)$.

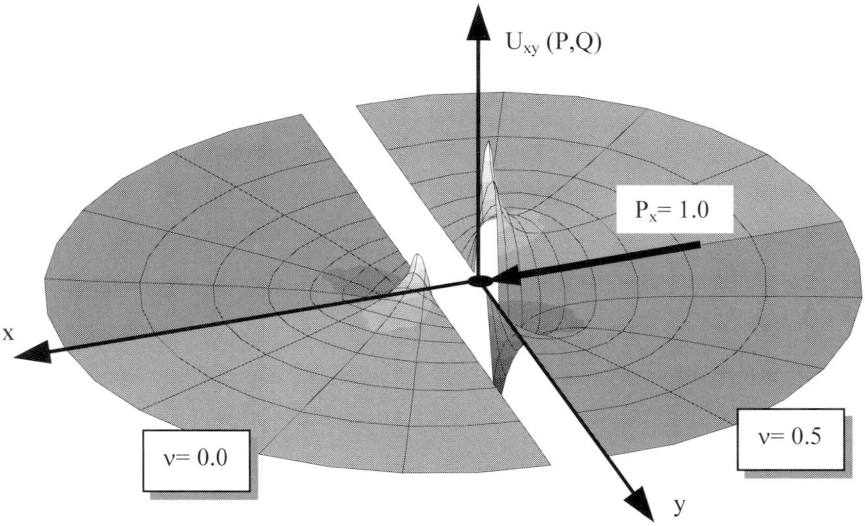

Figure 4.14 3-D Kelvin solution: variation of displacements in y-direction due to P_x= 1.0 for Poissons ratio of 0.0 (left figure) and 0.5 (right figure)

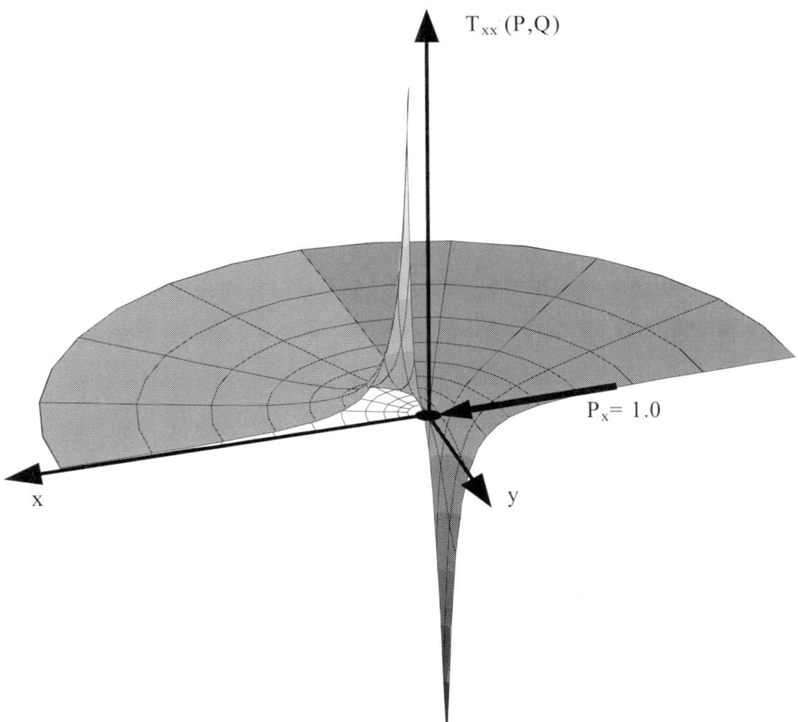

Figure 4.15 3-D Kelvin solution: variation of T_{xx} for $\mathbf{n}$ ={1,0,0}. This is equivalent to σ_x

```
FUNCTION UK(dxr,r,E,ny,Cdim)
!--------------------------------------------
!
!   FUNDAMENTAL SOLUTION FOR DISPLACEMENTS
!   isotropic material (Kelvin solution)
!
!--------------------------------------------
IMPLICIT NONE
REAL,INTENT(IN)    :: dxr(:)          !   r,x,r,y,r,z
REAL,INTENT(IN)    :: r               !   r
REAL,INTENT(IN)    :: E               !   Young's modulus
REAL,INTENT(IN)    :: ny              !   eff. Poisson's ratio
INTEGER,INTENT(IN):: Cdim             !   Cartesian dimension
REAL:: UK(Cdim,Cdim)                  !   Function returns array
REAL:: G,c,c1,onr,clog,conr           !   Temps
G= E/(2.0*(1+ny))
c1= 3.0 - 4.0*ny
SELECT CASE (Cdim)
CASE (2)          !     Plane strain solution
   c= 1.0/(8.0*Pi*G*(1.0 - ny))
   clog= -c1*LOG(r)
   UK(1,1)= c*(clog + dxr(1)*dxr(1))
   UK(2,2)= c*(clog + dxr(2)*dxr(2))
   UK(1,2)= c*dxr(1)*dxr(2)
   UK(2,1)= UK(1,2)
CASE(3)           !     Three-dimensional solution
   c= 1.0/(16.0*Pi*G*(1.0 - ny))
   conr=c/r
   UK(1,1)= conr*(c1 + dxr(1)*dxr(1))
   UK(2,2)= conr*(c1 + dxr(2)*dxr(2))
   UK(3,3)= conr*(c1 + dxr(3)*dxr(3))
   UK(1,2)= conr*dxr(1)*dxr(2)
   UK(1,3)= conr*dxr(1)*dxr(3)
   UK(2,1)= UK(1,2)
   UK(2,3)= conr*dxr(2)*dxr(3)
   UK(3,1)= UK(1,3)
   UK(3,2)= UK(2,3)
CASE DEFAULT
END SELECT
RETURN
END FUNCTION UK
```

Function TK requires one more parameter to be specified: the vector normal to the boundary (normal vector).

```
FUNCTION TK(dxr,r,Vnor,ny,Cdim)
!-------------------------------------------
!   FUNDAMENTAL SOLUTION FOR TRACTIONS
!   isotropic material (Kelvin solution)
!-------------------------------------------
IMPLICIT NONE
REAL,INTENT(IN)        :: dxr(:)           !  r,x,r,y,r,z
REAL,INTENT(IN)        :: r                !  r
REAL,INTENT(IN)        :: Vnor(:)          !  normal vector
REAL,INTENT(IN)        :: ny               !  eff. Poisson's ratio
INTEGER,INTENT(IN)     :: Cdim             !  Cartesian dimension
REAL                   :: TK(Cdim,Cdim)    !  Function returns
array
REAL                   :: c2,c3,costh,Conr !  Temps
c3= 1.0 - 2.0*ny
Costh= DOT_PRODUCT (Vnor,dxr)
SELECT CASE (Cdim)
CASE (2)            !    plane strain
   c2= 1.0/(4.0*Pi*(1.0 - ny))
   Conr= c2/r
   TK(1,1)= -(Conr*(C3 + 2.0*dxr(1)*dxr(1))*Costh)
   TK(2,2)= -(Conr*(C3 + 2.0*dxr(2)*dxr(2))*Costh)
   DO i=1,2
    DO j=1,3
     IF(i /= j) THEN
      TK(i,j)= -(Conr*(2.0*dxr(i)*dxr(j)*Costh &
         - c3*(Vnor(j)*dxr(i) - Vnor(i)*dxr(j))))
     END IF
    END DO
   END DO
CASE(3)             !    Three-dimensional
   c2= 1.0/(8.0*Pi*(1.0 - ny))
   Conr= c2/r**2
   TK(1,1)= -Conr*(C3 + 3.0*dxr(1)*dxr(1))*Costh
   TK(2,2)= -Conr*(C3 + 3.0*dxr(2)*dxr(2))*Costh
   TK(3,3)= -Conr*(C3 + 3.0*dxr(3)*dxr(3))*Costh
   DO i=1,3
    DO j=1,3
     IF(i /= j) THEN
      TK(i,j)= -Conr*(3.0*dxr(i)*dxr(j)*Costh &
           - c3*(Vnor(j)*dxr(i) - Vnor(i)*dxr(j)))
     END IF
    END DO
   END DO
CASE DEFAULT
END SELECT
END FUNCTION TK
```

Fundamental solutions for anisotropic material exist, but are rather complicated[3]. Further details are discussed in Chapter 18.

4.4. CONCLUSIONS

In this chapter we have dealt with the description of the material response in a mathematical way and have derived solutions for the equations governing the problem for simple loading. The solutions are for point sources, or loads, in an infinite domain. It has been shown that the implementation of these fundamental solutions into a F90 function is fairly straightforward. A particular advantage of the new facilities in F90 is that two-and three-dimensional solutions can be implemented in one FUNCTION, with the parameter Cdim determining the dimensionality of the result.

The Kelvin fundamental solution is not the only one which may be used for a boundary element analysis. Indeed, any solution may be used, including ones which satisfy some boundary conditions explicitly. For example, we may include the zero boundary traction conditions at the ground surface. Green's functions for a point load in a semi-infinite domain have been worked out, for example, by Melan in two dimensions[4] and Mindlin in three dimensions[5]. Also Bonnet[1] presents a solution for bonded half-spaces where two different materials may be considered implicitly in the solution. The fundamental solutions just derived will form the basis for the methods discussed in the next chapter.

4.5. REFERENCES

1. Bonnet, M, (1995) Boundary Integral Equation Methods for Solids and Fluids. Wiley, Chichester.
2. Sokolnikoff I.(1956) Mathematical Theory of Elesticity, McGraw-Hill, New York.
3. Tonon F, Pan E. and Amadei B. (2000) Green's functions and BEM formulations for 3-D anisotropic media. *Computers and Structures*, **79** (5):469-482.
4. Melan, E. (1932) Der Spannungszustand der durch eine Einzelkraft im Inneren beanspruchten Halbscheibe. *Z. Angew. Math. & Mech,* **12**, 343-346.
5. Mindlin R.D. (1936) Force at a point in the interior of a semi-infinite solid. *Physics* **7**: 195-202.

5

Boundary Integral Equations

There is nothing more practical
than a good theory

I. Kant

5.1 INTRODUCTION

As explained previously, the basic idea of the boundary element method comes from Trefftz[1], who suggested that in contrast to the method of Ritz, only functions satisfying the differential equations exactly should be used to approximate the solution inside the domain. If we use these functions it means, of course, that we only need to approximate the actual boundary conditions. This approach, therefore, has some considerable advantages:

- The solutions obtained inside the domain satisfy the differential equations exactly, approximations (or errors) only occur due to the fact that boundary conditions are only satisfied approximately.
- Since functions are defined globally, there is no need to subdivide the domain into elements.
- The solutions also satisfy conditions at infinity, therefore, there is no problem dealing with infinite domains, where the FEM has to use mesh truncation or approximate infinite elements.

The disadvantage is that we need solutions of differential equations to be as simple as possible, if we want to reduce computation time. The most suitable solutions are ones involving concentrated sources or loads in infinite domains. As we know from the previous chapter, these solutions also have some rather nasty properties, such as singularities. The integration of these functions will require special consideration.

The original method proposed by Trefftz is not suitable for writing general purpose programs as its accuracy is not satisfactory and, as will be seen later, convergence of the method cannot be assured. However, because of the inherent simplicity of the method, it serves well to explain some of the basic principles involved. Therefore, we will first introduce this method on a simple example in heat flow.

However, we will actually develop our programs using the direct method, which gets its name from the fact that no *fictitious* source or forces need to be computed, as in the Trefftz method, but that unknowns at the boundary are obtained directly. In the development of the integral equations we will use the theorem of Betti, which is better known to engineers than the Greens theorem.

5.2 TREFFTZ METHOD

To introduce the Trefftz method let us look at a simple two-dimensional example in heat flow. Consider an infinite homogeneous domain having conductivity k, where heat (q_0) flows only in the vertical (y) direction (Figure 5.1a).

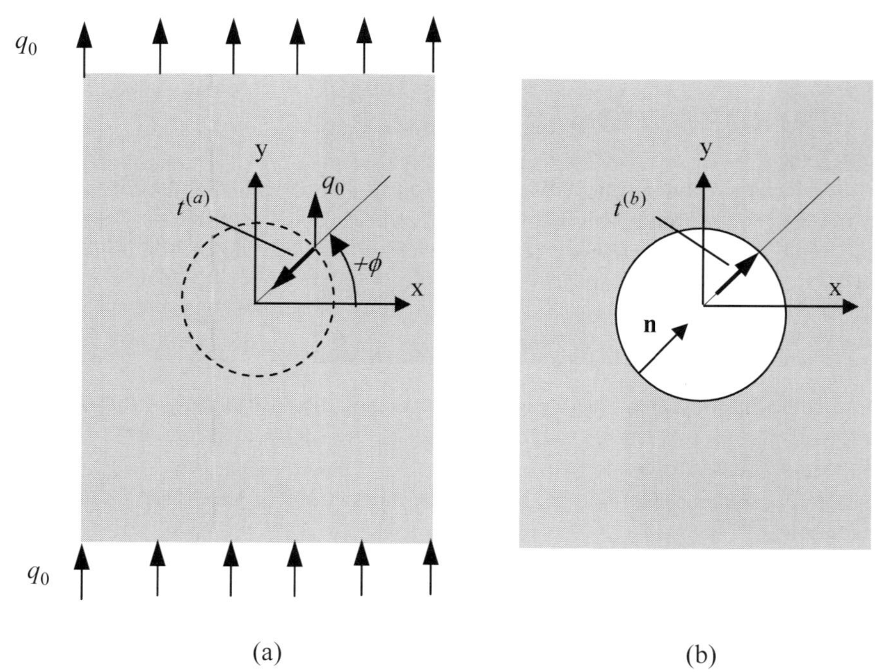

Figure 5.1 Heat flow in an infinite domain, case (a) and (b)

According to the Fourier law introduced in Chapter 4 we can write

$$\frac{\partial u}{\partial x} = 0 \quad \text{and} \quad \frac{\partial u}{\partial y} = -\frac{q_0}{k} \tag{5.1}$$

Solving the differential equations for u, the temperature at a point Q with coordinates x,y is obtained as

$$u^{(a)}(Q) = -\frac{q_0}{k} y + C \tag{5.2}$$

If we assume the temperature at the centre of the circle to be zero, then C= 0.

We now place a cylindrical isolator in the flow and compute how the flow pattern and temperature distribution changes. The isolator prevents flow to occur in a direction perpendicular to its boundary, which is computed by

$$t = -k\frac{\partial u}{\partial \mathbf{n}} = -k(n_x \frac{\partial u}{\partial x} + n_y \frac{\partial u}{\partial y}) = 0 \tag{5.3}$$

Where $\mathbf{n}$ $\{n_x\ ,\ n_y\}$is the vector normal to the boundary of the isolator (*outward normal*). Note that the positive direction of this vector is pointing from the infinite domain into the isolator. For the solution (5.2) just obtained, we find that this condition is not satisfied, because the flow in the direction normal to the isolator boundary (marked with a dotted line in Figure 5.1a) is computed as:

$$t^{(a)} = n_y q_0 = -q_0 \sin\phi \tag{5.4}$$

If we want to find out how the isolator changes the flow/temperature distribution, then we can think of the problem as divided into two parts: the first being the trivial one, whose solution we just obtained, the second being one where the solution is obtained for the following boundary condition:

$$t^{(b)} = -t^{(a)} = q_0 \sin\phi \tag{5.5}$$

If the two solutions for the flow normal to the boundary of the isolator are added then:

$$t = t^{(b)} + t^{(a)} = 0 \tag{5.6}$$

i.e. the boundary condition that no flow occurs normal to the isolator is satisfied. The final solution for the temperature is therefore

$$u(Q) = u^{(a)}(Q) + u^{(b)}(Q) \tag{5.7}$$

We now solve the boundary value problem (b) by the Trefftz method. To apply the Trefftz method, we quite arbitrarily select N points on the boundary of the isolator, where we wish to satisfy the boundary conditions, equation (5.5) and another set of points, where we apply *fictitious* sources. The reason these are called *fictitious* is that they are not actually present, but can be thought of as parameters of the global approximation functions. We have to be careful with the location of these points and this will be the major drawback of the method. The source points must be placed in such a way, that they do not influence the results. In our case, the best place is inside the isolator. Also, we must not place points P too close to the boundary points Q, because, as we know, when P approaches Q, the fundamental solutions become singular. In Figure 5.2 we show an example of the choice of locations for load points P_i and boundary points Q_i. We place points Q at quarter points on the boundary of the isolator, with radius R_Q and points P at a circle, with radius R_P inside the isolator.

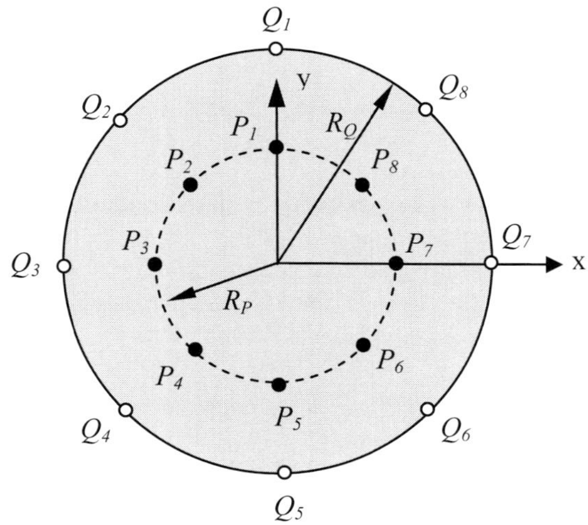

Figure 5.2 Points P for fictitious loads and Q, where boundary conditions are to be satisfied

In the Trefftz method, we attempt to satisfy the given boundary conditions, by adjusting the magnitude of the fictitious sources F_i applied at P_i. Noting that the fundamental solutions for the flow in direction **n**, which we derived in the last chapter, is $T(P,Q)$, the boundary condition at point Q_1 can be satisfied by

$$t^{(b)}(Q_1) = \sum_{i=1}^{8} T(P_i, Q_1) F_i \tag{5.8}$$

Here $T(P_i, Q_1)$ is the flow in direction $\mathbf{n}(Q_1)$ at point Q_1 due to a source at P_i. This is also sometimes referred to as an *influence coefficient*. We can now write a similar equation for each boundary point Q_i, a total of 8 equations:

$$t^{(b)}(Q_1) = \sum_{i=1}^{8} T(P_i, Q_1) F_i \;\; ; \;\; t^{(b)}(Q_2) = \sum_{i=1}^{8} T(P_i, Q_2) F_i \quad \text{etc.} \tag{5.9}$$

We obtain a system of simultaneous equations, which we can solve for unknown *fictitious* sources F_i. Obviously, the number of fictitious sources depends on the number of equations we can write and hence, on the number of boundary points Q_i. It is convenient, therefore, to have the same number of source points as we have field points. Once we have solved the system of simultaneous equations and calculated the *fictitious* sources F_i,, then the temperature at any point Q on the boundary of the isolator and in the domain (but outside the isolator) is given by

$$u(Q) = u^{(a)}(Q) + u^{(b)}(Q) \quad \text{where} \quad u^{(b)}(Q) = \sum_1^8 U(P_i, Q)\,\mathrm{F_i} \tag{5.10}$$

The flow at a point Q in x and y-directions may be obtained by

$$q_x = q_x^{(b)} \quad ; \quad q_y = q_0 + q_y^{(b)}$$
$$\text{where}$$
$$q_x^{(b)} = -k\frac{\partial u^{(b)}(Q)}{\partial x} = -k\sum_1^8 \frac{\partial U(P_i,Q)}{\partial x}\mathrm{F}_i \tag{5.11}$$
$$q_y^{(b)} = -k\frac{\partial u^{(b)}(Q)}{\partial y} = -k\sum_1^8 \frac{\partial U(P_i,Q)}{\partial y}\mathrm{F}_i$$

5.3 PROGRAM 5.1: FLOW AROUND CYLINDER, TREFFTZ METHOD

The program shown here allows us to numerically solve the problem of flow around a cylinder, with a variable number of source points and this allows the reader to get a better understanding of the Trefftz method and its limitations. We activate the Laplace_lib, which contains the fundamental solutions of the Laplace equation governing our problem and the Utility_lib containing the subroutine for solving equations by the USE statement. Next, we read some information about the problem, such as heat inflow, conductivity, number of source/boundary points and radius of the cylinder. We finally, quite arbitrarily, specify that the source points are located on a circle with radius *Rp*, which has to be smaller than the radius of the cylinder. We can later do numerical experiments on the effect of distance between source and boundary points on accuracy of results. Since the size of the arrays for storing the equation system is dependent on the number of source points specified, we allocate them at run time. Next, we loop over all boundary points (DO loop Field_points) and all source points (DO loop Source_points) to generate the matrix of *influence coefficients* and the right hand side. The points Q and P are assumed to be equally distributed over the circle. The system of equations is solved next with utility program SOLVE. The values of temperature are computed at boundary points and interior points, the coordinates of which are specified by the input. Both involve a summation of influences (i.e., fundamental solutions multiplied with the fictitious source intensities).

```
PROGRAM Trefftz
!---------------------------------
!   Program to compute the heat flow past a cylindrical isolator
!   in a 2-D infinite domain using the Trefftz method
!---------------------------------
USE Laplace_lib ; USE Utility_lib
IMPLICIT NONE                          ! declare all variables
REAL              ::  q                  ! inflow/outflow
REAL              ::  k                  ! Thermal conductivity
INTEGER           ::  npnts              ! Number of points P,Q
REAL              ::  rq                 ! radius of isolator
REAL              ::  rp                 ! radius of source points
REAL(KIND=8),ALLOCATABLE ::  Lhs(:,:) ! left hand side
REAL(KIND=8),ALLOCATABLE ::  Rhs(:)   ! right hand side
REAL(KIND=8),ALLOCATABLE ::  F(:)     ! fictitious sources
REAL              ::  dxr(2)             ! r,x, r,y
REAL              ::  vnorm(2)           ! normal vector
REAL              ::  Delth,Thetq,Thetp,xq,yq,xp,yp,xi,yi,r,uq
INTEGER           ::  npq,npp,ninpts,nin
OPEN(UNIT=10,FILE='INPUT.DAT',STATUS='OLD',ACTION='READ')
OPEN(UNIT=11,FILE='OUTPUT.DAT',STATUS='UNKNOWN',ACTION='WRITE')
READ(10,*) q,k,npnts,rq,rp
WRITE(11,*) ' Program 2: heat flow past a cylinder Trefftz
method'
WRITE(11,*) '  Heat inflow/outflow=  ',q
WRITE(11,*) '  Thermal conductivity= ',k
WRITE(11,*) '  Number of Points P,Q= ',npnts
WRITE(11,*) '  Radius of Isolator=   ',rq
WRITE(11,*) '  Radius of Sources =   ',rp
ALLOCATE (Lhs(npnts,npnts),Rhs(npnts),F(npnts)) !
Delth= 2*Pi/npnts ! increment in angle theta between points
Thetq= Pi/2.0     ! angle theta to first field point Q1
Field_points: &
DO npq= 1,npnts
   Rhs(npq)= q * SIN(Thetq)     ! right hand side
   xq= rq*COS(Thetq)            ! x-coordinate of field point
   yq= rq*SIN(Thetq)            ! y-coordinate of field point
   vnorm(1)= -COS(Thetq)        ! normal vector to Q
   vnorm(2)= -SIN(Thetq)
   Thetq= Thetq + Delth         ! angle to next field point Q
   Thetp= Pi/2.0                ! angle to first source point P1
   Source_points:  &
   DO npp= 1,npnts
      xp= rp*COS(Thetp)         ! x-coordinate of source point
      yp= rp*SIN(Thetp)         ! y-coordinate of source point
      dxr(1)= xp-xq
```

```
      dxr(2)= yp-yq
      r= SQRT(dxr(1)**2 + dxr(2)**2)    ! dist. field/source pnt
      dxr= dxr/r                        ! normalise vector dxr
      Lhs(npq,npp)= T(r,dxr,vnorm,2)    !
      Thetp= Thetp + Delth         ! angle to next point P
    END DO  &
    Source_points
END DO &
Field_points
Lhs= - Lhs    !Multiplication with "-" to avoid negative pivots
Rhs= - Rhs    !
!  Solve system of equations: calculate F out of Lhs and Rhs
CALL Solve(Lhs,Rhs,F)
!   Postprocessing - Boundary values of temperature
WRITE(11,*)  ''
WRITE(11,*)  'Temperatures at Boundary points:'
Thetq= Pi/2.0  ! angle to first field point Q1
Field_points1: &
DO npq= 1,npnts
   uq= 0.0
   xq= rq* COS(Thetq)             ! x-coordinate of field point
   yq= rq* SIN(Thetq)             ! y-coordinate of field point
   Thetq= Thetq + Delth           ! angle to next field point Q
   Thetp= Pi/2.0                  ! angle to first source point P1
   Source_points1: &
   DO npp= 1,npnts
     xp= rp* COS(Thetp)           ! x-coordinate of source point
     yp= rp* SIN(Thetp)           ! y-coordinate of source point
     dxr(1)= xp-xq
     dxr(2)= yp-yq
     r= SQRT(dxr(1)**2 + dxr(2)**2)
     uq= uq + U(r,k,2)*F(npp)
     Thetp= Thetp + Delth         ! angle to next source point P
   END DO &
   Source_points1
   uq=uq-q/k* yq
   WRITE(11,*) 'Temperature at field point',npq,' =',uq
 END DO  &
 Field_points1
 !   Postprocessing - Interior points
 WRITE(11,*)  ''
 WRITE(11,*)  'Temperatures at interior points:'
 READ(10,*) ninpts                ! read number of interior points
 Int_points:  &
 DO nin= 1,ninpts
    READ(10,*) xi,yi              ! coordinates of interior points
    uq= 0.0
    Thetp= Pi/2.0                 ! angle to first source point P1
    Source_points2:  &
    DO npp= 1,npnts
       xp= rp* COS(Thetp)         ! x-coordinate of source point
```

```
        yp= rp*SIN(Thetp)        ! y-coordinate of source point
        dxr(1)= xp-xi
        dxr(2)= yp-yi
        r= SQRT(dxr(1)**2 + dxr(2)**2)
        uq= uq + U(r,k,2)*F(npp)
        Thetp= Thetp + Delth     ! angle to next source point P
    END DO &
    Source_points2
    uq=uq-q/k*yi
    WRITE(11,*) 'Temperature at x=',xi,', y=',yi,' =',uq
  END DO &
  Int_points
  STOP
END PROGRAM Trefftz
```

INPUT DATA for program Trefftz

1.0 Problem specification

q,k, npnts, rq,rp

q ... Heat inflow
k ... Thermal conductivity
npnts ... Number of points P,Q
rq ... Radius of isolator
rp ... Radius of sources

2.0 Interior point specification

Npoints — Number of interior points

3.0 Interior point coordinates (Npoints cards)

x,y — x,y coordinates of interior points

5.3.1 Sample input and output

Here we show an example of the input for an isolator of radius 1.0 with 32 points P and Q, where the source points P are situated along a circle with a radius 0.7.

File INPUT.DAT

```
1.0 1.0 32 1.0 0.7
18
0. -5.
0. -4.5
0. -4.
0. -3.5
0. -3.
0. -2.5
0. -2.
0. -1.5
```

```
0. -1.
0. 1.
0. 1.5
0. 2.
0. 2.5
0. 3.
0. 3.5
0. 4.
0. 4.5
0. 5.
```

File OUTPUT.DAT

```
Program 2 : heat flow past a cylinder with Trefftz method
Heat inflow/outflow=        1.00000
Thermal conductivity=       1.00000
Number of Points P,Q=             32
Radius of Isolator=         1.00000
Radius of Sources =         0.700000

Temperatures at Boundary points:
Temperature at field point             1 =    -1.99996
….
Temperature at field point            32 =    -1.96546

Temperatures at interior points:
Temperature at x=    0.000000    , y=    -5.00000      =     5.19999
….
Temperature at x=    0.000000    , y=     5.00000      =    -5.19999
```

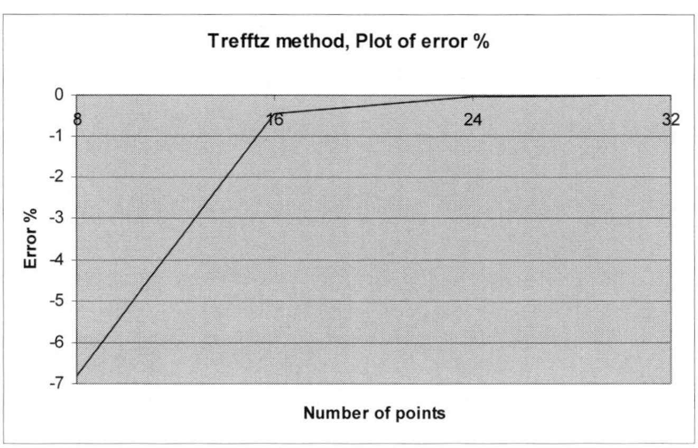

Figure 5.3 Plot of error in computing the temperature versus the number of points P

The error in the computation of the temperature at the top of the circular isolator (point Q_1) is plotted in Figure 5.3. It can be seen that very accurate results can be obtained with 24 elements.

5.4 DIRECT METHOD

As we have seen from the simple example, the Trefftz method is not suitable for general purpose programming. The method is not very user-friendly because, in addition to specifying points where boundary conditions are to be satisfied, we have to specify a second set of points where fictitious forces are to be applied. This is certainly not acceptable, especially if we want to go into three-dimensional problems. In addition, the convergence of the method can not be guaranteed for a general case as the number of points Q and P are increased.

5.4.1 Theorem of Betti and integral equations

An alternative to the Trefftz method is the direct method. Here we use the well known Betti theorem, rather elegantly to get rid of the need to compute *fictitious* sources or forces. We also abolish the need for an additional set of points, by placing the source points P to coincide with field points Q.

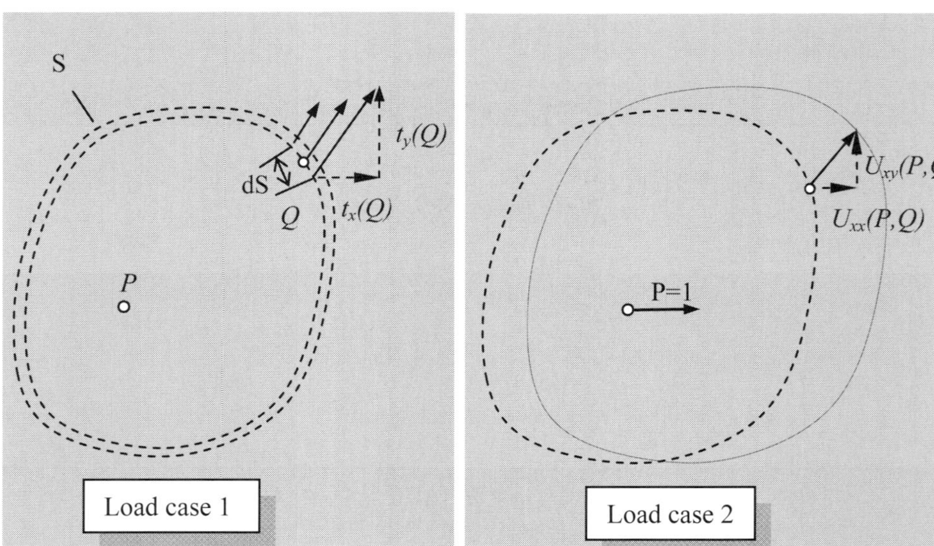

Figure 5.4 Application of Betti's theorem, tractions of load case 1 and displacements of load case 2 for computing W_{12}

This means that the method will become more complicated than Trefftz's, because we will now have to solve a set of integral equations and to cope with integrals, which are singular. The direct method, however, is much more user-friendly than Trefftz's method and has the advantage that convergence can be guaranteed. We explain the direct method with an example in elasticity, as engineers associate the Betti theorem with that type of problem. However, we will see that the integral equations can be derived for potential problems in a similar way.

Consider an infinite domain with two types of 'loading': *load case* number 1 we assume to be the case we want to solve and *load case* number 2, a case where only a unit load in the x-direction is specified at a point *P* (see Figure 5.4). Along a dotted line we show for load case 1 the stresses defined as forces per unit length of the line (*dS*). These are the *tractions* at point Q, with components $t_x(Q)$ and $t_y(Q)$. For load case 2, we show the displacements at point *Q* on *S*, which are the fundamental solutions $U_{xx}\ (P,Q)$ and $U_{xy}(P,Q)$.

As already mentioned in Chapter 4, we must cut through the continuum to show stresses. Here we cut along a dotted line, which forms a closed contour and which has been chosen quite arbitrarily. By this cut, the continuum is divided into two parts: the interior and exterior domains. Note that for the following derivation it does not matter which domain is considered and, therefore, the integral equations are valid for infinite as well as finite domains.

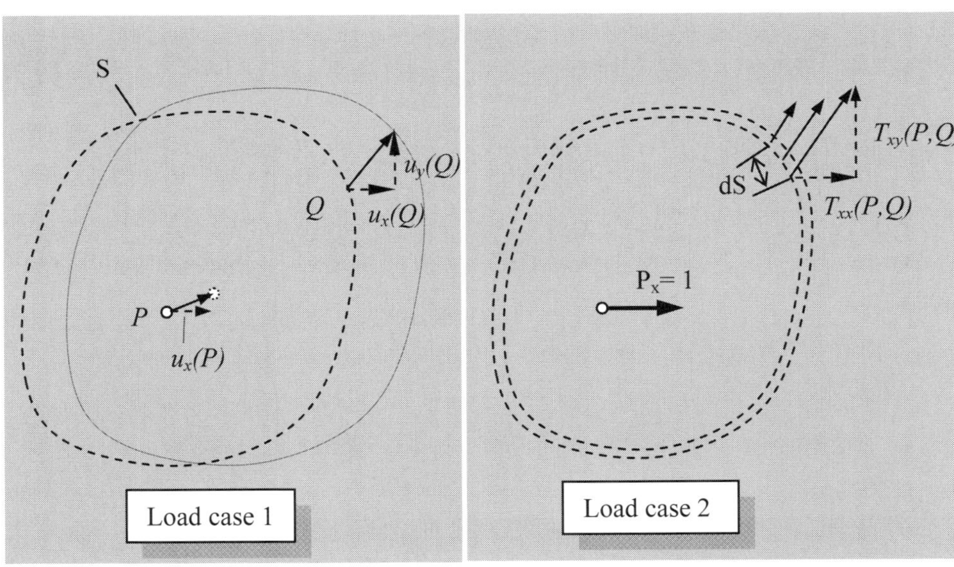

Figure 5.5 Application of Betti's theorem, displacements of load case 1 and tractions for load case 2 for computing W_{21}

The theorem of Betti states that the work done by the load of case 1 along the displacements of case 2 must equal the work done by the loads of case 2 along the displacements of case 1.

If we assume that there are no body forces acting in the domain (these will be introduced later), the work done by the first set of tractions and displacements is (Fig 5.4)

$$W_{12} = \int_S \left[t_x(Q) U_{xx}(P,Q) + t_y(Q) U_{xy}(P,Q) \right] dS \tag{5.12}$$

The work done by the second set of tractions/forces and displacements is (Fig 5.5)

$$W_{21} = \int_S \left[u_x(Q) T_{xx}(P,Q) + u_y(Q) T_{xy}(P,Q) \right] dS + 1 u_x(P) \tag{5.13}$$

The theorem of Betti states that $W_{12} = W_{21}$ and this gives the first integral equation

$$\begin{aligned} u_x(P) = & \int_S \left[t_x(Q) U_{xx}(P,Q) + t_y(Q) U_{xy}(P,Q) \right] dS \\ & - \int_S \left[u_x(Q) T_{xx}(P,Q) + u_y(Q) T_{xy}(P,Q) \right] dS \end{aligned} \tag{5.14}$$

A second integral equation can be obtained by placing the unit load in y direction

$$\begin{aligned} u_y(P) = & \int_S \left[t_x(Q) U_{yx}(P,Q) + t_y(Q) U_{yy}(P,Q) \right] dS \\ & - \int_S \left[u_x(Q) T_{yx}(P,Q) + u_y(Q) T_{yy}(P,Q) \right] dS \end{aligned} \tag{5.15}$$

Using matrix algebra we can combine equations (5.14) and (5.15)

$$\mathbf{u}(P) = \int_S \mathbf{U}(P,Q)\mathbf{t}(Q)\, dS - \int_S \mathbf{T}(P,Q)\mathbf{u}(Q)\, dS \tag{5.16}$$

where

$$\mathbf{u}(Q) = \begin{Bmatrix} u_x \\ u_y \end{Bmatrix} \quad , \quad \mathbf{U}(P,Q) = \begin{bmatrix} U_{xx} & U_{xy} \\ U_{yx} & U_{yy} \end{bmatrix}$$

$$\mathbf{t}(Q) = \begin{Bmatrix} t_x \\ t_y \end{Bmatrix} \quad , \quad \mathbf{T}(P,Q) = \begin{bmatrix} T_{xx} & T_{xy} \\ T_{yx} & T_{yy} \end{bmatrix} \tag{5.17}$$

Equations (5.16) represent for the two-dimensional problem discussed here a system of two integral equations which relate tractions **t** and displacements **u** at the boundary directly, thereby removing the need to compute *fictitious* forces.

For three-dimensional problems, three integral equations (5.15) can be obtained where S is a surface and

$$\mathbf{u}(Q)=\begin{Bmatrix} u_x \\ u_y \\ u_z \end{Bmatrix} \quad , \quad \mathbf{U}(P,Q)=\begin{bmatrix} U_{xx} & U_{xy} & U_{xz} \\ U_{yx} & U_{yy} & U_{yz} \\ U_{zx} & U_{zy} & U_{zz} \end{bmatrix} \tag{5.18}$$

$$\mathbf{t}(Q)=\begin{Bmatrix} t_x \\ t_y \\ t_z \end{Bmatrix} \quad , \quad \mathbf{T}(P,Q)=\begin{bmatrix} T_{xx} & T_{xy} & T_{xz} \\ T_{yx} & T_{yy} & T_{yz} \\ T_{zx} & T_{zy} & T_{zz} \end{bmatrix} \tag{5.19}$$

It can be shown that the Betti theorem can also be arrived at in a mathematical way, by using the divergence theorem and Green's symmetric identity[2]. Using this more general mathematical approach, it can be shown that for potential problems, the following single integral equation is obtained

$$u(P)= \int_S t(Q)U(P,Q)dS(Q)- \int_S u(Q)T(P,Q)dS(Q) \tag{5.20}$$

where *u(Q)* and *t(Q)* are the temperature/potential and the normal derivative respectively at point *Q* on *S,* and *U(P,Q)* and *T(P,Q)* are the fundamental solutions at *Q* for a source at point *P.* The integration is carried out over a line *S* for two-dimensional problems or a surface *S* for three-dimensional problems.

5.4.2 Limiting values of integrals as *P* coincides with *Q*

We have now succeeded to avoid computing the fictitious forces but have not succeeded yet in making the method more user-friendly since, we still need two sets of points: points *P* where the unit sources/loads are applied and points *Q* where we have to satisfy boundary conditions. Ideally, we would like to have only one set of points on the line where the points *Q* are specified. The problem is that some integrals in (5.16) or (5.20) only exist in the sense of a limiting value as *P* approaches *Q*.

This is explained in Figure 5.6 for two-dimensional potential problems. Here, we examine what happens when points *P* and *Q* coincide. We define a region of exclusion around point *P,* with radius ε and integrate around it. The integrals in equation (5.20) can now be split up into integrals over $S\text{-}S_\varepsilon$, that is, the part of the curve without the exclusion zone and into integrals over s_ε , that is, the part of the circular exclusion. As ε is taken to zero it does not matter if we integrate over s_ε or S_ε . The right hand side of equation (5.20) is written as:

$$\int_S t\cdot U\cdot dS - \int_S u\cdot T\cdot dS = \int_{S-S_\varepsilon} t\cdot U\cdot dS - \int_{S-S_\varepsilon} u\cdot T\cdot dS + \int_{s_\varepsilon} t\cdot U\cdot dS - \int_{s_\varepsilon} u\cdot T\cdot dS \tag{5.21}$$

We examine the integrals over s_ε further. For a smooth surface at P, using polar coordinates, as shown, we change the integration limits of the first integral to 0 and π and substitute for the fundamental solution *U*. Furthermore, as in the limit *P* will be coincident with *Q*, we can assume *t(Q)=t(P)* and *u(Q)= u(P)*. Then we have

$$\int_{s_\varepsilon} t(Q)U(P,Q)dS(Q) = t(P)\int_0^\pi \frac{1}{2\pi k} ln\left(\frac{1}{\varepsilon}\right)\varepsilon\, d\phi = t(P)\pi\frac{1}{k} ln\left(\frac{1}{\varepsilon}\right)\varepsilon \tag{5.22}$$

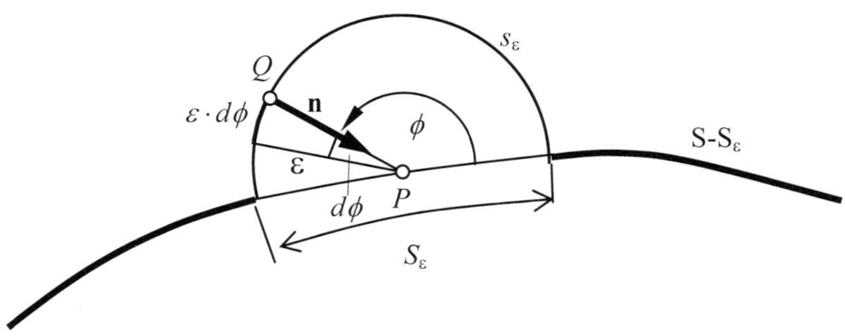

Figure 5.6 Diagram explaining the limiting value of integrals for two-dimensional potential problems

The integral approaches zero as ε approaches zero. Therefore

$$\lim_{\varepsilon\to 0} \int_{s_\varepsilon} t(Q)U(P,Q)dS(Q) = 0 \tag{5.23}$$

The second integral becomes

$$\int_{s_\varepsilon} u(Q)T(P,Q)dS(Q) = u(P)\int_0^\pi \frac{\cos\theta}{2\pi\varepsilon}\varepsilon\, d\phi = u(P)\int_0^\pi \frac{-1}{2\pi} d\phi = -\frac{1}{2}u(P) \tag{5.24}$$

As ε cancels out we do not have to take the limiting value of this integral. The integral equation that has to be used for the case where the source points are located on the continuous line *S*, is given by

$$\frac{1}{2}u(P) = \lim_{\varepsilon\to 0}\left[\int_{S-S_\varepsilon} t(Q)U(P,Q)dS(Q) - \right]\int_{S-S_\varepsilon} u(Q)T(P,Q)dS(Q) \tag{5.25}$$

For a three-dimensional problem, we take the zone of exclusion to be a sphere, as shown in Figure 5.7.

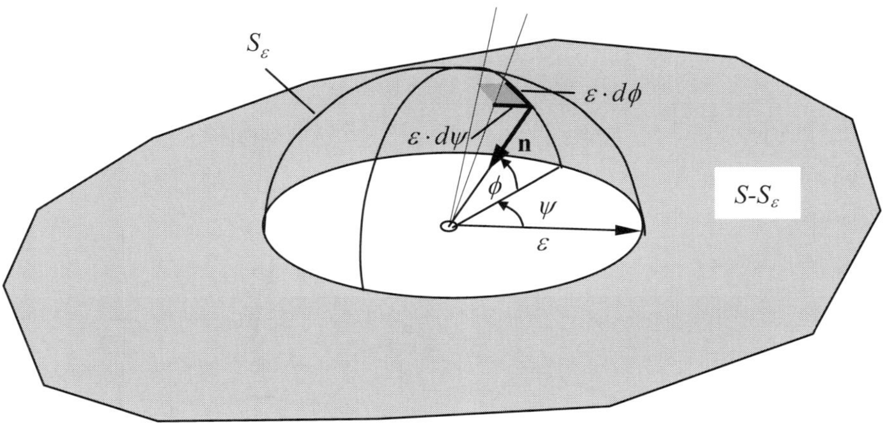

Figure 5.7 Computation of integrals for the case that P=Q, three-dimensional case

In this case the first integral also approaches zero as ε approaches zero. The second integral can be computed as

$$\int_{S_\varepsilon} u(Q)T(P,Q)dS(Q) = u(P)\int_0^{2\pi}\int_0^{\pi}\frac{\cos\theta}{4\pi\varepsilon^2}\,\varepsilon\, d\phi\, \varepsilon\, d\psi = -\frac{1}{2}u(P) \tag{5.26}$$

which for smooth surfaces gives the same result as before. Obviously, the same limiting procedure can be made for elasticity problems. If P=Q the integral equation (5.16) can be rewritten as

$$\frac{1}{2}\mathbf{u}(P) = \lim_{\varepsilon\to 0}\left[\int_{S-S\varepsilon}\mathbf{U}(P,Q)\mathbf{t}(Q)dS(Q) - \int_{S-S\varepsilon}\mathbf{T}(P,Q)\mathbf{u}(Q)dS(Q)\right] \tag{5.27}$$

If the boundary is not smooth but has a corner, as shown in Figure 5.8, then equation (5.24) has to be modified. The integration limits are changed and now depend on the angle γ :

$$\int_{S_\varepsilon} u(Q)T(P,Q)dS(Q) = u(P)\int_0^{\gamma}\frac{\cos\theta}{2\pi\varepsilon}\varepsilon d\phi = u(P)\int_0^{\gamma}\frac{-1}{2\pi}d\phi = -\frac{\gamma}{2\pi}u(P) \tag{5.28}$$

A more general integral equation can be written for potential problems

$$cu(P) = \lim_{\varepsilon\to 0}\left[\oint_{S-S_\varepsilon} t(Q)U(P,Q)dS(Q) - \oint_{S-S_\varepsilon} u(Q)T(P,Q)\, dS(Q)\right] \tag{5.29}$$

The reader may verify that

$$c = 1 - \frac{\gamma}{2\pi} \quad \textit{for} \quad 2-D \quad \textit{and} \quad c = 1 - \frac{\gamma}{4\pi} \quad \textit{for} \quad 3-D \tag{5.30}$$

where γ is defined as the angle subtended at P by s_ε.

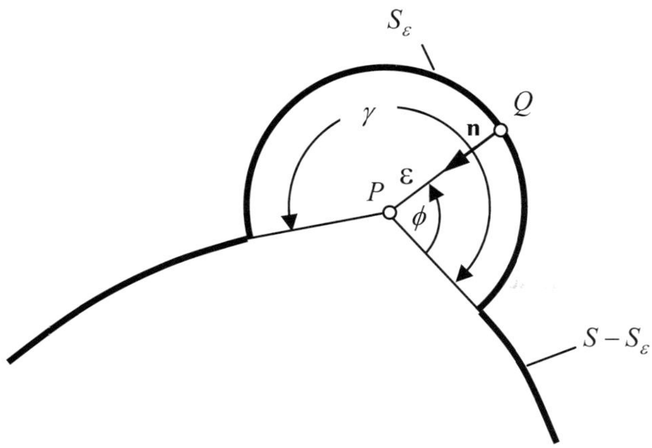

Figure 5.8 Limiting value of integral when P is located on a corner

For two and three-dimensional elasticity problems we may write a more general form of equation (5.25)

$$c\,\mathbf{Iu}(P) = \lim_{\varepsilon \to 0}\left[\int_{S-S\varepsilon} \mathbf{U}(P,Q)\mathbf{t}(Q)\,dS(Q) - \int_{S-S\varepsilon} \mathbf{T}(P,Q)\mathbf{u}(Q)\,dS(Q)\right] \tag{5.31}$$

where c is as previously defined and **I** is a 2x2 or 3x3 unit matrix.

5.4.3 Solution of integral equations

Using the direct method, a set of integral equations has been produced that relates the temperature/potential to the normal gradient, or the displacement to the traction at any point Q on the boundary. Since we are now able to place the source points coincidental with the points where the boundary conditions are to be satisfied, we no longer need to be concerned about these points. Indeed, in the direct method, the fictitious sources no longer play a role.

To use integral equations for the solution of boundary value problems we consider only one of the two regions created by cutting along the dotted line in Figure 5.4: the interior or the exterior region, as shown in Figure 5.9. With respect to the integral equations, the only difference between them is the direction of the outward normal **n**, which is assumed to point away from the solid. The interior region is a *finite* region, the exterior an *infinite* region.

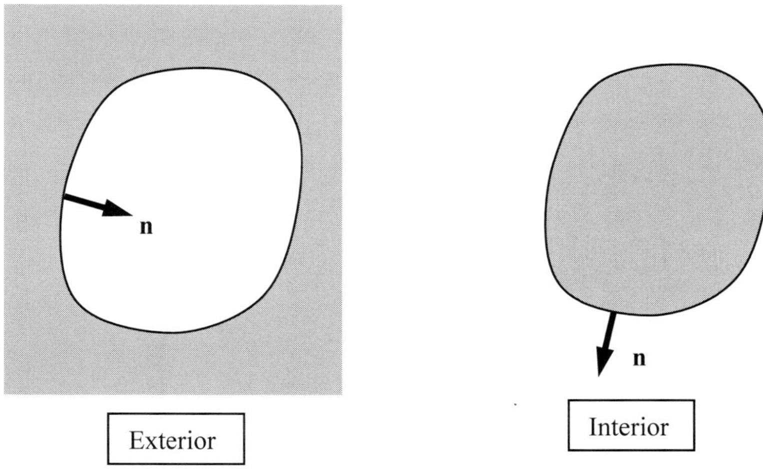

Figure 5.9 Exterior and interior regions obtained by separating the domain along dotted line

For potential problems, we obtain one integral equation per source point *P*. For elasticity problems, we get two or three integral equations per source point, depending on the dimensionality of the problem. Theoretically, if we want to satisfy the boundary conditions exactly at all points on the boundary, we would need an infinite number of points *P*=*Q*. In practice, we will solve the integral equations numerically and attempt to either satisfy the boundary conditions at a limited number of points *Q*, or specify that some norm of the error in satisfaction of the boundary conditions is a minimum.

For a boundary value problem, either *u* or *t* is specified and the other is the unknown to be determined by solving the integral equations. The boundary condition where potential *u* or displacement **u** is specified, is also known as the *Dirichlet* boundary condition, whereas the specification of flow *t* or traction **t** is referred to as a *Neuman* boundary condition.

Before we deal with the numerical solution of the integral equations, we must discuss the integrals a little further. As indicated, limiting values of the integrals have to be taken, as the region of exclusion around point *P* is reduced to zero.

The fundamental solutions or *kernels* of integrals *T* and *U* have different types of singularities, which affect this limiting process. The kernel *U* varies according to $ln r$ in two dimensions and with $1/r$ in three dimensions and is known as **weakly singular**. As we see later, the integration of this function poses no great problems. Kernel *T* has a $1/r$ singularity in two dimensions and a $1/r^2$ singularity in three dimensions. This is also known as **strongly singular**. The integral of this function only exists in the sense of a *Cauchy principal value*. We will discuss this further in the chapter on numerical implementation. In the simplest case, we may solve the integral equations by dividing the boundary for two-dimensional problems into straight line segments over which the values of *u* and *t* are assumed to be constant. We assume points *P* to be located at the centre of each segment.

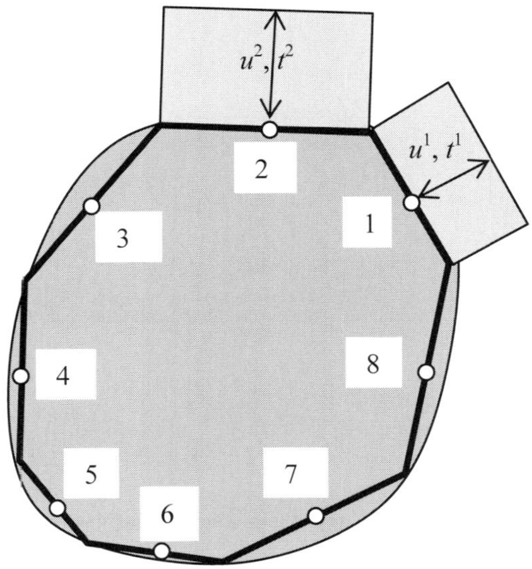

Figure 5.10 Solution of integral equations by linear segments

In the example shown in Fig 5.10 we assume the solution of a two-dimensional potential problem with eight segments, where either u or t is specified on the boundary. We see that this very simple *discretisation* into constant elements violates the continuity conditions between elements. However, we will see by numerical experiments, that the method converges, that is, exact results are obtained, as the number of elements tends to infinity. The integrals can now be evaluated over each element separately and the contributions added, that is, equation (5.25) can be re-written as eight equations

$$\frac{1}{2}u^e + \sum_{e=1}^{8} \Delta T_i^e u^e = \sum_{e=1}^{8} \Delta U_i^e t^e \quad for \quad i = 1,2\ldots 8 \tag{5.32}$$

Where u^e and t^e is the temperature and flow at the centre of element e. Note that as there is a smooth surface at the centres of the elements (at points P_i) c is assigned 1/2. The integrals over the segments are defined as

$$\Delta T_i^e = \int_{S_e} T(P_i, Q)\, dS_e(Q) \quad , \quad \Delta U_i^e = \int_{S_e} U(P_i, Q)\, dS_e(Q) \tag{5.33}$$

Using matrix notation, equation (5.32) can be written as

$$[\,\Delta T\,]\{\,u\,\} = [\,\Delta U\,]\{\,t\,\} \tag{5.34}$$

where

$$[\Delta T]=\begin{bmatrix} \frac{1}{2}+\Delta T_1^1 & \Delta T_1^2 & \dots \\ \Delta T_2^1 & \frac{1}{2}+\Delta T_2^2 & \dots \\ \dots & \dots & \dots \end{bmatrix} \quad , \quad [\Delta U]=\begin{bmatrix} \Delta U_1^1 & \Delta U_1^2 & \dots \\ \Delta U_2^1 & \Delta U_2^2 & \dots \\ \dots & \dots & \dots \end{bmatrix} \tag{5.35}$$

and

$$\{u\}=\begin{Bmatrix} u^1 \\ u^2 \\ . \end{Bmatrix} \quad , \quad \{t\}=\begin{Bmatrix} t^1 \\ t^2 \\ . \end{Bmatrix} \tag{5.36}$$

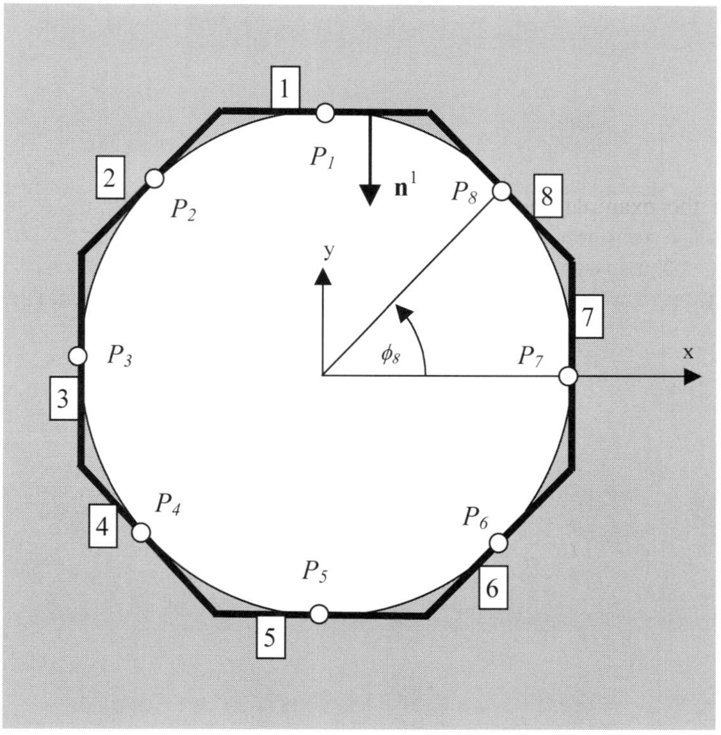

Figure 5.11 Discretisation into linear elements for problem of flow past cylinder

If we consider the solution of the heat flow problem, which we solved by the Trefftz method, then we have a problem where flow $\{t\}_0$ is specified at the boundary and temperatures are unknown (Figure 5.11).

This means that the system of equations can be written

$$[\Delta T]\{u\}=\{F\} \quad with \quad \{F\}=[\Delta U]\{t\}_0 \tag{5.37}$$

where vector $\{t\}_0$ is given by

$$\{t\}_0 = q_0\begin{Bmatrix} -n_y^1 \\ -n_y^2 \\ . \end{Bmatrix} = q_0\begin{Bmatrix} \sin\phi_1 \\ \sin\phi_2 \\ . \end{Bmatrix} \tag{5.38}$$

The integrals which have to be evaluated analytically are

$$\begin{aligned} \Delta T_i^e &= \int_{S_e} T(P_i,Q)\, dS_e(Q) = \int_{S_e} \frac{\cos\theta}{2\pi\, r}\, dS_e \\ \Delta U_i^e &= \int_{S_e} U(P_i,Q)\, dS_e(Q) = \int_{S_e} \frac{1}{2\pi} ln\frac{1}{r}\, dS_e \end{aligned} \tag{5.39}$$

The integrals can be evaluated using a local coordinate system $\bar{x},\bar{y}$ through point P and polar coordinates, as shown[3] in Figure 5.12 where θ is defined anticlockwise from a line perpendicular to the element e with start node A and end node B.

The angle θ is computed as follows: a unit vector from A to B is defined as:

$$\mathbf{v}_{AB} = \frac{1}{L}\begin{Bmatrix} x_A - x_B \\ y_A - y_B \end{Bmatrix} \tag{5.40}$$

The vector normal to element **n** is computed by taking the vector x-product of $\mathbf{V}_{AB}$ with the z-axis. This gives

$$\mathbf{n} = \frac{1}{L}\begin{Bmatrix} y_A - y_B \\ -(x_A - x_B) \end{Bmatrix} \tag{5.41}$$

The cosine and sine of θ are then computed by

$$\cos\theta = \mathbf{n}\bullet\mathbf{r}\frac{1}{r} \tag{5.42}$$

and

$$\sin\theta = \mathbf{v}_{AB}\bullet\mathbf{r}\frac{1}{r} \tag{5.43}$$

θ can be computed by

$$\theta = \cos^{-1}\left(\mathbf{n} \bullet \mathbf{r}\frac{1}{r}\right) SIGN\,(\sin\theta) \tag{5.44}$$

The first integral is evaluated as:

$$\Delta T_i^e = \int_{\theta_A}^{\theta_B} \frac{\cos\theta}{2\pi\, r}\frac{rd\theta}{\cos\theta} = \int_{\theta_A}^{\theta_B} \frac{1}{2\pi}\, d\theta = \frac{\theta}{2\pi}\Bigg|_{\theta_A}^{\theta_B} = \frac{1}{2\pi}\left(\theta_A - \theta_B\right) \tag{5.45}$$

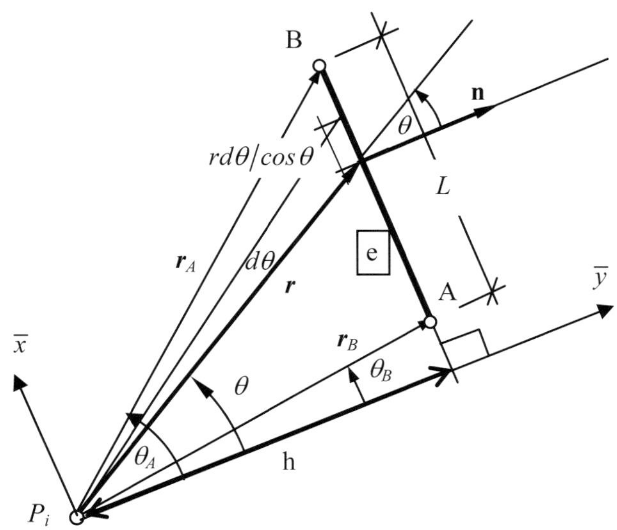

Figure 5.12 Polar coordinate system used for the analytic evaluation of integral ΔT_i^e

If P_i is at the centre of element e then we have to take the Cauchy principal value of the integral. As shown in Figure 5.13, the integration is carried out over the region of exclusion. The reader may verify that because of the anti-symmetry of T shown in Figure 4.4 we obtain $\Delta T_i^i = 0$

The second integral is computed as

$$\Delta U_i^e = \int_{\theta_A}^{\theta_B} \frac{1}{2\pi\, k}\ln\frac{1}{r}\frac{rd\theta}{\cos\theta} = -\int_{\theta_A}^{\theta_B} \frac{1}{2\pi\, k}\ln\left(\frac{h}{\cos\theta}\right)\frac{hd\theta}{\cos^2\theta} =$$
$$-\frac{h}{2\pi k}\left[\tan\theta\left(\ln\left(\frac{h}{\cos\theta}\right)-1\right)+\theta\right]_{\theta_B}^{\theta_A} \tag{5.46}$$

where $r = h/\cos\theta$ has been substituted

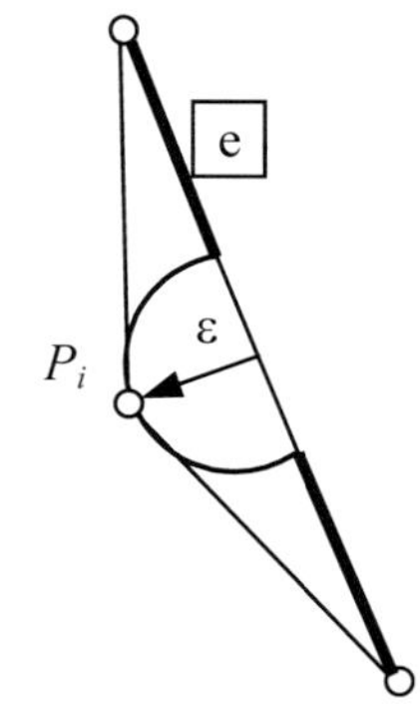

Figure 5.13 Cauchy principal value computation as P_i approaches the centre of element e

For programming purposes, it is convenient to write this expression in terms of r and θ

$$\Delta U_i^e = -\frac{1}{2\pi\, k}\Big[\; r\sin\theta\big(\ln r - 1\big) + \theta\, r\cos\theta \Big]_{\theta_B, r_B}^{\theta_A, r_A} \tag{5.47}$$

If P_i is at the centre of element e of length L then we have

$$h \to 0 \quad , \quad \theta_B = -\theta_A \to \pi/2 \quad , \quad r_B = r_A = L/2 \tag{5.48}$$

and the diagonal coefficient is computed as

$$\Delta U_2^2 = -\frac{L}{2\pi k}\left(\ln\left(\frac{L}{2}\right) - 1\right) \tag{5.49}$$

5.5 COMPUTATION OF RESULTS INSIDE THE DOMAIN

The solution of the integral equation only provides values of u and t on the boundary of the domain. Since we have defined *global shape functions* in the form of fundamental solutions, the results at any point inside the domain can be readily computed. In contrast to the FEM, where results at all nodes or Gauss points are computed as part of the solution, we compute the interior results as a post-processing exercise. To compute, for example, the temperature/potential at a point P_a inside the domain, we simply rewrite equation (5.20)

$$u(P_a) = \int_S t(Q)U(P_a,Q)dS(Q) - \int_S u(Q)T(P_a,Q)dS(Q) \tag{5.50}$$

or in discretised form using line segments

$$u(P_a) = \sum_{e=1}^{8} \Delta T^e(P_a) u^e - \sum_{e=1}^{8} \Delta U^e(P_a) t^e \tag{5.51}$$

where

$$\Delta T^e = \int_{S_e} T(P_a, Q)\, dS_e(Q) \quad , \quad \Delta U^e = \int_{S_e} U(P_a, Q)\, dS_e(Q) \tag{5.52}$$

The flows at P_a in x- and y-directions are computed by taking derivatives of (5.50)

$$\begin{aligned} q_x(P_a) &= -k\frac{\partial u}{\partial x}(P_a) = -k\left(\int_S t(Q)\frac{\partial U}{\partial x}(P_a, Q)dS(Q) - \int_S u(Q)\frac{\partial T}{\partial x}(P_a, Q)dS(Q)\right) \\ q_y(P_a) &= -k\frac{\partial u}{\partial y}(P_a) = -k\left(\int_S t(Q)\frac{\partial U}{\partial y}(P_a, Q)dS(Q) - \int_S u(Q)\frac{\partial T}{\partial y}(P_a, Q)dS(Q)\right) \end{aligned} \tag{5.53}$$

where the derivatives of U have been presented previously and the derivatives of T are given for two-dimensional problems as

$$\begin{aligned} \frac{\partial T}{\partial x} &= \frac{\partial}{\partial x}\left[n_x \frac{\partial U}{\partial x} + n_y \frac{\partial U}{\partial y}\right] \\ \frac{\partial T}{\partial y} &= \frac{\partial}{\partial y}\left[n_x \frac{\partial U}{\partial x} + n_y \frac{\partial U}{\partial y}\right] \end{aligned} \tag{5.54}$$

For constant boundary elements, equation (5.53) can be replaced by

$$\begin{aligned} q_x(P_a) &= -k\left(\sum_{e=1}^{E} \Delta S_{xa}^e\, t^e - \sum_{e=1}^{E} \Delta R_{xa}^e\, u^e\right) \\ q_y(P_a) &= -k\left(\sum_{e=1}^{E} \Delta S_{ya}^e\, t^e - \sum_{e=1}^{E} \Delta R_{ya}^e\, u^e\right) \end{aligned} \tag{5.55}$$

where the integrals

$$\begin{aligned} \Delta S_{xa}^e &= \int_{S_e} \frac{\partial U}{\partial x}(P_a, Q)dS(Q) \quad ; \quad \Delta S_{ya}^e = \int_{S_e} \frac{\partial U}{\partial y}(P_a, Q)dS(Q) \\ \Delta R_{xa}^e &= \int_{S_e} \frac{\partial T}{\partial x}(P_a, Q)dS(Q) \quad ; \quad \Delta R_{ya}^e = \int_{S_e} \frac{\partial T}{\partial y}(P_a, Q)dS(Q) \end{aligned} \tag{5.56}$$

can be evaluated analytically over element e.

Using the notation in Figure 5.12 with node P_i replaced by P_a we can evaluate the integrals analytically in terms of the local coordinates $\overline{x},\overline{y}$.

$$\begin{aligned}
\Delta S^e_{\overline{x}a} &= \frac{1}{2\pi k}\left(\theta_B - \theta_A \right) \\
\Delta S^e_{\overline{y}a} &= \frac{1}{2\pi k}\ln\left(\cos\theta_B / \cos\theta_A\right) \\
\Delta R^e_{\overline{x}a} &= \frac{1}{2\pi h}\left(\cos\theta_B \sin\theta_B - \cos\theta_A \sin\theta_A \right) \\
\Delta R^e_{\overline{y}a} &= \frac{1}{2\pi h}\left(\cos^2\theta_B - \cos^2\theta_A \right)
\end{aligned} \tag{5.57}$$

The contribution of element e to the flux in $\overline{x},\overline{y}$ -direction is given as:

$$\begin{aligned}
q^e_{\overline{x}}\left(P_a\right) &= -k\left(\Delta S^e_{\overline{x}a}\, t^e - \Delta R^e_{\overline{x}a}\, u^e\right) \\
q^e_{\overline{y}}\left(P_a\right) &= -k\left(\Delta S^e_{\overline{y}a}\, t^e - \Delta R^e_{\overline{y}a}\, u^e\right)
\end{aligned} \tag{5.58}$$

This has to be transformed into global directions x,y by

$$\begin{aligned}
q^e_x\left(P_a\right) &= q^e_{\overline{x}} n_x - q^e_{\overline{y}} n_y \\
q^e_y\left(P_a\right) &= q^e_{\overline{x}} n_y + q^e_{\overline{y}} n_x
\end{aligned} \tag{5.59}$$

where n_x ,n_y are the components of the vector normal to element e.

The final fluxes are computed by summing all element contributions

$$q_x\left(P_a\right) = \sum_{e=1}^{E} q^e_x \quad ; \quad q_y\left(P_a\right) = \sum_{e=1}^{E} q^e_y \tag{5.60}$$

5.6 PROGRAM 5.2: FLOW AROUND CYLINDER, DIRECT METHOD

We can now write a computer program for the solution of the flow around a cylinder problem, which was previously solved with the Trefftz method. The input section of the program is very similar to Program 5.1, except that no source points have to be specified. The circle is divided into *nseg* straight line segments. At the centre of each segment the boundary condition t_0 is specified. The coefficient matrices, equation (5.35), are set up with the results of the analytical integration, as computed in section 5.4.3. In setting up the coefficient matrices we distinguish between diagonal and off-diagonal coefficients. The diagonal coefficients are computed for the case where points P_i are coincidental with the centre of the segment (also sometimes called *self effects*).

```
PROGRAM Direct_Method
!---------------------------------
!   Program to compute the heat flow  past a cylindrical
!   isolator in an 2-D infinite domain using the direct BE
!   method with constant line segments
!---------------------------------
USE Utility_lib                  !   subroutine to solve equations
REAL          ::  q              !   inflow/outflow
REAL          ::  k              !   Thermal conductivity
INTEGER       ::  nseg           !   Number of segments
REAL          ::  rq             !   radius of isolator (inner)
REAL          ::  rqo            !   radius of isolator (outer)
REAL(KIND=8),ALLOCATABLE :: Lhs(:,:) ! [DT]
REAL(KIND=8),ALLOCATABLE :: F(:)   ! {F}
REAL(KIND=8),ALLOCATABLE :: u(:)   ! Temp at segment centers
REAL,ALLOCATABLE :: Rhs(:,:)       ! [DU]
REAL,ALLOCATABLE :: t0(:)          !  Applied flows
REAL,ALLOCATABLE :: xA(:,:),xB(:,:)  !   Start/end coords of seg
REAL,ALLOCATABLE :: xS(:,:)          !   Coords of points Pi
REAL,ALLOCATABLE :: Ve(:,:),Vn(:,:)  !  Vectors A-B and n
REAL :: vrA(2),vrB(2)    !  Vectors to point A and B of seg
REAL :: lens             !  Length of segment
C= 0.5/Pi
OPEN(UNIT=10,FILE='INPUT.DAT',STATUS='OLD',ACTION='READ')
OPEN(UNIT=11,FILE='OUTPUT.DAT',STATUS='UNKNOWN',ACTION='WRITE')
READ(10,*) q,k,nseg,rq
WRITE(11,*) 'Heat flow past a cylinder (direct BE method)'
WRITE(11,*) 'Input values:'
WRITE(11,*) ' Heat inflow/outflow= ',q
WRITE(11,*) ' Thermal conductivity=',k
WRITE(11,*) ' Radius of Isolator=  ',rq
WRITE(11,*) ' Number of segments=  ',nseg
ALLOCATE (Lhs(nseg,nseg),Rhs(nseg,nseg),F(nseg))
ALLOCATE (xA(2,nseg),xB(2,nseg),t0(nseg),u(nseg))
ALLOCATE (xS(2,nseg),ve(2,nseg),vn(2,nseg))
C1=0.5/(Pi*k)
Delth= 2.0*Pi/nseg      !     increment in angle theta
rqo=rq/COS(Delth/2.0)   !     outer radius of isolator
Thet= (Pi-Delth)/2.0
!  Compute start/end  coordinates of segments
xA(1,1)= rqo*COS(Thet)
xA(2,1)= rqo*SIN(Thet)
Segments: &
DO ns= 1,nseg-1
 Thet= Thet + Delth
 xB(1,ns)= rqo*COS(Thet)
 xB(2,ns)= rqo*SIN(Thet)
 xA(1,ns+1)= xB(1,ns)
 xA(2,ns+1)= xB(2,ns)
END DO &
```

```
Segments
xB(1,nseg)= xA(1,1)
xB(2,nseg)= xA(2,1)
!   Compute centre coordinates of segments (coll. point coords)
Segments1: &
DO ns= 1,nseg
 xS(1,ns)= (xB(1,ns) + xA(1,ns))/2.0
 xS(2,ns)= (xB(2,ns) + xA(2,ns))/2.0
END DO &
Segments1
!   Compute applied tractions at centers of elements
Thet= Pi/2.0
Segments2: &
DO ns= 1,nseg
 t0(ns)= q* SIN(Thet)
 Thet= Thet + Delth
END DO  &
Segments2
!    Assemble matrices DT and DU
Segments3: &
DO ns=1,nseg
 lens= dist(xA(:,ns),xB(:,ns),2)
 !    Vector parallel and normal to segment A-B
 dx= xA(1,ns) - xB(1,ns)
 dy= xA(2,ns) - xB(2,ns)
 ve(1,ns)= dx/lens
 ve(2,ns)= dy/lens
 vn(1,ns)= ve(2,ns)
 vn(2,ns)=-ve(1,ns)
 Points_Pi: &
 DO np=1,nseg
  rA= Dist(xA(:,ns),xS(:,np),2)
  rB= Dist(xB(:,ns),xS(:,np),2)
  vrA(1)= xA(1,ns)- xS(1,np)
  vrA(2)= xA(2,ns)- xS(2,np)
  vrB(1)= xB(1,ns)- xS(1,np)
  vrB(2)= xB(2,ns)- xS(2,np)
  COSThA= DOT_PRODUCT(vn(:,ns),vrA)/rA
  COSThB= DOT_PRODUCT(vn(:,ns),vrB)/rB
  SINThA= DOT_PRODUCT(ve(:,ns),vrA)/rA
  SINThB= DOT_PRODUCT(ve(:,ns),vrB)/rB
  ThetA= ACOS(COSThA)*SIGN(1.0,SinThA)
  ThetB= ACOS(COSThB)*SIGN(1.0,SinThB)
  IF(np == ns) THEN          ! Diagonal coefficients
   Lhs(np,np)=  0.5
   Rhs(np,np)= lens*C1*(LOG(lens/2.0)-1.0)
  ELSE                       ! off-diagonal coeff.
   Lhs(np,ns)= C*(ThetB-ThetA)
   Rhs(np,ns)= C1*(rB*SINThB*(LOG(rB)-1)+ThetB*rB*COSThB &
                 - rA*SINThA*(LOG(rA)-1)-ThetA*rA*COSThA)
  END IF
```

```
 END DO  &
 Points_Pi
END DO &
Segments3
F= MATMUL(Rhs,t0)    !    compute right hand side vector
CALL Solve(Lhs,F,u)  !    solve system of equations
!    output computed temperatures
WRITE(11,*) 'Temperatures at segment centers:'
Segments4: &
DO ns= 1,nseg
 WRITE(11,'(A,I5,A,F10.3)') &
' Segment',ns,'  T=',u(ns)-q/k*xS(2,ns)
END DO &
Segments4
DEALLOCATE (xS)
!   Compute Temperatures and flows at interior points
READ(10,*,IOSTAT=IOS) NPoints
IF(NPoints == 0 .OR. IOS /= 0) THEN
 PAUSE 'program Finshed'
 STOP
END IF
ALLOCATE (xS(2,NPoints))  !   re-use array Xs
WRITE(11,*) &
'Temperatures(T) and flow (q-x,q-y) at interior points:'
DO n=1,NPoints
 READ(10,*) xS(1,n),xS(2,n)
END DO
Interior_points: &
DO np=1,Npoints
    up= 0.0
    qx= 0.0
    qy= 0.0
    Segments5 : &
    DO ns=1,nseg
       rA= Dist(xA(:,ns),xS(:,np),2)
       rB= Dist(xB(:,ns),xS(:,np),2)
       vrA(1)= xA(1,ns)- xS(1,np)
       vrA(2)= xA(2,ns)- xS(2,np)
       vrB(1)= xB(1,ns)- xS(1,np)
       vrB(2)= xB(2,ns)- xS(2,np)
       COSThA= -DOT_PRODUCT(vn(:,ns),vrA)/rA
       COSThB= -DOT_PRODUCT(vn(:,ns),vrB)/rB
       SINThA= -DOT_PRODUCT(ve(:,ns),vrA)/rA
       SINThB= -DOT_PRODUCT(ve(:,ns),vrB)/rB
       H= RA*CosThA
       ThetA= ACOS(COSThA)*SIGN(1.0,SinThA)
       ThetB= ACOS(COSThB)*SIGN(1.0,SinThB)
       IF(ThetB-ThetA > Pi) ThetA= 2.0*Pi + ThetA ! θB-θA < 180°
       dT= C*(ThetB-ThetA)
       dU= C1*(rB*SINThB*(LOG(rB)-1)+ThetB*rB*COSThB &
             - rA*SINThA*(LOG(rA)-1)-ThetA*rA*COSThA)
```

```
        dSx= C/k*(ThetB-ThetA)
        Fact= CosthB/CosthA
        IF(Fact > 0.0) THEN
         dSy= -C/k*LOG(Fact)
        ELSE
         dSy= 0.
        END IF
        dRx= -C/H*(costhB*SINThB - cosThA*sinThA)
        dRy= C/H*(costhB**2 - cosThA**2)
        up= up + dU*t0(ns) - dT*u(ns)
        qxp= -k*(dSx*t0(ns)-dRx*u(ns))          !   q-x'
        qyp= -k*(dSy*t0(ns)-dRy*u(ns))          !   q-y'
        qx= qx + qxp*vn(1,ns) - qyp*vn(2,ns)
        qy= qy + qxp*vn(2,ns) + qyp*vn(1,ns)
    END DO &
    Segments5
    Up= Up - q/k*xS(2,np)   !   superimpose solutions
    qy= qy + q
    WRITE(11,'(5(A,F10.3))') &
  'x=',xS(1,np),', y=',xS(2,np),', T=',up,',  q-x=',qx,',  q-
y=',qy
END DO  &
Interior_points
STOP
END PROGRAM Direct_Method
```

INPUT DATA for program Direct_method

1.0 Problem specification

q,k, nseg, rq	q	Heat inflow
	k	Thermal conductivity
	nseg	Number of segments
	rq	Radius of isolator

4.0 Interior point specification

Npoints	Number of interior points

5.0 Interior point coordinates (Npoints cards)

x,y	x,y coordinates of interior points

5.6.1 Sample input and output

Here we show the input file for the calculation of the problem in Figure 5.11, with 16 segments and interior points along a horizontal and vertical line and the output file generated by Program 5.2.

File INPUT.DAT :

```
1.0 1.0 16 1.0
28
0. 1.
...
1. 0.
...
10. 0.
```

File OUTPUT.DAT:

```
Heat flow past a cylinder (direct BE method)
Input values:
Heat inflow/outflow=       1.00000
Thermal conductivity=      1.00000
Radius of Isolator=        1.00000
Number of segments=              16
Temperatures at segment centers:
 Segment    1  T=     -2.026
 ....
 Segment   16  T=     -1.872
Temperatures(T) and flow (q-x,q-y) at interior points:
x=0.000, y=1.000, T=    -2.026,  q-x=     0.000,  q-y=     0.032
.....
x=10.000 y=0.000, T=     0.000,  q-x=     0.000,  q-y=     1.010
```

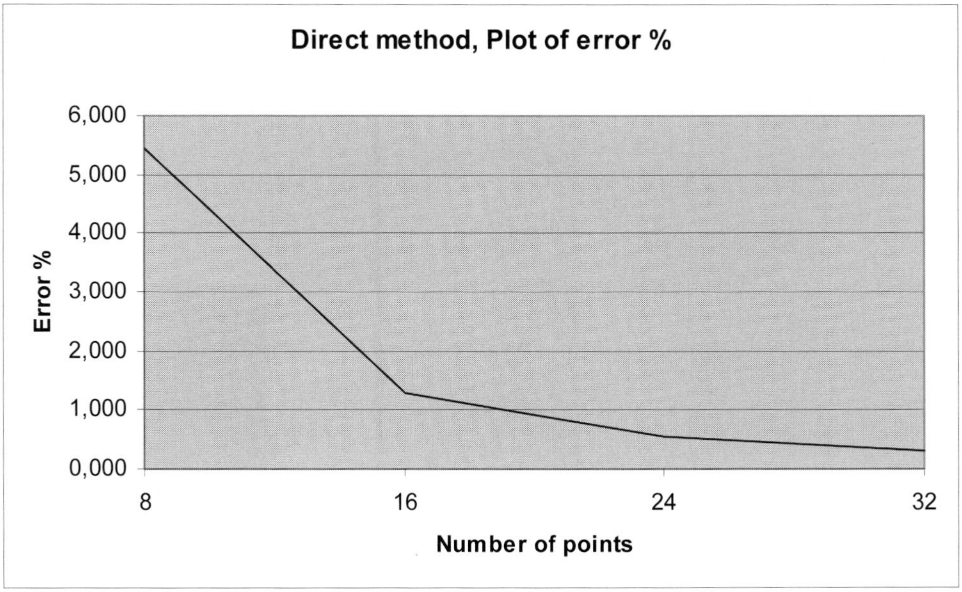

Figure 5.14 Error in the temperature at segment 1 for different no of elements (points)

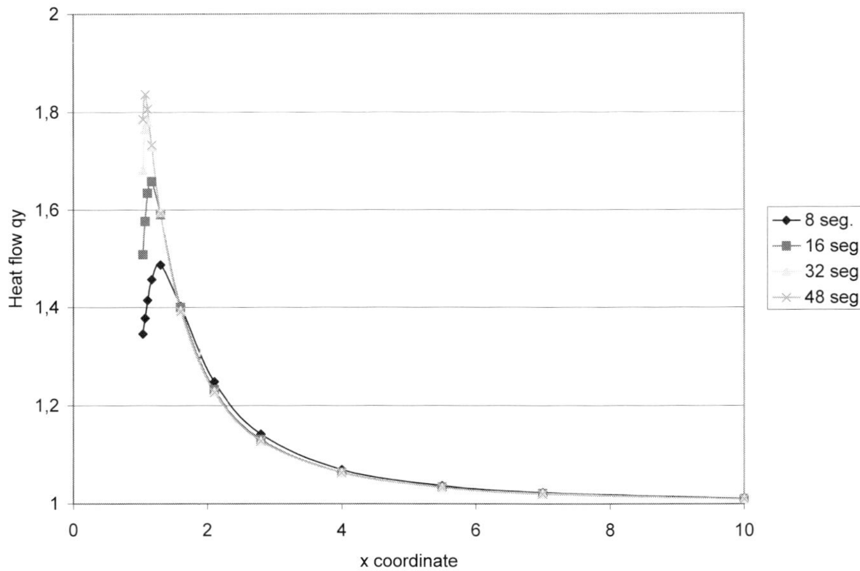

Figure 5.15 Heat flow in vertical direction along horizontal line results for different meshes

Figure 5.16 Flow past a cylindrical isolator: contour lines of temperature

The error in the temperature at segment 1 versus the number of elements (collocation points) is plotted in Figure 5.14. It can be seen that the error falls below 1% for 24 elements. A plot of the heat flow in vertical direction along a horizontal line depending on the number of segments is shown in Figure 5.15. The theoretical value of q_y should approach the value of 2.0 exactly on the boundary. It can be seen that as we get very near to the boundary the values are significantly in error and that this error depends on the element size adjacent to the interior point. As we will see later, this is typical of the boundary element method and will be more pronounced when numerical integration is used. However with the higher order elements introduced next we will see that results exactly on the boundary can be computed with an alternative method. Figure 5.16 and 5.17 finally show the graphical display of the results as it may be produced by a postprocessor. Figure 5.16 shows the contours of the temperature distribution whereas in Figure 5.17 the flow vectors are depicted by arrows whose magnitude depends on the value of heat flow. It can be seen that the temperature contours align normal to the boundary as they should and that the flow vectors approach zero values at the bottom and the top of the circular isolator.

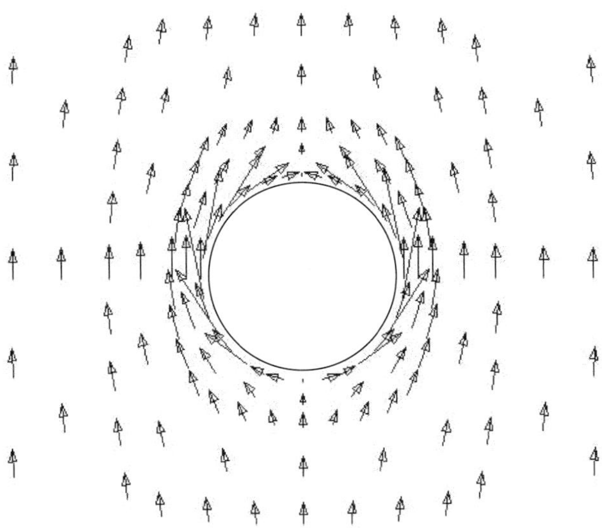

Figure 5.17 Flow past a cylindrical isolator: flow vectors

5.7 CONCLUSIONS

In this chapter we have introduced the Trefftz and boundary integral equation methods. Although we found that the Trefftz method is not suitable for general purpose programming it can be used to demonstrate the basic principles involved, because of its simplicity. A short program can be written and used for numerical experiments. As the original idea by Trefftz, conceived in the days before computers, was not found to be suitable, improvements to the method were sought. This lead initially to the so called indirect method, where sources were assumed to be distributed instead of concentrated at a point. This allowed, with similar limiting procedures as shown in this chapter, the placing of source points on the same contour as the boundary points, therefore alleviating the need for two sets of points. We have not discussed this method here as it has been largely superseded by the direct method, which avoids the computation of *fictitious* sources/forces altogether.

Using the well known theorem of Betti, we developed boundary integral equations relating tractions to displacements, or temperatures/potentials to normal gradients. We found that using a limiting procedure, source points can be placed on the boundary to be coincidental with the points where we satisfy given boundary conditions, thereby rendering the method usable for general purpose programming. However, we find that evaluating some boundary integrals causes difficulties, since the integrands tend to infinity at certain points. Some of the integrals exist only in the sense of a principal value. Indeed, the advanced mathematics involved, which may have prevented matching the success of the FEM in the early days, stems from the difficulty in evaluating these integrals. If simple elements, that is, line segments, such as the ones used here for solving the 2-D heat flow problem are defined, where the known boundary condition and the unknown are assumed to be constant, then the integration can be carried out analytically. For 3-D elasticity triangular elements with constant variation have been proposed, but the analytical evaluation of the integrals becomes rather involved. However, even for the simple heat flow example, we find that these constant elements are not very accurate and many elements are needed to model a smooth surface.

To the author's best knowledge, it was Lachat and Watson[4] who first thought of the idea of introducing isoparametric boundary elements of the same type as the ones already in use in the FEM at that time. These are commonly attributed to Ergatoudis[5], although the basic concept can be found in old mathematics books. The method, previously known as the *Boundary Integral Equation* method, became the *Boundary Element Method (BEM)*. Analytical integration is no longer a feasible way of computing the coefficients of the system of equations and we have to revert to numerical integration. For engineers, who usually find no pleasure in writing pages of analytical evaluation of integrals, this of course was a godsend. Using the Gauss integration method introduced in Chapter 3, the evaluation of the integrals can now be reduced to evaluating sums. However, because of the nature of the integrals we must be very careful that the accuracy is adequate. In contrast to the FEM, where less may be better, (i.e., the application of reduced integration for the evaluation of element stiffness) we will find that the BEM is much less forgiving when it comes to the accuracy of the integrals. The boundary element method using higher order isoparametric elements, is the method used almost exclusively in modern general purpose computer programs. We will therefore deal, in some depth, with the numerical implementation of the method in the next chapter.

5.8 EXERCISES

Exercise 5.1

Use Program 5.1 (Trefftz method) to find out the influence of the following on the accuracy of results of the heat flow example in Fig 5.1:

(a) when the distance between source points P and field points Q is reduced to ½ and ¼ of the value used in section 5.3.

(b) when the number of points *P,Q* is increased to twice and three times the value used in section 5.

Exercise 5.2

Expand Program 5.1 (Trefftz method), so that in addition to temperatures flow vectors **q** may be computed at interior points.

Exercise 5.3

Use program 5.2 (Direct_method), to compute the heat flow problem solved by the Trefftz method. Investigate the influence of the number of segments on results by using 8, 16 and 32 segments. Plot the norm of the error to show convergence.

Exercise 5.4

Modify program 5.1, so that potential problems for general boundary shapes can be analysed, by allowing points *P* and *Q* to be specified as input instead of being generated automatically. Test the program by analysing the flow past an elliptical isolator.

Exercise 5.5

Modify program 5.2, so that potential problems for general boundary shapes can be analysed, by allowing boundary segments and boundary conditions to be specified as input. Test the program by analysing the flow past an elliptical isolator.

5.9 REFERENCES

1. Trefftz, E. (1926) Ein Gegenstück zum Ritzschen Verfahren. *Proc. 2nd Int. Congress in Applied Mechanics,* Zürich, p.131.
2. Beer G. and Watson J.O. (1995) Introduction to Finite and Boundary Element Methods for Engineers. J. Wiley.
3. Banerjee P.K. (1994) Boundary Element Methods in Engineering Science. McGraw Hill.
4. Lachat, J.C. and Watson, J.O. (1976) Effective numerical treatment of boundary integral equations. *Int. J. Num. Meth. Eng.* **10**: 991-1005.
5. Ergatoudis J.G., Irons B.M. and Zienkiewicz O.C. (1968) Curved, Isoparametric 'Quadilateral' Elements for Finite Element Analysis, *Int. J. Solids & Struct.* **4**: 31-42.

6

Boundary Element Methods – Numerical Implementation

There is nothing more powerful
than an idea whose time has come

V. Hugo

6.1 INTRODUCTION

In the previous chapter we derived boundary integral equations relating the known boundary conditions to the unknowns. For practical problems, these integral equations can only be solved numerically. The simplest numerical implementation is using line elements, where the knowns and unknowns are assumed to be constant inside the element. In this case, the integral equation can be written as the sum of integrals over elements. The integrals over the elements can then be evaluated analytically. In the previous chapter we have presented constant elements for the solution of two-dimensional potential problems only. The analytical evaluation over elements would become quite cumbersome for two- and three-dimensional elasticity problems. Constant elements were used in the early days of the development, where the method was known under the name *Boundary Integral Equation* (BIE) Method[1]. This is similar to the development of the FEM, where triangular and tetrahedral elements, with exact integration, were used in the early days. In 1968, Ergatoudis and Irons[2] suggested that isoparametric finite elements and numerical integration could be used to obtain better results, with fewer elements. The concept of higher order elements and numerical integration is very appealing to engineers because it alleviates the need for tedious analytical integration and, more importantly, it allows the writing of general purpose software with a choice of element types. Indeed, this concept will allow us to develop one single program to solve two- and three-dimensional problems in elasticity and potential flow, or any other problem for which we can supply a fundamental solution (see Chapter 18).

The idea of using isoparametric concepts for boundary elements seems to have been first introduced by Lachat and Watson[3] and this prompted a change of name of the method to *Boundary Element Method.* This chapter is about the numerical implementation of isoparametric boundary elements, using the basic concepts that were already discussed in detail in Chapter 3.

6.2 DISCRETISATION WITH ISOPARAMETRIC ELEMENTS

We consider the numerical solution of the boundary integral equations using isoparametric elements where linear or quadratic functions are assumed for the variation of the known and the unknown boundary values. Recalling from Chapter 3, we have for a one-dimensional isoparametric element and for potential problems the following interpolations

$$
\begin{aligned}
\mathbf{x}(\xi) &= \sum N_n(\xi)\mathbf{x}_n^e \quad \textit{Geometry} \\
u(\xi) &= \sum N_n(\xi)u_n^e \quad \textit{Temperature / Potential} \\
t(\xi) &= \sum N_n(\xi)t_n^e \quad \textit{Flux}
\end{aligned}
\tag{6.1}
$$

Consider the example in Figure 6.1, where the boundary of a two-dimensional potential problem is divided into linear isoparametric elements.

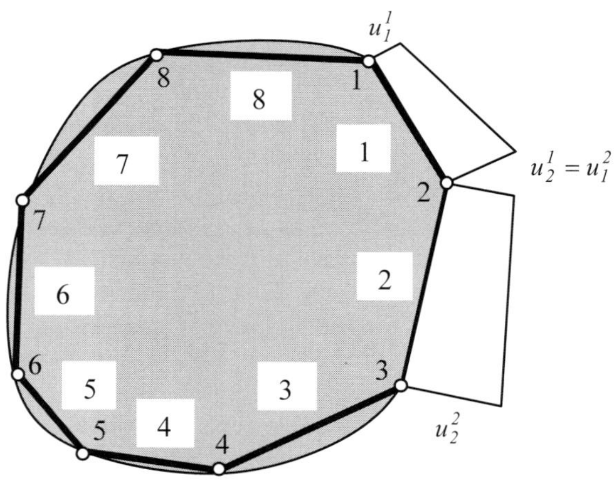

Figure 6.1 Discretisation of two-dimensional problem into linear boundary elements

Equations (6.1) are based on a local numbering as explained in Chapter 3. In order to enforce continuity conditions, we also define a global numbering of the nodes. That is, we define a global vector containing the potentials/temperatures at all nodes.

$$\{\mathbf{u}\} = \begin{Bmatrix} u_1 \\ u_2 \\ \vdots \end{Bmatrix} \tag{6.2}$$

The relation between local and global numbering is known as the element *connectivity* or *incidences*. For example, element 1 has connectivity vector {1,2}, which means that the values of u for the two nodes of the element appear at the first and second position in the global vector $\{\mathbf{u}\}$. Although we usually wish to enforce continuity of u, this is not necessary for t, the boundary flux, which may be discontinuous.

We now consider the numerical treatment of the integral equation.

$$cu(P) = \lim_{\varepsilon \to 0} \left[\int_{S-S_\varepsilon} t(Q)U(P,Q)dS(Q) - \int_{S-S_\varepsilon} u(Q)T(P,Q)dS(Q) \right] \tag{6.3}$$

Substituting equations (6.1) for $t(Q)$ and $u(Q)$ and splitting the integrals into a sum of integrals over elements gives (leaving out the limiting value process, which we now implicitly assume)

$$cu(P) = \sum_{e=1}^{E} \int_{S_e} \left(\sum_{n=1}^{N} N_n t_n^e \right) U(P,\xi)dS(\xi) - \sum_{e=1}^{E} \int_{S_e} \left(\sum_{n=1}^{N} N_n u_n^e \right) T(P,\xi)dS(\xi) \tag{6.4}$$

where E is the total number of elements and N is the number of nodes per element. The process is generally known as *discretisation of the integral equation.* Since t^e_n and u^e_n, being nodal values are constant with respect to the integration, they can be taken out of the integral and equation (6.4) can be rewritten

$$cu(P) = \sum_{e=1}^{E} \sum_{n=1}^{N} t_n^e \int_{S_e} N_n(\xi) U(P,\xi)dS(\xi) - \sum_{e=1}^{E} \sum_{n=1}^{N} u_n^e \int_{S_e} N_n(\xi) T(P,\xi)dS(\xi) \tag{6.5}$$

The integration has now been changed to a sum of integrations of Kernel shape function products over elements. We will deal with this in detail later.

Theoretically, Betti's theorem should be valid for any location P and, therefore, we can write equation (6.5) for an infinite number of points P_i. In practice, we select a limited number of points only. Since for potential problems either t or u must be known on the boundary, there will be as may unknowns as there are nodes. In the simplest numerical method, also known as *point collocation*, we therefore obtain the necessary integral equations by placing points P_i in turn at all the nodes of the mesh.

$$\begin{aligned} cu(P_i) = & \sum_{e=1}^{E}\sum_{n=1}^{N} t_n^e \int_{S_e} N_n(\xi) U(P_i,\xi) dS(\xi) \\ & - \sum_{e=1}^{E}\sum_{n=1}^{N} u_n^e \int_{S_e} N_n(\xi) T(P_i,\xi) dS(\xi) \quad i = 1,2 \ldots I \end{aligned} \tag{6.6}$$

where I is the total number of nodes, which has to equal to the number of unknowns.

This would mean, however, that the theorem by Betti is only satisfied for certain locations of P. In an alternative approach we seek to minimise the error in the satisfaction of the Betti theorem. This approach is also known by the term *weighted residual methods*, because weighting functions are used in the minimisation of the residual error. In the most popular method, the *Galerkin* method, the interpolation functions are used as weighting functions. The *Galerkin* method will not be discussed here because it is more complicated and it is not clear if the additional complexity and increased numerical work will result in a significant increase in accuracy[4].

Equation (6.6) can be re-written as

$$cu(P_i) + \sum_{e=1}^{E}\sum_{n=1}^{N} \Delta T_{ni}^e u_n^e = \sum_{e=1}^{E}\sum_{n=1}^{N} \Delta U_{ni}^e t_n^e \quad i = 1,2....I \tag{6.7}$$

where

$$\Delta U_{ni}^e = \int_{S_e} N_n(\xi) U(P_i,\xi) dS(\xi) \quad , \quad \Delta T_{ni}^e = \int_{S_e} N_n(\xi) T(P_i,\xi) dS(\xi) \tag{6.8}$$

where S_e is the element length and ξ is the intrinsic coordinate.

For elasticity problems, the integral equation which has to be discretised is given as

$$c\,\mathbf{Iu}(P) = \lim_{\varepsilon \to 0}\left[\int_{S-S\varepsilon} \mathbf{U}(P,Q)\mathbf{t}(Q) dS(Q) - \int_{S-S\varepsilon} \mathbf{T}(P,Q)\mathbf{u}(Q) dS(Q) \right] \tag{6.9}$$

In discretised form this equation is written as

$$\mathbf{cu}(P_i) + \sum_{e=1}^{E}\sum_{n=1}^{N} \Delta \mathbf{T}_{ni}^e \mathbf{u}_n^e = \sum_{e=1}^{E}\sum_{n=1}^{N} \Delta \mathbf{U}_{ni}^e \mathbf{t}_n^e \tag{6.10}$$

where, for two-dimensional problems

$$\Delta \mathbf{U}_{ni}^{e} = \int_{S_e} N_n(\xi)\,\mathbf{U}(P_i,\xi)\,dS(\xi) \quad , \quad \Delta \mathbf{T}_{ni}^{e} = \int_{S_e} N_n(\xi)\,\mathbf{T}(P_i,\xi)\,dS(\xi) \tag{6.11}$$

For three-dimensional problems

$$\begin{aligned} \Delta \mathbf{U}_{ni}^{e} &= \int_{S_e} N_n(\xi,\eta)\,\mathbf{U}(P_i,\xi,\eta)\,dS(\xi,\eta) \\ \Delta \mathbf{T}_{ni}^{e} &= \int_{S_e} N_n(\xi,\eta)\,\mathbf{T}(P_i,\xi,\eta)\,dS(\xi,\eta) \end{aligned} \tag{6.12}$$

where S_e is the element area and ξ,η are the intrinsic coordinates.

Since there are two or three integral equations per location P_i, we now get $2I$ or $3I$ equations depending on the Cartesian dimension. As we will see later in the section on assembly, Equation (6.10) can be written in matrix form, where coefficients are assembled in a similar way as in the FEM. For this it is convenient to store the coefficients for element into arrays $[\Delta\mathrm{U}]^{e}$ and $[\Delta\mathrm{T}]^{e}$. For potential problems we have for example

$$\begin{array}{c} \rightarrow \quad elem \quad nodes \\ [\Delta U]^{e} = \begin{bmatrix} \Delta U_{11} & \Delta U_{21} & \cdots \\ \Delta U_{12} & \Delta U_{22} & \cdots \\ \vdots & \vdots & \cdots \end{bmatrix} \downarrow \; coll.\ pnts \end{array} \tag{6.13}$$

The arrays are of size NxI, where N is the number of element nodes and I is the number of collocation points. For elasticity problems, the arrays are of size $2N$x$2I$, for two-dimensional problems and $3N$x$3I$, for three-dimensional problems. In the following section we will deal with the numerical integration of Kernel shape function products over elements.

6.3 INTEGRATION OF KERNEL SHAPE FUNCTION PRODUCTS

The evaluation of integrals (6.8) or (6.12) over isoparametric elements is probably the most crucial aspect of the numerical implementation of BEM and this is much more involved than in the FEM. The problem lies in the fact that the functions which have to be integrated exhibit singularities at certain points in the elements. Here we first discuss the treatment of "improper" integrals that exist as Cauchy principal values and then discuss the numerical treatment of the other integrals.

6.3.1 Singular integrals

How an integral can be evaluated depends on the type of singularity. In general, we can say that a *weakly singular* integral (functions of order *lnr* for 2-D and *1/r* for 3-D problems) can be evaluated using numerical integration, that is the Gauss Quadrature discussed in Chapter 3. However, care has to be taken that an appropriate accuracy is maintained, by choosing the number of integration points as a function of the closeness of the collocation point to the region of integration. Theoretically the integrals of functions which are *strongly singular* (functions of order *1/r* for 2-D problems and $1/r^2$ for 3-D problems) are improper integrals and only exist as *Cauchy* principal values[5]. However, we can show that for integration on a flat surface approaching the collocation point the symmetric part of the kernel is zero and the anti-symmetric part approaches $+\infty$ on one side and $-\infty$ on the other side of the point. If we assume a flat integration region extending equal distances to the left and right of point P_i, the integral of the anti-symmetric part also becomes zero.

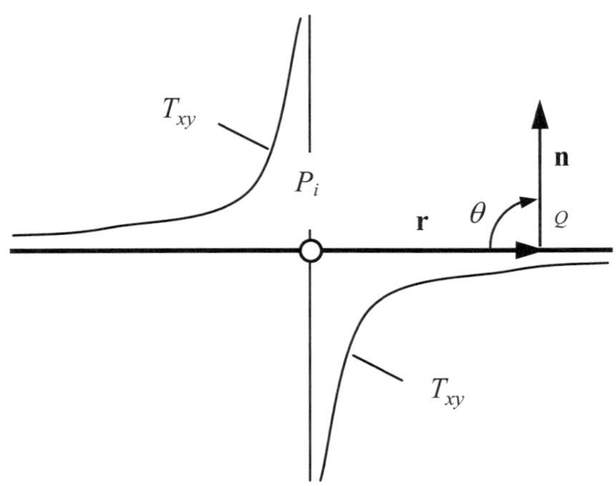

Figure 6.2 Variation of T_{xy} over a flat boundary

To explain this, consider a problem in 2-D elasticity with a flat surface at point P_i as shown in Figure 6.2. For this problem the angle between vector **r** and **n** is 90^0 and therefore $\cos\theta$ is zero.

According to Equation (4.64)

$$T_{xx} = \frac{C_2}{r}\left[C_3 + 2r_{,x}{}^2\right]\cos\theta = 0$$
$$T_{xy} = \frac{C_2}{r}\left[2r_{,x}r_{,y}\cos\theta + C_3\left(n_x r_{,y} - n_y r_{,x}\right)\right] = \frac{C}{r}\left(n_x r_{,y} - n_y r_{,x}\right) \tag{6.14}$$

From the distribution of the anti-symmetric part of T_{xy} , shown in Figure 6.2 we can see that given the restrictions stated above the integral of T_{xy} will give zero value. As a consequence, the diagonal coefficients only contain the "free term" c as computed in Equation (5.24). One could devise a scheme whereby we assume a flat boundary very near the collocation point, extending equally in both directions and use normal Gauss integration over parts, which exclude this flat region, so that we do not have to worry about computing the *Cauchy* principal value of the integral. However, the implementation of this is not trivial and we still have to deal with the determination of the "free term" which for corners and edges in a 3-D analysis, is also not trivial.

Two general approaches exist for the determination of the Cauchy principal value integral. One is a mathematical approach, by Guiggiani and Casalini[6], the other is based on simple engineering considerations. Since the second is simpler to implement, it will be the one used for the programs in this book.

6.3.2 Rigid body motion

The concept is based on the fact that we do not need to actually compute the integrals because the coefficients may be determined from the fact that for a pure rigid body translation of an elastic domain, there must be no change in shape of the body and therefore, applied tractions must be zero. We note that strongly singular integrals arise only for the T kernel and only if the collocation point P_i coincides with one of the element nodes. Let us rewrite equation (6.10)

$$\mathbf{cu}(P_i)+\sum_{e=1}^{E}\sum_{\substack{n=1\\g(n)=i}}^{N}\Delta\mathbf{T}_{ni}^{e}\,\mathbf{u}_{n}^{e}+\sum_{e=1}^{E}\sum_{\substack{n=1\\g(n)\neq i}}^{N}\Delta\mathbf{T}_{ni}^{e}\,\mathbf{u}_{n}^{e}=\sum_{e=1}^{E}\sum_{n=1}^{N}\Delta\mathbf{U}_{ni}^{e}\,\mathbf{t}_{n}^{e} \tag{6.15}$$

where $g(n)$ stands for the global node number of a node with the local node number n, therefore separating in the first sum all the terms that involve a strongly singular integration. To generate a rigid body translation for a two-dimensional finite domain we substitute $u_x = 1$ and $u_y = 0$ (translation in x-direction) and $u_x = 0$ and $u_y = 1$ (translation in y-direction) for all nodes and set all tractions to zero. For a plane problem we now can write 2 sets of equations

$$\mathbf{c}(P_i)\begin{pmatrix}1\\0\end{pmatrix}+\sum_{e=1}^{E}\sum_{\substack{n=1\\g(n)=i}}^{N}\Delta\mathbf{T}_{ni}^{e}\begin{pmatrix}1\\0\end{pmatrix}+\sum_{e=1}^{E}\sum_{\substack{n=1\\g(n)\neq i}}^{N}\Delta\mathbf{T}_{ni}^{e}\begin{pmatrix}1\\0\end{pmatrix}=0 \tag{6.16}$$

and

$$\mathbf{c}(P_i)\begin{pmatrix}0\\1\end{pmatrix}+\sum_{e=1}^{E}\sum_{\substack{n=1\\g(n)=i}}^{N}\Delta\mathbf{T}_{ni}^{e}\begin{pmatrix}0\\1\end{pmatrix}+\sum_{e=1}^{E}\sum_{\substack{n=1\\g(n)\neq i}}^{N}\Delta\mathbf{T}_{ni}^{e}\begin{pmatrix}0\\1\end{pmatrix}=0 \tag{6.17}$$

The equations can now be solved for the strongly singular terms including the free term. For the first set of equations we get:

$$\mathbf{c}(P_i)\begin{pmatrix}1\\0\end{pmatrix}+\sum_{e=1}^{E}\sum_{\substack{n=1\\g(n)=i}}^{N}\Delta\mathbf{T}_{ni}^{e}\begin{pmatrix}1\\0\end{pmatrix}=-\sum_{e=1}^{E}\sum_{\substack{n=1\\g(n)\neq i}}^{N}\Delta\mathbf{T}_{ni}^{e}\begin{pmatrix}1\\0\end{pmatrix} \tag{6.18}$$

The consequence of 6.18 is that the strongly singular terms, including the free term can be determined by simply summing up all coefficients - except the terms to be evaluated - of one equation and changing the sign of the sum. The advantage of this scheme is that not only do we avoid the strongly singular integration, but we also get the free term at no additional expense.

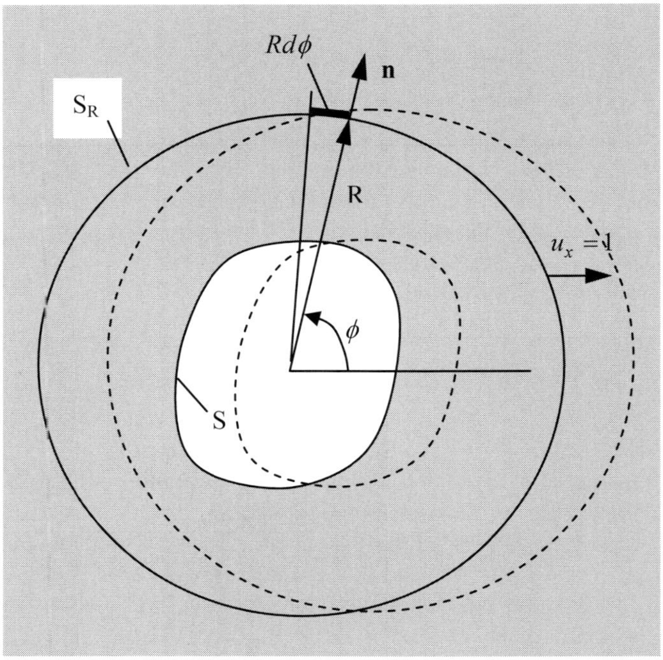

Figure 6.3 Rigid body translation in x-direction of a domain with radius R

For an infinite domain we cannot apply a rigid body translation. However, if we consider a two-dimensional domain to be bounded by an auxiliary surface, i.e., a circle of radius R (see Figure 6.3), where R is approaching infinity, then we may apply a rigid body translation.

We must consider now - in addition to the integrals which extend over the boundary of the problem S - also the ones over the boundary S_R , the auxiliary surface, that is

$$\mathbf{c}(P_i)\begin{pmatrix}1\\0\end{pmatrix}+\sum_{e=1}^{E}\sum_{\substack{n=1\\g(n)=i}}^{N}\Delta\mathbf{T}_{ni}^{e}\begin{pmatrix}1\\0\end{pmatrix}=-(\sum_{e=1}^{E}\sum_{\substack{n=1\\g(n)\neq i}}^{N}\Delta\mathbf{T}_{ni}^{e}\begin{pmatrix}1\\0\end{pmatrix}+\int_{S_R}\mathbf{T}(P_i,Q)dS) \tag{6.19}$$

The integral over S_R is known as the *azimuthal integral*[4]. Substituting $\cos\theta=-1$ and $r_{,x}=\cos\phi$ for two-dimensional elasticity problems, typical integrals are given by (see Figure. 6.3)

$$\int_{S_R} T_{xx}(P,Q)\,dS=\int_0^{2\pi}\frac{C_2}{R}\left(C_3+2\cos\phi^2\right)(-1)\,Rd\phi=-1 \tag{6.20}$$

and

$$\int_{S_R} T_{xy}(P,Q)dS=\int_0^{2\pi}\frac{C_2}{R}[2\cos\phi\, sin\phi(-1)-C_3(sin\phi\, cos\phi-cos\phi\, sin\phi)]Rd\phi=0 \tag{6.21}$$

The *azimuthal integral* of matrix **T** can therefore be written as

$$\int_{S_R}\mathbf{T}\,(P,Q)dS=-\mathbf{I} \tag{6.22}$$

where **I** is a 2x2 unit matrix.

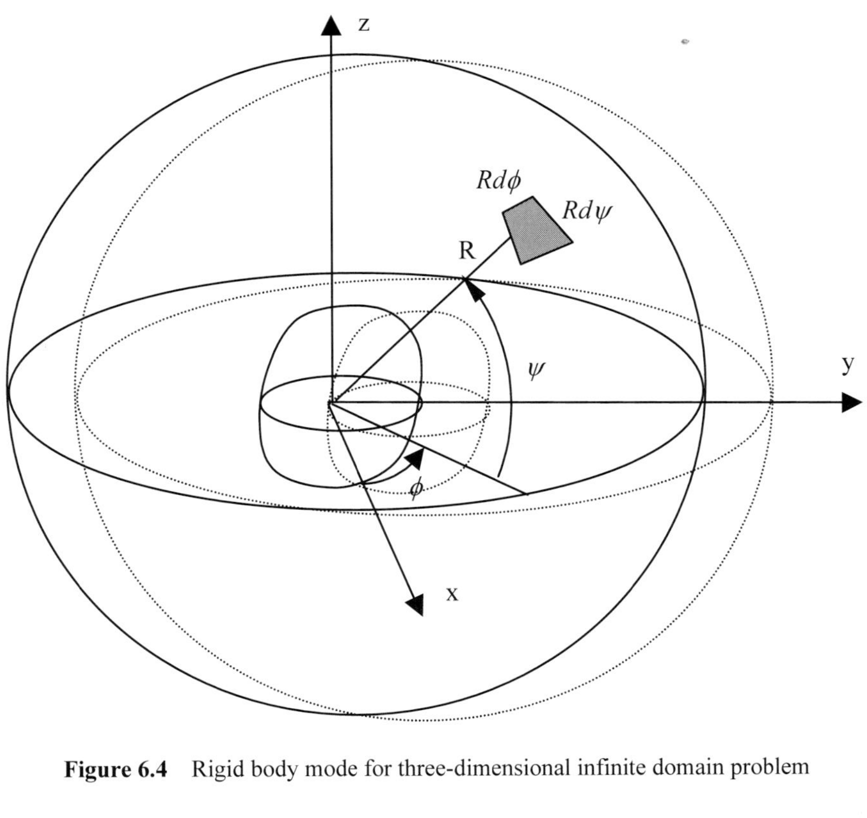

Figure 6.4 Rigid body mode for three-dimensional infinite domain problem

We see that since R cancels out, the integral is valid for any radius of the circle, including a radius of infinity, so the method of computing the strongly singular terms by rigid body translation is also valid for infinite domains.

For three-dimensional elasticity problems, the infinite domain is assumed to be a sphere of radius R. Typical values of the *azimuthal integral* are (see Figure 6.4):

$$\int_{S_R} T_{xx}(P,Q)dS = \int_0^{2\pi}\int_0^{2\pi}\frac{C_2}{R^2}\left(C_3 + 3\cos^2\phi\right)(-1)Rd\psi Rd\phi = -1 \tag{6.23}$$

and

$$\begin{aligned}&\int_{S_R} T_{xy}(P,Q)dS = \\ &\int_0^{2\pi}\int_0^{2\pi}\frac{C_2}{R^2}\left[3\cos\phi\sin\phi(-1) - C_3\left(\cos\phi\sin\phi - \sin\phi\cos\phi\right)\right]Rd\psi Rd\phi = 0\end{aligned} \tag{6.24}$$

so equation (6.22) is equally valid for three-dimensional problems, except that **I** is a 3x3 unit matrix.

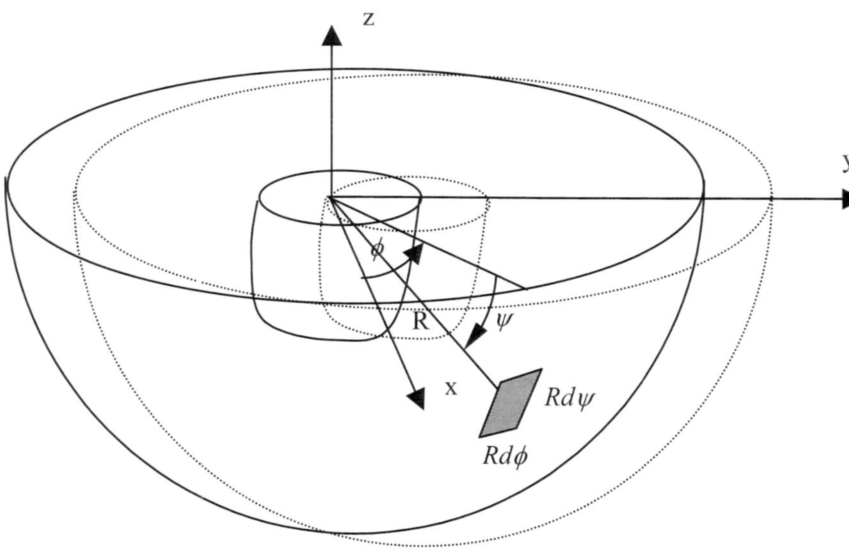

Figure 6.5 Rigid body mode for semi-infinite domain problem

For the case where the domain is semi-infinite, then the integration limits of the integral are from 0 to π and we have

$$\int_{S_R} T_{xx}(P,Q)\,dS = \int_0^{\pi}\int_0^{2\pi} \frac{C_2}{R^2}\left(C_3 + 3\cos^2\phi\right)(-1)\,Rd\psi\,Rd\phi = -\frac{1}{2} \tag{6.25}$$

$$\int_{S_R} \mathbf{T}\,(P,Q)\,dS = -\tfrac{1}{2}\mathbf{I} \tag{6.26}$$

For potential problems, we may consider a concept similar to the rigid body motion, by assuming that for uniform temperature at all nodes of the boundary and no internal heat generation, there can be no heat flow.

For a finite region we have

$$c(P_i) + \sum_{e=1}^{E}\sum_{\substack{n=1\\ g(n)=i}}^{N} \Delta T_{ni}^{e} = -\sum_{e=1}^{E}\sum_{\substack{n=1\\ g(n)\neq i}}^{N} \Delta T_{ni}^{e} \tag{6.27}$$

6.3.3 Numerical integration

It has already been mentioned that it is very important to maintain an adequate accuracy of the numerical integration. If this is not done, then significant errors may be introduced in the solution. In Chapter 3 we introduced the numerical integration by Gauss and pointed out that in this method the function to be integrated is approximated by a polynomial.

Here, we attempt to find an error bound for the integration of functions of type (1/r), ($1/r^2$) and ($1/r^3$) - which are not polynomials - depending on the number of Gauss points. Obviously, when point P_i is very close to the integration region, then the function varies very rapidly and higher and higher order polynomials are needed to approximate the function to be integrated and the number of required integration points has to increase. The error estimate that is introduced next allows us to ensure that the error made by numerical integration is nearly constant, regardless of the proximity of point P_i.

The Gauss integration formula in one dimension is (see Chapter 3)

$$\int_{-1}^{1} f(\xi) = \sum_{n=1}^{N} W_n f(\xi_n) \tag{6.28}$$

where N is the number of integration points.

Stroud and Secrest[8] provide a formula for the upper bound of error ε

$$\varepsilon \le 2\frac{4}{(2)^{2N}(2N)!}\left|\frac{\partial^{2N}}{\partial\xi^{2N}} f(\xi)\right| \tag{6.29}$$

Considering the integration over an element of length L with point P_i located at a distance R on the side (Figure 6.6), and taking $f(\xi)= 1/r$ we obtain

$$\left|\frac{\partial^{2N}}{\partial \xi^{2N}} f(\xi)\right| = 2\frac{(2N)!\,\mathrm{L}^{2N}}{(2)^{2N}\, r^{2N+1}} \tag{6.30}$$

and for the integration error

$$\varepsilon \leq \frac{4}{(4\mathrm{r}/\mathrm{L})^{2N}} \tag{6.31}$$

Therefore the integration error is a function of the distance r from point P_i to the integration region. Various schemes have been proposed [4,7] for determining the number of Gauss points on the basis of equation (6.31).

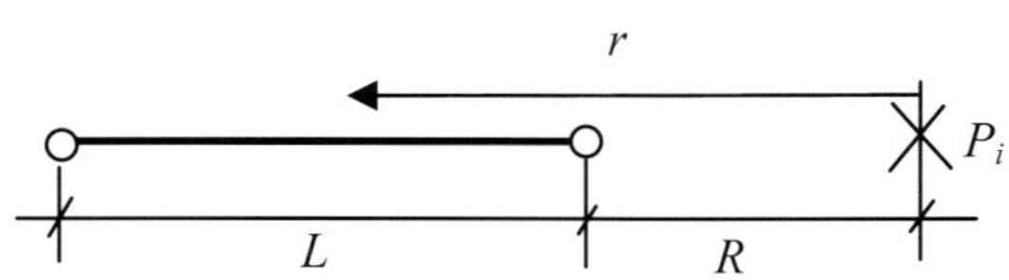

Figure 6.6 Integration over one-dimensional element

However, the actual functions to be integrated are more complicated than has been assumed above, because they involve products of the fundamental solution with the shape function and the *Jacobian*. In addition, it makes a difference if the P_i is located at the edge of the element as shown in Figure 6.6, or if it is located on the side. Finally the shape of the element (curved or straight) will also have an influence. For two-dimensional problems the integrals to be evaluated can be simplified to

$$I_{ni} = \int_{-1}^{1} N_n(\xi)\frac{1}{r^i(\xi)}\left|J(\xi)\right| d\xi \tag{6.32}$$

The idea is to determine the error in the integration as a function of the location of P_i if we integrate with a large number of Gauss points first to determine the actual value of the integral and then lower the number of Gauss points. If we do this for a large number of possible locations of points P_i then we can obtain contours of error for a given number of Gauss points. Figure 6.7 for example shows the contours of integration error 10^{-3} for a curved iso-parametric element for different integrals and for 4 Gauss Points. The contours can be interpreted in such a way that if point P_i lies on the contour then the error is exactly 10^{-3}, if it is outside it is less, if on the inside it is greater. It can be seen that near the Gauss points P_i can be placed closer to the element. To devise a table for the required number of Gauss points one may take an envelope that ensures that for all points on or inside the envelope the accuracy of integration is assured. This has been proposed by Eberwien et al [7].

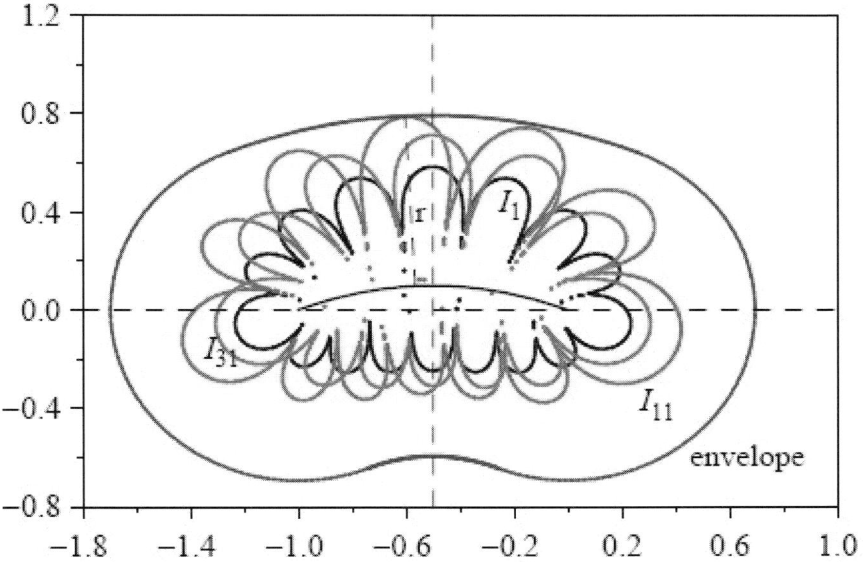

Figure 6.7 Contours showing the location of points P_i where the integration with 4 Gauss points gives an error of 10^{-3} (second subscript of I indicates that this is for a 1/r singularity)

Table 6.1 Number of Gauss points (Eberwien et al[7])

N	R/L		
	$O(1/r)$	$O(1/r^2)$	$O(1/r^3)$
3	1.4025	2.3187	3.4170
4	0.6736	0.9709	1.2908

The result is summarised in Table 6.1 as limiting values of R/L for an integration order of 4 and 5. Experience showed that the minimum number of integration points should not be lower than 3 and that it is more efficient to keep the maximum integration order low. This means that we have to subdivide the region of integration, so that the minimum ratios of R/L according to Table 6.1, are obeyed.

Cases where the point is very close to the element occur when there is a drastic change in element size, or the boundary surfaces are very close to each other, for example, in the case of a thin beam. Care has to be taken not to go to extremes with the value of R/L, because we must avoid cases where points P_i are too unevenly distributed since Betti's reciprocal theorem is only satisfied at these points.

We convert Table 6.1 into a FUNCTION Ngaus which returns the number of Gauss points according to the value of R/L.

```
INTEGER FUNCTION Ngaus(RonL,ne,RLIM)
!------------------------------------------------------
!    Function returns number of Gauss points needed
!    to integrate a function o(1/r^ne)
!    according to Eberwien et al.
!-------------------------------------------------------
REAL , INTENT(IN)          ::  RonL  !   R/L
INTEGER , INTENT(IN)       ::  ne    !   order of Kernel (1,2,3)
REAL, INTENT(OUT)          ::  Rlim(2)  !   array to store values of
table
SELECT CASE(ne)
 CASE(1)
  Rlim= (/1.4025, 0.7926/)
 CASE(2)
  Rlim= (/4.1029, 1.6776/)
 CASE(3)
  Rlim= (/3.4170, 1.2908/)
 CASE DEFAULT
END SELECT
DO   N=1,2      !   Determine minimum no of Gauss points needed
   IF(RonL >= Rlim(N)) THEN
        Ngaus= N+2
        EXIT
   END IF
END DO
IF(Ngaus == 0) THEN  ! Point is too close to the surface
     Ngaus=5     ! this value will trigger subdivision
END IF
RETURN
END FUNCTION Ngaus
```

6.3.4 Numerical integration over one-dimensional elements

In the integration of Kernel-shape function products care has to be taken because in some cases the function has a singularity or is discontinuous over the element depending on the location of P_i. Therefore, we have to distinguish integration schemes for the case where P_i is one of the element nodes and where it is not.
The integrals which have to be evaluated over the isoparametric element, shown in Figure 6.8, are for potential problems

$$\Delta U_{ni}^{e} = \int_{-1}^{1} N_n(\xi) U(P_i,\xi) J(\xi) d\xi \quad , \quad \Delta T_{ni}^{e} = \int_{-1}^{1} N_n(\xi) T(P_i,\xi) J(\xi) d\xi \tag{6.33}$$

where $U(P_i, \xi)$ and $T(P_i, \xi)$ are the fundamental solutions at $Q(\xi)$ for a source at point P_i, $J(\xi)$ is the *Jacobian* and $N_n(\xi)$ are linear or quadratic shape functions.

When point P_i is not one of the element nodes, both integrals can be evaluated by Gauss Quadrature and the integrals in equation (6.33) can be replaced by two sums

$$\Delta T_{ni}^{e} \approx \sum_{m=1}^{M} N_n(\xi_m) T(P_i, \xi_m) J(\xi_m) W_m$$
$$\Delta U_{ni}^{e} \approx \sum_{m=1}^{M} N_n(\xi_m) U(P_i, \xi_m) J(\xi_m) W_m \qquad (6.34)$$

where the number of integration points M is determined as a function of the proximity of P_i to the integration region as explained previously. If P_i is close to the integration region a subdivision will be necessary.

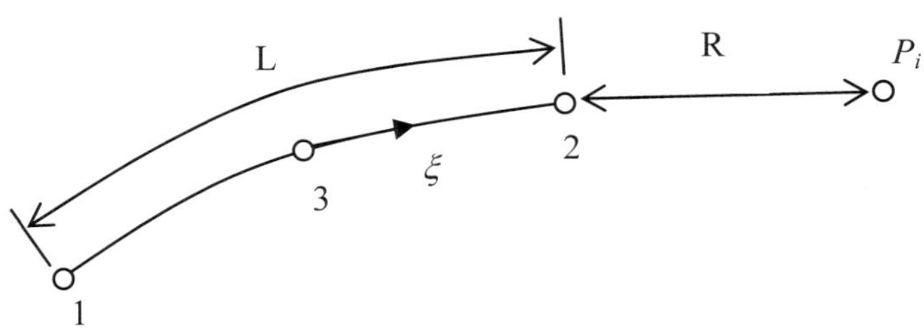

Figure 6.8 One dimensional element, integration where P_i is not one of the element nodes

When P_i is one of the element nodes, functions U and T tend to infinity within the integration region. Consider the two cases in Figure 6.9:

(a) P_i is located at point 1 and n in the equation (6.33) is 2:
This means that although Kernels T and U tend to infinity as point 1 is approached, the shape function tends to zero, so the integral of product $N_n(\xi)U(P_i,\xi)$ and $N_n(\xi)T(P_i,\xi)$ tend to a finite value. Thus, for the case where P_i is not at node n of the element, the integral can be evaluated with the formulae (6.34) without any problems.

(b) P_i is located at point 2 and n in the equation (6.33) is 2:
In this case, Kernels T and U tend to infinity and the shape function to unity and products $N_n(\xi)U(P_i,\xi)$ and $N_n(\xi)T(P_i,\xi)$ also tend to infinity. Since Kernel U has a singularity of order $ln(1/r)$, the first product cannot be integrated using Gauss

Quadrature. The integral of the second product only exists as a Cauchy principal value. However, these are the diagonal terms of the coefficient matrix that can be evaluated using equation (6.18), (6.19) or (6.27).

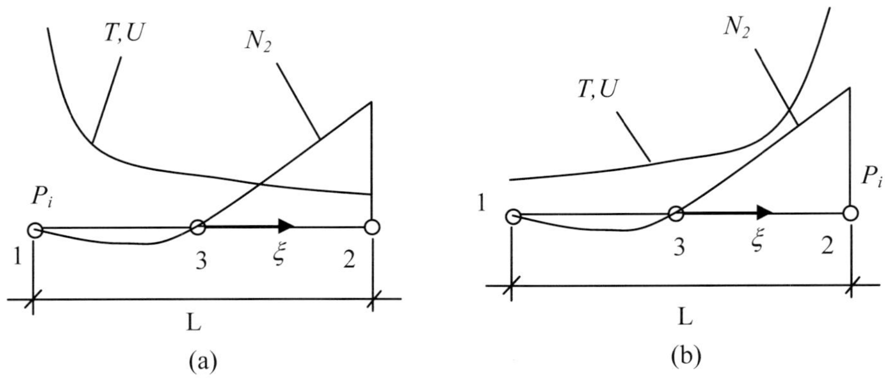

Figure 6.9 Integration when P_i is one of the element nodes

For the integration of the product with *ln(1/r),* we can use a modified Gauss Quadrature called *Gauss-Laguerre*[8] integration

$$\int_0^1 f(\bar{\xi}) ln(\frac{1}{\bar{\xi}}) d\xi \approx \sum_{m=1}^{M} W_m f(\bar{\xi}_m) \tag{6.35}$$

where M is the number of integration points.

The weights and coordinates are given by the Subroutine Gauss_Laguerre_coor, which is listed at the end of this section. Note that for this integration scheme $\bar{\xi} = 0$ at the singular point and the limits are from 0 to 1, so a change in coordinates has to be made before equation (6.35) can be applied.

This change in coordinate is given by (see Figure 6.10):

$$\begin{aligned} \xi &= 2\bar{\xi} - 1 \quad when \quad P_i \quad is \quad at \quad node \quad 1 \\ \xi &= 1 - 2\bar{\xi} \quad when \quad P_i \quad is \quad at \quad node \quad 2 \end{aligned} \tag{6.36}$$

For the case where we integrate over a quadratic element, the integrand is discontinuous if P_i is located at the midside node. The integration has to be split into two regions, one over $-1<\xi<0$, the other over $0<\xi<1$. For the computation of product $N_n(\xi)U(P_i,\xi)$, the intrinsic coordinates for the 2 sub-regions are computed by (see Figure 6.10):

$$\begin{aligned} \xi &= -\bar{\xi} \quad for \quad subregion \quad 1 \\ \xi &= \;\;\bar{\xi} \quad for \quad subregion \quad 2 \end{aligned} \tag{6.37}$$

To evaluate the first integral in equation (6.8) we must substitute for r as a function of ξ. For a linear element we may simply write $r= J\xi$ and obtain

$$\begin{aligned}\Delta U_{ni}^{e} &= \int_0^1 N_n(\xi)\frac{1}{2\pi k}\ln\left(\frac{1}{J\xi}\right)J\left|\frac{d\xi}{d\bar{\xi}}\right|d\bar{\xi} = \\ &\int_0^1 N_n(\xi)\frac{1}{2\pi k}\ln\left(\frac{1}{\xi}\right)J\left|\frac{d\xi}{d\bar{\xi}}\right|d\bar{\xi} + \int_0^1 N_n(\xi)\frac{1}{2\pi k}\ln\left(\frac{1}{J}\right)J\left|\frac{d\xi}{d\bar{\xi}}\right|d\bar{\xi}\end{aligned} \tag{6.38}$$

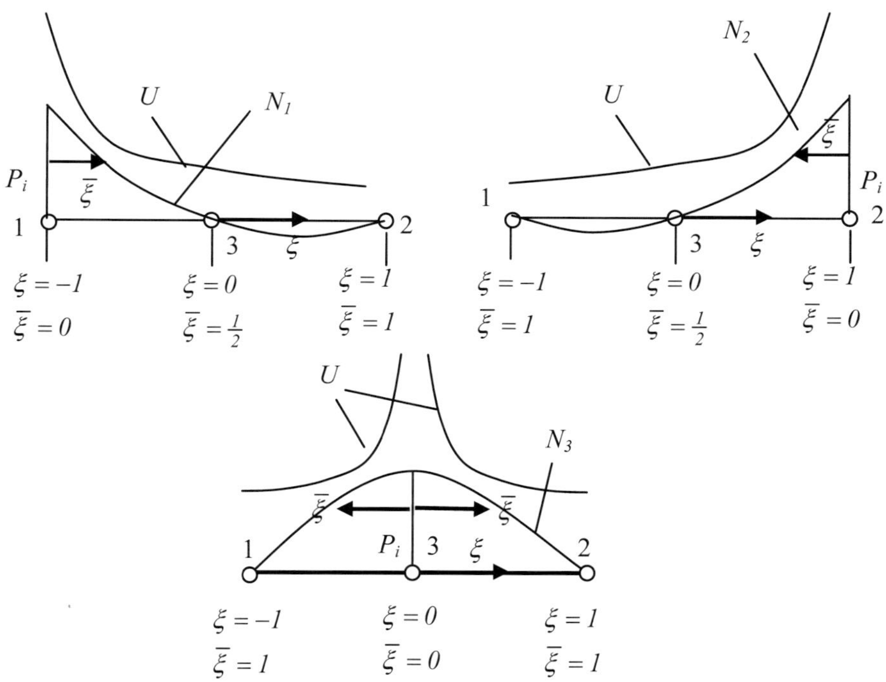

Figure 6.10 Integration when P_i and n coincide

The first integral may be evaluated with Gauss-Laguerre:

$$\begin{aligned}\Delta U_{ni}^{e} &= \int_0^1 N_n(\xi)\frac{1}{2\pi k}\ln\left(\frac{1}{\xi}\right)J(\xi)\frac{d\xi}{d\bar{\xi}}d\bar{\xi} = \\ &\approx \sum_{m=1}^{M} N_n(\xi_m)\frac{1}{2\pi k}J(\xi_m)W_m\frac{d\xi}{d\bar{\xi}}\end{aligned} \tag{6.39}$$

whereas the second part is integrated with normal Gauss Quadrature. The Jacobian $\partial\xi/\partial\bar{\xi}$ can be easily obtained by differentiation of equations (6.36) and (6.37). The second integral in (6.38) can be evaluated using normal Gauss Quadrature. For quadratic elements, the substitution for r in terms of ξ is more complicated. One may approximately substitute $r = a\xi$, where a is the length of a straight line between the end nodes of the element. This should give a small error for elements which are nearly straight. A more accurate computation r as a function of ξ is presented by Eberwien[7].

A SUBROUTINE which provides the coordinates and weights for a Gauss Laguerre integration is given below.

```
SUBROUTINE Gauss_Laguerre_coor(Cor,Wi,Intord)
!-----------------------------------
!  Returns Gauss_Laguerre coordinates and Weights
!  for 1 to 4 Gauss points
!-----------------------------------
IMPLICIT NONE
REAL, INTENT(OUT)   :: Cor(8)  !  Gauss point coordinate
REAL, INTENT(OUT)   :: Wi(8)   !  weigths
INTEGER,INTENT(IN)  :: Intord  !  integration order
SELECT CASE (Intord)
CASE (1)
    Cor(1)= 0.5 ; Wi(1) = 1.0
CASE(2)
    Cor(1)= .112008806 ; Cor(2)=.602276908
    Wi(1) = .718539319 ; Wi(2) =.281460680
CASE(3)
    Cor(1)= .063890793 ; Cor(2)= .368997063 ; Cor(3)= .766880303
    Wi(1) = .513404552 ; Wi(2) = .391980041 ; Wi(3) =.0946154065
CASE(4)
    Cor(1)= .0414484801 ; Cor(2)=.245274914 ; Cor(3)=.556165453
    Cor(4)= .848982394
    Wi(1) = .383464068 ; Wi(2) =.386875317 ; Wi(3) =.190435126
    Wi(4) = .0392254871
CASE DEFAULT
CALL Error_Message('Gauss points not in range 1-8')
END SELECT
END SUBROUTINE
```

6.3.5 Subdivision of region of integration

In some cases, when point P_i is near the element, the number of Gauss points required will exceed 4 in table 6.1. In this case it is necessary to subdivide the element into sub regions of integration. A simple approach is to subdivide the element into equal subdivisions depending on the value of R/L. If according to the R/L value the maximum number of Gauss points available is exceeded, the element is subdivided into K regions where

$$K = INT\left[\left(R/L\right)_{\min} / \left(R/L\right)\right] \tag{6.40}$$

Where *INT* means a rounding up of the result and $\left(R/L\right)_{\min}$ is the minimum value of *R/L* for 4 Gauss points in table 6.1.

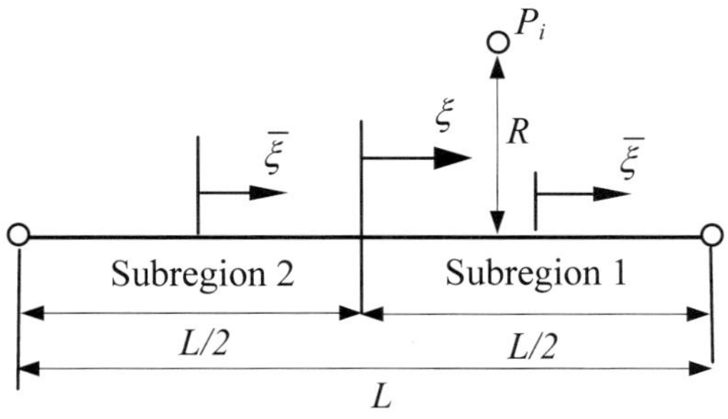

Figure 6.11 Subdivision of integration region

Note that for each sub region of integration the coordinates of the Gauss points have to be defined in a local coordinate system $\bar{\xi}$, whereas the shape functions are functions of ξ . For one-dimensional boundary elements the Gauss formula (6.34) is replaced by

$$\begin{aligned}\Delta T_{ni}^{e} &\approx \sum_{k=1}^{K}\sum_{m=1}^{M(k)} N_n\left(\xi_m\right) T\left(P_i, \xi_m\right) J\left(\xi_m\right) \cdot \bar{J}\, W_m \\ \Delta U_{ni}^{e} &\approx \sum_{k=1}^{K}\sum_{m=1}^{M(k)} N_n\left(\xi_m\right) U\left(P_i, \xi_m\right) J\left(\xi_m\right) \cdot \bar{J}\, W_m\end{aligned} \tag{6.41}$$

where *K* is the number of sub regions and *M(k)* is the number of Gauss points for sub region *k*. The relationship between ξ and $\bar{\xi}$ is given by

$$\xi = \frac{1}{2}(\xi_1 + \xi_2) + \frac{\bar{\xi}}{K} \tag{6.42}$$

where ξ_1 and ξ_2 are the start and end coordinates of the sub region. In the example shown in 6.11 this is (0 , 1) for sub region 1 and (-1 , 0) for sub region 2. If a uniform subdivision is assumed the *Jacobian* $\bar{J}$ for the transformation from ξ to $\bar{\xi}$ is for all regions.

$$\bar{J} = \frac{\partial \xi}{\partial \bar{\xi}} = \frac{1}{K} \tag{6.43}$$

The proposed scheme is not very efficient since the sub regions will have different minimum distances R to P_i and therefore should have different integration order also. A more efficient method would be to provide more subdivisions near P_i and less further away.

6.3.6 Implementation for plane problems

A SUBROUTINE Integ2P is shown below which integrates the Kernel/shape function products over one-dimensional isoparametric elements for potential problems.

```
SUBROUTINE Integ2P (Elcor,Inci,Nodel,Ncol,xP,k,dUe,dTe)
!-------------------------------------------------
!   Computes  Element contributions[dT]e and [dU]e
!   for 2-D potential problems
!   by numerical integration
!-------------------------------------------------
IMPLICIT NONE
REAL, INTENT(IN):: Elcor(:,:)    ! Element coordinates
INTEGER, INTENT(IN)  :: Inci(:)  ! Element Incidences
INTEGER, INTENT(IN)  :: Nodel ! No. of Element Nodes
INTEGER , INTENT(IN):: Ncol  ! Number of points Pi
REAL, INTENT(IN)   :: xP(:,:) ! Array with coll. point coords.
REAL, INTENT(IN)   :: k         ! Permeability/Conductivity
REAL, INTENT(OUT) :: dUe(:,:),dTe(:,:)
REAL :: epsi= 1.0E-4      !    Small value for comparing coords
REAL ::Eleng,Rmin,RonL,Glcor(8),Wi(8),Ni(Nodel),Vnorm(2),GCcor(2)
REAL :: UP,Jac,dxr(2),TP,r,pi,c1,c2,xsi,eta,dxdxb
REAL :: RLIM(2),xsi1,xsi2,RJACB
INTEGER :: i,m,n,Mi,nr,ldim,cdim,nreg,ndiv,ndivs
pi=3.14159265
ldim= 1
cdim=ldim+1
CALL Elength(Eleng,Elcor,Nodel,ldim) ! Element Length
!-------------------------------------------------------------
!   Integration off-diagonal coeff.  -> normal Gauss Quadrature
!--------------------------------------------------------------
  dUe= 0.0 ; dTe= 0.0        ! Clear arrays for summation
Colloc_points:&
DO i=1,Ncol
  Rmin= Min_dist(Elcor,xP(:,i),Nodel,ldim,inci)! Distance to Pi
  RonL= Rmin/Eleng   ! R/L
  Mi= Ngaus(RonL,1,RLIM)! Number of Gauss points for (1/r) sing.
  IF(Mi == 5) THEN     !  check if subdivisions are required
    NDIVS= INT(RLIM(2)/RonL))+1
    RJACB= 1/NDIVS
    Mi=4
  ELSE
    NDIVS=1
     RJACB=1.0
  END IF
```

```
    Call Gauss_coor(Glcor,Wi,Mi)        ! Assign coords/Weights
     Xsi1=-1
    Subregions: &
    DO NDIV=1,NDIVS
     IF(NDIVS > 1) THEN
      Xsi2= Xsi1+2/NDIVS
     Gauss_points:&
     DO m=1,Mi
      xsi= Glcor(m)
      IF(NDIVS > 1) Xsi= 0.5*(Xsi1+Xsi2)+xsi/NDIVS
       CALL Serendip_func(Ni,xsi,eta,ldim,Nodel,Inci)
       Call Normal_Jac(Vnorm,Jac,xsi,eta,ldim,Nodel,Inci,elcor)
       CALL Cartesian(GCcor,Ni,ldim,elcor) ! Coords of Gauss pt
       r= Dist(GCcor,xP(:,i),cdim)         ! Dist. P,Q
       dxr= (GCcor-xP(:,i))/r              ! rx/r , ry/r
       UP= U(r,k,cdim) ; TP= T(r,dxr,Vnorm,cdim)     ! Kernels
       Node_points: &
       DO n=1,Nodel
          IF(Dist(Elcor(:,n),xP(:,i),cdim) < epsi) EXIT ! Pi is n
          dUe(i,n)= dUe(i,n) + Ni(n)*UP*Jac*Wi(m)*RJACB
          dTe(i,n)= dTe(i,n) + Ni(n)*TP*Jac*Wi(m)*RJACB
       END DO &
       Node_points
    END DO &
    Gauss_points
  END DO &
  Subregions
END DO &
Colloc_points
!-------------------------------
!     Diagonal terms of dUe
!-------------------------------
c1= 1/(2.0*pi*k)
Colloc_points1: &
DO i=1,Ncol
 Node_points1: &
 DO n=1,Nodel
   IF(Dist(Elcor(:,n),xP(:,i),cdim) > Epsi) CYCLE ! Pi not n
   Nreg=1
   IF(n == 3) nreg= 2
!-------------------------------------------
!    Integration of logarithmic term
!--------------------------------------------
   Subregions: &
   DO nr=1,Nreg
    Mi= 4
    Call Gauss_Laguerre_coor(Glcor,Wi,Mi)
    Gauss_points1: &
    DO m=1,Mi
     SELECT CASE (n)
     CASE (1)
```

```
      xsi= 2.0*Glcor(m)-1.0
      dxdxb= 2.0
     CASE (2)
      xsi= 1.0 -2.0*Glcor(m)
      dxdxb= 2.0
     CASE (3)
      dxdxb= 1.0
      IF(nr == 1) THEN
       xsi= -Glcor(m)
      ELSE
       xsi= Glcor(m)
      END IF
     CASE DEFAULT
     END SELECT
     CALL Serendip_func(Ni,xsi,eta,1,Nodel,Inci)
     Call Normal_Jac(Vnorm,Jac,xsi,eta,1,Nodel,Inci,elcor)
     dUe(i,n)= dUe(i,n) + Ni(n)*c1* Jac* dxdxb* Wi(m)
    END DO &
    Gauss_points1
   END DO &
   Subregions
!-----------------------------------------
!    Integration of non logarithmic term
!------------------------------------------
     Mi= 2
    Call Gauss_coor(Glcor,Wi,Mi)  ! Assign coords/Weights
    Gauss_points2: &
    DO m=1,Mi
     SELECT CASE (n)
     CASE (1:2)
      c2=-LOG(Eleng)*c1
     CASE (3)
      c2=LOG(2/Eleng)*c1
     CASE DEFAULT
     END SELECT
     xsi= Glcor(m)
     CALL Serendip_func(Ni,xsi,eta,ldim,Nodel,Inci)
     Call Normal_Jac(Vnorm,Jac,xsi,eta,ldim,nodel,Inci,elcor)
         dUe(i,n)= dUe(i,n) + Ni(n)*c2* Jac* Wi(m)
    END DO &
    Gauss_points2
  END DO &
  Node_points1
END DO &
Colloc_points1
RETURN
END SUBROUTINE Integ2P
```

The above integration scheme is equally applicable to elasticity problems, except that when integrating functions with Kernel **U** when P_i is one of the nodes of the element we have to consider that only U_{xx} and U_{yy} have a logarithmic and non-logarithmic part. The logarithmic part is integrated with *Gauss-Laguerre,* for example:

$$\begin{aligned}\Delta U_{xxni}^{e} &= \int_{0}^{1} N_n(\xi)\,\frac{(1-\nu)(3-4\nu)}{4\pi\, E\,(1-\nu)}\ln\left(\frac{1}{\xi}\right)(P_i,\xi)\,J(\xi)\left|\frac{d\xi}{d\bar{\xi}}\right| d\bar{\xi} \\ &\approx \sum_{m=1}^{M} N_n(\xi_m)\,\frac{(1-\nu)(3-4\nu)}{4\pi\, E\,(1-\nu)}\,J(\xi_m)\,W_m\left|\frac{d\xi}{d\bar{\xi}}\right|\end{aligned} \tag{6.44}$$

The non-logarithmic part is integrated using Gauss Quadrature.

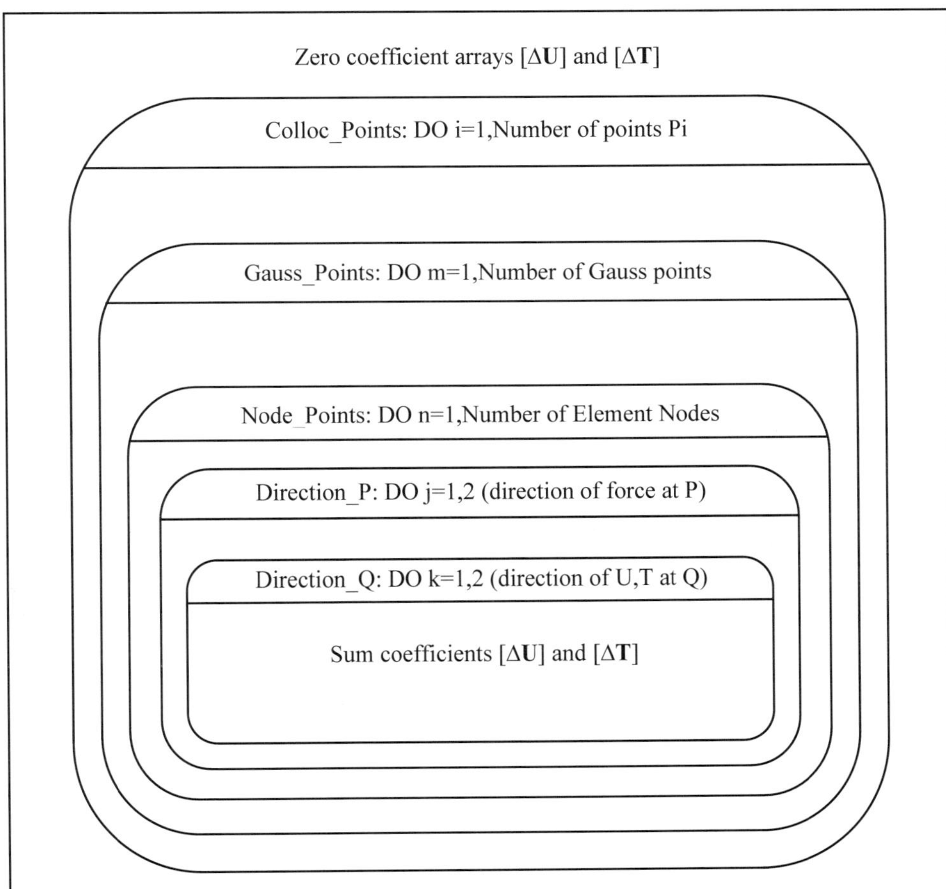

Figure 6.12 Structure chart for SUBROUTINE Integ2E

A SUBROUTINE for integrating over one-dimensional elements for elasticity is written. The main differences to the previous subroutine are that the Kernels **U** and **T** are now 2x2 matrices and we have to add two more Do-loops for the direction of the load at P_i and the direction of the displacement/traction at $Q(\xi)$. A structure chart of SUBROUTINE Integ2E is shown in Figure 6.12, where for the sake of clarity, the subdivision of the region of integration is not shown.

For the implementation of symmetry, as will be discussed in Chapter 7 two additional parameters are used: ISYM and NDEST. The first parameter contains the symmetry code, the second is an array that is used to eliminate variables which have zero value, because they are situated on a symmetry plane.

Note that the storage of coefficients is by degree of freedom number rather than node number. There are two columns per node and two rows per collocation point. The storage of the element coefficients $[\Delta U]^e$ is as follows:

$$\begin{array}{c} \rightarrow \; element \quad nodes \\ [\Delta U]^e = \begin{bmatrix} \Delta U_{xx11} & \Delta U_{xy11} & \Delta U_{xx21} & \Delta U_{xy21} & \cdots \\ \Delta U_{yx11} & \Delta U_{yy11} & \Delta U_{yx21} & \Delta U_{yy21} & \cdots \\ \Delta U_{xx12} & \Delta U_{xy12} & \Delta U_{xx22} & \Delta U_{xy22} & \cdots \\ \vdots & \vdots & \vdots & \vdots & \ddots \end{bmatrix} \downarrow coll.\ pnts \end{array} \tag{6.45}$$

```
SUBROUTINE
Integ2E(Elcor,Inci,Nodel,Ncol,xP,E,ny,dUe,dTe,Ndest,Isym)
!--------------------------------------------------
!    Computes  [dT]e and [dU]e for 2-D elasticity problems
!    by numerical integration
!--------------------------------------------------
IMPLICIT NONE
REAL, INTENT(IN)     :: Elcor(:,:)      !   Element coordinates
INTEGER, INTENT(IN)  :: Ndest(:,:)    !  Node destination vector
INTEGER, INTENT(IN)  :: Inci(:)         !   Element Incidences
INTEGER, INTENT(IN)  :: Nodel           !   No. of Element Nodes
INTEGER , INTENT(IN):: Ncol             !   Number of points Pi
INTEGER , INTENT(IN):: Isym
REAL, INTENT(IN)     :: E,ny            !   Elastic constants
REAL, INTENT(IN)     :: xP(:,:)         !   Coloc. Point coords
REAL(KIND=8), INTENT(OUT)   :: dUe(:,:),dTe(:,:)
REAL :: epsi= 1.0E-4  !    Small value for comparing coords
REAL :: Eleng,Rmin,RonL,Glcor(8),Wi(8),Ni(Nodel),Vnorm(2),GCcor(2)
REAL :: Jac,dxr(2),UP(2,2),TP(2,2), xsi, eta, r
&,dxdxb,Pi,C,C1,Rlim(2),Xsi1,Xsi2,RJacB
INTEGER :: i,j,k,m,n,Mi,nr,ldim,cdim,iD,nD,Nreg,NDIV,NDIVS,MAXDIVS
Pi=3.14159265359
C=(1.0+ny)/(4*Pi*E*(1.0-ny))
ldim= 1                                ! Element dimension
cdim=ldim+1
MAXDIVS=1
CALL Elength(Eleng,Elcor,nodel,ldim)  ! Element Length
  dUe= 0.0 ; dTe= 0.0                 ! Clear arrays for summation
```

```
Colloc_points: DO i=1,Ncol
Rmin= Min_dist1(Elcor,xP(:,i),Nodel,inci,ELeng,Eleng,ldim)
  RonL= Rmin/Eleng                         !  R/L
!----------------------------------------------------
!     Integration off-diagonal coeff.  -> normal Gauss Quadrature
!--------------------------------------------------------
  Mi= Ngaus(RonL,1,Rlim) !  Number of Gauss points for o(1/r)
  NDIVS= 1
  RJacB=1.0
  IF(Mi == 5) THEN       !   Determine number of subdiv. In ξ
    IF(RonL > 0.0) NDIVS= INT(RLim(2)/RonL) + 1
    IF(NDIVS > MAXDIVS) MAXDIVS= NDIVS
    RJacB= 1.0/NDIVS
    Mi=4
  END IF
  Call Gauss_coor(Glcor,Wi,Mi)  ! Assign coords/Weights
  Xsi1=-1
  Subdivisions: DO NDIV=1,NDIVS
    Xsi2= Xsi1 + 2.0/NDIVS
    Gauss_points: DO m=1,Mi
    xsi= Glcor(m)
    IF(NDIVS > 1) xsi= 0.5*(Xsi1+Xsi2)+xsi/NDIVS
    CALL Serendip_func(Ni,xsi,eta,ldim,nodel,Inci)
    Call Normal_Jac(Vnorm,Jac,xsi,eta,ldim,nodel,Inci,elcor)
    CALL Cartesian(GCcor,Ni,ldim,elcor)     ! coords of Gauss pt
    r= Dist(GCcor,xP(:,i),cdim)             !  Dist. P,Q
    dxr= (GCcor-xP(:,i))/r               !  rx/r , ry/r
    UP= UK(dxr,r,E,ny,Cdim) ; TP= TK(dxr,r,Vnorm,ny,Cdim)
     Node_points: DO n=1,Nodel
      Direction_P: DO j=1,2
       IF(Isym == 0)THEN
         iD= 2*(i-1) + j
       ELSE
         iD= Ndest(i,j)                    !  line number in array
       END IF
       IF (id == 0) CYCLE
       Direction_Q: DO k= 1,2
        nD= 2*(n-1) + k                   !  column number in array
        IF(Dist(Elcor(:,n),xP(:,i),cdim) > epsi) THEN
          dUe(iD,nD)= dUe(iD,nD)  + Ni(n)*UP(j,k)*Jac*Wi(m)*RJacB
          dTe(iD,nD)= dTe(iD,nD)  + Ni(n)*TP(j,k)*Jac*Wi(m)*RJacB
         ELSE
          dUe(iD,nD)= dUe(iD,nD)  + &
                      Ni(n)*C*dxr(j)*dxr(k)*Jac*Wi(m)*RJacB
         END IF
        END DO Direction_Q
       END DO Direction_P
      END DO Node_points
     END DO Gauss_points
     Xsi1= Xsi2
    END DO Subdivisions
```

```
 END DO Colloc_points
!-----------------------------------------------------------
!    Integration diagonal coeff.
!---------------------------------------------------------------
 C= C*(3.0-4.0*ny)
Colloc_points1: DO i=1,Ncol
  Node_points1:  DO n=1,Nodel
    IF(Dist(Elcor(:,n),xP(:,i),cdim) > Epsi) CYCLE
    Nreg=1
    IF (n == 3) nreg= 2
    Subregions: DO nr=1,Nreg
      Mi= 4
      Call Gauss_Laguerre_coor(Glcor,Wi,Mi)
      Gauss_points1: DO m=1,Mi
        SELECT CASE (n)
         CASE (1)
          xsi= 2.0*Glcor(m)-1.0
          dxdxb= 2.0
         CASE (2)
          xsi= 1.0 -2.0*Glcor(m)
          dxdxb= 2.0
         CASE (3)
          dxdxb= 1.0
          IF(nr == 1) THEN
            xsi= -Glcor(m)
          ELSE
            xsi= Glcor(m)
          END IF
          CASE DEFAULT
        END SELECT
        CALL Serendip_func(Ni,xsi,eta,ldim,nodel,Inci)
        Call Normal_Jac(Vnorm,Jac,xsi,eta,ldim,nodel,Inci,elcor)
        Direction1: DO j=1,2
         IF(Isym == 0)THEN
           iD= 2*(i-1) + j
         ELSE
           iD= Ndest(i,j)                 !  line number in array
         END IF
         IF (id == 0) CYCLE
         nD= 2*(n-1) + j                 !  column number in array
         dUe(iD,nD)= dUe(iD,nD) + Ni(n)*C*Jac*dxdxb*Wi(m)
        END DO Direction1
       END DO Gauss_points1
      END DO Subregions
      Mi= 2
      Call Gauss_coor(Glcor,Wi,Mi)
      Gauss_points2: DO m=1,Mi
       SELECT CASE (n)
        CASE (1)
          C1=-LOG(Eleng)*C
        CASE (2)
```

```
          C1=-LOG(Eleng)*C
        CASE (3)
          C1=LOG(2/Eleng)*C
        CASE DEFAULT
       END SELECT
       xsi= Glcor(m)
       CALL Serendip_func(Ni,xsi,eta,ldim,nodel,Inci)
       Call Normal_Jac(Vnorm,Jac,xsi,eta,ldim,nodel,Inci,elcor)
       Direction2: DO j=1,2
         IF(Isym == 0)THEN
           iD= 2*(i-1) + j
         ELSE
            iD= Ndest(i,j)                  !  line number in array
         END IF
         IF (id == 0) CYCLE
         nD= 2*(n-1) + j                   !  column number in array
         dUe(iD,nD)= dUe(iD,nD) + Ni(n)*C1*Jac*Wi(m)
        END DO Direction2
       END DO Gauss_points2
    END DO Node_points1
 END DO Colloc_points1
 RETURN
 END SUBROUTINE Integ2E
```

6.3.7 Numerical integration for two-dimensional elements

Here we discuss numerical integration over two-dimensional isoparametric finite boundary elements. We find that the basic principles are very similar to integration over one-dimensional elements in that we separate the cases where P_i is not one of the nodes of an element and where it is. Starting with potential problems, the integrals which have to be evaluated (see Figure 6.13) are:

$$\Delta U_{ni}^{e} = \int_{-1}^{1}\int_{-1}^{1} N_n(\xi,\eta)\, U\big(P_i,Q(\xi,\eta)\big)\, J(\xi,\eta)\, d\xi\, d\eta$$
$$\Delta T_{ni}^{e} = \int_{-1}^{1}\int_{-1}^{1} N_n(\xi,\eta)\, T\big(P_i,Q(\xi,\eta)\big)\, J(\xi,\eta)\, d\xi\, d\eta \tag{6.46}$$

When P_i is not one of the element nodes, then the integrals can be evaluated using Gauss Quadrature in the ξ and η direction. This gives

$$\Delta U_{ni}^{e} \approx \sum_{m=1}^{M}\sum_{k=1}^{K} N_n(\xi_m,\eta_k)\, U\big(P_i,Q(\xi_m,\eta_k)\big)\, J(\xi_m,\eta_k) W_m W_k$$
$$\Delta T_{ni}^{e} \approx \sum_{m=1}^{M}\sum_{k=1}^{K} N_n(\xi_m,\eta_k)\, T\big(P_i,Q(\xi_m,\eta_k)\big)\, J(\xi_m,\eta_k) W_m W_k \tag{6.47}$$

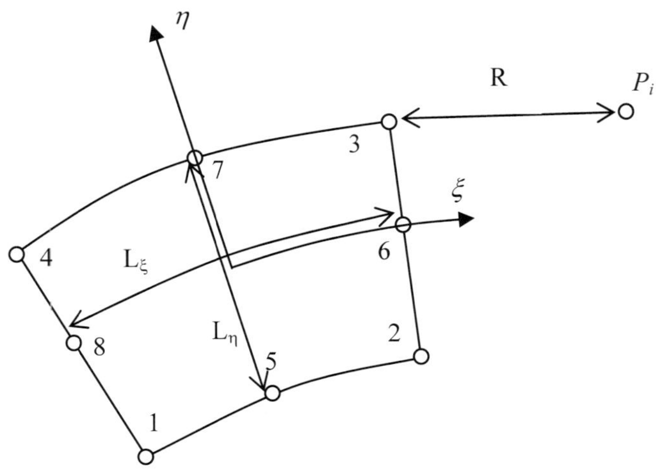

Figure 6.13 Two-dimensional isoparametric element

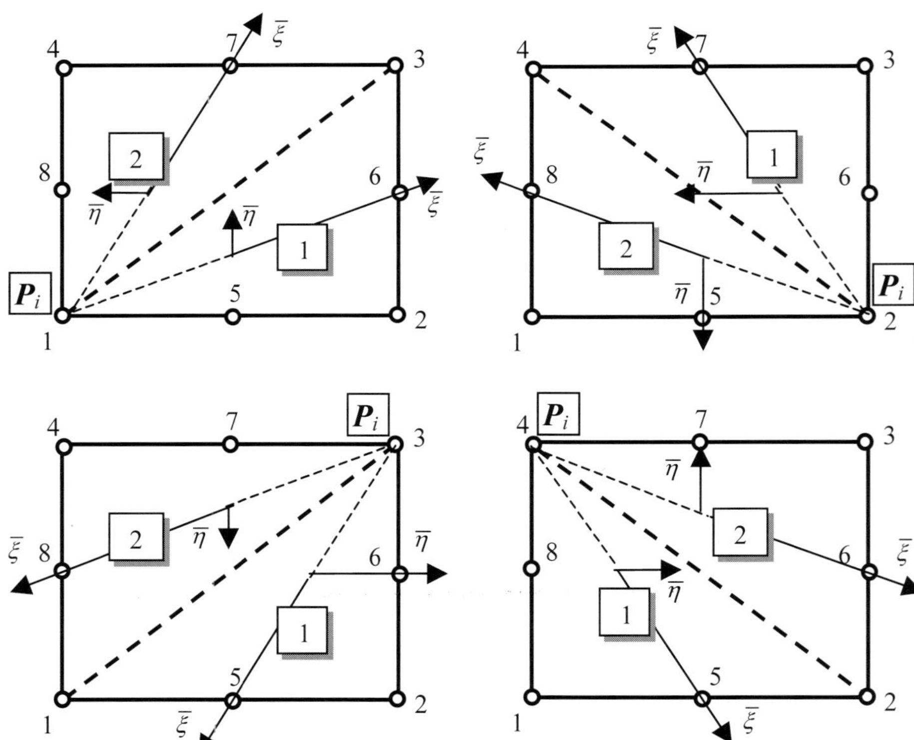

Figure 6.14 Sub-elements for numerical integration when P_i is a corner node of element

The number of integration points in ξ direction M is determined from Table 6.1, where L is taken as the size of the element in ξ direction, L_ξ, and the number of points in η direction K is determined by substituting for L the size of the element in η direction (L_η) in Figure 6.13.

When P_i is a node of the element but not node n, then Kernel U approaches infinity as $(1/r)$ but the shape function approaches zero, so product N_nU may be determined using Gauss Quadrature. Kernel T approaches infinity as $o(1/r^2)$ and cannot be integrated using the above scheme. When P_i is node n of the element, then product N_nU cannot be evaluated with Gauss Quadrature. The integral of the product N_nT only exists as a Cauchy principal value but this can be evaluated using equations (6.17) and (6.18).

For the evaluation of the second integral in equation (6.46), when P_i is a node of the element but not node n and for evaluating the first integral, when P_i is node n, we propose to split up the element into triangular subelements, as shown in Figures 6.14 and 6.15. For each subelement we introduce a local coordinate system that is chosen in such a way that the Jacobian of the transformation approaches zero at node P_i. Numerical integration formulae are then applied over two or three subelements depending if P_i is a corner or mid-side node.

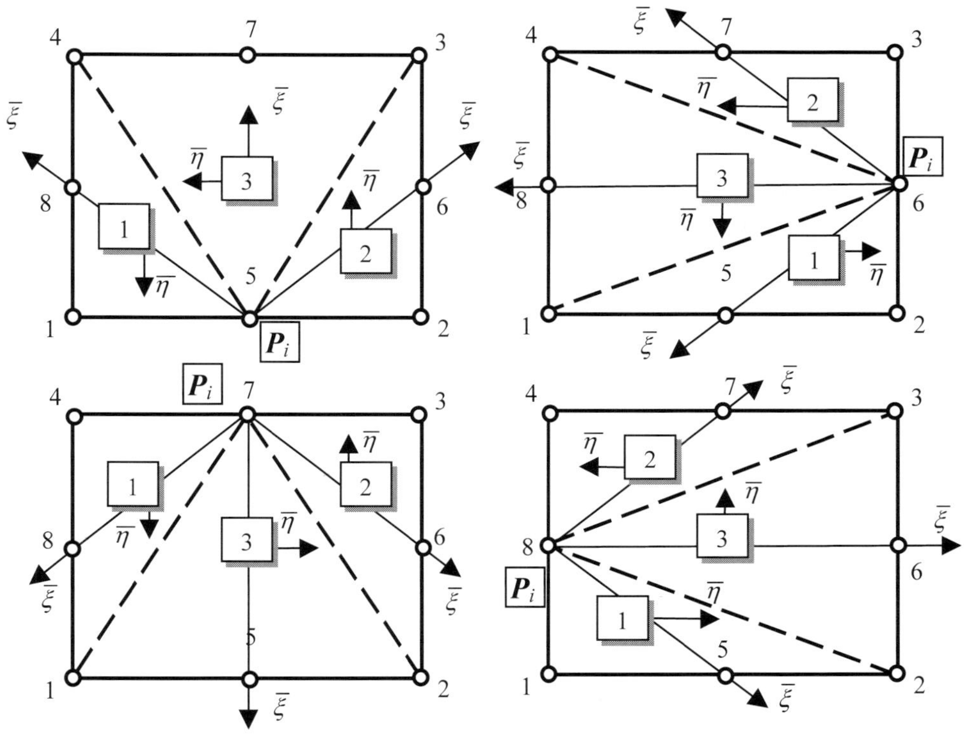

Figure 6.15 Sub-elements for numerical integration when P_i is a mid-side node of element

Using this scheme, the first integral in equation (6.46) is re-written as

$$\Delta \underset{g(n)=P_i}{U^e_{ni}} = \sum_{s=1}^{2(3)} \int_{-1}^{1}\int_{-1}^{1} N_n(\xi,\eta)\, U\big(P_i, Q(\xi,\eta)\big)\, J(\xi,\eta) \frac{\partial(\xi,\eta)}{\partial(\bar{\xi},\bar{\eta})} d\bar{\xi}\, d\bar{\eta} \tag{6.48}$$

The equation for numerical evaluation of the integral using Gauss Quadrature is given by

$$\Delta \underset{g(n)=P_i}{U^e_{ni}} \approx \sum_{s=1}^{2(3)} \sum_{m=1}^{M} \sum_{k=1}^{K} N_n\left(\bar{\xi}_m,\bar{\eta}_k\right) U\left(P_i, Q\left(\bar{\xi}_m,\bar{\eta}_k\right)\right) J\left(\bar{\xi}_m,\bar{\eta}_k\right) \bar{J}\left(\bar{\xi}_m,\bar{\eta}_k\right) W_m W_k \tag{6.49}$$

where $\bar{J}(\bar{\xi},\bar{\eta})$ is the *Jacobian* of the transformation from $\bar{\xi},\bar{\eta}$ to ξ,η.

The transformation from local element coordinates to subelement coordinates is given by

$$\xi = \sum_{n=1}^{3} \bar{N}_n(\bar{\xi},\bar{\eta})\xi_{l(n)} \quad , \quad \eta = \sum_{n=1}^{3} \bar{N}_n(\bar{\xi},\bar{\eta})\eta_{l(n)} \tag{6.50}$$

where *l(n)* is the local number of the nth subelement node and the shape functions are given by

$$\bar{N}_1 = \tfrac{1}{4}\left(1+\bar{\xi}\right)\left(1-\bar{\eta}\right) \quad , \quad \bar{N}_2 = \tfrac{1}{4}\left(1+\bar{\xi}\right)\left(1+\bar{\eta}\right) \; ; \; \bar{N}_3 = \tfrac{1}{2}\left(1-\bar{\xi}\right) \tag{6.51}$$

Tables 6.2 and 6.3 give the local node numbers *l(n)* in equation (6.50), depending on the number of the subelement and the position of P_i.

Table 6.2 Local node number *l(n)* of subelement nodes, P_i at corner nodes

P_i at node	Subelement 1			Subelement 2		
	n=1	*n=2*	*n=3*	*n=1*	*n=2*	*n=3*
1	2	3	1	3	4	1
2	3	4	2	4	1	2
3	1	2	3	4	1	3
4	1	2	4	2	3	4

The Jacobian matrix of the transformation (6.50) is given by

$$\bar{\mathbf{J}} = \begin{bmatrix} \dfrac{\partial\xi}{\partial\bar{\xi}} & \dfrac{\partial\eta}{\partial\bar{\xi}} \\ \dfrac{\partial\xi}{\partial\bar{\eta}} & \dfrac{\partial\eta}{\partial\bar{\eta}} \end{bmatrix} \tag{6.52}$$

where

$$\frac{\partial \xi}{\partial \bar{\xi}} = \sum_{n=1}^{3} \frac{\partial \bar{N}_n}{\partial \bar{\xi}}(\bar{\xi},\bar{\eta})\,\xi_n \quad , \quad \frac{\partial \xi}{\partial \bar{\eta}} = \sum_{n=1}^{3} \frac{\partial \bar{N}_n}{\partial \bar{\eta}}(\bar{\xi},\bar{\eta})\,\eta_n$$
$$\frac{\partial \eta}{\partial \bar{\xi}} = \sum_{n=1}^{3} \frac{\partial \bar{N}_n}{\partial \bar{\xi}}(\bar{\xi},\bar{\eta})\,\eta_n \quad , \quad \frac{\partial \eta}{\partial \bar{\eta}} = \sum_{n=1}^{3} \frac{\partial \bar{N}_n}{\partial \bar{\eta}}(\bar{\xi},\bar{\eta})\,\eta_n \tag{6.53}$$

Table 6.3 Local node number *l(n)* of subelement nodes, P_i at mid-side nodes

P_i at node	Subelement 1			Subelement 2			Subelement 3		
	n= 1	*n= 2*	*n= 3*	*n= 1*	*n= 2*	*n= 3*	*n= 1*	*n= 2*	*n= 3*
5	4	1	5	2	3	5	3	4	5
6	1	2	6	3	4	6	4	1	6
7	4	1	7	2	3	7	1	2	7
8	1	2	8	3	4	8	2	3	8

The *Jacobian* is given by

$$\bar{J} = det\left|\bar{\mathbf{J}}\right| = \frac{\partial \xi}{\partial \bar{\xi}}\frac{\partial \eta}{\partial \bar{\eta}} - \frac{\partial \eta}{\partial \bar{\xi}}\frac{\partial \xi}{\partial \bar{\eta}} \tag{6.54}$$

The reader may verify that for $\bar{\xi} = -1$ the *Jacobian* is zero. Without modification, the integration scheme is applicable to elasticity problems. In equations (6.46) we simply replace the scalars *U* and *T* with matrices **U** and **T**.

6.3.8 Subdivision of region of integration

As for the plane problems we need to implement a subdivision scheme for the integration. In the simplest implementation we subdivide elements into sub regions as shown in Figure 6.16. The number of sub regions N_ξ in ξ and N_η in η direction is determined by

$$N_\xi = INT[(R/L)_{\min}/(R/L_\xi)] \; ; \; N_\eta = INT[(R/L)_{\min}/(R/L_\eta)] \tag{6.55}$$

Equation 6.47 is replaced by

$$\Delta U^e_{ni} \approx \sum_{l=1}^{N_\xi}\sum_{j=1}^{N_\eta}\sum_{m=1}^{M(l)}\sum_{k=1}^{K(j)} N_n\left(\bar{\xi}_m,\bar{\eta}_k\right) U\left(P_i,Q\left(\bar{\xi}_m,\bar{\eta}_k\right)\right) J\cdot\bar{J}\,W_m W_k$$
$$\Delta U^e_{ni} \approx \sum_{l=1}^{N_\xi}\sum_{j=1}^{N_\eta}\sum_{m=1}^{M(l)}\sum_{k=1}^{K(j)} N_n\left(\bar{\xi}_m,\bar{\eta}_k\right) T\left(P_i,Q\left(\bar{\xi}_m,\bar{\eta}_k\right)\right) J\cdot\bar{J}\,W_m W_k \tag{6.56}$$

where *M(l)* and *K(j)* are the number of Gauss points in ξ and η direction for the sub region.

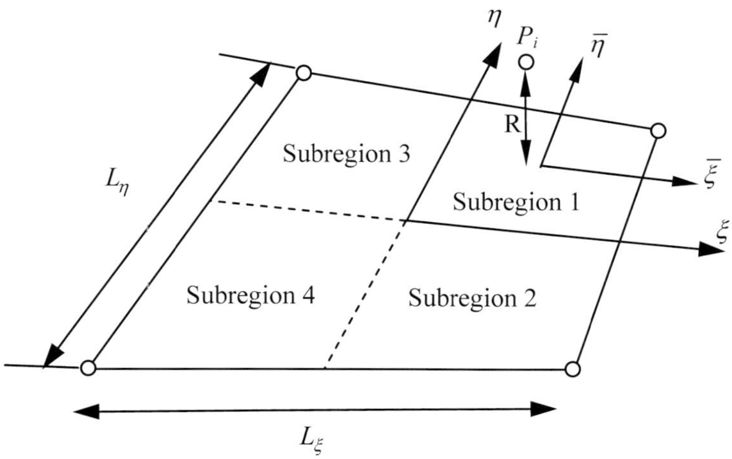

Figure 6.16 Subdivision of two-dimensional element

The relationship between global and local coordinates is defined as

$$\xi = \frac{1}{2}(\xi_1 + \xi_2) + \frac{\bar{\xi}}{N_\xi}$$
$$\eta = \frac{1}{2}(\eta_1 + \eta_2) + \frac{\bar{\eta}}{N_\eta} \tag{6.57}$$

where ξ_1, ξ_2 and η_1, η_2 define the sub-region. The *Jacobian* is given by

$$\left|\bar{J}\right| = \frac{\partial \xi}{\partial \bar{\xi}} \frac{\partial \eta}{\partial \bar{\eta}} = \frac{1}{N_\xi \cdot N_\eta} \tag{6.58}$$

6.3.9 Infinite elements

Since the integration over infinite elements is carried out in the (finite) local coordinate space no special integration scheme need to be introduced for infinite "decay" elements. However, special consideration has to be given to infinite "plane strain" elements[9]. This is explained on an example of an infinitely long cavity (tunnel) in Figure 6.17. For a "plane strain" element there is no change of the value of the variable in the infinite direction and Equation 6.12 becomes.

$$\Delta \mathbf{U}_i^e = \int_{-1}^{1}\int_{-1}^{1} \mathbf{U}(P_i,\xi,\eta)\, J(\xi,\eta)\, d\xi\, d\eta \;\; ; \;\; \Delta \mathbf{T}_i^e = \int_{-1}^{1}\int_{-1}^{1} \mathbf{T}(P_i,\xi,\eta)\, J(\xi,\eta)\, d\xi\, d\eta \tag{6.59}$$

For a two-dimensional element the sides of the element going to infinity must be parallel, so J is $o(r^2)$. $\mathbf{T}(P_i,Q)$ is $o(1/r^2)$ so the product $\mathbf{T}(P_i,Q)\cdot J$ is $o(1)$ and may be integrated using Gauss Quadrature. However, $\mathbf{U}(P_i,Q)$ is $o(1/r)$, the product $\mathbf{U}(P_i,Q)\cdot J$ is $o(r)$ and therefore the integral, with η going to infinity, does not exist. However we may replace the integral

$$\Delta\mathbf{U}_i^e = \int_{-1}^{1}\int_{-1}^{1}\mathbf{U}(P_i,Q(\xi,\eta))\cdot J\cdot d\xi\cdot d\eta \tag{6.60}$$

with

$$\Delta\mathbf{U}_i^e = \int_{-1}^{1}\int_{-1}^{1}(\mathbf{U}(P_i,Q)-\mathbf{U}(P_i,Q'))\cdot J\cdot d\xi\cdot d\eta \tag{6.61}$$

where Q' is a point dropped from Q to a "plane strain" axis, as shown in Figure 6.17.

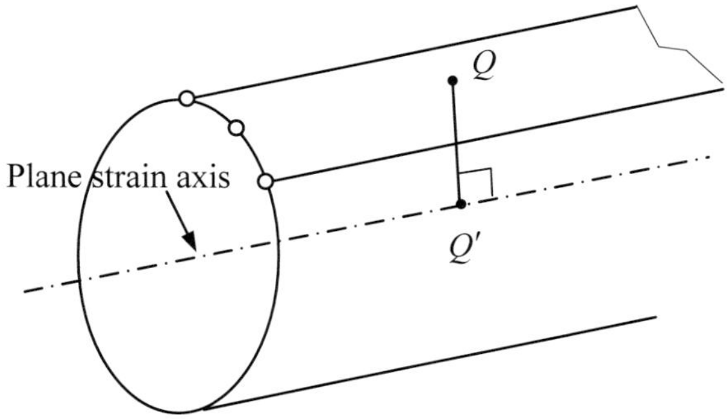

Figure 6.17 A cavity (tunnel) with the definition of the "plane strain" axis

Replacement of 6.60 with 6.61 has no effect on the satisfaction of the integral equation because tractions must integrate to zero around the cavity.

6.3.10 Implementation for three-dimensional problems

A sub-program, which calculates the element coefficient arrays $[\Delta U]^e$ and $[\Delta T]^e$ for potential problems, or $[\Delta\mathbf{U}]^e$ and $[\Delta\mathbf{T}]^e$ for elasticity problems, can be written based on the theory discussed. The diagonal coefficients of $[\Delta T]^e$ cannot be computed by integration over elements of Kernel-shape function products. As has already been discussed in section 6.3.2, these can be computed from the consideration of rigid body modes. The implementation will be discussed in the next chapter. In Subroutine Integ3 we distinguish between elasticity and potential problems by the input variable Ndof (number of degrees of freedom per node), which is set to 1 for potential problems and to 3 for elasticity problems.

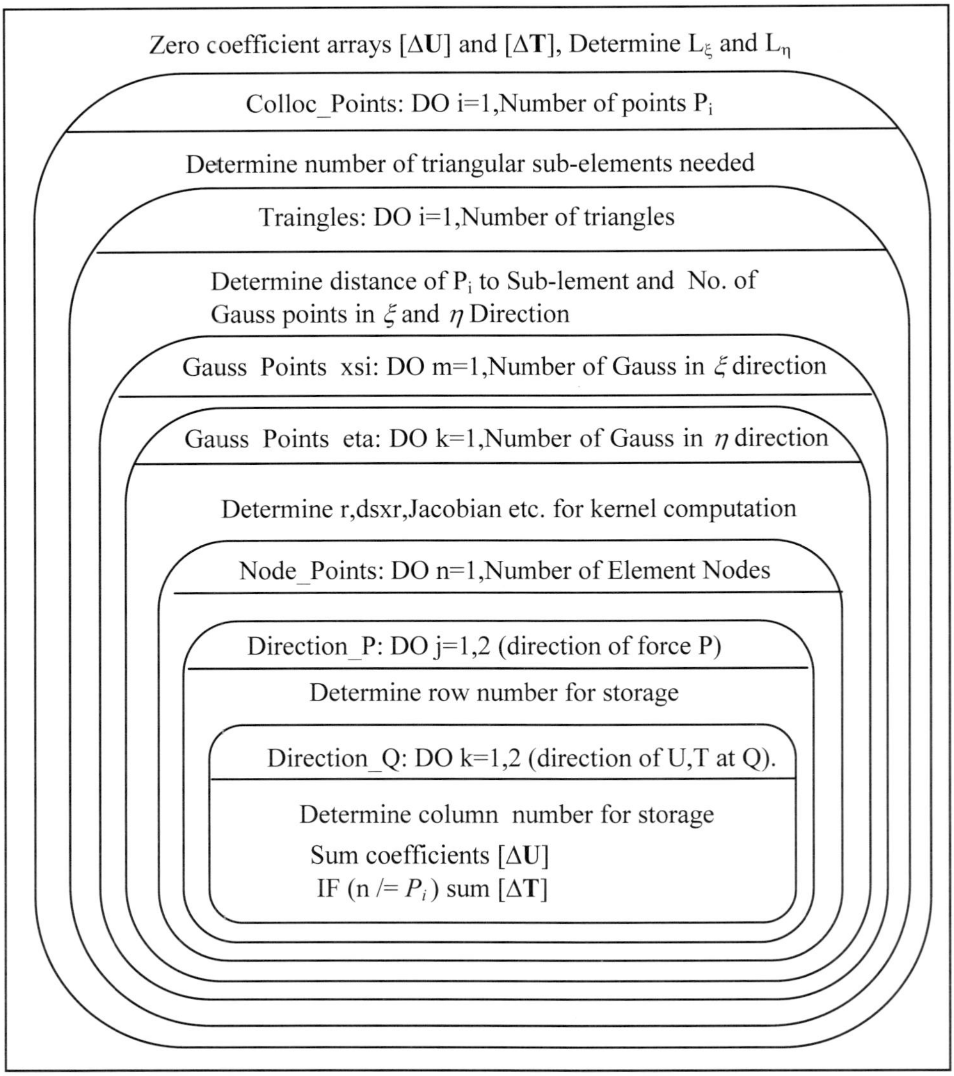

Figure 6.18 Structure chart for computation of [ΔT] and if [ΔU] if P_i is one of the element nodes

Subroutine INTEG3 is divided into two parts. The first part deals with integration when P_i is not one of the nodes of the element over which the integration is made. Gauss integration in two directions is used here. The integration of ΔT^e_{ni} and ΔU^e_{ni} is carried out concurrently. It should actually be treated separately, because the functions to be integrated have different degrees of singularity and, therefore, require a different number of Gauss points. For simplicity, both are integrated using the number of Gauss points required for the higher order singularity. Indeed, the subroutine presented has not been programmed very efficiently but, for the purpose of this book, simplicity was the paramount factor. Additional improvements in efficiency can, for example, be made by carefully examining if the operations in the DO loops actually depend on the DO loop variable. If they do not, then that operation should be taken outside of the corresponding DO loop. Substantial savings can be made here for a program that involves up to seven implied DO loops and which has to be executed for all boundary elements.

The second part of the SUBROUTINE deals with the case where P_i is one of the nodes of the element which we integrate over. To deal with the singularity of the integrand the element has to be subdivided into 2 or 3 triangles, as explained previously. Since there are a lot of implied DO loops involved, we show a structure chart of this part of the program in Figure 6.18.

A subdivision of the integration region has been implemented, but in order to improve clarity of the structure chart is not shown there. The subdivision of integration involves two more DO loops.

```
SUBROUTINE
Integ3(Elcor,Inci,Nodel,Ncol,xPi,Ndof,E,ny,ko,dUe,dTe,Ndest,Isym)
!-----------------------------------------------------
!    Computes  [dT]e and [dU]e for 3-D problems
!-----------------------------------------------------
IMPLICIT NONE
REAL, INTENT(IN)      :: Elcor(:,:)     !   Element coordinates
INTEGER, INTENT(IN)  :: Ndest(:,:)     !   Node destination vector
INTEGER, INTENT(IN)  :: Inci(:)        !   Element Incidences
INTEGER, INTENT(IN)  :: Nodel          !   No. of Element Nodes
INTEGER , INTENT(IN) :: Ncol           !   Number of points Pi
REAL , INTENT(IN)     :: xPi(:,:)       !   coll. points coords.
INTEGER , INTENT(IN) :: Ndof      !   Number DoF /node (1 or 3)
INTEGER , INTENT(IN) :: Isym
REAL , INTENT(IN)     :: E,ny      !   Elastic constants
REAL , INTENT(IN)     :: ko
REAL(KIND=8) , INTENT(OUT)  :: dUe(:,:),dTe(:,:)
REAL :: Elengx,Elenge,Rmin,RLx,RLe,Glcorx(8),Wix(8),Glcore(8)&
,Wie(8),Weit,r
REAL :: Ni(Nodel),Vnorm(3),GCcor(3),dxr(3),Jac,Jacb,xsi,eta,xsib&
,etab,Rlim(2)
REAL :: Xsi1,Xsi2,Eta1,Eta2,RJacB,RonL
REAL :: UP(Ndof,Ndof),TP(Ndof,Ndof)    !   for storing kernels
INTEGER :: i,m,n,k,ii,jj,ntr,Mi,Ki,id,nd,lnod,Ntri,NDIVX&
,NDIVSX,NDIVE,NDIVSE,MAXDIVS
INTEGER :: ldim= 2        !   Element dimension
INTEGER :: Cdim= 3        !   Cartesian dimension
ELengx=&
Dist((Elcor(:,3)+Elcor(:,2))/2.,(Elcor(:,4)+Elcor(:,1))/2.,Cdim)
ELenge=&
Dist((Elcor(:,2)+Elcor(:,1))/2.,(Elcor(:,3)+Elcor(:,4))/2.,Cdim)
dUe= 0.0 ; dTe= 0.0                    ! Clear arrays for summation
```

```
!-----------------------------------------------------------------
!      Part 1 : Pi is not one of the element nodes
!-----------------------------------------------------------------
Colloc_points: DO i=1,Ncol
 IF(.NOT. ALL(Inci /= i)) CYCLE       !  Check if inci contains i
 Rmin= Min_dist1(Elcor,xPi(:,i),Nodel,inci,ELengx,Elenge,ldim)
 Mi= Ngaus(Rmin/Elengx,2,Rlim)    !  Number of G.P. in xsi dir.
 RonL= Rmin/Elengx
 NDIVSX= 1 ; NDIVSE= 1
 RJacB=1.0
 IF(Mi == 5) THEN     !  Subdivision in ξ required
   IF(RonL > 0.0) NDIVSX= INT(RLim(2)/RonL) + 1
   Mi=4
 END IF
 Call Gauss_coor(Glcorx,Wix,Mi)
 Ki= Ngaus(Rmin/Elenge,2,Rlim)
 RonL= Rmin/Elenge
 IF(Ki == 5) THEN   !  Subdivision in η required
   IF(RonL > 0.0) NDIVSE= INT(RLim(2)/RonL) + 1
   Ki=4
 END IF
 IF(NDIVSX > 1 .OR. NDIVSE>1) RJacB= 1.0/(NDIVSX*NDIVSE)
 Call Gauss_coor(Glcore,Wie,Ki)
 Xsi1=-1.0
   Subdivisions_xsi: DO NDIVX=1,NDIVSX
    Xsi2= Xsi1 + 2.0/NDIVSX
    Eta1=-1.0
    Subdivisions_eta: DO NDIVE=1,NDIVSE
     Eta2= Eta1 + 2.0/NDIVSE
     Gauss_points_xsi: DO m=1,Mi
      xsi= Glcorx(m)
      IF(NDIVSX > 1) xsi= 0.5*(Xsi1+Xsi2)+xsi/NDIVSX
      Gauss_points_eta: DO k=1,Ki
       eta= Glcore(k)
       IF(NDIVSE > 1) eta= 0.5*(Eta1+Eta2)+eta/NDIVSE
       Weit= Wix(m)*Wie(k)*RJacB
       CALL Serendip_func(Ni,xsi,eta,ldim,nodel,Inci)
       Call Normal_Jac(Vnorm,Jac,xsi,eta,ldim,nodel,Inci,elcor)
       CALL Cartesian(GCcor,Ni,ldim,elcor)
       r= Dist(GCcor,xPi(:,i),Cdim)            !  Dist. P,Q
       dxr= (GCcor-xPi(:,i))/r                 !  rx/r , ry/r
       IF(Ndof .EQ. 1) THEN
        UP= U(r,ko,Cdim) ; TP= T(r,dxr,Vnorm,Cdim) !  Pot. problem
       ELSE
        UP= UK(dxr,r,E,ny,Cdim) ; TP= TK(dxr,r,Vnorm,ny,Cdim)
       END IF
        Direction_P: DO ii=1,Ndof
         IF(Isym == 0)THEN
          iD= Ndof*(i-1) + ii      !  line number in array
         ELSE
          iD= Ndest(i,ii)                   !  line number in array
```

```
        END IF
        IF (id == 0) CYCLE
         Direction_Q: DO jj=1,Ndof
           Node_points: DO n=1,Nodel
             nD= Ndof*(n-1) + jj   !  column number in array
             dUe(iD,nD)= dUe(iD,nD)  + Ni(n)*UP(ii,jj)*Jac*Weit
             dTe(iD,nD)= dTe(iD,nD)  + Ni(n)*TP(ii,jj)*Jac*Weit
           END DO Node_points
          END DO Direction_Q
         END DO Direction_P
        END DO Gauss_points_eta
       END DO Gauss_points_xsi
       Eta1= Eta2
      END DO Subdivisions_eta
      Xsi1= Xsi2
     END DO Subdivisions_xsi
   END DO Colloc_points
!-----------------------------------------------------------------
!     Part 1 : Pi is one of the element nodes
!-----------------------------------------------------------------
Colloc_points1: DO i=1,Ncol
 lnod= 0
 DO n= 1,Nodel          !   Determine which local node is Pi
   IF(Inci(n) .EQ. i) THEN
    lnod=n
   END IF
 END DO
 IF(lnod .EQ. 0) CYCLE          !  None -> next Pi
 Ntri= 2
 IF(lnod > 4) Ntri=3           !  Number of triangles
 Triangles: DO ntr=1,Ntri
   CALL Tri_RL(RLx,RLe,Elengx,Elenge,lnod,ntr)
   Mi= Ngaus(RLx,2,Rlim)
   IF(Mi == 5) Mi=4   !  Triangles are not sub-divided
   Call Gauss_coor(Glcorx,Wix,Mi)
   Ki= Ngaus(RLe,2,Rlim)
   IF(Ki == 5) Ki=4
   Call Gauss_coor(Glcore,Wie,Ki)
   Gauss_points_xsi1: DO m=1,Mi
    xsib= Glcorx(m)
    Gauss_points_eta1: DO k=1,Ki
     etab= Glcore(k)
     Weit= Wix(m)*Wie(k)
     CALL Trans_Tri(ntr,lnod,xsib,etab,xsi,eta,Jacb)
     CALL Serendip_func(Ni,xsi,eta,ldim,nodel,Inci)
     Call Normal_Jac(Vnorm,Jac,xsi,eta,ldim,nodel,Inci,elcor)
     Jac= Jac*Jacb
     CALL Cartesian(GCcor,Ni,ldim,elcor)
     r= Dist(GCcor,xPi(:,i),Cdim)
     dxr= (GCcor-xPi(:,i))/r
     IF(Ndof .EQ. 1) THEN
```

```
    UP= U(r,ko,Cdim) ; TP= T(r,dxr,Vnorm,Cdim) !  Potential
   ELSE
    UP= UK(dxr,r,E,ny,Cdim) ; TP= TK(dxr,r,Vnorm,ny,Cdim)
   END IF
   Direction_P1: DO ii=1,Ndof
    IF(Isym == 0)THEN
     iD= Ndof*(i-1) + ii      !  line number in array
    ELSE
     iD= Ndest(i,ii)                 !  line number in array
    END IF
    IF (id == 0) CYCLE
    Direction_Q1:  DO jj=1,Ndof
     Node_points1: DO n=1,Nodel
       nD= Ndof*(n-1) + jj     !  column number in array
       dUe(iD,nD)= dUe(iD,nD) + Ni(n)*UP(ii,jj)*Jac*Weit
       IF(Inci(n) /= i) THEN  !   diagonal elements of dTe
        dTe(iD,nD)= dTe(iD,nD) + Ni(n)*TP(ii,jj)*Jac*Weit
       END IF
      END DO Node_points1
     END DO Direction_Q1
    END DO Direction_P1
   END DO Gauss_points_eta1
  END DO Gauss_points_xsi1
 END DO Triangles
END DO Colloc_points1
RETURN
END SUBROUTINE Integ3
```

6.4 CONCLUSIONS

In this chapter we have discussed in some detail, numerical methods which can be used to perform the integration of Kernel-shape function products over boundary elements. Because of the nature of these functions, special integration schemes had to be devised, so that the precision of integration is similar for all locations of P_i relative to the boundary element over which the integration is carried out. If this is not taken into consideration, results obtained from a BEM analysis will be in error and, in extreme cases, meaningless.

The number of integration points which has to be used to obtain a given precision of integration is not easy to determine. Whereas error estimates have been worked out by several researchers based on mathematical theory, so far they are only applicable to regular meshes and not to isoparametric elements of arbitrary curved shape. The scheme proposed here for working out the number of integration points has been developed on a semi-empirical basis, but has been found to work well.

We have now developed a library of subroutines which we will need for the writing of a general purpose computer program. All that is needed is the assembly of coefficient matrices from element contributions, to specify the boundary conditions and to solve the system of equations.

6.5 EXERCISES

Exercise 6.1

Check the accuracy of integration, using Subroutine Integ2P for a one-dimensional boundary element by performing the numerical integration of

$$\Delta U_{ni}^{e} = \int_{S_e} N_n(\xi) U(P_i,\xi) dS(\xi) \quad , \quad \Delta T_{ni}^{e} = \int_{S_e} N_n(\xi) T(P_i,\xi) dS(\xi) \tag{6.62}$$

(where U and T are the Kernels for potential problems) for a straight line element with linear shape function located parallel to the x-axis of length 2 for n=1, with two different locations of P_i (shown in Figure 6.19):

(a) along the element

(b) perpendicular to the centre of the element

Do this for values of R/L= 0.5, 0.1 , 0.05 and compare with the analytical solution using aprogram such as MATHEMATICA.

Modify Subroutine Integ2P to disable the subdivision of the region of integration and compare the results with the analytical solution

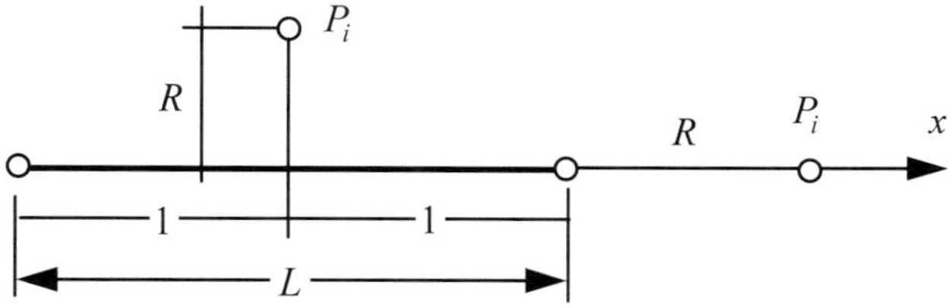

Figure 6.19 2-D problem for checking accuracy of numerical integration

Exercise 6.2

Check the integration schemes proposed for two-dimensional boundary elements using Subroutine INTEG3 by performing the integration of equations

$$\begin{aligned} \Delta U_{ni}^{e} &= \int_{S_e} N_n(\xi,\eta)\, U\big(P_i, Q(\xi,\eta)\big)\, dS(\xi) \\ \Delta T_{ni}^{e} &= \int_{S_e} N_n(\xi,\eta)\, T\big(P_i, Q(\xi,\eta)\big)\, dS(\xi) \end{aligned} \tag{6.63}$$

for a square element with linear shape function of size 2x2, as shown in Figure 6.20 for n=1 with two different locations of P_i:

(c) In the same plane as the element

(d) perpendicular to the centre of the element

for values of R/L= 0.5, 0.1 , 0.05. Compare with the analytical solution using MATHEMATICA or similar.

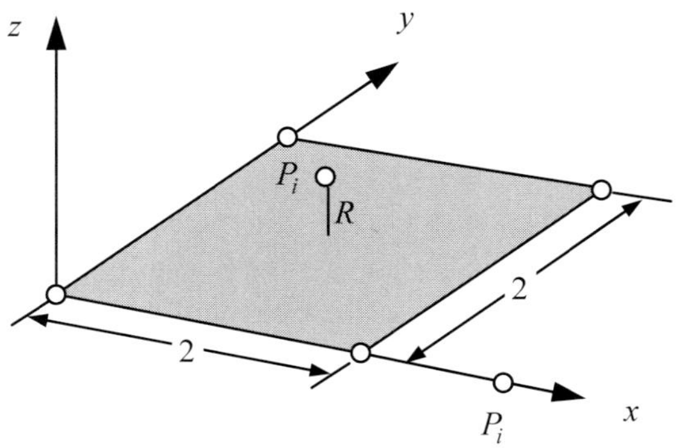

Figure 6.20 3-D problem for checking the accuracy of numerical integration

6.6 REFERENCES

1. Deist, F.H. and Georgiadis, E. (1973) A computer system for three-dimensional stress analysis in elastic media. *Rock Mechanics* **5**: 189-202.
2. Ergatoudis, J.G.,Irons, B.M.and Zienkiewicz O.C. (1968) Three dimensional analysis of arch dams and their foundations. *Proc. Symp. Arch Dams.* Inst. Civ. Eng.
3. Lachat, J.C. and Watson, J.O. (1976) Effective numerical treatment of boundary integral equations. *Int. J. Num. Meth. Eng.* **10**: 991-1005.
4. Beer G. and Watson J.O. (1991) Introduction to Finite and Boundary Element Methods for Engineers. J.Wiley.
5. Kreyszig E. (1999) Advanced Engineering Mathematics. J.Wiley.
6. Guigiani M and Casalini P.(1987) Direct computation of Cauchy principal value integrals in advanced boundary elements. *Int. J. Num. Meth. Eng.* **24**:1711-1720.
7. Eberwien U., Duenser C., Moser W. (2005) Efficient calculation of internal results in 2D elasticity BEM. *Engineering Analysis with Boundary Elements* **29**: 447-453.
8. Stroud, A.H. and Secrest, D. (1966) Gaussian Quadrature Formulas. Prentice-Hall, Englewood Cliffs.
9. Beer G and Watson J.O. (1989) Infinite Boundary Elements. *Int. J. Num. Meth. Eng.* **28**: 1233-1247.

7

Assembly and Solution

Intellektuelle Erkenntnisse sind papier
(Intellectual findings are just paper)
H. Hesse

7.1 INTRODUCTION

The previous chapter, dealing with the numerical integration of Kernel-Shape function products, addressed probably the most important aspect of the boundary element method. We note that this is much more involved than the integration used in the FEM for determining the element stiffness matrix.

In the current chapter we will find that subsequent steps in solving the integral equations are fairly straightforward and similar to the FEM, especially with respect to the assembly of element contributions into the global coefficient matrix. In this book we will discuss two approaches: one where we first assemble the coefficient matrices and then solve for the boundary unknowns and another where the assembly and solution are intermixed. The analogy to this in the FEM is the direct solution and the Element by Element solution using iterative solvers[1]. We will see that the iterative solution approach is very suitable to parallel processing. The assembly and direct solution is discussed in this chapter and the iterative solution strategy in the subsequent chapter, where also aspects of parallelisation will be discussed.

The system of equations will be different to the FEM, as we have to deal with non-symmetric and fully populated coefficient matrices. The fact that the equation system is fully populated has been claimed to be one of the main drawbacks of the method. However, because the system of equations obtained is always significantly smaller, it more than compensates for this and, as we will see later, computation times required for the solution are usually much lower than the FEM. We will also see in chapter 11, that if we introduce the concept of multiple boundary element regions sparsity is introduced to the system of equations.

At the end of this chapter we will have all the procedures necessary for a general purpose program, which can solve steady state problems in potential flow and elasticity. The program, however, will only give us values of the unknown at the boundary. As already pointed out, a special feature of the BEM is that results at any point inside the domain can be computed with greater accuracy as a postprocessing exercise. This topic will be dealt with in Chapter 9.

7.2 ASSEMBLY OF SYSTEM OF EQUATIONS

We start with potential problems. In the previous section we discussed the computation of element contributions to equation (6.7), that is

$$cu(P_i)+\sum_{e=1}^{E}\sum_{n=1}^{N}\Delta T_{ni}^{e}\, u_{n}^{e}=\sum_{e=1}^{E}\sum_{n=1}^{N}\Delta U_{ni}^{e}\, t_{n}^{e} \tag{7.1}$$

We recall the notation used

$$\begin{array}{c} \textit{Element number} \\ \downarrow \\ \Delta T^{e}_{\underset{\uparrow}{n}i\leftarrow} \;\; \textit{Collocation point} \\ \textit{Node Number} \end{array} \tag{7.2}$$

For the solution of the system of equations it is convenient to replace the double sums by a matrix multiplication of the type

$$[\Delta T]\{u\}=[\Delta U]\{t\} \tag{7.3}$$

where vectors $\{u\}$, $\{t\}$ contain potential/temperature and fluxes respectively for all nodes in global numbering system

$$\{u\}=\{u_1,u_2,\cdots\}^{\mathrm{T}} \tag{7.4}$$

and $[\Delta T]$, $[\Delta U]$ are global coefficient matrices assembled by *gathering* element contributions. In the global coefficient arrays, rows correspond to collocation points P_i and columns to the global node number. The gathering process is very similar to the assembly process in the FEM, except that whole columns are added. For the gathering process we need the Connectivity or Incidences of element *e*, which refer to the global node numbers of the element.

Referring to the simple 2-D mesh with linear elements in Figure 7.1 the incidences of are given in Table 7.1.

Table 7.1 Connectivity (Incidences)

Element	*Node 1*	*Node 2*
1	1	2
2	2	3
3	3	4
4	4	5
5	5	6
6	6	7
7	7	1

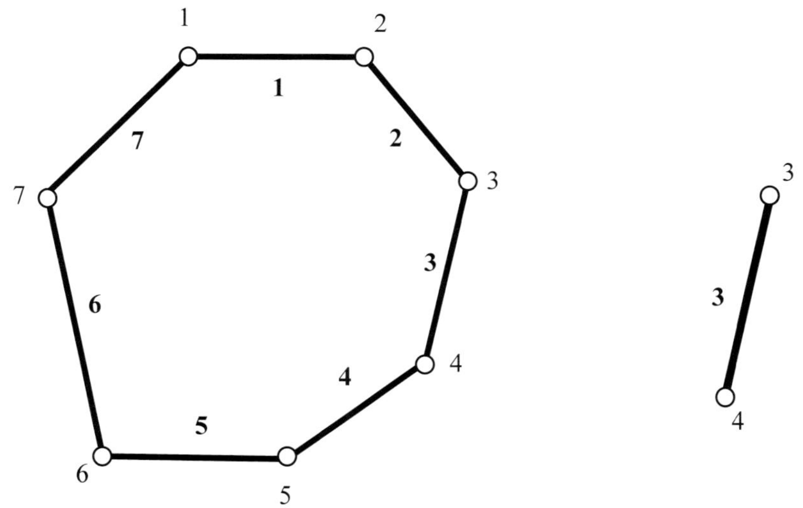

Figure 7.1 2-D BE mesh for explaining assembly (potential problems)

For example, to assemble the contributions of element 3 with connectivity (/3,4/), columns of the coefficient matrix $[\Delta T]^3$ are added to the global matrix $[\Delta T]$

$$\begin{array}{c} \rightarrow \quad Node \quad numbers \\ \begin{array}{ccccccc} 1 & 2 & \mathbf{3} & \mathbf{4} & 5 & 6 & 7 \end{array} \end{array}$$

$$[\Delta T] = \begin{bmatrix} \cdot & \cdot & \Delta T_{11}^{3} & \Delta T_{21}^{3} & \cdot & \cdot & \cdot \\ \cdot & \cdot & \Delta T_{12}^{3} & \Delta T_{22}^{3} & \cdot & \cdot & \cdot \\ \cdot & \cdot & \Delta T_{13}^{3} & \Delta T_{23}^{3} & \cdot & \cdot & \cdot \\ \cdot & \cdot & \Delta T_{14}^{3} & \Delta T_{24}^{3} & \cdot & \cdot & \cdot \\ \cdot & \cdot & \Delta T_{15}^{3} & \Delta T_{25}^{3} & \cdot & \cdot & \cdot \\ \cdot & \cdot & \Delta T_{16}^{3} & \Delta T_{26}^{3} & \cdot & \cdot & \cdot \\ \cdot & \cdot & \Delta T_{17}^{3} & \Delta T_{27}^{3} & \cdot & \cdot & \cdot \end{bmatrix} \begin{array}{c} 1 \\ 2 \\ 3 \\ 4 \quad \downarrow \quad P_i \\ 5 \\ 6 \\ 7 \end{array} \tag{7.5}$$

For elasticity problems there is more than one unknown per node, so columns are numbered according to the degree of freedom, rather than node number. For two-dimensional elasticity problems, each node has two degrees of freedom and the incidences of element 3 are expanded to *destinations* shown in table 7.2.

Table 7.2 *Destinations*

Element	*Node 1*		*Node 2*	
	x	*y*	*x*	*y*
1	1	2	3	4
2	3	4	5	6
3	5	6	7	8
4	7	8	9	10
5	9	10	11	12
6	11	12	13	14
7	13	14	1	2

For element 3 the destination vector is (/5,6,7,8/) and the assembly is

$$\begin{array}{ccccccccc} & & & & \rightarrow \textit{Destination} & \textit{Numbers} & & & \\ 1 & 2 & 3 & 4 & 5 & 6 & 7 & 8 & \cdots \end{array}$$

$$[\Delta \mathbf{T}] = \begin{bmatrix} \cdot & \cdot & \cdot & \cdot & \Delta T^3_{xx11} & \Delta T^3_{yx11} & \Delta T^3_{xx21} & \Delta T^3_{yx21} & \cdot & \cdot & \cdot & \cdot & \cdot & \cdot \\ \cdot & \cdot & \cdot & \cdot & \Delta T^3_{xy11} & \Delta T^3_{yy11} & \Delta T^3_{xy21} & \Delta T^3_{yy21} & \cdot & \cdot & \cdot & \cdot & \cdot & \cdot \\ \cdot & \cdot & \cdot & \cdot & \Delta T^3_{xx12} & \Delta T^3_{yx12} & \Delta T^3_{xx22} & \Delta T^3_{yx22} & \cdot & \cdot & \cdot & \cdot & \cdot & \cdot \\ \cdot & \cdot & \cdot & \cdot & \Delta T^3_{xy12} & \Delta T^3_{yy12} & \Delta T^3_{xy22} & \Delta T^3_{yy22} & \cdot & \cdot & \cdot & \cdot & \cdot & \cdot \\ \vdots & \vdots & \vdots & \vdots & \vdots & \vdots & \vdots & \vdots & \vdots & \vdots & \vdots & \vdots & \vdots & \vdots \end{bmatrix} \tag{7.6}$$

Note that destination numbers are now used for numbering the columns.

Coming back to potential problems and assuming that, as in the introductory example solved with the Trefftz method, the flux t is known on all boundary nodes and solution u is required, we assemble the left hand side, perform the matrix multiplication on the right and solve the system of equations. Alternatively, multiplication $[\Delta U]\{t\}$ can be made element by element at the assembly level, without explicitly creating the matrix $[\Delta U]$, therefore saving on storage space. This would also allow us to consider discontinuous distribution of normal gradients or tractions. For the simple example in Figure 7.1, equation (7.1) can be replaced by

$$[\Delta T]\{u\} = \{F\} \tag{7.7}$$

where the coefficients of the right hand side vector $\{F\}$ are given by

$$F_i = \sum_{e=1}^{7} \sum_{n=1}^{N} \Delta U_{ni}^{e} \, t_n^{e} \tag{7.8}$$

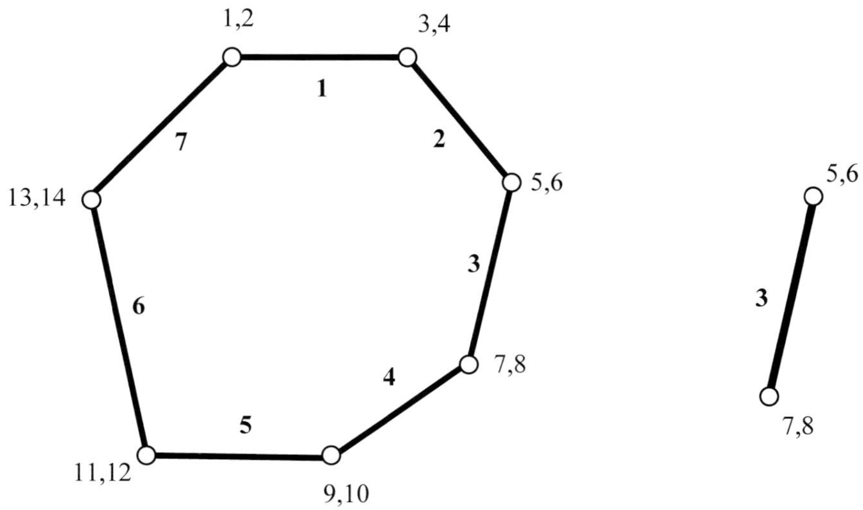

Figure 7.2 2-D mesh for explaining assembly (2-D elasticity problems)

Often, however, we have a mixed boundary value problem where u is prescribed on some portion of the boundary and t on the other. We must therefore exchange columns so that coefficients which multiply with unknowns are on the left hand side and coefficients which multiply with known values are on the right. We consider the simple example in Figure 7.3, where temperatures u are prescribed along element 4 and flow values are prescribed on elements 1,2 and 3. Note that since the outward normals are different at the corner nodes, the flow values are discontinuous there, i.e., different for the element left and right of the node. However, there can only be one temperature value at a node.

Writing equations (7.1) in longhand we obtain:

$$\begin{aligned}
&cu(P_i) + \Delta T_{1i}^{1} u_1^{1} + \Delta T_{2i}^{1} u_2^{1} + \Delta T_{1i}^{2} u_1^{2} + \Delta T_{2i}^{2} u_2^{2} + \Delta T_{1i}^{3} u_1^{3} + \Delta T_{2i}^{3} u_2^{3} \\
&+ \Delta T_{1i}^{4} u_1^{4} + \Delta T_{2i}^{4} u_2^{4} = \\
&\Delta U_{1i}^{1} t_1^{1} + \Delta U_{2i}^{1} t_2^{1} + \Delta U_{1i}^{2} t_1^{2} + \Delta U_{2i}^{2} t_2^{2} + \Delta U_{1i}^{3} t_1^{3} + \Delta U_{2i}^{3} t_2^{3} + \Delta U_{1i}^{4} t_1^{4} + \Delta U_{2i}^{4} t_2^{4}
\end{aligned} \tag{7.9}$$

The assembly procedure has to be modified, so that we put all known values on the right hand side and all unknown ones on the left side of the equation.

Known values are

$$\begin{aligned} &t_1^1 \; ; \; t_2^1 \; ; \; t_1^2 \; ; \; t_2^2 \; ; \; t_1^3 \; ; \; t_2^3 \\ &u_1 \; ; \; u_4 \end{aligned} \tag{7.10}$$

Unknown values are

$$\begin{aligned} &u_2^1 = u_1^2 = u_2 \quad ; \quad u_2^2 = u_1^3 = u_3 \\ &t_1^4 \; ; \; t_2^4 \end{aligned} \tag{7.11}$$

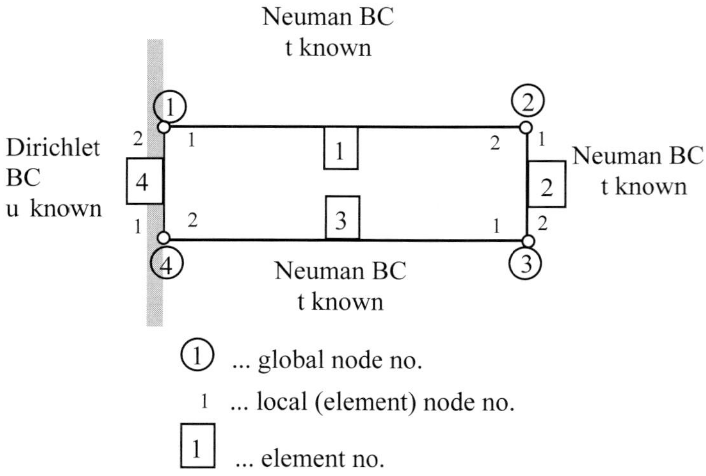

Figure 7.3 Example of two-dimensional potential problem with mixed boundary conditions

After placing unknown values on the left and known values on the right, equation (7.9) is written as

$$\begin{aligned} &cu(P_i) + (\Delta T_{2i}^1 + \Delta T_{1i}^2)\, u_2 + (\Delta T_{2i}^2 + \Delta T_{1i}^3)\, u_3 \\ &- \Delta U_{1i}^4\, t_1^4 - \Delta U_{2i}^4\, t_2^4 = \\ &\Delta U_{1i}^1\, t_1^1 + \Delta U_{2i}^1 t_2^1 + \Delta U_{1i}^2\, t_1^2 + \Delta U_{2i}^2 t_2^2 + \Delta U_{1i}^3\, t_1^3 + \Delta U_{2i}^3\, t_2^3 \\ &- \Delta T_{1i}^4\, u_1 - \Delta T_{2i}^4\, u_4 \end{aligned} \tag{7.12}$$

In equation (7.12) the global numbering for the nodes has been implemented. This equation can now be written for source points P_i located at nodes 1,2,3,4 as a matrix equation (7.13). The diagonal elements involving ΔT are highlighted by brackets. As explained in 6.3.1., we compute and assemble these diagonal coefficients by considering "rigid body modes".

$$\begin{array}{cccc} 1 & 2 & 3 & 4 \end{array}$$
$$\begin{bmatrix} -\Delta U_{11}^4 & \Delta T_{21}^1 + \Delta T_{11}^2 & \Delta T_{21}^2 + \Delta T_{11}^3 & -\Delta U_{21}^4 \\ -\Delta U_{12}^4 & \left[\Delta T_{22}^1 + \Delta T_{12}^2\right] & \Delta T_{22}^2 + \Delta T_{12}^3 & -\Delta U_{22}^4 \\ -\Delta U_{13}^4 & \Delta T_{23}^1 + \Delta T_{13}^2 & \left[\Delta T_{23}^2 + \Delta T_{13}^3\right] & -\Delta U_{23}^4 \\ -\Delta U_{14}^4 & \Delta T_{24}^1 + \Delta T_{14}^2 & \Delta T_{24}^2 + \Delta T_{14}^3 & -\Delta U_{24}^4 \end{bmatrix} \begin{Bmatrix} t_1^4 \\ u_2 \\ u_3 \\ t_2^4 \end{Bmatrix} = \{F\} \tag{7.13}$$

The coefficients F_i of the right hand side vector $\{F\}$ is computed as

$$F_i = \Delta U_{1i}^1\, t_1^1 + \Delta U_{2i}^1 t_2^1 + \Delta U_{1i}^2\, t_1^2 + \Delta U_{2i}^2 t_2^2 + \Delta U_{1i}^3\, t_1^3 + \Delta U_{2i}^3\, t_2^3$$
$$-\Delta T_{1i}^4\, u_1 - \Delta T_{2i}^4\, u_4 \tag{7.14}$$

For problems in elasticity, the assembly process for mixed boundary value problems is similar but, since the assembly is by degrees of freedom rather than node numbers, boundary conditions will also depend on the direction. An example of this is given in Figure 7.4.

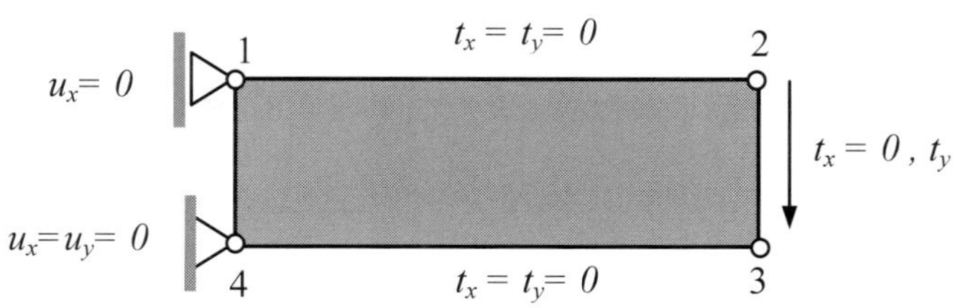

Figure 7.4 Example of discontinuous boundary condition in elasticity: fixed beam

To summarise the assembly process we note that element contributions are assembled into the global matrix by gathering the coefficients according to incidences or destinations (in the case of elasticity problems). Depending on the boundary codes defined at a particular node, the coefficients ΔT_{ni}^e and ΔU_{ni}^e are assembled either into the left or right hand side. So the information that is needed for the assembly is the *Connectivity* or *Destination* vector and the boundary code for each element. Note that the boundary code is defined locally for each element and can have two values at a particular node.

7.2.1 Symmetry

In many cases it is possible to take into account the symmetry of a problem and thereby considerably reduce the amount of analysis effort. In the FEM such conditions are simply implemented by generating only part of the mesh and providing the appropriate boundary conditions at the plane(s) of symmetry. In the BEM we can take a different approach, alleviating the need to have boundary elements on the symmetry plane. For example, for the problem shown in Figure 7.5, of a circular excavation in an infinite domain, nodes do not exist on the plane of symmetry.

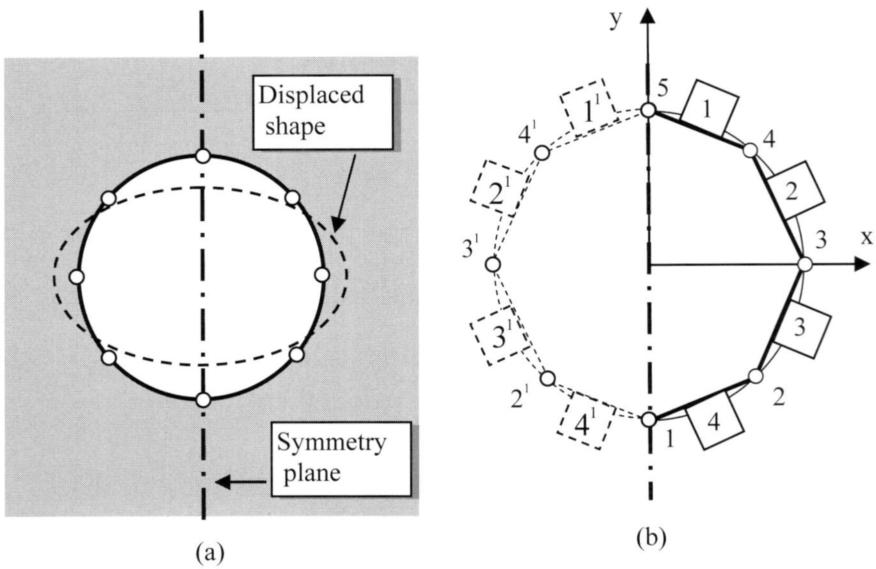

Figure 7.5 Example with one plane of symmetry and mesh used to explain implementation

The approach in dealing with symmetry conditions in the BEM will be explained here. Consider the simple mesh for the analysis of a circular excavation consisting of a total of 8 elements shown in Figure 7.5b. The idea is to input only elements on the right hand side of the symmetry plane (elements 1 to 4) and to automatically generate the elements on the left (elements 1^1 to 4^1). The incidences of all elements are:

Element	i	j	Element	i	j
1	4	5	1^1	5	4^1
2	3	4	2^1	4^1	3^1
3	2	3	3^1	3^1	2^1
4	1	2	4^1	2^1	1

Note that the sequence of nodes for all mirrored elements is reversed. This is important because it affects the direction of the outward normal vector **n**. The coordinates of nodes at the left of the symmetry plane can be computed from those on the right by:

Node	x	y
2^1	$-x_2$	y_2
3^1	$-x_3$	y_3
4^1	$-x_4$	y_4

Substantial savings in computational effort can be made, if during assembly we consider that the unknowns on the left hand side of the symmetry plane can be determined from the ones on the right. For potential problems we simply have $u_2^{1}= u_2$, $u_3^{1}= u_3$ and $u_4^{1}= u_4$.

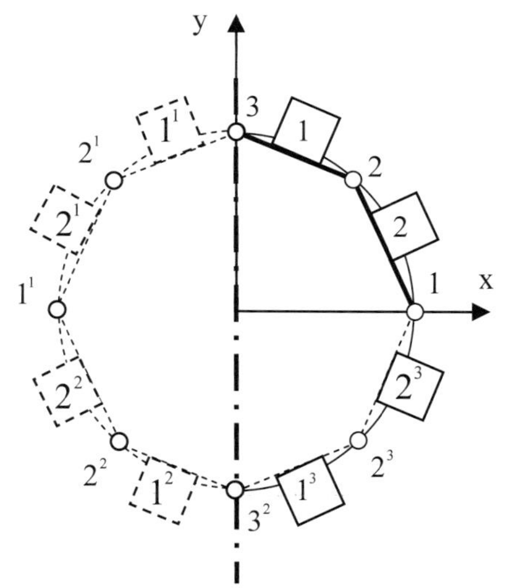

Figure 7.6 Example with two planes of symmetry

For elasticity problems we have (see displaced shape in Figure 7.5 (a))

Node	u_x	u_y
2^1	$-u_{x2}$	u_{y2}
3^1	$-u_{x3}$	u_{y3}
4^1	$-u_{x4}$	u_{y4}

For assembly of the system of equations this means that the coefficients of the mirrored nodes are assembled in the same location as for the un-primed nodes.

For elasticity problems, the negative signs of the x-component of the displacement have to be considered during assembly. If this assembly procedure is used, then the number of unknowns for the problem is reduced to the nodes on the right hand side of the symmetry plane and on the plane itself. The only additional computational effort will be the integration of the Kernel shape function products over the mirrored elements. If conditions of symmetry exist about the x and y axis, then the elements are "mirrored" twice, as shown in Figure 7.6.

The incidence vectors are now

Element	i	j	Element	i	j
1	2	3	1^2	2^2	3^2
2	1	2	2^2	1^2	2^2
1^1	3	2^1	1^3	3^1	2^3
2^1	2^1	1^1	2^3	2^3	1

Note that for all elements, except 1^2 and 2^2, the incidences are reversed. The coordinates of the "mirrored" nodes are

Node	x	y
1^1	$-x_1$	y_1
3^2	x_3	$-y_3$
2^1	$-x_2$	y_2
2^2	$-x_2$	$-y_2$
2^3	x_2	$-y_2$

For potential problems we have $u_2{}^1 = u_2{}^2 = u_2{}^3 = u_2$, $u_1{}^1 = u_1$ and $u_3{}^1 = u_3$. For elasticity problems the displacements at the primed nodes are given by

Node	u_x	u_y
1^1	$-u_{x1}$	0
3^1	0	$-u_{y3}$
2^1	$-u_{x2}$	u_{y2}
2^2	$-u_{x2}$	$-u_{y2}$
2^3	u_{x2}	$-u_{y2}$

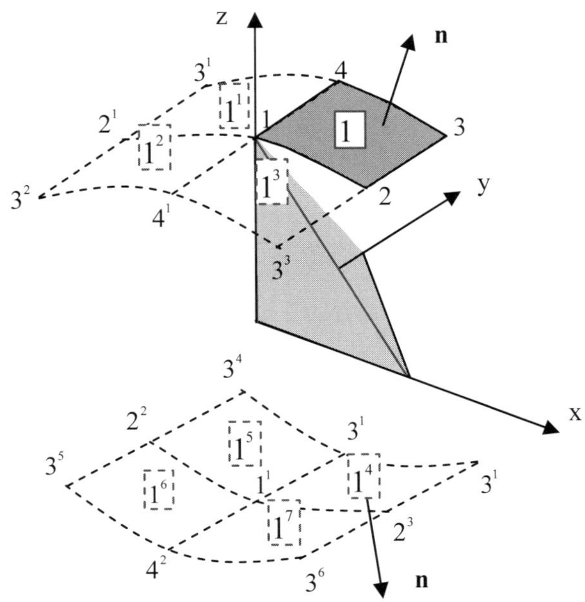

Figure 7.7 Three-dimensional BE mesh with 3 planes of symmetry

The method can be extended to three-dimensional problems. Up to three planes of symmetry are possible and an element has to be projected seven times. For the mesh in Figure 7.7, we determine the incidences for the mirrored elements keeping a consistent outward normal, as shown (anti-clockwise numbering of element 1).

Element	i	j	k	l
1	1	2	3	4
1^1	1	4	3^1	2^1
1^2	1	2^1	3^2	4^1
1^3	1	4^1	3^3	2
1^4	1^1	4^3	3^7	2^3
1^5	1^1	2^2	3^4	4^3
1^6	1^1	4^2	3^5	2^2
1^7	1^1	2^3	3^6	4^2

We note that for mirror image number n= 1,3,4 and 6, the incidences have to be reversed to maintain a consistent outward normal, as shown.

We now discuss the computer implementation of up to 3 symmetry planes. We specify a symmetry code

Symmetry code m	Symmetry about	No. of mirrored elements
1	y-z plane	1
2	y-z and x-z plane	3
3	All 3 planes	7

For the mirrored nodes we can compute coordinates **x**, displacements **u** and tractions **t** from the original nodes by

$$\mathbf{x}^n = \mathbf{T}^n\mathbf{x} \quad ; \quad \mathbf{u}^n = \mathbf{T}^n\mathbf{u} \quad ; \quad \mathbf{t}^n = \mathbf{T}^n\mathbf{t} \tag{7.15}$$

where superscript n denotes the mirror image number, as used in Figures 7.5 to 7.7.

Transformation matrices **T** are computed as follows

First we define three matrices $\mathbf{T}_m$

$$\mathbf{T}_1 = \begin{bmatrix} -1 & 0 & 0 \\ 0 & 1 & 0 \\ 0 & 0 & 1 \end{bmatrix} \quad ; \quad \mathbf{T}_2 = \begin{bmatrix} 1 & 0 & 0 \\ 0 & -1 & 0 \\ 0 & 0 & 1 \end{bmatrix} \quad ; \quad \mathbf{T}_3 = \begin{bmatrix} 1 & 0 & 0 \\ 0 & 1 & 0 \\ 0 & 0 & -1 \end{bmatrix} \tag{7.16}$$

In terms of these, the transformation matrices are defined as

$$\begin{aligned} &\mathbf{T}^1 = \mathbf{T}_1 \quad ; \quad \mathbf{T}^2 = \mathbf{T}_2 \quad ; \quad \mathbf{T}^3 = \mathbf{T}_1\mathbf{T}_2 \quad ; \\ &\mathbf{T}^4 = \mathbf{T}_3 \quad ; \quad \mathbf{T}^5 = \mathbf{T}_1\mathbf{T}_3 \quad ; \quad \mathbf{T}^6 = \mathbf{T}_2\mathbf{T}_3 \quad ; \quad \mathbf{T}^7 = \mathbf{T}_1\mathbf{T}_2\mathbf{T}_3 \end{aligned} \tag{7.17}$$

For implementation we provide an additional loop for each element which, depending on the symmetry code, is executed 1,2,4 or 8 times. For no symmetry (code 0) we consider only the original element. For code 1 (symmetry about y-z plane) we consider one mirrored image of the element. For symmetry codes 2 and 3, three and seven mirrored images of the element are considered.

7.2.2 Subroutine MIRROR

Subroutine MIRROR has been written to generate elements across symmetry planes. It returns the incidence, destination and coordinate vector of the mirrored element, as well as multiplication factors for the assembly. In the subroutine we assume that if points are on the symmetry plane, then they have exactly zero coordinate and one must ensure that this is actually the case. We note that for subroutines INTEG to work for the mirrored elements, we must change the mirrored node numbers to some arbitrary value. Here we have chosen to add the maximum node number to the incidences. Note that some numbers of the original node must not be changed if they lie on a specified symmetry plane. Table 7.3 gives an overview of the node number that must not be changed if it lies on a specified symmetry plane.

Table 7.3 List of node numbers that should not be changed during "mirroring"

Node is on	Mirror image (n) which must not be changed
x-z plane	3
y-z plane	1
x-y plane	4
2 planes	1 to 3

Destination vectors of the mirrored elements used for the assembly remain the same as for the original element, except that if incidences are reversed to maintain a consistent outward normal, then the destination vector must also be reversed. Subroutine Reverse is used and it has been put in the Utility Library. Note that in the implementation of the transformation matrices (equation 7.17), only the diagonal terms are considered.

```
SUBROUTINE Mirror(Isym,nsy,Nodes,Elcor,Fac,Incie,Ldeste &
                 Elres_te,Elres_ue,,Nodel,Ndof,Cdim)
!---------------------------------------------
!    Creates mirror image of an element
!---------------------------------------------
INTEGER, INTENT(IN)  ::  Isym        ! symmetry indicator
INTEGER, INTENT(IN)  ::  nsy         ! symmetry count
INTEGER, INTENT(IN)  ::  nodes       ! highest node no
REAL, INTENT(IN OUT)::   Elcor(:,:)  ! Coords (will be modified)
REAL, INTENT(OUT)    ::  Fac(:)      ! Multiplication factors
INTEGER, INTENT(IN OUT)::  Incie(:)! Incidences   (will be
INTEGER, INTENT(IN OUT):: Ldeste(:)! Destinations  modified)
REAL, INTENT(IN OUT)    ::  Elres_te(:) ! Element tractions
REAL, INTENT(IN OUT)    ::  Elres_ue(:) ! Element displacements
INTEGER, INTENT(IN)  ::  Nodel       ! Nodes per element
INTEGER, INTENT(IN)  ::  Ndof        ! d.o.F. per Node
INTEGER, INTENT(IN)  ::  Cdim        ! Cartesian dimension
REAL   :: TD(3)  ! Transformation vector (diagonal elements of T)
INTEGER :: n,m,Ison1,Ison2,Ison3,i
Fac(1:nodel*ndof)= 1.0
IF(nsy == 1) RETURN
SELECT CASE (nsy-1)
    CASE(1)
      TD=(/-1.0,1.0,1.0/)
    CASE(2)
      TD=(/-1.0,-1.0,1.0/)
```

```
    CASE(3)
      TD=(/1.0,-1.0,1.0/)
    CASE(4)
      TD=(/1.0,1.0,-1.0/)
    CASE(5)
      TD=(/-1.0,1.0,-1.0/)
    CASE(6)
      TD=(/1.0,-1.0,-1.0/)
    CASE(7)
      TD=(/-1.0,-1.0,-1.0/)
END SELECT
!     generate coordinates and incidences
Nodes0: &
DO n=1,nodel
 Elcor(:,n)= Elcor(:,n)*TD(m)
 !   Check if point is on any symmetry plane
 Ison1= 0 ; Ison2= 0 ; Ison3= 0
 IF(Elcor(1,n)==0.0) Ison1=1
 IF(Elcor(2,n)==0.0) Ison2=1
 IF(Cdim > 2 .AND. Elcor(3,n)==0.0) Ison3=1
 !   only change incidences for unprimed nodes
 IF(ison1==1 .AND. nsy-1 ==1) CYCLE
 IF(ison2==1 .AND. nsy-1 ==3) CYCLE
 IF(ison1+ison2+ison3 > 1 .AND. nsy-1<4) CYCLE
 Incie(n)= Incie(n) + Nodes
END DO &
Nodes0
!     generate multiplication factors, elast. Problems only
IF(Ndof > 1) THEN
 I=0
 Nodes1: &
 DO n=1,nodel
   Degrees_of_freedom: &
   DO m=1,Ndof
     I=I+1
     Fac(I)= TD(m)
   END DO &
   Degrees_of_freedom
 END DO &
 Nodes1
END IF
!   Reverse destination vector for selected elements
SELECT CASE (nsy-1)
 CASE (1,3,4,6)
  CALL &
 Reverse(Incie,elcor,ldeste,Elres_te,Elres_ue,Ndof,Cdim,nodel)
 CASE DEFAULT
END SELECT
RETURN
END SUBROUTINE Mirror
```

7.2.3 Subroutine Assembly

A sub-program for assembling the coefficient matrices using a vector of incidences or destinations, as well as information about the type of boundary and symmetry condition, is presented. The information about the boundary condition is supplied for each node or each degree of freedom of an element and the code is 0 for *Neuman* and 1 for the *Dirichlet* condition. Care has to be taken where the boundary condition is discontinuous. For example, in Figure 7.3, both temperature and flow values are known at the first node of element 1, but only temperature is known at the second node of element 4 (both nodes equal 1 in global numbering). For the assembly we must therefore specify a global code, in addition to a boundary code for each element. Then, if *Neuman* BC is specified and the global code is *Dirichlet*, both ΔT and ΔU are assembled on the right hand side. In the parameter list vectors, Elres_u and Elres_t are introduced. These will eventually contain all results of an element. At the stage when the SUBROUTINE is called, however, they contain only known (prescribed) values with all other values being zero. SUBROUTINE Assembly can be used for the assembly of two or three-dimensional problems in potential flow or elasticity. The incidence vector in potential flow problems and the destination vector in elasticity problems is defined as LDEST.

```
SUBROUTINE Assembly(Lhs,Rhs,DTe,DUe,Ldest,BCode,Ncode &
                  ,Elres_ue,Elres_te,Diag,Ndofe,Ndof,Nodel,Fac)
!---------------------------------------------
!  Assembles Element contributions DTe , DUe
!  into global matrix Lhs and vector Rhs
!  Also sums off-diagonal coefficients
!  for the computation of diagonal coefficients
!---------------------------------------------
REAL(KIND=8)              :: Lhs(:,:),Rhs(:)     !  Global arrays
REAL(KIND=8), INTENT(IN)::  DTe(:,:),DUe(:,:)    !  Element arrays
INTEGER , INTENT(IN)  :: LDest(:) ! Element destination vector
INTEGER , INTENT(IN)  :: BCode(:) ! Boundary code(local)
INTEGER , INTENT(IN)  :: NCode(:) ! Boundary code (global)
INTEGER , INTENT(IN)  :: Ndofe    ! D.o.F´s / Elem
INTEGER , INTENT(IN)  :: Ndof     ! D.o.F´s / Node
INTEGER , INTENT(IN)  :: Nodel      ! Nodes/Element
REAL , INTENT(IN)     :: Elres_ue(:) ! vector u for element
REAL , INTENT(IN)     :: Elres_te(:) ! vector t for element
REAL , INTENT(IN)     :: Fac(:)      ! Mult. factor for symmetry
REAL(KIND=8) :: Diag(:,:) ! Array containing diagonal coeff of DT
INTEGER :: n,Ncol
DoF_per_Element:&
DO m=1,Ndofe
  Ncol=Ldest(m)        !   Column number
  IF(BCode(m) == 0) THEN     !    Neumann BC
    Rhs(:) = Rhs(:) + DUe(:,m)*Elres_te(m)*Fac(m)
!    The assembly of dTe depends on the global BC
    IF (NCode(Ldest(m)) == 0) THEN
      Lhs(:,Ncol)=  Lhs(:,Ncol) + DTe(:,m)*Fac(m)
```

```
   END IF
   IF (NCode(Ldest(m)) == 1) THEN
     Rhs(:) = Rhs(:) - DTe(:,m) * Elres_ue(m)*Fac(m)
   END IF
  END IF
  IF(BCode(m) == 1) THEN    !   Dirichlet BC
     Lhs(:,Ncol) = Lhs(:,Ncol) - DUe(:,m)*Fac(m)
     Rhs(:)= Rhs(:) - DTe(:,m) * Elres_ue(m)*Fac(m)
  END IF
END DO &
DoF_per_Element
!      Sum of off-diagonal coefficients
DO n=1,Nodel
  DO k=1,Ndof
   l=(n-1)*Ndof+k
   Diag(:,k)= Diag(:,k) - DTe(:,l)*Fac(m)
  END DO
END DO
RETURN
END SUBROUTINE Assembly
```

Element contributions to the coefficient matrix, which have been computed numerically with SUBROUTINE Integ, have zero values for coefficients ΔT^{e}_{ni} when $g(n)$ is point I, because these coefficients are not computed using numerical integration. In the assembled matrix $[\Delta T]$, these coefficients correspond to diagonal elements. Equation (6.17) or (6.18) can be applied to compute these coefficients.

For example, the assembled diagonal coefficient ΔT_{ii} is given by

$$\Delta T_{ii} = -\sum_{e=1}^{E} \sum_{\substack{n=1 \\ g(n)\neq i}}^{N} \Delta T^{e}_{ni} + A \tag{7.18}$$

where A is the *azimuthal* integral (see section 6.3.2). The double sum is computed by SUBROUTINE Assembly and stored in an array Diag for later use.

7.3 SOLUTION OF SYSTEM OF EQUATIONS

After assembly and adding the *azimuthal* integral as required, a system of simultaneous equations is obtained. The difference to the system of equations obtained for the FEM is that it is not symmetric and fully populated. The non-symmetry of the coefficient matrix had engineers, who were used to symmetric stiffness matrices, baffled for a while. The question was why, since we have used the theorem by Betti, which Maxwell used to prove reciprocity and therefore the symmetry of the stiffness matrix, are we not getting a symmetric coefficient matrix? The answer lies in the fact that we are not solving the integral equations exactly but by numerical approximation. Instead of enforcing the Betti theorem at an infinite number of points, as we should, we select a limited number of points which are nodal points of the mesh. It can be shown that as the fineness of the mesh increases, the coefficient matrices become more and more symmetric. Indeed, in the limit with an infinite number of elements, full symmetry should be attained. Another fact that has been discovered is that if boundary elements with linear functions are used, then the non-symmetry is much less pronounced than if elements with quadratic variation are used. The reason for this is not quite clear.

The fact that coefficient matrices are fully populated makes things easier in the sense that we do not need to worry about sparse solvers at this stage. We will see later that when we introduce multiple regions, for example, to cater for non-homogenities or to model cracks or faults, we will also introduce sparseness.

The lack of sparseness of course means that no savings can be made by using special schemes, such as band or skyline storage. The number of degrees of freedom, however, should be considerably smaller, especially for soil or rock mechanics problems where the domain can be assumed to extend to infinity.

7.3.1 Gauss elimination

The Gauss elimination method is probably the oldest and most used for solving the system of equations. Consider the following system of equations

$$\mathbf{A}\mathbf{u} = \mathbf{F} \tag{7.19}$$

The solution for unknowns **u** involve two steps

STEP 1: Reduction

Here we introduce zeroes below the diagonal elements, so that we end up with an upper triangular coefficient matrix.

For example, consider the *n-th* and *i-th* equation of a system

$$\begin{aligned} a_{nn}u_n + \ldots + a_{nj}u_j + \ldots &= F_n \\ a_{in}u_n + \ldots + a_{ij}u_j + \ldots &= F_i \end{aligned} \tag{7.20}$$

To introduce a zero in the *n-th* column of the *i-th* equation, we subtract $(a_{in}/\, a_{nn})$ times the equation *n* from equation *i* :

$$\begin{aligned} a_{nn}u_n + \ldots + a_{nj}u_j + \ldots &= F_n \\ 0 \qquad + \ldots + a^*_{ij}u_j + \ldots &= F^*_i \end{aligned} \tag{7.21}$$

where

$$\begin{aligned} a^*_{ij} &= a_{ij} - \frac{a_{in}}{a_{nn}} a_{nj} \\ F^*_i &= F_i - \frac{a_{in}}{a_{nn}} F_n \end{aligned} \tag{7.22}$$

The procedure, which is sometimes referred to as *elimination of variable n,* can be visualised as a repeated modification of the coefficients a to a^*, sometimes referred to as *starring* operation. We continue doing the procedure for all the equations until all the coefficients of **A** below the diagonal are zero.

STEP 2: Backsubstitution

The results may now be obtained by computing the unknown from the last equation, which involves one unknown only. The formula for computing the n-th unknown is given by

$$u_n = -\frac{1}{a_{nn}} \sum_{i=n+1}^{N} a_{ni} u_i - F_n \tag{7.23}$$

The above procedure is easily converted to a subroutine. Subroutine SOLVE shown here assumes that coefficient matrix Lhs can be stored in memory.

```
SUBROUTINE Solve(Lhs,Rhs,u)
!---------------------------------------------
!    Solution of system of equations
!    by Gauss Elimination
!---------------------------------------------
REAL(KIND=8) :: Lhs(:,:)       !    Equation Left hand side
REAL(KIND=8) :: Rhs(:)         !    Equation right hand side
REAL(KIND=8) ::   u(:)         !    Unknown
INTEGER             M          !    Size of system
REAL(KIND=8) ::  FAC
M= UBOUND(Rhs,1)
!  Reduction
Equation_n: &
DO n=1,M-1
  IF(Lhs(n,n) < 1.0E-10 .AND. Lhs(n,n) > -1.0E-10) THEN
    CALL Error_Message('Singular Matrix')
  END IF
  Equation_i: &
  DO I= n + 1,M
     FAC= Lhs(i,n)/Lhs(n,n)
     Lhs(i, n+1 : M)= Lhs(i, n+1 : M) - Lhs(n, n+1 : M)*FAC
     Rhs(i)= Rhs(i) - Rhs(n)*FAC
  END DO  &
  Equation_i
END DO &
Equation_n
!    Backsubstitution
Unknown_n: &
DO n= M,1,-1
  u(n)= -1.0/Lhs(n,n)*(SUM(Lhs(n,n + 1:M)*u(n + 1:M)) - Rhs(n))
END DO Unknown_n
RETURN ;  END SUBROUTINE Solve
```

As already mentioned previously for the solution of equations involving many subtractions, it is necessary to use REAL (KIND=8) for the arrays, to avoid an accumulation of round-off error. For a 3-D elasticity problem involving 1000 nodes, the space required for storing the coefficient matrix in REAL (KIND=8) is 72 Mbytes. For the solution on small computers this space may not be available and special algorithms must be devised, where part of the matrix is written onto disk. Methods for the partitioned solution of large systems are presented, for example by Beer and Watson[2].

For the reduction of the system of equations we need three implied DO-loops. In the implementation the innermost DO-loop is written implicitly using the new feature available in FORTRAN 90. The innermost DO-loop involves one multiplication and one subtraction and is executed (M – n) times, where M is the number of unknowns. The DO loop above it involves a division and is also executed (M – n) times. Finally the outermost DO-loop is executed M – 1 times. It can be shown, therefore, that the total number of operations required is $2/3M^3 + ½M^2 + 1/6M$. For large systems the first term is dominant

This means that, for example, for a problem in three-dimensional elasticity involving 1000 nodes, approx. $2x10^{10}$ operations are necessary for the reduction. If we want to analyse these problems in a reasonable time there is clearly a need for more efficient solvers. Recently there has been a resurgence of iterative solvers[1]. The advantage of these solvers is that the number of operations and hence the solution time is only proportional to M^2 and that they can be adapted easily to run on parallel computers. This will be discussed in the next chapter.

7.3.2 Scaling

When we look at the fundamental solutions for elasticity we note that kernel **U** contains the modulus of elasticity whereas **T** does not. Depending on the chosen units used we expect a large difference in values. As we have seen at the beginning of this chapter, if there is a mixed boundary value problem then there is a mixture of **U** and **T** terms in the assembled coefficient matrix. This may cause problems in the solution of equations, since very small terms would be subtracted from very large ones. Additionally, we note that for 2-D problems kernel U varies with $ln(1/r)$ which gives $-\infty$ value as $r \to \infty$.

For the above reasons scaling of the data is recommended. Scaling is applied in such a way that all tractions are divided by E and all coordinates by the largest difference between coordinates (which results in a scaled problem size of unity).

7.4 PROGRAM 7.1: GENERAL PURPOSE PROGRAM, DIRECT METHOD, ONE REGION

We now have developed all necessary tools for writing a general purpose computer program for computing two and three-dimensional problems in potential flow and elasticity. The first part of the program reads input data. There are three types of data:

job specification, geometry and boundary data. They are read in by calling three separate subroutines **Jobin**, **Geomin** and **BCinput**. The job information consists of the Cartesian dimension of the problem (2-D or 3-D), type of region (finite or infinite), whether it is a potential or elasticity problem, type of elements used (linear or quadratic), properties, that is conductivity for potential problems and modulus of elasticity and Poisson's ratio for elasticity problems and number of elements/nodes. The geometrical information consists of the coordinates of nodes and element incidences. Finally the boundary conditions are input. In the program we assume that all nodes have *Neuman* boundary condition with zero prescribed value by default. All nodes with *Dirichlet* boundary conditions and all nodes having *Neumann* BC, with non-zero prescribed values have to be input. After the specification of the BC´s element, destination vectors can be set up by a call to Subroutine **Destination** contained in the utility library. As explained previously destinations are the addresses of the coefficients in the global arrays. Note that for symmetry it is of advantage to exclude those degrees of freedom which have zero value and a node destination vector (Ndest) has been included to consider this. As explained previously a global boundary code vector is needed to cater for the case where the boundary code is discontinuous at a node. Scaling, as described above, is applied by a call to **SUBROUTINE Scal**.

The assembly is made by calling **SUBROUTINE Assemb**. Since the diagonal coefficients are not computed using numerical integration but are determined using the 'rigid body mode' method, all off-diagonal coefficients are summed and, if the region is infinite, the azimuthal integral is added. Diagonal coefficients are stored in a vector **Diag** and the boundary condition codes will determine if these are assembled into the left or right hand side. The system of equations is solved next. Using the element destination vector, results **Elres_u** and **Elres_t** are gathered from global vector **u1**. As will seen later, it is convenient for postprocessing to store results element by element.

```
PROGRAM General_purpose_BEM
!---------------------------------------------------
!     General purpose BEM program
!     for solving elasticity and potential problems
!---------------------------------------------------
USE Utility_lib ; USE Elast_lib ; USE Laplace_lib
USE Integration_lib
IMPLICIT NONE
INTEGER, ALLOCATABLE :: Inci(:,:)   !  Element Incidences
INTEGER, ALLOCATABLE :: BCode(:,:), NCode(:) !  Element BC´s
INTEGER, ALLOCATABLE :: Ldest(:,:) !  Element dest. vector
INTEGER, ALLOCATABLE :: Ndest(:,:) !  Node destination vector
REAL, ALLOCATABLE :: Elres_u(:,:)  !  Results , u
REAL, ALLOCATABLE :: Elres_t(:,:)  !  Results , t
REAL, ALLOCATABLE :: Elcor(:,:)    !  Element coordinates
REAL, ALLOCATABLE :: xP(:,:)       !  Node co-ordinates
REAL(KIND=8), ALLOCATABLE :: dUe(:,:),dTe(:,:),Diag(:,:)
REAL(KIND=8), ALLOCATABLE :: Lhs(:,:),F(:)
REAL(KIND=8), ALLOCATABLE :: u1(:) !  global vector of unknown
CHARACTER (LEN=80) :: Title
INTEGER :: Cdim,Node,m,n,Istat,Nodel,Nel,Ndof,Toa
```

```
INTEGER :: Nreg,Ltyp,Nodes,Maxe,Ndofe,Ndofs,ndg,NE_u,NE_t
INTEGER :: nod,nd,i,j,k,l,DoF,Pos,Isym,nsym,nsy
REAL,ALLOCATABLE     :: Fac(:)      !  Factors for symmetry
REAL,ALLOCATABLE     :: Elres_te(:),Elres_ue(:)
INTEGER,ALLOCATABLE :: Incie(:)    !  Incidences 1 element
INTEGER,ALLOCATABLE :: Ldeste(:)   !  Destination vector
REAL :: Con,E,ny,Scat,Scad
!-----------------------------------------------------
!   Read job information
!-----------------------------------------------------
OPEN (UNIT=1,FILE='INPUT',FORM='FORMATTED') !   Input
OPEN (UNIT=2,FILE='OUTPUT',FORM='FORMATTED')!   Output
Call Jobin(Title,Cdim,Ndof,Toa,Nreg,Ltyp,Con,E,ny,&
      Isym,nodel,nodes,maxe)
Nsym= 2**Isym    !   number of symmetry loops
ALLOCATE(xP(Cdim,Nodes))    !  Array for node coordinates
ALLOCATE(Inci(Maxe,Nodel)) !  Array for incidences
CALL Geomin(Nodes,Maxe,xp,Inci,Nodel,Cdim)
Ndofe= Nodel*Ndof    !     Total degrees of freedom of element
ALLOCATE(BCode(Maxe,Ndofe))  !     Element Boundary codes
ALLOCATE(Elres_u(Maxe,Ndofe),Elres_t(Maxe,Ndofe))
CALL BCinput(Elres_u,Elres_t,Bcode,nodel,ndofe,ndof)
ALLOCATE(Ldest(maxe,Ndofe))  ! Elem. destination vector
ALLOCATE(Ndest(Nodes,Ndof))
!-------------------------------------------------------
!  Determine Node and Element destination vectors
!-------------------------------------------------------
CALL Destination(Isym,Ndest,Ldest,xP,&
            Inci,Ndofs,nodes,Ndof,Nodel,Maxe)
!------------------------------------------
!     Determine global Boundary code vector
!---------------------------------------------
ALLOCATE(NCode(Ndofs))
DoF_o_System: &
DO nd=1,Ndofs
 DO Nel=1,Maxe
  DO m=1,Ndofe
   IF (nd == Ldest(Nel,m) .and. NCode(nd) == 0) THEN
    NCode(nd)= NCode(nd)+BCode(Nel,m)
   END IF
  END DO
 END DO
END DO &
DoF_o_System
CALL Scal(E,xP(:,:),Elres_u(:,:),Elres_t(:,:),Cdim,Scad,Scat)
ALLOCATE(dTe(Ndofs,Ndofe),dUe(Ndofs,Ndofe)) ! Elem. coef.
ALLOCATE(Diag(Ndofs,Ndof))               ! Diagonal coefficients
ALLOCATE(Lhs(Ndofs,Ndofs),F(Ndofs),u1(Ndofs)) ! global arrays
ALLOCATE(Elcor(Cdim,Nodel))               !  Elem. Coordinates
ALLOCATE(Fac(Ndofe))                      !  Factor symmetry
ALLOCATE(Incie(Nodel))                    !  Element incidences
```

```
ALLOCATE(Ldeste(Ndofe))                    !  Element destination
ALLOCATE(Elres_te(Ndofe),Elres_ue(Ndofe))
!-----------------------------------------------------------------
!  Compute element coefficient matrices
!-----------------------------------------------------------------
Lhs(:,:) = 0.0; F(:) = 0.0; u1(:) = 0.0
Elements_1:&
DO Nel=1,Maxe
  Symmetry_loop:&
  DO nsy= 1,Nsym
    Elcor(:,:)= xP(:,Inci(Nel,:))   ! gather element coordinates
    Incie= Inci(nel,:)              ! incidences
    Ldeste= Ldest(nel,:)            ! and destinations
    Fac(1:nodel*ndof)= 1.0
    Elres_te(:)=Elres_t(Nel,:)
    IF(Isym > 0) THEN
      CALL Mirror(Isym,nsy,Nodes,Elcor,Fac,&
                 Incie,Ldeste,Elres_te,Elres_ue &
                 ,nodel,ndof,Cdim)
    END IF
    IF(Cdim == 2) THEN
      IF(Ndof == 1) THEN
        CALL Integ2P(Elcor,Incie,Nodel,Nodes&
                    ,xP,Con,dUe,dTe,Ndest,Isym)
      ELSE
        CALL Integ2E(Elcor,Incie,Nodel,Nodes&
                    ,xP,E,ny,dUe,dTe,Ndest,Isym)
      END IF
    ELSE
      CALL Integ3(Elcor,Incie,Nodel,Nodes,xP,Ndof &
               ,E,ny,Con,dUe,dTe,Ndest,Isym)
    END IF
    CALL Assembly(Lhs,F,DTe,DUe,Ldeste,BCode(Nel,:),Ncode &
               ,Elres_u(Nel,:),Elres_te,Diag&
              ,Ndofe,Ndof,Nodel,Fac)
  END DO &
  Symmetry_loop
END DO &
Elements_1
!-------------------------------------------------------------
!  Add azimuthal integral for infinite regions
!-------------------------------------------------------------
IF(Nreg == 2) THEN
  DO m=1, Nodes
    DO n=1, Ndof
      IF(Ndest(m,n) == 0)CYCLE
      k=Ndest(m,n)
      Diag(k,n) = Diag(k,n) + 1.0
    END DO
  END DO
END IF
```

```
!---------------------------------------------------------------
!  Add Diagonal coefficients
!---------------------------------------------------------------
Collocation_points: &
DO m=1,Ndofs
  Nod=0
  DO n=1, Nodes
    DO l=1,Ndof
      IF (m == Ndest(n,l))THEN
        Nod=n
        EXIT
      END IF
    END DO
    IF (Nod /= 0)EXIT
  END DO
  DO k=1,Ndof
    DoF=Ndest(Nod,k)
    IF(DoF /= 0) THEN
      IF(NCode(DoF) == 1) THEN
        Nel=0
        Pos=0
        DO i=1,Maxe
          DO j=1,Ndofe
            IF(DoF == Ldest(i,j))THEN
              Nel=i
              Pos=j
              EXIT
            END IF
          END DO
          IF(Nel /= 0)EXIT
        END DO
        F(m) = F(m) - Diag(m,k) * Elres_u(Nel,Pos)
      ELSE
        Lhs(m,DoF)= Lhs(m,DoF) + Diag(m,k)
      END IF
    END IF
  END DO
END DO &
Collocation_points
!-----------------------------------------------------------
!   Solve system of equations
!-----------------------------------------------------------
CALL Solve(Lhs,F,u1)
CLOSE(UNIT=2)
OPEN (UNIT=2,FILE='BERESULTS',FORM='FORMATTED')
!   Gather Element results from global result vector u1
Elements_2:&
DO nel=1,maxe
  D_o_F1:   &
  DO nd=1,Ndofe
    IF(Ldest(nel,nd) /= 0)THEN
```

```
      IF(NCode(Ldest(nel,nd)) == 0) THEN
        Elres_u(nel,nd) = Elres_u(nel,nd) + u1(Ldest(nel,nd))
      ELSE
        Elres_t(nel,nd) = Elres_t(nel,nd) + u1(Ldest(nel,nd))
      END IF
    END IF
  END DO &
  D_o_F1
  Elres_u(nel,:)= Elres_u(nel,:) * Scad
  Elres_t(nel,:)= Elres_t(nel,:) / Scat
  WRITE(2,'(24F12.5)') (Elres_u(nel,m), m=1,Ndofe)
  WRITE(2,'(24F12.5)') (Elres_t(nel,m), m=1,Ndofe)
END DO &
Elements_2
END PROGRAM
```

To make the program more readable and easier to modify, the reading of the input has been delegated to subroutines. This also gives the reader some freedom to determine the input FORMAT and implement simple mesh-generation facilities.

```
SUBROUTINE Jobin(Title,Cdim,Ndof,Toa,Nreg,Ltyp,Con,E,ny &
                ,Isym,nodel,nodes,maxe)
!-------------------------------------------------
!    Subroutine to read in basic job information
!-------------------------------------------------
CHARACTER(LEN=80), INTENT(OUT):: Title
INTEGER, INTENT(OUT) :: Cdim,Ndof,Toa,Nreg,Ltyp,Isym,nodel
INTEGER, INTENT(OUT) :: Nodes,Maxe
REAL, INTENT(OUT)    :: Con,E,ny
READ(1,'(A80)') Title
WRITE(2,*)'Project:',Title
READ(1,*) Cdim
WRITE(2,*)'Cartesian_dimension:',Cdim
READ(1,*) Ndof          !     Degrees of freedom per node
IF(NDof == 1) THEN
  WRITE(2,*)'Potential Problem'
ELSE
  WRITE(2,*)'Elasticity Problem'
END IF
IF(Ndof == 2)THEN
  READ(1,*) Toa ! Analysis type (plane strain= 1,plane stress= 2)
  IF(Toa == 1)THEN
    WRITE(2,*)'Type of Analysis: Solid Plane Strain'
  ELSE
    WRITE(2,*)'Type of Analysis: Solid Plane Stress'
  END IF
END IF
READ(1,*) Nreg          !    Type of region
IF(NReg == 1) THEN
  WRITE(2,*)'Finite Region'
```

```
ELSE
  WRITE(2,*)'Infinite Region'
END IF
READ(1,*) Isym        !   Symmetry code
SELECT CASE (isym)
      CASE(0)
        WRITE(2,*)'No symmetry'
      CASE(1)
        WRITE(2,*)'Symmetry about y-z plane'
      CASE(2)
        WRITE(2,*)'Symmetry about y-z and x-z planes'
      CASE(3)
        WRITE(2,*)'Symmetry about all planes'
END SELECT
READ(1,*) Ltyp          !   Element type
IF(Ltyp == 1) THEN
 WRITE(2,*)'Linear Elements'
ELSE
 WRITE(2,*)'Quadratic Elements'
END IF
!     Determine number of nodes per element
IF(Cdim == 2) THEN      !   Line elements
 IF(Ltyp == 1) THEN
  Nodel= 2
 ELSE
  Nodel= 3
 END IF
ELSE                    !   Surface elements
 IF(Ltyp == 1) THEN
   Nodel= 4
 ELSE
   Nodel= 8
 END IF
END IF
!   Read properties
IF(Ndof == 1) THEN
  READ(1,*) Con
  WRITE(2,*)'Conductivity=',Con
ELSE
  READ(1,*) E,ny
  IF(ToA == 2) ny = ny/(1+ny)      ! Solid Plane Stress
  WRITE(2,*)'Modulus:',E
  WRITE(2,*)'Poissons ratio:',ny
END IF
READ(1,*) Nodes
WRITE(2,*)'Number of Nodes of System:',Nodes
READ(1,*) Maxe
WRITE(2,*)'Number of Elements of System:', Maxe
RETURN
END SUBROUTINE Jobin
```

```
SUBROUTINE Geomin(Nodes,Maxe,xp,Inci,Nodel,Cdim)
!-------------------------------------
!   Inputs mesh geometry
!-------------------------------------
INTEGER, INTENT(IN) ::Nodes     !   Number of nodes
INTEGER, INTENT(IN) ::Maxe     !   Number of elements
INTEGER, INTENT(IN) ::Nodel     !   Number of Nodes of elements
INTEGER, INTENT(IN) ::Cdim     !   Cartesian Dimension
REAL, INTENT(OUT)   ::xP(:,:)   !   Node co-ordinates
REAL            :: xmax(Cdim),xmin(Cdim),delta_x(Cdim)
INTEGER, INTENT(OUT)::Inci(:,:) !   Element incidences
INTEGER             :: Node,Nel,M,n
!-------------------------------------------------------
!  Read Node Co-ordinates from Inputfile
!-------------------------------------------------------
DO Node=1,Nodes
 READ(1,*) (xP(M,Node),M=1,Cdim)
 WRITE(2,'(A5,I5,A8,3F8.2)') 'Node ',Node,&
          '  Coor  ', (xP(M,Node),M=1,Cdim)
END DO
!-------------------------------------------------------
!  Read Incidences from Inputfile
!-------------------------------------------------------
WRITE(2,*)''
WRITE(2,*)'Incidences: '
WRITE(2,*)''
Elements_1:&
DO Nel=1,Maxe
 READ(1,*) (Inci(Nel,n),n=1,Nodel)
 WRITE(2,'(A3,I5,A8,24I5)')'EL ',Nel,'  Inci  ',Inci(Nel,:)
END DO &
Elements_1
RETURN
END SUBROUTINE Geomin
```

```
SUBROUTINE BCInput(Elres_u,Elres_t,Bcode,nodel,ndofe,ndof)
!-------------------------------------------
!    Reads boundary conditions
!-------------------------------------------
REAL,INTENT(OUT)      :: Elres_u(:,:)   !  Element results , u
REAL,INTENT(OUT)      :: Elres_t(:,:)   !  Element results , t
INTEGER,INTENT(OUT) :: BCode(:,:)     !  Element BC´s
INTEGER,INTENT(IN)  :: nodel          !  Nodes per element
INTEGER,INTENT(IN)  :: ndofe          !  D.o.F. per Element
INTEGER,INTENT(IN)  :: ndof           !  D.o.F per Node
INTEGER :: NE_u,NE_t
WRITE(2,*)''
WRITE(2,*)'Elements with Dirichlet BC´s: '
WRITE(2,*)''
Elres_u(:,:)=0  ! Default prescribed values for u = 0.0
```

```
BCode = 0        ! Default BC= Neumann Condition
READ(1,*)NE_u
IF(NE_u > 0) THEN
  Elem_presc_displ: &
  DO n=1,NE_u
   READ(1,*) Nel,(Elres_u(Nel,m),m=1,Ndofe)
   BCode(Nel,:)=1
   WRITE(2,*)'Element ',Nel,'  Prescribed values: '
   Na= 1
   Nodes: &
   DO M= 1,Nodel
     WRITE(2,*) Elres_u(Nel,na:na+ndof-1)
     Na= na+Ndof
   END DO &
   Nodes
  END DO &
  Elem_presc_displ
END IF
WRITE(2,*)''
WRITE(2,*)'Elements with Neuman BC´s: '
WRITE(2,*)''
Elres_t(:,:)=0   !   Default prescribed values = 0.0
READ(1,*)NE_t
Elem_presc_trac:  &
DO n=1,NE_t
  READ(1,*) Nel,(Elres_t(Nel,m),m=1,Ndofe)
  WRITE(2,*)'Element ',Nel,'  Prescribed values: '
  Na= 1
  Nodes1: &
  DO M= 1,Nodel
    WRITE(2,*) Elres_t(Nel,na:na+ndof-1)
    Na= na+Ndof
  END DO &
  Nodes1
END DO &
Elem_presc_trac
RETURN
END SUBROUTINE BCInput
```

7.4.1 User's manual

The input data which have to be supplied in file INPUT are described below. Free field input is used, that is, numbers are separated by blanks. However, all numbers, including zero entries must be specified.

The input is divided into two parts. First, general information about the problem is read in, then the mesh geometry is specified. The problem may consist of linear and quadratic elements, as shown in Figure 7.8. The sequence in which node numbers have to be entered when specifying incidences is also shown. Note that this order determines the direction of the outward normal, which has to point away from the material. For 3-D elements, if node numbers are entered in an anticlockwise direction, the outward normal points towards the viewer.

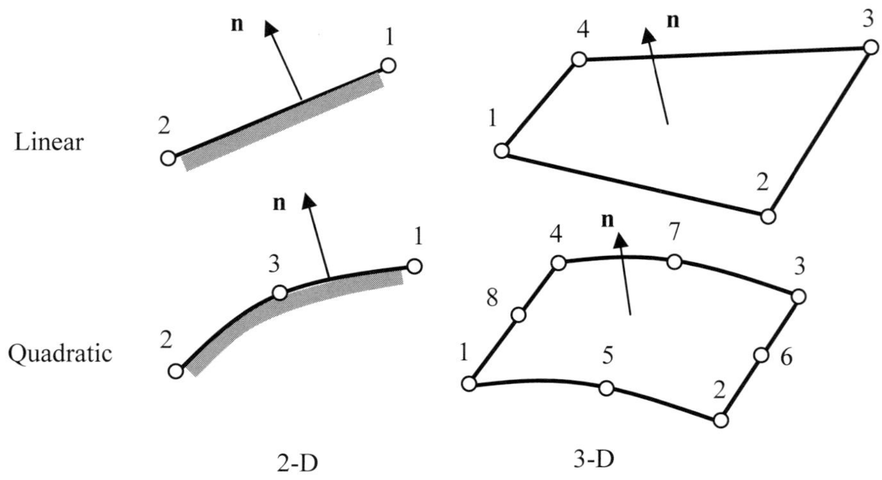

Figure 7.8 Element library

INPUT DATA SPECIFICATION FOR General_purpose-BEM program

1.0	Title specification **TITLE**	 Project Title (max 60 characters)
2.0	Cartesian dimension of problem **Cdim**	 Cartesian dimension 2= two-dimensional problem 3= three-dimensional problem
3.0	Problem type specification **Ndof**	 Degree of freedom per node 1= potential problem 2,3= elasticity problem
4.0	Analysis type (Only input for Ndof= 2 !!) **Toa**	 Type of analysis 1= Plane strain 2= plane stress
5.0	Region type specification **Nreg**	 Region code 1= finite region 2= infinite region

6.0 Symmetry specification	
ISym	Symmetry code 0= no symmetry 1= symmetry about y-z plane 2= symmetry about y-z and x-z planes 3= symmetry about all 3 planes
7.0 Element type specification	
Ltyp	Element type 1= linear 2= quadratic
8.0 Material properties	
C1, C2	Material properties **C**1= k (conductivity) for Ndof=1 = E (Elastic Modulus) for Ndof=2,3 **C**2= Poisson´s ratio for Ndof=2,3
9.0 Node specification	
Nodes	Number of nodes
10.0Element specification	
Maxe	Number of elements
11.0 Loop over nodes	
x, y, (z)	Node coordinates
12.0 Loop over all elements	
Inci (1:Element nodes)	Global node numbers of element nodes
13.0 *Dirichlet* boundary conditions	
NE_u Number of elements with *Dirichlet* BC	
14.0 Prescribed values for *Dirichlet* BC for NE_u elements	
Nel, Elres_u(1 : Element D.o.F.)	Specification of boundary condition **Nel** = Element number to be assigned BC **Elres_u** = Prescribed values for all degrees of Freedom of element: all d.o.F first node; all d.o.F second node etc.
15.0 *Neumant* boundary conditions	
NE_t	Number of elements with *Neuman* BC Only specify for no-zero prescribed values.
16.0 Prescribed values for *Neuman* BC for NE_t elements	
Nel, Elres_t(1 : Element D.o.F.)	Specification of boundary condition **Nel** = Element number to be assigned BC **Elres_t** = Prescribed values for all degrees of Freedom of element: all d.o.F first node; all d.o.F second node etc.

7.4.2 Sample input file

For the example of the heat flow past a cylindrical isolator, which was solved with the Trefftz method and the direct method with constant elements, we present the input file for an analysis with 8 linear elements and no symmetry (Figure 7.9).

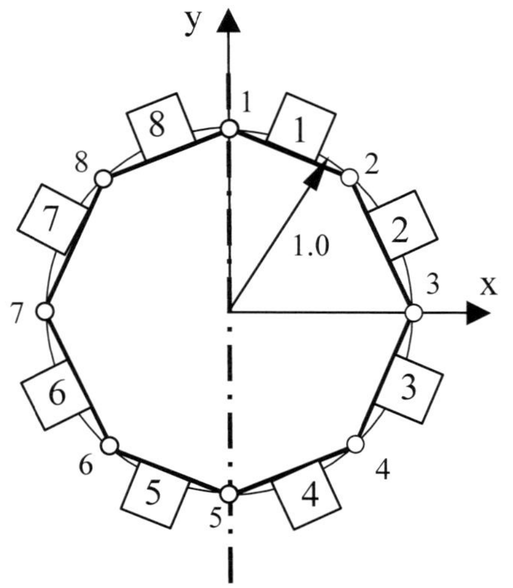

Figure 7.9 Discretisation of cylindrical isolator used for sample inputfile

File INPUT

```
Flow past cylindrical isolator, 8 linear elements
2     ! Cdim , 2-D problem
1     ! Ndof , potential problem
2     ! Nreg , Infinite region
0     ! ISym , no symmetry
1     ! Ltyp , linear element
1.00 ! C1 , Conductivity
 8    ! Nodes
 8    ! Number of Elements
0.0 1.000    !    Coordinates
0.707 0.707
1.0 0.0
0.707 -0.707
0.0 -1.0
-0.707 -0.707
-1.0 0.0
-0.707 0.707
 1 2              !     Incidences
 2 3
```

```
 3 4
 4 5
 5 6
 6 7
 7 8
 8 1
 0              !  no Dirichlet BC' s
 8              !  Neuman BC' c
 1  1.000  0.707
 2  0.707  0.000
 3  0.000 -0.707
 4 -0.707 -1.000
 5 -1.000 -0.707
 6 -0.707  0.000
 7  0.000  0.707
 8  0.707  1.000
```

7.5 CONCLUSIONS

In this chapter we have developed a general purpose program, which can be used to solve any problem in elasticity and potential flow, or if we substitute the appropriate fundamental solutions, any problem at all. This versatility has been made possible through the use of isoparametric elements and numerical integration. In essence, the boundary element method has borrowed here ideas from the finite element method and, in particular, the ideas of Ergatoudis, who first suggested the use of parametric elements and numerical integration.

Indeed, there are also other similarities with the FEM in that the system of equations is obtained by assembling element contributions. In the assembly procedure we have found that the treatment of discontinous boundary conditions, as they are encountered often in practical applications, needs special attention and will change the assembly process.

The implementation of the program is far from efficient. If one does an analysis of runtime spent in each part of the program, one will realise that the computation of the element coefficient matrices will take a significant amount of time. This is because, as pointed out in Chapter 6, the order of DO loops in the numerical integration is not optimised to reduce the number of calculations. Also in the implementation, all matrices must be stored in RAM, and this may severely restrict the size of problems which can be solved.

We have noted that the system of equations obtained is fully populated, that is, the coefficient matrix contains no zero elements. This is in contrast to the FEM, where systems are sparsely populated, i.e., containing a large number of zeroes. The other difference with the FEM is that the stiffness matrix is not symmetric. This has been claimed as one of the disadvantages of the method. However, this is more than compensated by the fact that the size of the system is significantly smaller.

The output from the program consists only of the values of the unknown at the boundary. The unknown are either the temperature/displacement or the flow normal to the boundary/boundary stresses. The computation of the complete flow vectors/stress tensor at the boundary, as well as the computation of values inside the domain is discussed in Chapter 9.

7.6 EXERCISES

Exercise 7.1
Using Program 7.1 compute the problem of flow past a cylindrical isolator, find out the influence of the following on the accuracy of results:

(a) when linear and quadratic boundary elements are used.
(b) when the number of elements is 8,16 and 32.

Plot the error in the computation of maximum temperature against number of elements.

Exercise 7.2
Modify the problem computed in Exercise 7.1 by changing the shape of the isolator, so that it has an elliptical shape, with a ratio vertical to horizontal axis of 2.0. Comment on the changes in the boundary values due to the change in shape.

Exercise 7.3
Using Program 7.1, compute the problem of a circular excavation in a plane strain infinite pre-stressed domain Figure 7.10, find out the influence of the following on the accuracy of results:

(a) when linear and quadratic boundary elements are used.
(b) when the number of elements is 8,16 and 32.

Plot the error against the number of elements.

Hint: *This is the elasticity problem equivalent to the heat flow problem in Exercise 7.1. The problem is divided into two:*

1. Continuum with no hole and the initial stresses only
2. Continuum with a hole and Neuman boundary conditions. The boundary conditions are computed in such a way that when stresses at the boundary of problem 1 are added to the ones at problem 2, zero values of boundary tractions are obtained. To compute the boundary tractions equivalent to the initial stresses use equation (4.28).

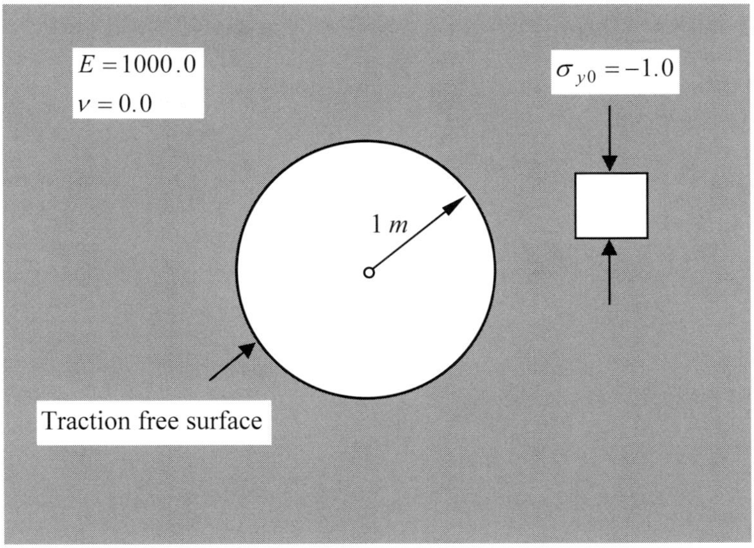

Figure 7.10 Circular excavation in an infinite domain

Exercise 7.4
Modify the problem computed in Exercise 7.3 by changing the shape of the excavation, so that it has an elliptical shape with a ratio vertical to horizontal axis of 2.0. Comment on the changes in the deformations due to the change in shape.

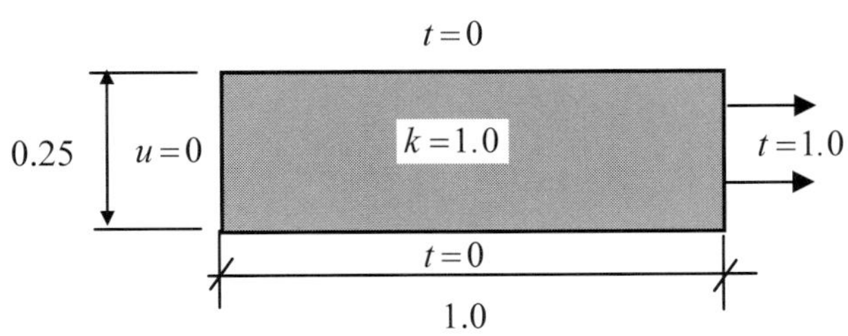

Figure 7.11 Potential problem with boundary conditions

Exercise 7.5
Using program 7.1, compute the potential problem of the beam depicted in Figure 7.12. Assume k=1.0 and a prescribed temperature of 0.0 at the left end and a prescribed flux of 1.0 at the right end. Construct two meshes, one with linear and one with quadratic boundary elements. Comment on the results.

Exercise 7.6
Using program 7.1, analyse the problem of the cantilever beam depicted in Figure 7.12. Plot the displaced shape and distribution of the normal and shear tractions at the fixed end. Construct two meshes, one with linear and the other with quadratic boundary elements. Comment on the results.

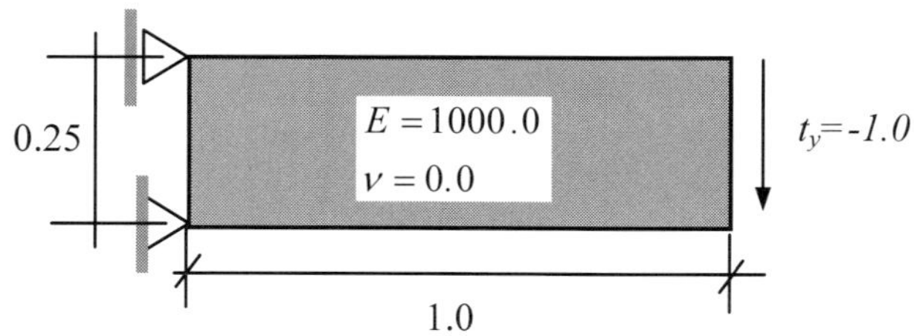

Figure 7.12 Example of cantilever beam

Exercise 7.7
Using program 7.1, compute the problem of the cantilever beam depicted in Figure 7.12. but apply a vertical movement of unity to the top support instead of traction at the free end. Plot the displaced shape and verify that this is just a rigid body rotation of the beam.

7.7 REFERENCES

1. Smith I M and Griffiths D.V. (2004) Programming the Finite Element Method. J.Wiley
2. Beer G and Watson J.O. (1992) Introduction to Finite and Boundary Elements for Engineers. J.Wiley.

8

Element-by-element techniques and Parallel Programming

I am a little world
made cunningly of elements
Donne

8.1 INTRODUCTION

In the previous Chapter we considered "traditional" techniques of assembly and solution involving element matrix assembly (additive) followed by Gaussian elimination performed on the resulting non-symmetric, fully-populated, linear equation system. We noted that computer storage requirements for the element matrix coefficients become demanding, as do processing requirements, for large numbers of elements particularly in three dimensions.

Typical single processor storage capacity, at the time of writing, is about 2Gb or roughly 200 million 64-bit locations. Therefore, the number of assembled boundary element equations that can be handled by one processor is approximately 14,000, implying a 3-D model with less than about 5000 nodes.

Consider a cubical cavity (cavern) in an infinite elastic medium with each of its 6 faces meshed by n*n boundary elements. For linear elements, the number of nodes (equations) is close to 6n**2 (18n**2) and for quadratic elements 18n**2 (54n**2), so a linear element mesh would be restricted to about 30*30 elements per face and a quadratic one to about 16*16, if no symmetries can be exploited.

8.1 THE ELEMENT-BY-ELEMENT CONCEPT

This arose in finite element work, probably first in "explicit" time marching analyses, where a solution, say **u**, Δt units of time after a previous one, say **v**, can simply be obtained by a matrix*vector multiplication of the form

$$\mathbf{u} = (\mathbf{M} - \Delta t\mathbf{K})\mathbf{v} = \mathbf{A}\cdot\mathbf{v} \tag{8.1}$$

where **M** and **K** are the system "mass" and "stiffness" matrices respectively. The above product $\mathbf{A}\cdot\mathbf{v}$ can be carried out "piece-by-piece", as long as the sum of the "pieces" adds up to **A**. For example

$$\mathbf{A} = \begin{pmatrix} 1 & 2 \\ 3 & 4 \end{pmatrix} = \begin{pmatrix} 1 & 0 \\ 3 & 0 \end{pmatrix} + \begin{pmatrix} 0 & 2 \\ 0 & 4 \end{pmatrix} = \begin{pmatrix} .5 & 1 \\ 2 & 3 \end{pmatrix} + \begin{pmatrix} .5 & 1 \\ 1 & 1 \end{pmatrix} \tag{8.2}$$

or any other suitable partitioning. Then taking, for example, the last partitioning

$$\mathbf{A}\cdot\mathbf{v} = \begin{pmatrix} 1 & 2 \\ 3 & 4 \end{pmatrix}\begin{pmatrix} v_1 \\ v_2 \end{pmatrix} = \begin{pmatrix} .5 & 1 \\ 2 & 3 \end{pmatrix}\begin{pmatrix} v_1 \\ v_2 \end{pmatrix} + \begin{pmatrix} .5 & 1 \\ 1 & 1 \end{pmatrix}\begin{pmatrix} v_1 \\ v_2 \end{pmatrix} = \begin{pmatrix} v_1 & 2v_2 \\ 3v_1 & 4v_2 \end{pmatrix} \tag{8.3}$$

gives the same result as without partitioning. In boundary element or finite element work, element assembly involves just such a partitioning where the appropriate "pieces" of **A** are the element matrices themselves. So whenever a matrix*vector multiplication is needed a global **A** need never be assembled at all and instead the product computed as:

$$\mathbf{u} = \sum \mathbf{A}^e \cdot \mathbf{v}^e \tag{8.4}$$

where $\mathbf{A}^e$ is the element matrix and $\mathbf{v}^e$ is the appropriate part of **v**, gathered for element e as described in Section 7.2. In essence the idea is to replace the double sum in Equation 7.1 by a single sum and a matrix*vector multiplication.

$$c\mathbf{u} + \sum_{e=1}^{E} \Delta\mathbf{T}^e\mathbf{u}^e = \sum_{e=1}^{E} \Delta\mathbf{U}^e\mathbf{t}^e \tag{8.5}$$

To extend this idea to solving sets of linear equations we have to look for a solution technique at the heart of which is a matrix*vector product like equation (8.1). Fortunately a whole class of iterative methods for equation solution is of this type. For example there are the "gradient" methods for symmetric systems, typified by the preconditioned conjugate gradient method (PCG) or the generalised minimum residual methods for non-symmetric systems, for example GMRES, which are appropriate for boundary element equations.

Following Chapter 7 we can write the final system of equations in the following form

$$\{\mathbf{R}\}_0 = \{\mathbf{F}\} - [\mathbf{K}_m] \cdot \{\mathbf{u}\}_0 \tag{8.6}$$

Where $\{\mathbf{R}\}_0$ is the "residual" or error for a first trial solution of $\{\mathbf{u}\}$ namely $\{\mathbf{u}\}_0$. In elasticity problems with only Neumann boundary conditions, for example, $[\mathbf{K}_m]$ would be the assembled matrix $[\Delta\mathbf{T}]$ and $\{\mathbf{u}\}$ a vector of displacements. However, we note that in an element by element (EBE) iterative solution the system of equations (8.6) need never be actually assembled.

Set $\{\mathbf{P}\}_0 = \{\mathbf{R}\}_0$; $\{\mathbf{u}\}_0 = \{0\}$ or some intial estimate

choose $\{\hat{\mathbf{R}}_0\}$ such that $\{\mathbf{R}_0\}^T \{\hat{\mathbf{R}}_0\} \neq 0$

FOR $k = 1,2,3....niter$ **DO**

$$\{\mathbf{Q}\}_{k-1} = [\mathbf{K}_m]\{\mathbf{P}\}_{k-1}$$

$$\{\mathbf{u}\}_{k-1/2} = \{\mathbf{u}\}_{k-1} + \alpha_{k-1}\{\mathbf{P}\}_{k-1} \quad ; \quad \alpha_{k-1} = \frac{\{\mathbf{R}\}_{k-1}^T \{\hat{\mathbf{R}}_0\}}{\{\mathbf{Q}\}_{k-1}^T \{\hat{\mathbf{R}}_0\}}$$

$$\{\mathbf{R}\}_{k-1/2} = \{\mathbf{R}\}_{k-1} + \alpha_{k-1}\{\mathbf{Q}\}_{k-1}$$

$$\{\mathbf{S}\}_{k-1} = [\mathbf{K}_m]\{\mathbf{R}\}_{k-1/2}$$

$$\{\mathbf{u}\}_k = \{\mathbf{u}\}_{k-1/2} + \omega_k\{\mathbf{R}\}_{k-1/2} \quad ; \quad \omega_{k-1} = \frac{\{\mathbf{R}\}_{k-1/2}^T \{\mathbf{S}\}_{k-1/2}}{\{\mathbf{S}\}_{k-1/2}^T \{\mathbf{S}\}_{k-1/2}}$$

$$\{\mathbf{R}\}_k = \{\mathbf{R}\}_{k-1/2} - \omega_k\{\mathbf{S}\}_{k-1/2}$$

$$\{\mathbf{P}\}_k = \{\mathbf{R}\}_k + \beta_k\{\mathbf{P}\}_{k-1} - \beta_k\omega_k\{\mathbf{Q}\}_{k-1} \quad ; \beta_k = \frac{\alpha_{k-1}\{\mathbf{R}\}_k^T \{\hat{\mathbf{R}}_0\}}{\omega_k\{\mathbf{R}\}_{k-1}^T \{\hat{\mathbf{R}}_0\}}$$

IF (converged) **EXIT**

END DO

Figure 8.1 Pseudo-code for BiCGStab

For the BEM we could choose any of the GMRES-type class of solution techniques. In particular, we select the BiCGStab algorithm, which has been shown to be effective in Finite Element work[1]. It follows the two-stage ("Bi") procedure shown in Figure 8.1 being dominated by two matrix*vector products such as

$$\{\mathbf{Q}\}_{k-1} = [\mathbf{K}_m]\{\mathbf{P}\}_{k-1} \tag{8.7}$$

carried out on an element by element basis. All other operations involve vector dot-products.

8.1.1 Element-by-element storage requirements

We are not now interested in storing the fully assembled Ndofs*Ndofs system of Program 7.1, but rather the element level arrays dTe and dUe which are of size Ndofs*Ndofe (assuming both have to be stored). In fact our storage requirements will be somewhat greater than for the assembled system, but of course we look forward to employing a parallel environment in which this storage will be distributed across the number of parallel processors available, npes.

Returning to our cubical "cavern" mesh, for linear elements the dUe (and dTe) boundary element coefficient matrices will need 12*18*n**2/npes locations (24*54*n**2/npes for quadratic elements). In this way we can solve much larger problems given that a sufficient number of processors is available. A very significant additional advantage when we come to parallel processing is that the time-consuming computation of dUe and dTe will also take place in parallel. Since this part of the computation involves no communication between processors it is an example of "perfectly" parallelisable code and with 1000 processors we shall compute the element matrix coefficients 1000 times faster than in serial mode.

Before going on to parallel processing, we go through two intermediate stages. Starting from Program 7.1 we first make the (very small) alterations so that we retain traditional assembly, but solve the resulting equations iteratively using the BiCGStab(l) algorithm (Program 8.1). Then we illustrate the change to an element-by-element iterative solution strategy (Program 8.2) and finally progress to the fully parallelised version (Program 8.3).

Example analyses illustrate the efficiency of parallelism in terms of processing speed and problems involving up to 60,000 boundary elements are solved.

8.2 PROGRAM 8.1 : REPLACING DIRECT BY ITERATIVE SOLUTION

```
PROGRAM General_purpose_BEM
!---------------------------------------------------------
!    General purpose BEM program
!    for solving elasticity and potential problems
!    This version iterative equation solution by BiCGStab(l)
!---------------------------------------------------------
USE bem_lib        !  contains precision
IMPLICIT NONE      !  Ndof changed to N_dof
INTEGER, ALLOCATABLE :: Inci(:,:)  !  Element Incidences
INTEGER, ALLOCATABLE :: BCode(:,:), NCode(:) !  Element BC´s
INTEGER, ALLOCATABLE :: Ldest(:,:) !  Element destination vector
INTEGER, ALLOCATABLE :: Ndest(:,:) !  Node destination vector
```

```
REAL(iwp), ALLOCATABLE :: Elres_u(:,:)  !  Element results , u
REAL(iwp), ALLOCATABLE :: Elres_t(:,:)  !  Element results , t
REAL(iwp), ALLOCATABLE :: Elcor(:,:)    !  Element coordinates
REAL(iwp), ALLOCATABLE :: xP(:,:)       !  Node co-ordinates
REAL(iwp), ALLOCATABLE :: dUe(:,:),dTe(:,:),Diag(:,:)
REAL(iwp), ALLOCATABLE :: Lhs(:,:),F(:)
REAL(iwp), ALLOCATABLE :: u1(:)     !  global vector of unknowns
CHARACTER (LEN=80) :: Title
INTEGER :: Cdim,Node,m,n,Istat,Nodel,Nel,N_dof,Toa
INTEGER :: Nreg,Ltyp,Nodes,Maxe,Ndofe,Ndofs,ndg,NE_u,NE_t
INTEGER :: nod,nd,i,j,k,l,DoF,Pos,Isym,nsym,nsy,its,ell
REAL(iwp),ALLOCATABLE     :: Fac(:)     !  Factors for symmetry
REAL(iwp),ALLOCATABLE     :: Elres_te(:),Elres_ue(:)
INTEGER,ALLOCATABLE :: Incie(:)   !  Incidences for one element
INTEGER,ALLOCATABLE :: Ldeste(:)  !  Destination vector 1 elem
REAL(iwp) :: Con,E,ny,Scat,Scad,tol,kappa
!--------------------------------------------------------
!   Read job information
!--------------------------------------------------------
OPEN (UNIT=11,FILE='prog81.dat',FORM='FORMATTED') !  Input
OPEN (UNIT=12,FILE='prog81.res',FORM='FORMATTED') !  Output
Call Jobin(Title,Cdim,N_dof,Toa,Nreg,Ltyp,Con,E,ny,&
           Isym,nodel,nodes,maxe)
Nsym= 2**Isym   !   number of symmetry loops
ALLOCATE(xP(Cdim,Nodes))   !  Array for node coordinates
ALLOCATE(Inci(Maxe,Nodel)) !  Array for incidences
CALL Geomin(Nodes,Maxe,xp,Inci,Nodel,Cdim)
Ndofe= Nodel*N_dof   !    Total degrees of freedom of element
ALLOCATE(BCode(Maxe,Ndofe))       !    Element Boundary codes
ALLOCATE(Elres_u(Maxe,Ndofe),Elres_t(Maxe,Ndofe))
CALL BCinput(Elres_u,Elres_t,Bcode,nodel,ndofe,n_dof)
READ(11,*) tol,its,ell,kappa    ! data for bicgstab(l)
ALLOCATE(Ldest(maxe,Ndofe))  ! Elem. destination vector
ALLOCATE(Ndest(Nodes,N_dof))
!-------------------------------------------------------------
!     Determine Node destination vector and Element dest vector
!-------------------------------------------------------------
CALL Destination(Isym,Ndest,Ldest,xP,Inci,Ndofs,nodes,&
N_dof,Nodel,Maxe)
!----------------------------------------------
!     Determine global Boundary code vector
!----------------------------------------------
ALLOCATE(NCode(Ndofs))
NCode=0
DoF_o_System: &
DO  nd=1,Ndofs
     DO Nel=1,Maxe
         DO m=1,Ndofe
            IF (nd == Ldest(Nel,m) .and. NCode(nd) == 0) THEN
               NCode(nd)= NCode(nd)+BCode(Nel,m)
            END IF
```

```
         END DO
     END DO
END DO &
DoF_o_System
IF(N_dof ==1)E= Con
CALL Scal(E,xP(:,:),Elres_u(:,:),Elres_t(:,:),Cdim,Scad,Scat)
ALLOCATE(dTe(Ndofs,Ndofe),dUe(Ndofs,Ndofe))! Elem. coef. matrices
ALLOCATE(Diag(Ndofs,N_dof))                 ! Diagonal coefficients
ALLOCATE(Lhs(Ndofs,Ndofs),F(Ndofs),u1(Ndofs))    ! global arrays
ALLOCATE(Elcor(Cdim,Nodel))                 !  Elem. Coordinates
ALLOCATE(Fac(Ndofe))                  !  Factor for symmetric
elements
ALLOCATE(Incie(Nodel))                   !  Element incidences
ALLOCATE(Ldeste(Ndofe))                  !  Element destination
ALLOCATE(Elres_te(Ndofe),Elres_ue(Ndofe))    ! Tractions of
Element
!-----------------------------------------------------------------
!  Compute element coefficient matrices
!-----------------------------------------------------------------
Lhs(:,:) = 0.0_iwp; F(:) = 0.0_iwp; u1(:) = 0.0_iwp; Diag(:,:) =
0.0_iwp
Elements_1:&
DO Nel=1,Maxe
        Symmetry_loop:&
        DO nsy= 1,Nsym
           Elcor(:,:)= xP(:,Inci(Nel,:))!gather element coord
           Incie= Inci(nel,:)                  !   incidences
           Ldeste= Ldest(nel,:)                !   and destinations
           Fac(1:nodel*n_dof)= 1.0_iwp
           Elres_te(:)=Elres_t(Nel,:)
           IF(Isym > 0) THEN
               CALL Mirror(Isym,nsy,Nodes,Elcor,Fac,    &
           Incie,Ldeste,Elres_te,Elres_ue,nodel,n_dof,Cdim)
           END IF
           IF(Cdim == 2) THEN
             IF(N_dof == 1) THEN
                  CALL Integ2P(Elcor,Incie,Nodel,Nodes, &
                         xP,Con,dUe,dTe,Ndest,Isym)
             ELSE
                  CALL Integ2E(Elcor,Incie,Nodel,Nodes, &
                         xP,E,ny,dUe,dTe,Ndest,Isym)
             END IF
            ELSE
                CALL Integ3(Elcor,Incie,Nodel,Nodes,xP,N_dof &
                                 ,E,ny,Con,dUe,dTe,Ndest,Isym)
            END IF
            CALL Assembly(Lhs,F,DTe,DUe,Ldeste,BCode(Nel,:),&
        Ncode,Elres_u(Nel,:),Elres_te,Diag,Ndofe,N_dof,Nodel,Fac)
        END DO &
        Symmetry_loop
END DO &
```

```
Elements_1
!-------------------------------------------------------------
!  Add azimuthal integral for infinite regions
!-------------------------------------------------------------
IF(Nreg == 2) THEN
        DO m=1, Nodes
           DO n=1, N_dof
              IF(Ndest(m,n) == 0)CYCLE
              k=Ndest(m,n)
              Diag(k,n) = Diag(k,n) + 1.0_iwp
           END DO
        END DO
END IF
!-------------------------------------------------------------
!  Add Diagonal coefficients
!-------------------------------------------------------------
DO m=1,Ndofs              ! Loop over collocation points
   Nod=0
   DO n=1, Nodes
      DO l=1,N_dof
         IF (m == Ndest(n,l))THEN
             Nod=n
             EXIT
         END IF
      END DO
      IF (Nod /= 0)EXIT
   END DO
   DO k=1,N_dof
      DoF=Ndest(Nod,k)
      IF(DoF /= 0) THEN
         IF(NCode(DoF) == 1) THEN
            Nel=0    ;    Pos=0
            DO i=1,Maxe
               DO j=1,Ndofe
                  IF(DoF == Ldest(i,j))THEN
                    Nel=i   ;    Pos=j  ;     EXIT
                  END IF
               END DO
            IF(Nel /= 0)EXIT
            END DO
            F(m) = F(m) - Diag(m,k) * Elres_u(Nel,Pos)
         ELSE
           Lhs(m,DoF)= Lhs(m,DoF) + Diag(m,k)
         END IF
       END IF
    END DO
END DO
!-----------------------------------------------------------
!   Solve system of equations iteratively
!-----------------------------------------------------------
CALL bicgstab_l(Lhs,F,Ndofs,u1,0.0_iwp,tol,its,ell,kappa)
```

```
!   Gather Element results from global result vector u1
Elements_2:      &
DO nel=1,maxe,maxe - 1
        D_o_F1:          &
        DO nd=1,Ndofe
         IF(Ldest(nel,nd) /= 0)THEN
           IF(NCode(Ldest(nel,nd)) == 0) THEN
            Elres_u(nel,nd) = Elres_u(nel,nd) + u1(Ldest(nel,nd))
           ELSE
            Elres_t(nel,nd) = Elres_t(nel,nd) + u1(Ldest(nel,nd))
           END IF
          END IF
        END DO &
        D_o_F1
        Elres_u(nel,:)= Elres_u(nel,:) * Scad
        Elres_t(nel,:)= Elres_t(nel,:) / Scat
        WRITE(12,'(24F12.5)') (Elres_u(nel,m), m=1,Ndofe)
        WRITE(12,'(24F12.5)') (Elres_t(nel,m), m=1,Ndofe)
END DO &
Elements_2
END PROGRAM General_purpose_BEM
```

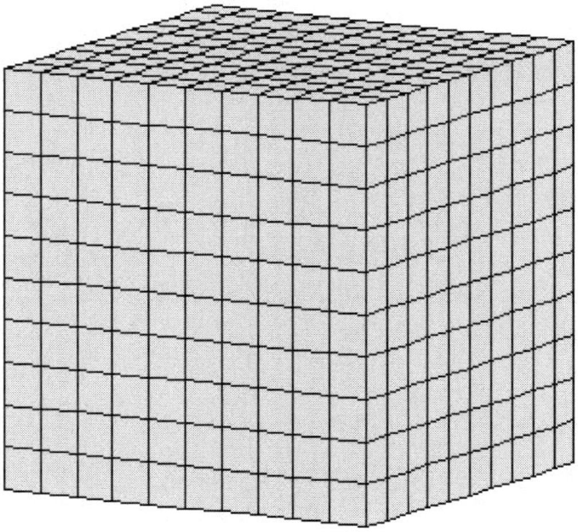

Figure 8.2 Boundary mesh for cubical cavity

The changes required to create Program 8.1 from Program 7.1 are minimal. Libraries **Utility_lib**, **Elast_lib**, **Laplace_lib** and **Integration_lib** are unchanged (apart from minimised output) and have been combined into a single library **bem_lib,** which also contains the iterative solution subroutine **bicgstab_l**. For reasons associated with global variable names used later in parallelised programs, the number of degrees of freedom per element, called **Ndof** in Program 7.1, has been changed to **N_dof**. In the iterative algorithm there are variables **ell** (**INTEGER**) and **kappa** (**REAL**) which have to be declared and input, and since the process must be terminated somehow there are declarations and input of an iteration termination counter **its** (**INTEGER**) and a convergence tolerance **tol** (**REAL**). The only other change is the replacement of the direct solver **solve** by the iterative one **bicgstab_l** ; further, because the sample input file (although for a "small" problem) takes an example with 600 elements, the output has been truncated to list only the first and last node, coordinate, element and so on. Therefore in the output **Elements_2** loop the counter increment is **maxe–1** rather than **1**.

8.2.1 Sample input file

The example chosen is of a cubical cavern in an infinite elastic medium loaded with a uniform traction on all 6 faces and meshed by 600 linear elements (see Figure 8.2).

```
Square excavation 3D
 3
 3
 2
 0
 1
0.1000E+04
0.0000E+00
   602
   600
10.000      0.000      0.000
10.000      1.000      0.000
 9.000      1.000      0.000
 9.000      0.000      0.000
 8.000      1.000      0.000
 8.000      0.000      0.000
...........................
10.000      5.000      9.000
10.000      6.000      9.000
10.000      7.000      9.000
10.000      8.000      9.000
10.000      9.000      9.000
  1    2    3    4
  4    3    5    6
  6    5    7    8
  8    7    9   10
 10    9   11   12
 12   11   13   14
......................
597  131  133  598
```

```
598  133  135  599
599  135  137  600
600  137  139  601
601  139  141  602
602  141  143  331
    0
    600
 1  0.0 0.0 10.0 0.0 0.0 10.0 0.0 0.0 10.0 0.0 0.0 10.0
 2  0.0 0.0 10.0 0.0 0.0 10.0 0.0 0.0 10.0 0.0 0.0 10.0
 3  0.0 0.0 10.0 0.0 0.0 10.0 0.0 0.0 10.0 0.0 0.0 10.0
 4  0.0 0.0 10.0 0.0 0.0 10.0 0.0 0.0 10.0 0.0 0.0 10.0
.......................................................
596 -10.0 0.0 0.0 -10.0 0.0 0.0 -10.0 0.0 0.0 -10.0 0.0 0.0
597 -10.0 0.0 0.0 -10.0 0.0 0.0 -10.0 0.0 0.0 -10.0 0.0 0.0
598 -10.0 0.0 0.0 -10.0 0.0 0.0 -10.0 0.0 0.0 -10.0 0.0 0.0
599 -10.0 0.0 0.0 -10.0 0.0 0.0 -10.0 0.0 0.0 -10.0 0.0 0.0
600 -10.0 0.0 0.0 -10.0 0.0 0.0 -10.0 0.0 0.0 -10.0 0.0 0.0
1.E-9  50  4  .7
```

Young's modulus is 1000.0 and Poisson's ratio zero. The iteration tolerance is 1.e-9, a maximum of 50 iterations is specified and the iterative algorithm parameters are set to ell=4 and kappa = 0.7.

8.2.2 Sample output file

```
    Project:
 Square excavation 3D

 Cartesian_dimension:            3
 Elasticity Problem
 Infinite Region
 No symmetry
 Linear Elements
 Modulus:    1000.00000000000
 Poissons ratio:   0.000000000000000E+000
 Number of Nodes of System:           602
 Number of Elements of System:           600
Node      1   Coor      10.00      0.00      0.00
Node    602   Coor      10.00      9.00      9.00

 Incidences:

EL      1  Inci       1    2    3    4
EL    600  Inci     602  141  143  331

 Elements with Dirichlet BC´s:
```

```
 Elements with Neuman BC's:

 Element            600   Prescribed values:
  -10.0000000000000                          0.000000000000000E+000
0.000000000000000E+000
  -10.0000000000000                          0.000000000000000E+000
0.000000000000000E+000

It took BiCGSTAB_L     4 iterations to converge

-0.00508        0.00508        0.00508       -0.00787        0.00123
0.00787        0.00079       -0.00079        0.02236       -0.00123
0.00787        0.00787
0.00000        0.00000       10.00000        0.00000        0.00000
10.00000        0.00000        0.00000       10.00000        0.00000
0.00000    10.00000
-0.02236     0.00079     0.00079     -0.00787     -0.00123
-0.00787    -0.00508    -0.00508     -0.00508     -0.00787
-0.00787    -0.00123
-10.00000        0.00000        0.00000      -10.00000        0.00000
0.00000     -10.00000        0.00000        0.00000     -10.00000
0.00000     0.00000
```

These results are the same as those produced by Program 7.1 to 5 decimal places.

8.3 PROGRAM 8.2 : REPLACING ASSEMBLY BY THE ELEMENT-BY-ELEMENT PROCEDURE

```
PROGRAM EBE_BEM
!-------------------------------------------------------
!     General purpose BEM program for solving elasticity problems
!     This version EBE with bicgstab(l)
!-------------------------------------------------------
USE bem_lib         ;     IMPLICIT NONE   ! N_dof replaces Ndof
INTEGER, ALLOCATABLE :: Inci(:,:)   !  Element Incidences
INTEGER, ALLOCATABLE :: BCode(:,:), NCode(:) !  Element BC's
INTEGER, ALLOCATABLE :: Ldest(:,:) !  Element destination vector
INTEGER, ALLOCATABLE :: Ndest(:,:) !  Node destination vector
REAL(iwp), ALLOCATABLE :: Elres_u(:,:)   !  Element results , u
REAL(iwp), ALLOCATABLE :: Elres_t(:,:)   !  Element results , t
REAL(iwp), ALLOCATABLE :: Elcor(:,:)     !  Element coordinates
REAL(iwp), ALLOCATABLE :: xP(:,:)        !  Node co-ordinates
REAL(iwp), ALLOCATABLE :: dUe(:,:),dTe(:,:),lhs(:,:),Diag(:,:)&
,pmul(:)
REAL(iwp), ALLOCATABLE :: km(:,:),qmul(:)
REAL(iwp), ALLOCATABLE :: store_dUe(:,:,:),store_dTe(:,:,:)
REAL(iwp), ALLOCATABLE :: F(:)      !  global RHS
REAL(iwp), ALLOCATABLE :: ul(:)     !  global vector of unknowns
```

```
CHARACTER (LEN=80) :: Title
INTEGER :: Cdim,Node,m,n,Istat,Nodel,Nel,N_dof,Toa
INTEGER :: Nreg,Ltyp,Nodes,Maxe,Ndofe,Ndofs,ndg,NE_u,NE_t
INTEGER :: nod,nd,i,j,k,l,DoF,Pos,Isym,nsym,nsy
INTEGER :: its,iters,ell
REAL(iwp),ALLOCATABLE    :: Fac(:)     !  Factors for symmetry
REAL(iwp),ALLOCATABLE    :: Elres_te(:),Elres_ue(:)
INTEGER,ALLOCATABLE :: Incie(:)   !  Incidences for one element
INTEGER,ALLOCATABLE :: Ldeste(:),g(:)
REAL(iwp)::Con,E,ny,Scat,Scad,tol,kappa,alpha,beta,rho,gama,
&omega,norm_r,r0_norm,error,one=1._iwp,zero=.0_iwp
LOGICAL:: converged
REAL(iwp),ALLOCATABLE::s(:),GG(:,:),Gamma(:),
&rt(:),y(:),y1(:),r(:,:),uu(:,:)
!-----------------------------------------------------
!   Read job information
!-----------------------------------------------------
OPEN (UNIT=11,FILE='prog82.dat',FORM='FORMATTED') !  Input
OPEN (UNIT=12,FILE='prog82.res',FORM='FORMATTED')!  Output
Call Jobin(Title,Cdim,N_dof,Toa,Nreg,Ltyp,Con,E,ny,&
          Isym,nodel,nodes,maxe)
Nsym= 2**Isym    !   number of symmetry loops
ALLOCATE(xP(Cdim,Nodes))    !  Array for node coordinates
ALLOCATE(Inci(Maxe,Nodel)) !  Array for incidences
CALL Geomin(Nodes,Maxe,xp,Inci,Nodel,Cdim)
Ndofe= Nodel*N_dof   !    Total degrees of freedom of element
ALLOCATE(BCode(Maxe,Ndofe))       !    Element Boundary codes
ALLOCATE(Elres_u(Maxe,Ndofe),Elres_t(Maxe,Ndofe))
CALL BCinput(Elres_u,Elres_t,Bcode,nodel,ndofe,n_dof)
READ(11,*) tol,its,ell,kappa ! BiCGStab  data
ALLOCATE(Ldest(maxe,Ndofe))  ! Elem. destination vector
ALLOCATE(Ndest(Nodes,N_dof))
!-------------------------------------------------------------
! Determine Node destination vector and Element dest vector
!-------------------------------------------------------------
CALL Destination(Isym,Ndest,Ldest,xP,Inci,Ndofs,nodes,&
                N_dof,Nodel,Maxe)
!-------------------------------------------
!     Determine global Boundary code vector
!-------------------------------------------
ALLOCATE(NCode(Ndofs))
NCode=0
DoF_o_System: DO  nd=1,Ndofs
    DO Nel=1,Maxe
        DO m=1,Ndofe
           IF (nd == Ldest(Nel,m)  .and. NCode(nd) == 0) THEN
               NCode(nd)= NCode(nd)+BCode(Nel,m)
           END IF
        END DO
    END DO
END DO DoF_o_System
```

```
IF(N_dof ==1)E= Con
CALL Scal(E,xP(:,:),Elres_u(:,:),Elres_t(:,:),Cdim,Scad,Scat)
ALLOCATE(dTe(Ndofs,Ndofe),dUe(Ndofs,Ndofe),lhs(Ndofs,Ndofe))!
Elem.coef.matrices
ALLOCATE(store_dTe(Maxe,Ndofs,Ndofe),store_dUe(Maxe,Ndofs,Ndofe))
! store els
ALLOCATE(Diag(Ndofs,N_dof))            ! Diagonal coefficients
ALLOCATE(F(Ndofs),u1(Ndofs))           ! global arrays
ALLOCATE(Elcor(Cdim,Nodel))           !  Elem. Coordinates
ALLOCATE(Fac(Ndofe))             !  Factor for symmetric elements
ALLOCATE(Incie(Nodel))                !  Element incidences
ALLOCATE(Ldeste(Ndofe),pmul(Ndofe),km(N_dof,N_dof)&
                      ,g(N_dof),qmul(N_dof))
ALLOCATE(Elres_te(Ndofe),Elres_ue(Ndofe)) ! Tractions of Element
!-----------------------------------------------------------------
!  Compute element coefficient matrices
!-----------------------------------------------------------------
Lhs=zero;  F  =  zero;  u1  =  zero;  Diag  =  zero;store_dUe  =  zero;
store_dTe = zero
Elements_1:&
DO Nel=1,Maxe
        Symmetry_loop:&
        DO nsy= 1,Nsym
                Elcor(:,:)= xP(:,Inci(Nel,:))
                Incie= Inci(nel,:)
                Ldeste= Ldest(nel,:)
                Fac(1:nodel*n_dof)= 1.0_iwp
                Elres_te(:)=Elres_t(Nel,:)
                IF(Isym > 0) THEN
                CALL Mirror(Isym,nsy,Nodes,Elcor,Fac&
                Incie,Ldeste,Elres_te,Elres_ue,nodel,n_dof,Cdim)
                END IF
                IF(Cdim == 2) THEN
                  IF(N_dof == 1) THEN
                   CALL Integ2P(Elcor,Incie,Nodel,Nodes,
                   &xP,Con,dUe,dTe,Ndest,Isym)
                  ELSE
                   CALL Integ2E(Elcor,Incie,Nodel,Nodes,
                   &xP,E,ny,dUe,dTe,Ndest,Isym)
                  END IF
                ELSE
                   CALL Integ3(Elcor,Incie,Nodel,Nodes,xP,N_dof &
                                   ,E,ny,Con,dUe,dTe,Ndest,Isym)
                END IF
!  Now build global F and diag but not LHS
        CALL rhs_and_diag(F,DTe,DUe,Ldeste,BCode(Nel,:),Ncode
        &,Elres_u(Nel,:),Elres_te,Diag,Ndofe,N_dof,Nodel,Fac)
        END DO &
        Symmetry_loop
        store_dUe(Nel,:,:) = dUe;    store_dTe(Nel,:,:) = dTe
END DO &
```

```
Elements_1
!--------------------------------------------------------------
!  Add azimuthal integral for infinite regions
!--------------------------------------------------------------
IF(Nreg == 2) THEN
        DO m=1, Nodes
                DO n=1, N_dof
                        IF(Ndest(m,n) == 0)CYCLE
                        k=Ndest(m,n)
                        Diag(k,n) = Diag(k,n) + 1.0_iwp
                END DO
        END DO
END IF
!---------------------------------------------------------------
!  Store active Diagonal coefficients
!---------------------------------------------------------------
DO m=1,Ndofs              ! Loop over collocation points
   Nod=0
   DO n=1, Nodes
      DO l=1,N_dof
         IF (m == Ndest(n,l))THEN
              Nod=n    ;      EXIT
         END IF
      END DO
      IF (Nod /= 0)EXIT
   END DO
   DO k=1,N_dof
      DoF=Ndest(Nod,k)
      IF(DoF /= 0) THEN
         IF(NCode(DoF) == 1) THEN
               Nel=0    ;     Pos=0
           DO i=1,Maxe
              DO j=1,Ndofe
                 IF(DoF == Ldest(i,j))THEN
                    Nel=i   ;    Pos=j   ;    EXIT
                 END IF
              END DO
              IF(Nel /= 0)EXIT
           END DO
           F(m) = F(m) - Diag(m,k) * Elres_u(Nel,Pos)
         END IF
      END IF
   END DO
END DO
!-----------------------------------------------------------
!   Solve system of equations  element by element
!-----------------------------------------------------------
ALLOCATE(s(ell+1),GG(ell+1,ell+1),Gamma(ell+1),&
rt(Ndofs),y(Ndofs),y1(Ndofs),r(Ndofs,ell+1),uu(Ndofs,ell+1))
!      initialisation phase
u1 = zero   ;    y = u1    ;    y1 = zero
```

```
Elements_2 : DO Nel = 1 , Maxe
     Dte = store_DTe(Nel,:,:);  DUe = store_DUe(Nel,:,:)
     Ldeste = Ldest(Nel,:); pmul = y(Ldeste)
     CALL form_lhs(lhs,DTe,DUe,Ldeste,BCode(Nel,:)&
                  ,Ncode,Ndofe,Fac)
     y1 = y1 + MATMUL(lhs,pmul)
END DO Elements_2
DO i = 1 , nodes
     CALL get_km(Cdim,i,y,Diag,g,qmul,km)
     y1(g) = y1(g) + MATMUL(km,qmul)
END DO
    y=y1; rt = F - y
    r=zero ; r(:,1) = rt  ;  uu = zero  ; gama = one  ; omega=one
    norm_r = norm(rt);r0_norm = norm_r;error = one ;  iters = 0
!     bicgstab(ell)  iterations
iterations : DO
   iters = iters + 1    ;            converged = error  < tol
   IF(iters==its.OR. converged) EXIT
   gama = - omega*gama  ;  y = r(:,1)
    DO j = 1 , ell
      rho = DOT_PRODUCT(rt,y)  ;  beta = rho/gama
      uu(:,1:j) = r(:,1:j) - beta * uu(:,1:j)   ;  y = uu(:,j)
      y1 = zero
      Elements_3: DO Nel = 1 , Maxe
        Dte = store_DTe(Nel,:,:);  DUe = store_DUe(Nel,:,:)
        Ldeste = Ldest(Nel,:); pmul = y(Ldeste)
        CALL form_lhs(lhs,DTe,DUe,Ldeste,BCode(Nel,:)&
                     ,Ncode,Ndofe,Fac)
        y1 = y1 + MATMUL(lhs,pmul)
      END DO Elements_3
      DO i = 1 , nodes
        CALL get_km(Cdim,i,y,Diag,g,qmul,km)
        y1(g) = y1(g) + MATMUL(km,qmul)
      END DO
      y=y1; uu(:,j+1) = y
      gama = DOT_PRODUCT(rt,y); alpha = rho/gama
      u1=u1+ alpha * uu(:,1)
      r(:,1:j) = r(:,1:j) - alpha * uu(:,2:j+1)
      y = r(:,j)
      y1 = zero
      Elements_4: DO Nel = 1 , Maxe
       Dte = store_DTe(Nel,:,:);  DUe = store_DUe(Nel,:,:)
       Ldeste = Ldest(Nel,:);  pmul = y(Ldeste)
       CALL form_lhs(lhs,DTe,DUe,Ldeste,BCode(Nel,:)&
                    ,Ncode,Ndofe,Fac)
        y1 = y1 + MATMUL(lhs,pmul)
      END DO Elements_4
      DO i = 1 , nodes
         CALL get_km(Cdim,i,y,Diag,g,qmul,km)
         y1(g) = y1(g) + MATMUL(km,qmul)
      END DO
```

```
        y=y1 ; r(:,j+1) = y
     END DO
     GG = MATMUL(TRANSPOSE(r),r)
     CALL form_s(gg,ell,kappa,omega,gamma,s)
     u1 = u1 - MATMUL(r,s);r(:,1)=MATMUL(r,Gamma)
     uu(:,1)=MATMUL(uu,Gamma)
     norm_r = norm(r(:,1))  ;  error = norm_r/r0_norm
END DO iterations
WRITE(12,'(/,A,I5,A,/)')"It took BiCGSTAB_L ",iters," iterations
to converge"
!   Gather Element results from global result vector u1
Elements_5:      &
DO nel=1,maxe , maxe - 1
   D_o_F1:        &
   DO nd=1,Ndofe
      IF(Ldest(nel,nd) /= 0)THEN
         IF(NCode(Ldest(nel,nd)) == 0) THEN
            Elres_u(nel,nd) = Elres_u(nel,nd) + u1(Ldest(nel,nd))
         ELSE
            Elres_t(nel,nd) = Elres_t(nel,nd) + u1(Ldest(nel,nd))
         END IF
      END IF
   END DO &
   D_o_F1
   Elres_u(nel,:)= Elres_u(nel,:) * Scad
   Elres_t(nel,:)= Elres_t(nel,:) / Scat
   WRITE(12,'(24F12.5)') (Elres_u(nel,m), m=1,Ndofe)
   WRITE(12,'(24F12.5)') (Elres_t(nel,m), m=1,Ndofe)
END DO &
Elements_5
END PROGRAM EBE_BEM
```

The essential difference between this program and the preceding one is that we do not now store the Ndofs*Ndofs array **Lhs**. However, in this version, storage is allocated to all the element **dUe** and **dTe** matrices as **store_dUe** and **store_dTe** respectively. The basic BiCGStab algorithm is unchanged, but has to be "unscrambled" from the **bicgstab_l** subroutine, because instead of a simple **MATMUL** operation to complete the matrix*vector product of equation (8.1) we have to carry this operation out element-by-element as in equation (8.4). Since the iterative algorithm is not now hidden in a subroutine there are additional declarations of **REAL**s **alpha**, **beta** etc, but these need not concern the user. The same applies to arrays **s**, **GG** etc.

The early part of the program, up to the call to subroutine **Scal**, remains unchanged. In the subsequent array allocations **lhs** is used for the element-sized "left hand side" in the element-by-element matrix*vector product, so **Lhs** can be deleted. The three-dimensional storage arrays are added, as are **pmul**, **km**, **g** and **qmul** which are necessary for the addition of diagonal components.

In the **Elements_1** loop, instead of calling subroutine **Assembly**, we call **rhs_and_diag,** which merely omits to form **Lhs,** but otherwise forms **F** and **Diag** as before. At the end of the loop, the element matrices are stored. The section adding the azimuthal integral for infinite regions is unchanged. In the next loop **F** is augmented from **Diag** as before, but of course there is no **Lhs**.

We now proceed to a new section for solving the equation system element-by-element. This contains two new subroutines **form_lhs** and **get_km**. The first of these combines element **dUe** and **dTe** matrices appropriately and the second gets the appropriate part of **Diag** for its addition into the matrix*vector multiplication. In the first case the "gather" vector is **Ldeste** and in the second it is **g**.

So in the BiCGStab process there are three matrix*vector products: **Elements_2**, to start the process and **Elements_3** and **Elements_4** in the "ell" loop. The usual value of **ell** is taken to be 4 but this can be changed by the user.

The results are collected from the global result vector **u1** in exactly the same manner as in the previous program.

8.3.1 Sample input file

The input file is precisely the same as for the previous program.

8.3.2 Sample output file

```
 Project:
 Square excavation 3D

 Cartesian_dimension:            3
 Elasticity Problem
 Infinite Region
 No symmetry
 Linear Elements
 Modulus:    1000.00000000000
 Poissons ratio:  0.000000000000000E+000
 Number of Nodes of System:          602
 Number of Elements of System:          600
Node      1   Coor      10.00     0.00     0.00
Node    602   Coor      10.00     9.00     9.00

 Incidences:

EL      1   Inci       1    2    3    4
EL    600   Inci     602  141  143  331

 Elements with Dirichlet BC´s:
 Elements with Neuman BC´s:
 Element             600    Prescribed values:
  -10.000    0.000E+000 0.000E+000
  -10.000    0.000E+000 0.000E+000
```

```
It took BiCGSTAB_L     4 iterations to converge

    -0.00508        0.00508        0.00508       -0.00787        0.00123
0.00787        0.00079       -0.00079        0.02236       -0.00123
0.00787     0.00787
     0.00000        0.00000       10.00000        0.00000        0.00000
10.00000       0.00000        0.00000       10.00000        0.00000
0.00000    10.00000
    -0.02236       0.00079        0.00079       -0.00787       -0.00123     -
0.00787      -0.00508      -0.00508       -0.00508       -0.00787       -
0.00787    -0.00123
   -10.00000              0.00000              0.00000         -10.00000
0.00000       0.00000    -10.00000       0.00000       0.00000     -
10.00000       0.00000       0.00000
```

The output can be seen to be the same as for Program 8.1 to 5 significant figures. The iterative algorithm converged in 4 iterations in this case.

8.4 PROGRAM 8.3 : PARALLELISING THE ELEMENT_BY_ELEMENT PROCEDURE

```
PROGRAM PARALLEL_BEM
!-------------------------------------------------------
!     General purpose BEM program for solving elasticity problems
!     This version parallel with bicgstab(l)
!-------------------------------------------------------
USE bem_lib_p; USE precision; USE timing; USE utility; USE
mp_module
USE global_variables1; USE gather_scatter6
IMPLICIT NONE  !  Ndof changed to N_dof
INTEGER, ALLOCATABLE :: Inci(:,:)  !  Element Incidences
INTEGER, ALLOCATABLE :: BCode(:,:), NCode(:) !  Element BC´s
INTEGER, ALLOCATABLE :: Ldest(:,:) !  Element destination vector
INTEGER, ALLOCATABLE :: Ndest(:,:) !  Node destination vector
REAL(iwp), ALLOCATABLE :: Elres_u(:,:)   !  Element results , u
REAL(iwp), ALLOCATABLE :: Elres_t(:,:)   !  Element results , t
REAL(iwp), ALLOCATABLE :: Elcor(:,:)     !  Element coordinates
REAL(iwp), ALLOCATABLE :: xP(:,:)        !  Node co-ordinates
REAL(iwp),ALLOCATABLE  :: &
dUe(:,:),dTe(:,:),lhs(:,:),Diag(:,:),pmul(:)
REAL(iwp), ALLOCATABLE :: km(:,:),qmul(:),Diag1(:,:)
REAL(iwp), ALLOCATABLE :: store_dUe_pp(:,:,:),store_dTe_pp(:,:,:)
REAL(iwp), ALLOCATABLE :: F(:),F1(:)      !  global RHS
REAL(iwp), ALLOCATABLE :: u1(:),y_cop(:) ! vector of unknowns
CHARACTER (LEN=80) :: Title
INTEGER :: Cdim,m,n,Nodel,Nel,N_dof,Toa,N_tot
```

```
INTEGER :: Nreg,Ltyp,Nodes,Maxe,Ndofe,Ndofs,ndg,NE_u,NE_t
INTEGER:: nod,nd,i,j,k,l,DoF,Pos,Isym,nsym,nsy,its,iters,ell
REAL(iwp),ALLOCATABLE    :: Fac(:)      !  Factors for symmetry
REAL(iwp),ALLOCATABLE    :: Elres_te(:),Elres_ue(:)
INTEGER,ALLOCATABLE :: Incie(:)    !  Incidences for one element
INTEGER,ALLOCATABLE :: Ldeste(:),g(:)
REAL(iwp)   ::   Con,E,ny,Scat,Scad,tol,kappa,alpha,beta,rho,gama,
&omega,norm_r,r0_norm,error,one=1._iwp,zero=.0_iwp
LOGICAL:: converged
REAL(iwp),ALLOCATABLE::s(:),GG(:,:),Gamma(:),rt(:),y(:),y1(:),r(:
,:),uu(:,:)
timest(1) = elap_time(); CALL find_pe_procs(numpe,npes)
!---------------------------------------------------
!   Read job information
!---------------------------------------------------
OPEN  (UNIT=11,FILE='prog83.dat',FORM='FORMATTED',ACTION='READ')  !
Input
IF(numpe==1)OPEN(UNIT=12,FILE='prog83.res',FORM='FORMATTED',ACTIO
N='WRITE')!O/P
IF(numpe==1) WRITE(12,*) "This job ran on ",npes," processors"
Call Jobin(Title,Cdim,N_dof,Toa,Nreg,Ltyp,Con,E,ny,&
           Isym,nodel,nodes,nels)
Nsym= 2**Isym   !   number of symmetry loops
ALLOCATE(xP(Cdim,Nodes))    !  Array for node coordinates
ALLOCATE(Inci(nels,Nodel)) !  Array for incidences
CALL Geomin(Nodes,nels,xp,Inci,Nodel,Cdim)
Ndofe= Nodel*N_dof    !    Total degrees of freedom of element
ALLOCATE(BCode(nels,Ndofe))        !    Element Boundary codes
ALLOCATE(Elres_u(nels,Ndofe),Elres_t(nels,Ndofe))
CALL BCinput(Elres_u,Elres_t,Bcode,nodel,ndofe,n_dof)
READ(11,*) tol,its,ell,kappa       ! BiCGStab data
ALLOCATE(Ldest(nels,Ndofe))   ! Elem. destination vector
ALLOCATE(Ndest(Nodes,N_dof))
!------------------------------------------------------------
! Determine Node destination vector and Element dest vector
!------------------------------------------------------------
CALL Destination(Isym,Ndest,Ldest,xP,Inci,Ndofs,&
nodes,N_dof,Nodel,nels)
!------------------------------------------------------------
!     Determine global Boundary code vector
!------------------------------------------------------------
ALLOCATE(NCode(Ndofs));   CALL   calc_nels_pp   !   elements   per
processor
IF(numpe==1) WRITE(12,*) "Elements on first processor ",nels_pp
NCode=0
DoF_o_System: &
DO  nd=1,Ndofs
    DO Nel=1,nels
        DO m=1,Ndofe
           IF (nd == Ldest(Nel,m) .and. NCode(nd) == 0) THEN
               NCode(nd)= NCode(nd)+BCode(Nel,m)
```

```
          END IF
        END DO
     END DO
END DO &
DoF_o_System
IF(N_dof ==1)E= Con
CALL Scal(E,xP(:,:),Elres_u(:,:),Elres_t(:,:),Cdim,Scad,Scat)
ALLOCATE(dTe(Ndofs,Ndofe),dUe(Ndofs,Ndofe),lhs(Ndofs,Ndofe))!
Elem.coef.matrices
ALLOCATE(store_dTe_pp(Ndofs,Ndofe,nels_pp),         &
         store_dUe_pp(Ndofs,Ndofe,nels_pp))  !  store  el  matrices
on procs
ALLOCATE(Diag(Ndofs,N_dof),Diag1(Ndofs,N_dof))!Diag cos
ALLOCATE(F(Ndofs),u1(Ndofs),F1(Ndofs))   ! global arrays
ALLOCATE(Elcor(Cdim,Nodel))              !  Elem. Coordinates
ALLOCATE(Fac(Ndofe))   !  Factor for symmetric elements
ALLOCATE(Incie(Nodel))                !  Element incidences
ALLOCATE(Ldeste(Ndofe),pmul(Ndofe),km(N_dof,N_dof),g(N_dof),qmul(
N_dof))!dest.
ALLOCATE(Elres_te(Ndofe),Elres_ue(Ndofe)) ! Tractions of Element
!-----------------------------------------------------------------
!  Compute element coefficient matrices
!-----------------------------------------------------------------
Lhs=zero;  F1  =  zero;  u1  =  zero;  Diag1  =  zero;      N_tot  =
Ndofs*N_dof
store_dUe_pp = zero; store_dTe_pp = zero    ; ielpe = iel_start
Elements_1:&
DO Nel=1,nels_pp
 Symmetry_loop:&
 DO nsy= 1,Nsym
  Elcor(:,:)= xP(:,Inci(ielpe,:))
  Incie= Inci(ielpe,:)
  Ldeste= Ldest(ielpe,:)
  Fac(1:nodel*n_dof)= 1.0_iwp
  Elres_te(:)=Elres_t(ielpe,:)
  IF(Isym > 0) THEN
   CALL Mirror(Isym,nsy,Nodes,Elcor,Fac,     &
            Incie,Ldeste,Elres_te,Elres_ue,nodel,n_dof,Cdim)
  END IF
  IF(Cdim == 2) THEN
   IF(N_dof == 1) THEN
    CALL Integ2P(Elcor,Incie,Nodel,Nodes,&
             xP,Con,dUe,dTe,Ndest,Isym)
   ELSE
    CALL Integ2E(Elcor,Incie,Nodel,Nodes, &
             xP,E,ny,dUe,dTe,Ndest,Isym)
   END IF
  ELSE
   CALL Integ3(Elcor,Incie,Nodel,Nodes,xP,N_dof &
            ,E,ny,Con,dUe,dTe,Ndest,Isym)
  END IF
```

```
!  Now build global F and diag but not LHS
   CALL rhs_and_diag(F1,DTe,DUe,Ldeste,BCode(ielpe,:),Ncode &
  ,Elres_u(ielpe,:),Elres_te,Diag1,Ndofe,N_dof,Nodel,Fac)
 END DO &
 Symmetry_loop              ;    ielpe = ielpe + 1
 store_dUe_pp(:,:,Nel) = dUe; store_dTe_pp(:,:,Nel) = dTe
END DO &
Elements_1
CALL MPI_ALLREDUCE(F1,F,Ndofs,MPI_REAL8,MPI_SUM&
                  ,MPI_COMM_WORLD,ier)
CALL MPI_ALLREDUCE(Diag1,Diag,N_tot,MPI_REAL8,MPI_SUM&
                  ,MPI_COMM_WORLD,ier)
!-------------------------------------------------------------
!  Add azimuthal integral for infinite regions
!-------------------------------------------------------------
IF(Nreg == 2) THEN
      DO m=1, Nodes
            DO n=1, N_dof
                  IF(Ndest(m,n) == 0)CYCLE
                  k=Ndest(m,n)
                  Diag(k,n) = Diag(k,n) + 1.0_iwp
            END DO
      END DO
END IF
!--------------------------------------------------------------
!  Store active Diagonal coefficients
!--------------------------------------------------------------
DO m=1,Ndofs               ! Loop over collocation points
   Nod=0
   DO n=1, Nodes
      DO l=1,N_dof
         IF (m == Ndest(n,l))THEN
             Nod=n   ;       EXIT
         END IF
      END DO
      IF (Nod /= 0)EXIT
   END DO
   DO k=1,N_dof
      DoF=Ndest(Nod,k)
      IF(DoF /= 0) THEN
         IF(NCode(DoF) == 1) THEN
              Nel=0    ;     Pos=0
           DO i=1,nels
              DO j=1,Ndofe
                 IF(DoF == Ldest(i,j))THEN
                    Nel=i   ;    Pos=j   ;    EXIT
                 END IF
              END DO
              IF(Nel /= 0)EXIT
           END DO
           F(m) = F(m) - Diag(m,k) * Elres_u(Nel,Pos)
```

```
        END IF
      END IF
   END DO
END DO
IF(numpe==1) WRITE(12,*) "Time before eq solution is ",&
elap_time()-timest(1)
!---------------------------------------------------------
!   Solve system of equations  element by element
!---------------------------------------------------------
ALLOCATE(s(ell+1),GG(ell+1,ell+1),Gamma(ell+1),y_cop(Ndofs),
& rt(Ndofs),y(Ndofs),y1(Ndofs),r(Ndofs,ell+1),uu(Ndofs,ell+1))
!      initialisation phase
u1 = zero ; y = u1 ; y_cop = y; y1 = zero ; neq = Ndofs
ielpe = iel_start
Elements_2 : DO Nel = 1 , nels_pp
   Dte = store_DTe_pp(:,:,Nel);  DUe = store_DUe_pp(:,:,Nel)
   Ldeste = Ldest(ielpe,:); pmul = y(Ldeste)
   CALL form_lhs(lhs,DTe,DUe,Ldeste,BCode(Nel,:),Ncode,Ndofe,Fac)
   y1 = y1 + MATMUL(lhs,pmul) ;  ielpe = ielpe + 1
END DO Elements_2
CALL MPI_ALLREDUCE(y1,y,neq,MPI_REAL8,MPI_SUM,MPI_COMM_WORLD,ier)
DO i = 1 , nodes
     CALL get_km(Cdim,i,y_cop,Diag,g,qmul,km)
     y(g) = y(g) + MATMUL(km,qmul)
END DO
rt = F - y
r=zero ; r(:,1) = rt   ;  uu = zero  ; gama = one  ; omega=one
norm_r = norm(rt)   ;   r0_norm = norm_r   ;  error = one
iters = 0
!     bicgstab(ell)  iterations
iterations : DO
   iters = iters + 1    ;            converged = error  < tol
   IF(iters==its.OR. converged) EXIT
   gama = - omega*gama  ;   y = r(:,1)
    DO j = 1 , ell
     rho = DOT_PRODUCT(rt,y)  ;  beta = rho/gama
     uu(:,1:j) = r(:,1:j) - beta * uu(:,1:j)   ;  y = uu(:,j)
     y1 = zero     ; y_cop = y  ;  ielpe = iel_start
     Elements_3: DO Nel = 1 , nels_pp
       Dte = store_DTe_pp(:,:,Nel);  DUe = store_DUe_pp(:,:,Nel)
       Ldeste = Ldest(ielpe,:); pmul = y(Ldeste)
       CALL form_lhs(lhs,DTe,DUe,Ldeste,BCode(Nel,:)&
                   ,Ncode,Ndofe,Fac)
       y1 = y1 + MATMUL(lhs,pmul)  ;  ielpe = ielpe + 1
      END DO Elements_3
      CALL MPI_ALLREDUCE(y1,y,neq,MPI_REAL8,MPI_SUM&
                       ,MPI_COMM_WORLD,ier)
      DO i = 1 , nodes
         CALL get_km(Cdim,i,y_cop,Diag,g,qmul,km)
         y(g) = y(g) + MATMUL(km,qmul)
      END DO
```

```
    uu(:,j+1) = y
    gama = DOT_PRODUCT(rt,y); alpha = rho/gama
    u1=u1+ alpha * uu(:,1)
    r(:,1:j) = r(:,1:j) - alpha * uu(:,2:j+1)
    y = r(:,j)
    y1 = zero    ; y_cop = y ;   ielpe = iel_start
    Elements_4: DO Nel = 1 , nels_pp
      Dte = store_DTe_pp(:,:,Nel);  DUe = store_DUe_pp(:,:,Nel)
      Ldeste = Ldest(ielpe,:);  pmul = y(Ldeste)
      CALL form_lhs(lhs,DTe,DUe,Ldeste,BCode(Nel,:)&
                   ,Ncode,Ndofe,Fac)
     y1 = y1 + MATMUL(lhs,pmul) ;   ielpe = ielpe + 1
    END DO Elements_4
    CALL MPI_ALLREDUCE(y1,y,neq,MPI_REAL8,MPI_SUM&
                      ,MPI_COMM_WORLD,ier)
    DO i = 1 , nodes
       CALL get_km(Cdim,i,y_cop,Diag,g,qmul,km)
       y(g) = y(g) + MATMUL(km,qmul)
    END DO
    r(:,j+1) = y
  END DO
  GG = MATMUL(TRANSPOSE(r),r)
  CALL form_s(gg,ell,kappa,omega,gamma,s)
  u1 = u1 - MATMUL(r,s);r(:,1)=MATMUL(r,Gamma)
  uu(:,1)=MATMUL(uu,Gamma)
  norm_r = norm(r(:,1))  ;  error = norm_r/r0_norm
END DO iterations
IF(numpe==1) WRITE(12,'(/,A,I5,A,/)')&
     "It took BiCGSTAB_L ",iters," iterations to converge"
!  Gather Element results from global result vector u1
Elements_5:    &
DO nel=1,nels , nels - 1
  D_o_F1:        &
  DO nd=1,Ndofe
    IF(Ldest(nel,nd) /= 0)THEN
      IF(NCode(Ldest(nel,nd)) == 0) THEN
         Elres_u(nel,nd) = Elres_u(nel,nd) + u1(Ldest(nel,nd))
      ELSE
         Elres_t(nel,nd) = Elres_t(nel,nd) + u1(Ldest(nel,nd))
      END IF
    END IF
  END DO &
  D_o_F1
  Elres_u(nel,:)= Elres_u(nel,:) * Scad
  Elres_t(nel,:)= Elres_t(nel,:) / Scat
  WRITE(12,'(24F12.5)') (Elres_u(nel,m), m=1,Ndofe)
  WRITE(12,'(24F12.5)') (Elres_t(nel,m), m=1,Ndofe)
END DO &
Elements_5
IF(numpe==1)  WRITE(12,*)  "This  analysis  took  ",  elap_time()  -
timest(1)
```

```
CALL shutdown()
END PROGRAM PARALLEL_BEM
```

Comparing this parallelised program with the previous one, we see a simple logical development. Extra libraries are **USE**d to include routines needed for parallel processing and the basic library is **bem_lib_p** rather than **bem_lib** but the user sees a basically unchanged coding. New arrays **F1** and **Diag1** are needed to hold the parts of **F** and **Diag** on the different processors until they are accumulated into **F** and **Diag**. **N_tot** is the total number of entries in **Diag**. Otherwise the declarations are unchanged.

The first sign of parallelisation is the need to establish the number of parallel processors being used (**npes**) and their "rank" (**numpe**). So **numpe** takes values 1 to **npes**. This is done by subroutine **find_pe_procs**. Before this, the clock is started to assess the time taken in different parts of the program (**timing**). To maintain simplicity, all processors read the input data although it might be better for one processor to do the reading and then "broadcast" to the other processors. The first processor (**numpe=1**) is used for output and the number of processors being used is listed. The next task is, given the number of elements in the analysis from **Jobin**, to calculate the number of elements to be allocated to each processor. This is done by simple division using **calc_nels_pp** so that there will be nearly the same number **nels_pp** on each processor. There is no need to employ other than a "naïve" serial distribution - first 10, second 10 etc. As a check the number of elements on the first processor is output.

The next section establishing **Ncode** remains unchanged, but in the subsequent array allocations each processor will only hold its own element matrices and so the storage arrays become **store_dUe_pp** and **store_dTe_pp** respectively. These could have (slightly) different sizes **nels_pp**. We can then proceed to compute the element coefficient matrices and store them. An important parameter established in the parallel libraries is **iel_start,** which is the number of the first element on each processor. So when counting round the elements, we count **Nel = 1 , nels_pp** rather than **Nel = 1 , Maxe** and counter **ielpe** replaces **Nel** to identify an element in parallel. In the call to **rhs_and_diag** we calculate only the parts of **F** and **Diag** which reside on that particular processor as **F1** and **Diag1**. Therefore, after the **Elements_1** loop we need MPI routine **MPI_ALLREDUCE** with parameter **MPI_SUM** to collect the contributions from all processors and add into **F** and **Diag.** The "azimuthal integral" and "Diagonal coefficients" sections of the program are unchanged and the analysis time after element calculation (which can be quite significant) is printed.

In the equation solution section, the parallel coding looks very similar to its serial counterpart. The element loops are over **nels_pp** rather than **Maxe** and one has to be careful to use element counter **ielpe** rather than **Nel** where appropriate. To compute a total vector **y**, the partial vectors **y1** have always to be collected using **MPI_ALLREDUCE**. Gathering the element results is unchanged and it is necessary to close down MPI using subroutine **shutdown**. By comparing Programs 8.2 and 8.3 the hope is that the serial Fortran programmer will see that the step to parallel analyses is indeed a small one.

8.4.1 Sample input file

The input is precisely the same as for the previous two programs.

8.4.2 Sample output file

```
 This job ran on                  4  processors
 Project:
 Square excavation 3D
 Cartesian_dimension:               3
 Elasticity Problem
 Infinite Region
 No symmetry
 Linear Elements
 Modulus:    1000.00000000000
 Poissons ratio:  0.000000000000000E+000
 Number of Nodes of System:              602
 Number of Elements of System:              600
Node      1  Coor      10.00      0.00     0.00
Node    602  Coor      10.00      9.00     9.00
 Incidences:
EL      1  Inci       1    2    3    4
EL    600  Inci     602  141  143  331
 Elements with Dirichlet BC's:
 Elements with Neuman BC's:
 Element            600    Prescribed values:
  -10.000 0.000E+000 0.000E+000
  -10.000 0.000E+000 0.000E+000
 Elements on first processor             150
 Time before eq solution is    1.85099999999875
It took BiCGSTAB_L      4 iterations to converge
    -0.00508       0.00508       0.00508      -0.00787       0.00123
0.00787       0.00079      -0.00079       0.02236      -0.00123
0.00787    0.00787
     0.00000       0.00000      10.00000       0.00000       0.00000
10.00000      0.00000       0.00000      10.00000       0.00000
0.00000    10.00000
    -0.02236      0.00079      0.00079     -0.00787     -0.00123    -
0.00787     -0.00508     -0.00508     -0.00508     -0.00787    -
0.00787    -0.00123
   -10.00000      0.00000       0.00000     -10.00000       0.00000
0.00000     -10.00000       0.00000       0.00000     -10.00000
0.00000     0.00000
 This analysis took    2.81362999999903
```

The output is again unchanged to five significant figures. The number of processors used was 4 and the speedup due to parallelisation, even on such a small problem, is shown in Table 8.1, the results being produced on a Bull computer

Table 8.1 Results for 600 element problem

Processors	Analysis time (seconds
1	8.3
2	4.6
4	2.8

8.4.3 Results from larger analyses

When the number of elements was increased to 9600 the results obtained using a Bull computer and the UK National HPCx system are shown in Table 8.2

Table 8.2 Results for 9600 element problem

Processors	Analysis time (seconds)	
	Bull	HPx
32	490	
64	260	
128	140	63
256		45
512		43

The scaling is satisfactory up to about a couple of hundred processors but not beyond. Recall that this analysis could not be run on a single processor because of storage limitations. Increasing the number of elements to 21600 leads to the following results on HPCx shown in Table 8.3

Table 8.3 Results for 21600 element problem

Processors	Analysis time (seconds)
512	288
1024	249

In this case there is no advantage in going beyond about 500 processors, but this quite large problem is solved in about 4 minutes – "coffee break time".

A concern when using iterative methods is that iteration counts may become "large" as problems get bigger. The counts for the above analyses are listed below in Table 8.4

Table 8.4 Convergence statistics for BiCGStab

Equations	Iterations to Converge
1800	4
28800	15
64800	49

It can be seen that the increase in iteration count with problem size is modest. A second concern about iterative methods relates to the conditioning of the system equations. When Poisson's Ratio was increased to 0.5 the iterations to convergence increased by only about 50%[2].

8.5 CONCLUSIONS

In this chapter we have illustrated how the basic program from the previous Chapter can be modified easily to use an iterative equation solution technique rather than a direct one. The purpose behind this is then to extend further to replace traditional element assembly by an element-by-element approach whereby no large system matrices are ever assembled at all. This opens up the prospect of a simple parallel processing strategy for boundary element methods. When this is done, solution times can be dramatically reduced and much larger problems solved. Finally readers should note that the new libraries referred to are not listed but are available for download from the web (see information in the preface).

8.6 REFERENCES

1. Smith, I.M. and Griffiths,D.V. (2004) Programming the Finite Element Method, J.Wiley
2. Smith, I.M. and Margetts,L. (2007) Parallel Boundary Element Analysis of Tunnels, *Proceedings EUROTUN 2007*, Vienna.

9

Postprocessing

Man soll auf alles achten, denn man kann alles deuten
(You should consider everything
because you can interpret everything)
H. Hesse

9.1 INTRODUCTION

In the previous Chapters we developed a general purpose computer program for the analysis of two and three-dimensional problems in elasticity and potential flow. This program only calculates the values of unknowns (temperature/displacements or boundary flow/tractions) at the nodes of boundary elements. In this chapter we will develop procedures for the calculation of other results which are of interest. These are the flow vector or the stress tensor at the boundary and at points inside the domain.

There are two types of approximations involved in a boundary element analysis. The first is that the distribution of temperature/displacement, or boundary flow/stress, is approximated at the boundary by shape functions defined locally for each element. The second approximation is that the theorem by Betti is only ensured to be satisfied at the nodal points on the boundary elements (collocation points).

Because we use fundamental solutions, the variation of temperature/displacements inside the domain is known in terms of boundary values. It is therefore possible to compute the results at any point inside the domain as a postprocessing exercise, after the analysis has been performed. This is in contrast to the FEM, where results are only available at points inside finite elements. Now the results at interior points can be part of a graphical postprocessor, with the option that the user may freely specify locations where results are required. Instead of using interpolation between values at nodal points of elements, we can use a direct procedure to determine the contour lines and this will be discussed later.

In the discussion on the computation of results we distinguish between values inside the domain and on the boundary. For the computation of results the integral equation for the displacement (5.16) or temperature/potential (5.20) can be used. The strain tensor is determined by taking the derivatives of the displacement solution. However, we note that the singularity of the Kernels increase by one order and tend to infinity as the boundary is approached. Special consideration has to be given to this case and therefore we will deal separately with the computation of values on the boundary and internal values.

Because of the increased singularity of the Kernels, special care has to be taken in the numerical integration when points are very close to the boundary and a subdivision of the integration region will become inevitable.

9.2 COMPUTATION OF BOUNDARY RESULTS

For the computation of postprocessed results on the boundary, we use a procedure which is essentially the same as the one used in the finite element method for computing results inside elements. Along a boundary element we know (after the solution), the variation of both u and t because of the approximation introduced in 6.2. We can therefore compute fluxes or stresses tangential to the boundary by differentiation of u. In the following we will discuss two and three-dimensional potential and elasticity problems separately.

9.2.1 Potential problems

The temperature distribution on a boundary element in terms of nodal values u_n^e is

$$u = \sum_{n=1}^{N} N_n u_n^e \tag{9.1}$$

For two-dimensional problems we define a vector $\mathbf{V}_\xi$ in the direction tangential to the boundary (Figure 9.1) where

$$\mathbf{V}_\xi = \frac{\partial}{\partial \xi}\mathbf{x} = \sum_{n=1}^{N} \frac{\partial N_n}{\partial \xi}\mathbf{x}_n^e \tag{9.2}$$

and $\mathbf{x}_n^e$ is a vector containing the coordinates of the element nodes. The flow in tangential direction is given by

$$q_{\bar{x}} = -k\frac{\partial u}{\partial \bar{x}} = -k\sum_{n=1}^{N} \frac{\partial N_n}{\partial \xi}\frac{\partial \xi}{\partial \bar{x}} u_n^e \tag{9.3}$$

where $\partial\xi/\partial\bar{x}$ is the inverse of the *Jacobian* given as

$$\frac{\partial \xi}{\partial \bar{x}} = \frac{1}{\sqrt{V_{\xi x}^2 + V_{\xi y}^2}} \tag{9.4}$$

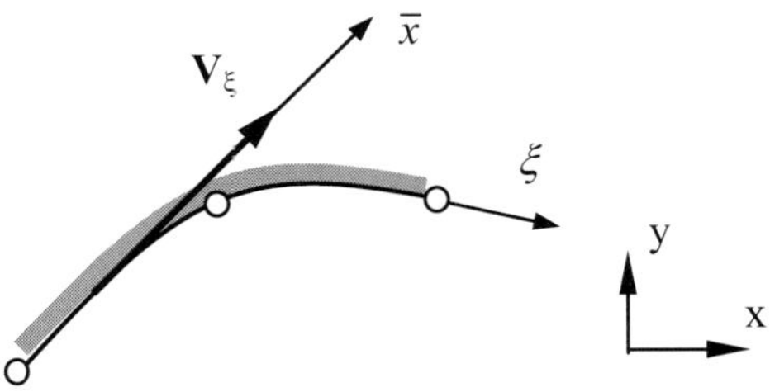

Figure 9.1 Local coordinate system for computing boundary values

For three-dimensional problems we construct, at a point inside the element, a local orthogonal coordinate system $\overline{x}$ and $\overline{y}$, with directions as specified by vectors $\mathbf{v}_1$ and $\mathbf{v}_2$ as outlined in section 3.9 (see Figure 9.2), where $\mathbf{v}_1$ is in the ξ direction.

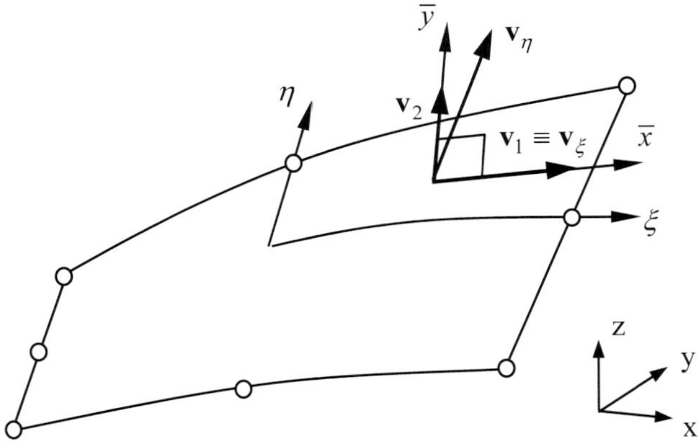

Figure 9.2 Local coordinate systems for surface element

The components of the flow vector in the local directions are

$$
\begin{aligned}
q_{\overline{x}} &= -k\frac{\partial u}{\partial \overline{x}} = -k\frac{\partial u}{\partial \xi}\frac{\partial \xi}{\partial \overline{x}} = -k\sum_{n=1}^{N}\frac{\partial N_n}{\partial \xi}\frac{\partial \xi}{\partial \overline{x}}u_n^e \\
q_{\overline{y}} &= -k\frac{\partial u}{\partial \overline{y}} = -k\left(\frac{\partial u}{\partial \xi}\frac{\partial \xi}{\partial \overline{y}} + \frac{\partial u}{\partial \eta}\frac{\partial \eta}{\partial \overline{y}}\right) = -k\left(\sum_{n=1}^{N}\frac{\partial N_n}{\partial \xi}\frac{\partial \xi}{\partial \overline{y}}u_n^e + \sum_{n=1}^{N}\frac{\partial N_n}{\partial \eta}\frac{\partial \eta}{\partial \overline{y}}u_n^e\right)
\end{aligned}
\quad (9.5)
$$

To compute the values of $\partial\xi / \partial\bar{x}$ etc. we consider a view normal to the surface of the element (Figure 9.3) in order to more clearly show the relationship between the skew and the orthogonal axes.

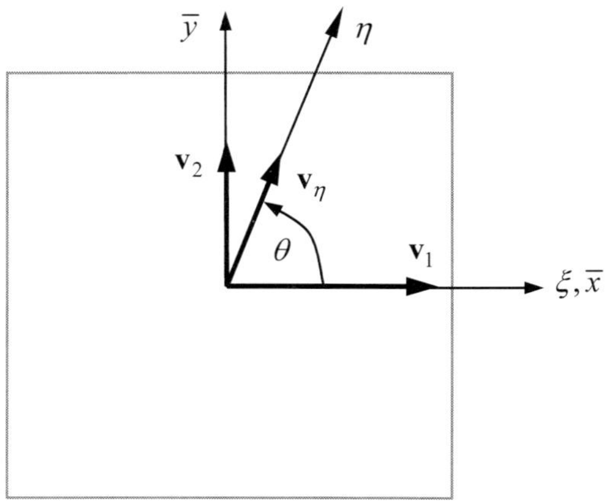

Figure 9.3 View normal to surface of element

The unit vectors $\mathbf{v}_\xi, \mathbf{v}_\eta$ in the direction ξ, η are computed by

$$\mathbf{v}_1 = \mathbf{v}_\xi = \frac{1}{J_\xi}\mathbf{V}_\xi = \frac{1}{J_\xi}\frac{\partial \mathbf{x}}{\partial \xi} \quad ; \quad \mathbf{v}_\eta = \frac{1}{J_\eta}\mathbf{V}_\eta = \frac{1}{J_\eta}\frac{\partial \mathbf{x}}{\partial \eta} \tag{9.6}$$

where J_ξ, J_η are stretch factors given by

$$J_\xi = \sqrt{V_{\xi x}^2 + V_{\xi y}^2 + V_{\xi z}^2} \quad ; \quad J_\eta = \sqrt{V_{\eta x}^2 + V_{\eta y}^2 + V_{\eta z}^2} \tag{9.7}$$

The relationship between $\bar{x}, \bar{y}$ and ξ, η is (see Figure 9.3)

$$\begin{aligned} \bar{x} &= J_\xi \cdot \xi + J_\eta \cdot \eta \cdot \cos\theta \\ \bar{y} &= J_\eta \cdot \eta \cdot \sin\theta \end{aligned} \tag{9.8}$$

where

$$\cos\theta = \mathbf{v}_\xi \bullet \mathbf{v}_\eta \quad ; \quad \sin\theta = \mathbf{v}_\eta \bullet \mathbf{v}_2 \tag{9.9}$$

Solving equations (9.8) for ξ, η

$$\xi = \frac{1}{J_\xi}\left(\bar{x} - \frac{\cos\theta}{\sin\theta}\bar{y}\right) \quad ; \quad \eta = \frac{1}{J_\eta \cdot \sin\theta}\bar{y} \tag{9.10}$$

and taking the derivatives we obtain

$$\frac{\partial \xi}{\partial x} = \frac{1}{J_\xi} \quad ; \quad \frac{\partial \xi}{\partial y} = -\frac{\cos\theta}{J_\xi \cdot \sin\theta} \quad ; \quad \frac{\partial \eta}{\partial y} = \frac{1}{J_\eta \cdot \sin\theta} \tag{9.11}$$

The theory just outlined may be programmed into a subroutine which computes boundary flows at a point on the boundary element with intrinsic coordinates ξ, *(*η*)*.

```
SUBROUTINE BFLOW(Flow,xsi,eta,u,Inci,Elcor,k)
!--------------------------------------------
!    Computes flow vectors in direction tangential to the
!    Boundary
!--------------------------------------------
REAL , INTENT(OUT)    :: Flow(:)  !  Flow vector
REAL , INTENT(IN)     :: xsi,eta  !  intrinsic coord. of point
REAL , INTENT(IN)     :: u(:,:)    !  Nodal temps/potentials
INTEGER, INTENT (IN) :: Inci(:)  !  Element Incidences
REAL, INTENT (IN)     :: Elcor(:,:)  !  Element coordinates
REAL, INTENT (IN)     :: k        !  Conductivity
REAL, ALLOCATABLE     :: Vxsi(:),Veta(:),DNi(:,:),V3(:)
INTEGER :: Nodes,Cdim,Ldim
REAL :: Jxsi,Jeta,v1(3),v2(3),CosT,SinT,DuDxsi,DuDeta,V3_L
REAL :: DxsiDx,DxsiDy,DetaDx,DetaDy
Nodes= UBOUND(ELCOR,2)  !  Number of nodes
Cdim= UBOUND(ELCOR,1)   !  Cartesian Dimension
Ldim= Cdim-1            !  Local (element) dimension
ALLOCATE (Vxsi(cdim),Dni(Nodes,Ldim),v3(cdim))
IF(ldim > 1) ALLOCATE (Veta(cdim))
!   Compute Vector(s) tangential to boundary surface
CALL Serendip_deriv(DNi,xsi,eta,ldim,nodes,inci)
Vxsi(1)= Dot_Product(Dni(:,1),Elcor(1,:))
Vxsi(2)= Dot_Product(Dni(:,1),Elcor(2,:))
IF(Cdim == 2) THEN
  CALL Vector_norm(Vxsi,Jxsi)
  Flow(1)= -k*Dot_product(Dni(:,1),u(:,1))/Jxsi
ELSE
  Vxsi(3)= Dot_Product(Dni(:,1),Elcor(3,:))
  CALL Vector_norm(Vxsi,Jxsi)
  Veta(1)= Dot_Product(Dni(:,2),Elcor(1,:))
  Veta(2)= Dot_Product(Dni(:,2),Elcor(2,:))
  Veta(3)= Dot_Product(Dni(:,2),Elcor(3,:))
  CALL Vector_norm(Veta,Jeta)
   v3= Vector_ex(Vxsi,Veta)
   Call Vector_norm(v2,v3_L)
    v1=Vxsi
    v2= Vector_ex(v3,v1)
    DuDxsi= Dot_Product(Dni(:,1),u(:,1))
    DuDeta= Dot_Product(Dni(:,2),u(:,1))
    CosT= DOT_Product(Vxsi,Veta)
```

```
    SinT= ABS(DOT_Product(V2,Veta))
    DxsiDx= 1/Jxsi
    DxsiDy= -CosT /(Jxsi* SinT)
    DetaDx= 0.0
    DetaDy= 1/(Jeta* SinT)
!    Flow in local coordinate directions
    Flow(1)= -k* DuDxsi* DxsiDx
    Flow(2)= -k* (DuDxsi* DxsiDy+DuDeta* DetaDy)
END IF
RETURN
END SUBROUTINE BFLOW
```

9.2.2 Elasticity problems

The computation of boundary values of stress for elasticity problems is similar to the one for potential problems. For elasticity, displacements **u** inside a boundary element are given in terms of nodal displacements $\mathbf{u}_n^e$ by

$$\mathbf{u} = \sum_{n=1}^{N} N_n \mathbf{u}_n^e \tag{9.12}$$

For two-dimensional problems, the strain in tangential direction is computed by

$$\varepsilon_{\bar{x}} = \frac{\partial u_{\bar{x}}}{\partial \bar{x}} = \left(\frac{\partial \mathbf{u}}{\partial \xi} \bullet \mathbf{v}_\xi \right) \frac{\partial \xi}{\partial \bar{x}} \tag{9.13}$$

where

$$\frac{\partial \mathbf{u}}{\partial \xi} = \begin{Bmatrix} \dfrac{\partial u_x}{\partial \xi} \\ \dfrac{\partial u_y}{\partial \xi} \end{Bmatrix} \tag{9.14}$$

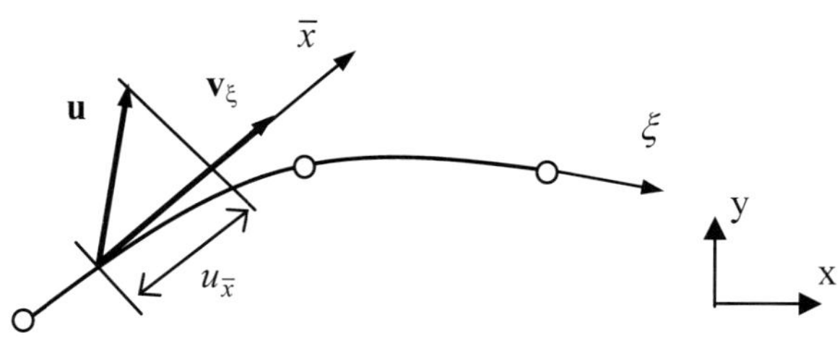

Figure 9.4 Computation of tangential strain

The derivatives of the displacements are given by

$$\frac{\partial u_x}{\partial \xi} = \sum_{n=1}^{N} \frac{\partial N_n}{\partial \xi} u_{xn}^e \quad , \quad \frac{\partial u_y}{\partial \xi} = \sum_{n=1}^{N} \frac{\partial N_n}{\partial \xi} u_{yn}^e \tag{9.15}$$

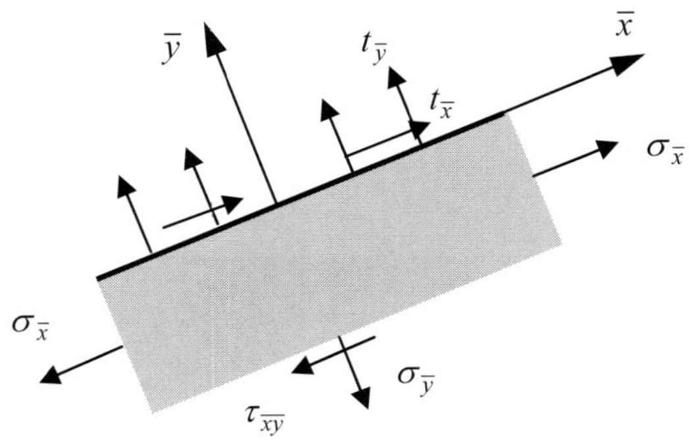

Figure 9.5 Computation of stresses for plane strain problems

The stresses in tangential direction are computed using Hooke`s law. For plane stress conditions we have (Figure 9.5):

$$\sigma_{\bar{x}} = E\varepsilon_{\bar{x}} + \nu t_{\bar{y}} \tag{9.16}$$

where $t_{\bar{y}}$ is the traction normal to the boundary.

For plane strain we have

$$\sigma_{\bar{x}} = \frac{1}{1-\nu}\left(\frac{E}{1+\nu}\varepsilon_{\bar{x}} + \nu\, t_{\bar{y}}\right) \tag{9.17}$$

The local stress pseudo-vector for plane stress is

$$\bar{\boldsymbol{\sigma}} = \begin{pmatrix} \sigma_{\bar{x}} \\ \sigma_{\bar{y}} \\ \sigma_z \\ \tau_{\overline{xy}} \end{pmatrix} = \begin{pmatrix} \sigma_{\bar{x}} \\ t_{\bar{y}} \\ 0 \\ t_{\bar{x}} \end{pmatrix} \tag{9.18}$$

and for plane strain

$$\bar{\boldsymbol{\sigma}} = \begin{pmatrix} \sigma_{\bar{x}} \\ t_{\bar{y}} \\ \nu(\sigma_{\bar{x}} + t_{\bar{y}}) \\ t_{\bar{x}} \end{pmatrix} \tag{9.19}$$

The stresses may be transformed into global directions using Eq (4.36). For three-dimensional problems we compute the strain components in local $\bar{x}, \bar{y}$ directions. These strains are obtained in the same way as for two-dimensional problems by projecting the displacement vector **u** onto the unit tangential vectors $\mathbf{v}_1$ and $\mathbf{v}_2$ and by taking the derivatives to $\bar{x}, \bar{y}$ (see Fig. 9.2). The strains in the local directions are given by

$$\begin{aligned}
\varepsilon_{\bar{x}} &= \frac{\partial u_{\bar{x}}}{\partial \bar{x}} = \left(\frac{\partial \mathbf{u}}{\partial \xi} \bullet \mathbf{v}_1 \right) \frac{\partial \xi}{\partial \bar{x}} \\
\varepsilon_{\bar{y}} &= \frac{\partial u_{\bar{y}}}{\partial \bar{y}} = \left(\frac{\partial \mathbf{u}}{\partial \xi} \bullet \mathbf{v}_2 \right) \frac{\partial \xi}{\partial \bar{y}} + \left(\frac{\partial \mathbf{u}}{\partial \eta} \bullet \mathbf{v}_2 \right) \frac{\partial \eta}{\partial \bar{y}} \\
\gamma_{\bar{x}\bar{y}} &= \frac{\partial u_{\bar{x}}}{\partial \bar{y}} + \frac{\partial u_{\bar{y}}}{\partial \bar{x}} = \left(\frac{\partial \mathbf{u}}{\partial \xi} \bullet \mathbf{v}_1 \right) \frac{\partial \xi}{\partial \bar{y}} + \left(\frac{\partial \mathbf{u}}{\partial \eta} \bullet \mathbf{v}_1 \right) \frac{\partial \eta}{\partial \bar{y}} + \left(\frac{\partial \mathbf{u}}{\partial \xi} \bullet \mathbf{v}_2 \right) \frac{\partial \xi}{\partial \bar{x}}
\end{aligned} \tag{9.20}$$

where

$$\frac{\partial \mathbf{u}}{\partial \xi} = \sum \frac{\partial N_n}{\partial \xi} \cdot \mathbf{u}_n^e \; ; \; \frac{\partial \mathbf{u}}{\partial \eta} = \sum \frac{\partial N_n}{\partial \eta} \cdot \mathbf{u}_n^e \tag{9.21}$$

The components of stress in the local orthogonal system are shown in Figure 9.6. According to Chapter 4, stresses in the tangential plane are related to strains by

$$\begin{aligned}
\varepsilon_{\bar{x}} &= \frac{1}{E}(\sigma_{\bar{x}} - \nu\sigma_{\bar{y}} - \nu\sigma_{\bar{z}}) \\
\varepsilon_{\bar{y}} &= \frac{1}{E}(\sigma_{\bar{y}} - \nu\sigma_{\bar{x}} - \nu\sigma_{\bar{z}}) \\
\gamma_{\bar{x}\bar{y}} &= \frac{1}{G}\tau_{\bar{x}\bar{y}}
\end{aligned} \tag{9.22}$$

Using (9.20) and equilibrium conditions with boundary tractions $t_{\bar{x}}, t_{\bar{y}}, t_{\bar{z}}$, as shown in Figure 9.6, the stresses may be computed as

$$\begin{aligned}
\sigma_{\bar{x}} &= C_1(\varepsilon_{\bar{x}} + \nu\varepsilon_{\bar{y}}) + C_2 t_{\bar{z}} \\
\sigma_{\bar{y}} &= C_1(\varepsilon_{\bar{y}} + \nu\varepsilon_{\bar{x}}) + C_2 t_{\bar{z}} \\
\sigma_{\bar{z}} &= t_{\bar{z}} \\
\tau_{\bar{x}\bar{y}} &= G\gamma_{\bar{x}\bar{y}} \quad \tau_{\bar{x}\bar{z}} = t_{\bar{x}} \quad \tau_{\bar{y}\bar{z}} = t_{\bar{y}}
\end{aligned} \tag{9.23}$$

where

$$C_1 = \frac{E}{1-\nu^2} \; ; \; C_2 = \frac{\nu}{1-\nu} \tag{9.24}$$

The stresses may be transformed into the global x, y, z coordinate system by applying the transformation (4.36). The theory is translated into SUBROUTINE Bstress, which computes the stress components in the tangential plane at a point with the intrinsic coordinates ξ, η on a boundary element. The subroutine is very similar to Bflow, except that in addition to **u**, we have to specify **t** in the list of input parameters and we must provide an indicator specifying whether plane stress or plane strain is assumed for a 2-D analysis.

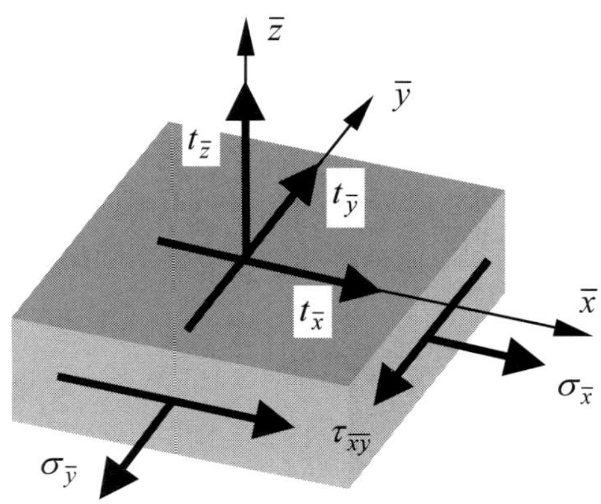

Figure 9.6 Stresses and boundary tractions at a boundary point

```
SUBROUTINE BStress(Stress,xsi,eta,u,t,Inci,Elcor,E,Ny,IPS)
!--------------------------------------------------
!    Computes stresses in a plane tangential to the
!    Boundary Element
!---------------------------------------------------
REAL , INTENT(OUT)     :: Stress(:)!  Stress vector
REAL , INTENT(IN)      :: xsi,eta  !  intrinsic coordinates
REAL , INTENT(IN)      :: u(:,:)   !  Nodal displacements
REAL , INTENT(IN)      :: t(:,:)   !  Nodal Tractions
INTEGER, INTENT (IN)   :: Inci(:)  !  Element Incidences
REAL, INTENT (IN)      :: Elcor(:,:)  !  Element coordinates
REAL, INTENT (IN)      :: E,Ny
INTEGER , INTENT (IN)  :: IPS
REAL, ALLOCATABLE      :: Vxsi(:),Veta(:),DNi(:,:),Ni(:),trac_GP(:)
REAL :: Jxsi,Jeta,v1(3),v2(3),v3(3),v3_L,CosT,SinT
REAL :: DxsiDx, DxsiDy, DetaDx, DetaDy
REAL :: C1,C2,G,tn,ts,ts1,ts2
REAL , ALLOCATABLE :: Dudxsi(:),Dudeta(:),Strain(:)
INTEGER :: Nodes, Cdim, Ldim
Nodes= UBOUND(Elcor,2)
Cdim= UBOUND(Elcor,1)
ldim= Cdim-1
ALLOCATE (Vxsi(cdim),Veta(cdim),Dni(Nodes,Ldim),Ni(Nodes))
```

```
ALLOCATE (Dudxsi(Cdim),Dudeta(Cdim),trac_GP(Cdim))
!   Compute Vector(s) tangential to boundary surface
CALL Serendip_deriv(DNi,xsi,eta,ldim,nodes,inci)
CALL Serendip_func(Ni,xsi,eta,ldim,nodes,inci)
trac_GP(1)= Dot_Product(Ni,t(:,1))
trac_GP(2)= Dot_Product(Ni,t(:,2))
Vxsi(1)= Dot_Product(Dni(:,1),Elcor(1,:))
Vxsi(2)= Dot_Product(Dni(:,1),Elcor(2,:))
IF(Cdim == 2) THEN
!  2-D Calculation
   ALLOCATE (Strain(1))
   CALL Vector_norm(Vxsi,Jxsi)
   V1(1:2)= Vxsi
   V3(1)=   Vxsi(2)
   V3(2)= - Vxsi(1)
   tn= Dot_Product(v3(1:2),trac_GP)
   ts= Dot_Product(v1(1:2),trac_GP)
   DuDxsi(1)= Dot_Product(Dni(:,1),u(:,1))
   DuDxsi(2)= Dot_Product(Dni(:,1),u(:,2))
   Strain(1)= Dot_Product(DuDxsi,V1(1:2))/Jxsi
!   Compute stresses in local directions
   IF(IPS == 2) THEN      ! plane stress
     Stress(1)= E*Strain(1) + ny*tn
     Stress(2)= tn
     Stress(3)= 0
     Stress(4)= ts
   ELSE                   ! Plane strain
     Stress(1)= 1/(1.0-ny)*(E/(1.0+ny)*Strain(1) + ny*tn)
     Stress(2)= tn
     Stress(3)= ny*(Stress(1)+ Stress(2))
     Stress(4)= ts
   END IF
   DEALLOCATE (Strain)
ELSE
!  3-D Calculation
   ALLOCATE (Strain(3))
   trac_GP(3)= Dot_Product(Ni,t(:,3))
   Vxsi(3)= Dot_Product(Dni(:,1),Elcor(3,:))
   CALL Vector_norm(Vxsi,Jxsi)
   Veta(1)= Dot_Product(Dni(:,2),Elcor(1,:))
   Veta(2)= Dot_Product(Dni(:,2),Elcor(2,:))
   Veta(3)= Dot_Product(Dni(:,2),Elcor(3,:))
   CALL Vector_norm(Veta,Jeta)
   v3= Vector_ex(Vxsi,veta)
   CALL Vector_norm(v3,v3_L)
   v1=Vxsi
   v2= Vector_ex(v3,v1)
   DuDxsi(1)= Dot_Product(Dni(:,1),u(:,1))
   DuDxsi(2)= Dot_Product(Dni(:,1),u(:,2))
   DuDxsi(3)= Dot_Product(Dni(:,1),u(:,3))
   DuDeta(1)= Dot_Product(Dni(:,2),u(:,1))
```

```
    DuDeta(2)= Dot_Product(Dni(:,2),u(:,2))
    DuDeta(3)= Dot_Product(Dni(:,2),u(:,3))
    CosT= DOT_Product(Vxsi,Veta)
    SinT= ABS(DOT_Product(V2,Veta))
    DxsiDx= 1/Jxsi
    DxsiDy= -CosT /(Jxsi*SinT)
    DetaDx= 0.0
    DetaDy= 1/(Jeta*SinT)
!    Strains
    Strain(1)= Dot_product(DuDxsi,v1)*DxsiDx
    Strain(2)= Dot_product(DuDxsi,v2)*DxsiDy&
             + Dot_product(DuDeta,v2)*DetaDy
    Strain(3)= Dot_product(DuDxsi,v1)*DxsiDy&
             + Dot_product(DuDeta,v1)*DetaDy&
             + Dot_product(DuDxsi,v2)*DxsiDx
    tn= Dot_Product(v3,trac_GP)
    ts1= Dot_Product(v1,trac_GP)
    ts2= Dot_Product(v2,trac_GP)
!    Compute stresses in local directions
    C1= E/(1.0-ny**2)   ;   C2= ny/(1.0-ny) ; G=E/(2*(1.0+ny))
    Stress(1)= C1*(Strain(1)+ny*strain(2))+ C2*tn
    Stress(2)= C1*(Strain(2)+ny*strain(1))+ C2*tn
    Stress(3)= tn
    Stress(4)= G*Strain(3)
    Stress(5)= ts2
    Stress(6)= ts1
  DEALLOCATE (Strain)
END IF
!  Transformation of local stresses in global stresses
CALL Stress_Transformation(v1,v2,v3,Stress,Cdim)
RETURN
End SUBROUTINE BStress
```

9.3 COMPUTATION OF INTERNAL RESULTS

For the computation of results which are not on the boundary the integral equation for the temperature/potential and the displacement is used.

9.3.1 Potential problems

To compute temperature/potential at a point P_a we rewrite equation (5.20)

$$u(P_a) = \int_S t(Q)\, U(P_a, Q)\, dS(Q) - \int_S u(Q)\, T(P_a, Q)\, dS(Q) \tag{9.25}$$

Flows at P_a in x-, y-and z-directions are given by

$$
\begin{aligned}
q_x(P_a) &= -k\frac{\partial u}{\partial x}(P_a) = -k\left(\int_S t(Q)\frac{\partial U}{\partial x}(P_a,Q)dS(Q) - \int_S u(Q)\frac{\partial T}{\partial x}(P_a,Q)dS(Q) \right) \\
q_y(P_a) &= -k\frac{\partial u}{\partial y}(P_a) = -k\left(\int_S t(Q)\frac{\partial U}{\partial y}(P_a,Q)dS(Q) - \int_S u(Q)\frac{\partial T}{\partial y}(P_a,Q)dS(Q) \right) \\
q_z(P_a) &= -k\frac{\partial u}{\partial z}(P_a) = -k\left(\int_S t(Q)\frac{\partial U}{\partial z}(P_a,Q)dS(Q) - \int_S u(Q)\frac{\partial T}{\partial z}(P_a,Q)dS(Q) \right)
\end{aligned}
\tag{9.26}
$$

where derivatives of U have been presented previously and derivatives of T are for two-dimensional problems

$$
\begin{aligned}
\frac{\partial T}{\partial x} &= \frac{\partial}{\partial x}\left[n_x \frac{\partial U}{\partial x} + n_y \frac{\partial U}{\partial y} \right] = \frac{\cos\theta}{2\pi r^2} r_{,x} \\
\frac{\partial T}{\partial y} &= \frac{\partial}{\partial y}\left[n_x \frac{\partial U}{\partial x} + n_y \frac{\partial U}{\partial y} \right] = \frac{\cos\theta}{2\pi r^2} r_{,y}
\end{aligned}
\tag{9.27}
$$

and for three-dimensional problems

$$
\begin{aligned}
\frac{\partial T}{\partial x} &= \frac{\partial}{\partial x}\left[n_x \frac{\partial U}{\partial x} + n_y \frac{\partial U}{\partial y} + n_z \frac{\partial U}{\partial z} \right] = \frac{\cos\theta}{4\pi r^3} r_{,x} \\
\frac{\partial T}{\partial y} &= \frac{\partial}{\partial y}\left[n_x \frac{\partial U}{\partial x} + n_y \frac{\partial U}{\partial y} + n_z \frac{\partial U}{\partial z} \right] = \frac{\cos\theta}{4\pi r^3} r_{,y} \\
\frac{\partial T}{\partial z} &= \frac{\partial}{\partial z}\left[n_x \frac{\partial U}{\partial x} + n_y \frac{\partial U}{\partial y} + n_z \frac{\partial U}{\partial z} \right] = \frac{\cos\theta}{4\pi r^3} r_{,z}
\end{aligned}
\tag{9.28}
$$

The derivatives of fundamental solutions T for three-dimensional space have a singularity of $1/r^2$ for 2-D and $1/r^3$ for 3-D problems and are therefore *hypersingular*. We now extend the Laplace_lib to include the derivatives of the fundamental solution.

```
FUNCTION dU(r,dxr,Cdim)
!--------------------------------
!   Derivatives of Fundamental solution for Potential problems
!   Temperature/Potential
!--------------------------------
REAL,INTENT(IN):: r     !  Distance between source and field point
REAL,INTENT(IN)::  dxr(:)!  Distances in x,y directions div. by r
REAL :: dU(UBOUND(dxr,1))     !   dU is array of same dim as dxr
INTEGER ,INTENT(IN):: Cdim    !   Cartesian dimension (2-D,3-D)
REAL :: C
SELECT CASE (CDIM)
    CASE (2)              !  Two-dimensional solution
```

```
     C=1/(2.0*Pi*r)
     dU(1)= C*dxr(1)
     dU(2)= C*dxr(2)
    CASE (3)             !  Three-dimensional solution
     C=1/(4.0*Pi*r**2)
     dU(1)= C*dxr(1)
     dU(2)= C*dxr(2)
     dU(3)= C*dxr(3)
    CASE DEFAULT
END SELECT
RETURN
END FUNCTION dU
FUNCTION dT(r,dxr,Vnorm,Cdim)
!-------------------------------
!  derivatives of the Fundamental solution for Potential problems
!  Normal gradient
!-------------------------------
INTEGER,INTENT(IN) :: Cdim      !   Cartesian dimension
REAL,INTENT(IN):: r  !   Distance between source and field point
REAL,INTENT(IN):: dxr(:)!Distances in Cartesian dir divided by R
REAL,INTENT(IN)::  Vnorm(:)     !   Normal vector
REAL :: dT(UBOUND(dxr,1))       !   dT is array of same dim as dxr
REAL :: C,COSTH
COSTH= DOT_PRODUCT (Vnorm,dxr)
SELECT CASE (Cdim)
    CASE (2)             !  Two-dimensional solution
     C= 1/(2.0*Pi*r**2)
     dT(1)= C*COSTH*dxr(1)
     dT(2)= C*COSTH*dxr(2)
    CASE (3)             !  Three-dimensional solution
     C= 3/(4.0*Pi*r**3)
     dT(1)= C*COSTH*dxr(1)
     dT(2)= C*COSTH*dxr(2)
     dT(3)= C*COSTH*dxr(3)
    CASE DEFAULT
END SELECT
RETURN
END FUNCTION dT
```

The discretised form of equation (9.25) is

$$u\left(P_a\right)=\sum_{e=1}^{E}\sum_{n=1}^{N}\Delta T_n^e(P_a)u_n^e-\sum_{e=1}^{E}\sum_{n=1}^{N}\Delta U_n^e(P_a)t_n^e \tag{9.29}$$

where u_n^e and t_n^e are the solutions obtained for the temperature/potential and boundary flow on node n on boundary element e and

$$\Delta T_n^e = \int_{S_e} T(P_a, Q) N_n \, dS_e(Q) \quad , \quad \Delta U_n^e = \int_{S_e} U(P_a, Q) \, N_n dS_e(Q) \tag{9.30}$$

The discretised form of equation (9.26) is given by

$$\mathbf{q}(P_a) = -k\left(\sum_{e=1}^{E} \sum_{n=1}^{N} \Delta \mathbf{S}_n^e \, t_n^e - \sum_{e=1}^{E} \sum_{n=1}^{N} \Delta \mathbf{R}_n^e \, u_n^e \right) \tag{9.31}$$

where

$$\mathbf{q} = \begin{Bmatrix} q_x \\ q_y \\ q_z \end{Bmatrix} \quad and \quad \Delta \mathbf{S}_n^e = \begin{Bmatrix} \Delta \mathrm{S}_{xn}^e \\ \Delta \mathrm{S}_{yn}^e \\ \Delta \mathrm{S}_{zn}^e \end{Bmatrix} \quad ; \quad \Delta \mathbf{R}_n^e = \begin{Bmatrix} \Delta \mathrm{R}_{xn}^e \\ \Delta \mathrm{R}_{yn}^e \\ \Delta \mathrm{R}_{zn}^e \end{Bmatrix} \tag{9.32}$$

The components of $\Delta\mathbf{S}$ and $\Delta\mathbf{R}$ are defined as

$$\begin{aligned} \Delta S_{xn}^e &= \int_{S_e} \frac{\partial U}{\partial x}(P_a, Q) N_n dS(Q) \quad ; \quad \Delta S_{yn}^e = \int_{S_e} \frac{\partial U}{\partial y}(P_a, Q) N_n dS(Q) \quad etc. \\ \Delta R_{xn}^e &= \int_{S_e} \frac{\partial T}{\partial x}(P_a, Q) N_n dS(Q) \quad ; \quad \Delta R_{yn}^e = \int_{S_e} \frac{\partial T}{\partial y}(P_a, Q) N_n dS(Q) \quad etc. \end{aligned} \tag{9.33}$$

The integrals can be evaluated numerically over element **e** using Gauss Quadrature, as explained in detail in Chapter 6. For 2-D problems this is

$$\begin{aligned} \Delta U_n^e &= \sum_{k=1}^{K} U(P_a, Q(\xi_k)) \, N_n(\xi_k) J(\xi_k) W_k \\ \Delta T_n^e &= \sum_{k=1}^{K} T(P_a, Q(\xi_k)) \, N_n(\xi_k) J(\xi_k) W_k \end{aligned} \tag{9.34}$$

and

$$\begin{aligned} \Delta S_{xn}^e &= \sum_{k=1}^{K} \frac{\partial U}{\partial x}(P_a, Q(\xi_k)) \, N_n(\xi_k) J(\xi_k) W_k \quad etc. \\ \Delta R_{xn}^e &= \sum_{k=1}^{K} \frac{\partial T}{\partial y}(P_a, Q(\xi_k)) \, N_n(\xi_k) J(\xi_k) W_k \quad etc. \end{aligned} \tag{9.35}$$

For 3-D problems the equations are

$$\Delta U_n^e = \sum_{m=1}^{M}\sum_{k=1}^{K} U(P_a, Q(\xi_k, \eta_m))\, N_n(\xi_k, \eta_m) J(\xi_k, \eta_m) W_k W_m$$
$$\Delta T_n^e = \sum_{m=1}^{M}\sum_{k=1}^{K} T(P_a, Q(\xi_k, \eta_m))\, N_n(\xi_k, \eta_m) J(\xi_k, \eta_m) W_k W_m \tag{9.36}$$

and

$$\Delta S_{xn}^e = \sum_{m=1}^{M}\sum_{k=1}^{K} \frac{\partial}{\partial x} U(P_a, Q(\xi_k, \eta_m)) \cdot N_n(\xi_k, \eta_m) \cdot J(\xi_k, \eta_m) W_k W_m \qquad etc. \tag{9.37}$$

The number of Gauss points in ξ and η direction M,K needed for accurate integration will again depend on the proximity of P_a to the element over which the integration is carried out. For computation of displacements, Kernel **T** has a singularity of $1/r$ for 2-D problems and $1/r^2$ for 3-D. Kernel **R** has a $1/r^2$ singularity for 2-D and a $1/r^3$ singularity for 3-D problems and the number of integration points is chosen according to Table 6.1.

9.3.2 Elasticity problems

The displacements at a point P_a inside the domain can be computed by using the integral equation for the displacement

$$\mathbf{u}(P_a) = \int_S \mathbf{U}(P_a, Q)\mathbf{t}(Q) dS - \int_S \mathbf{T}(P_a, Q)\mathbf{u}(Q) dS \tag{9.38}$$

The strains can be computed by using equation (4.31)

$$\boldsymbol{\varepsilon} = \mathbf{B}\mathbf{u}(P_a) = \int_S \mathbf{B}\mathbf{U}(P_a, Q)\mathbf{t}(Q) dS - \int_S \mathbf{B}\mathbf{T}(P_a, Q)\mathbf{u}(Q) dS \tag{9.39}$$

Finally, stresses can be computed by using equation (4.45)

$$\boldsymbol{\sigma} = \mathbf{D}\boldsymbol{\varepsilon} = \int_S \mathbf{DBU}(P_a, Q)\mathbf{t}(Q) dS - \int_S \mathbf{DBT}(P_a, Q)\mathbf{u}(Q) dS \tag{9.40}$$

or

$$\boldsymbol{\sigma} = \int_S \mathbf{S}(P_a, Q)\mathbf{t}(Q) dS - \int_S \mathbf{R}(P_a, Q)\mathbf{u}(Q) dS \tag{9.41}$$

where the derived fundamental solutions **S** and **R** are defined as

$$\mathbf{S} = \mathbf{DBU}(P_a, Q) \quad , \quad \mathbf{R} = \mathbf{DBT}(P_a, Q) \tag{9.42}$$

and the pseudo-stress vector $\boldsymbol{\sigma}$ is defined as

$$\boldsymbol{\sigma} = \begin{Bmatrix} \sigma_x \\ \sigma_y \\ \sigma_z \\ \tau_{xy} \\ \tau_{yz} \\ \tau_{xz} \end{Bmatrix} \quad for \quad 3-D \quad and \quad \boldsymbol{\sigma} = \begin{Bmatrix} \sigma_x \\ \sigma_y \\ \tau_{xy} \end{Bmatrix} \quad for \quad 2-D \tag{9.43}$$

Matrices **S** and **R** are of dimension 3x2 for two-dimensional problems and of dimension 6x3 for three-dimensional problems.

Matrix **S** is given by[1]

$$\mathbf{S} = \begin{bmatrix} S_{xxx} & S_{xxy} & S_{xxz} \\ S_{yyx} & S_{yyy} & S_{yyz} \\ S_{zzx} & S_{zzy} & S_{zzz} \\ S_{xyx} & S_{xyy} & S_{xyz} \\ S_{yzx} & S_{yzy} & S_{yzz} \\ S_{xzx} & S_{xzy} & S_{xzz} \end{bmatrix} \tag{9.44}$$

The coefficients of **S** are given by:

$$S_{ijk} = \frac{C_2}{r^n}\left[C_3(\delta_{ki} r_{,j} + \delta_{kj} r_{,i} - \delta_{ij} r_{,k}) + (n+1) r_{,i} r_{,j} r_{,k} \right] \tag{9.45}$$

Values x, y, z are substituted for i, j, k. Constants are defined in Table 9.1 for plane stress/strain and 3-D problems and

Matrix **R** is given by

$$\mathbf{R} = \begin{bmatrix} R_{xxx} & R_{xxy} & R_{xxz} \\ R_{yyx} & R_{yyy} & R_{yyz} \\ R_{zzx} & R_{zzy} & R_{zzz} \\ R_{xyx} & R_{xyy} & R_{xyz} \\ R_{yzx} & R_{yzy} & R_{yzz} \\ R_{xzx} & R_{xzy} & R_{xzz} \end{bmatrix} \tag{9.46}$$

where[1]

$$R_{kij} = \frac{C_5}{r^{n+1}} \begin{bmatrix} (n+1)\cos\theta(C_3\delta_{ij}r_{,k} + \nu(\delta_{ik}r_{,j} + \delta_{jk}r_{,i}) - C_6 r_{,i}r_{,j}r_{,k}) \\ +(n+1)\nu(n_i r_{,j}r_{,k} + n_j r_{,i}r_{,k}) \\ +C_3((n+1)n_k r_{,i}r_{,j} + n_j\delta_{ik} + n_i\delta_{jk}) - C_7 n_k\delta_{ij} \end{bmatrix} \tag{9.47}$$

x, y, z may be substituted for *i, j, k* and $cos\theta$ has been defined previously. Values of the constants are given in Table 9.1.

Table 9.1 Constants for fundamental solutions **S** and **R**

	Plane strain	Plane stress	3-D
n	1	1	2
C_2	1/4π(1−ν)	(1+ν)/4π	1/8π(1−ν)
C_3	1-2 ν	(1- ν)/(1+ ν)	1-2 ν
C_5	G/(2π(1-ν))	(1+ ν)G/2π	G/(4π(1-ν))
C_6	4	4	15
C_7	1-4 ν	(1-3 ν)/(1+ ν)	1-4 ν

For plane stress assumptions the stresses perpendicular to the plane are computed by $\sigma_z = 0$, whereas for plane strain $\sigma_z = \nu(\sigma_x + \sigma_y)$.

Subroutines for calculating Kernels **S** and **R** are added to the Elasticity_lib.

```
SUBROUTINE SK(TS,DXR,R,C2,C3)
!-------------------------------------------------------------
!   KELVIN SOLUTION FOR STRESS
!   TO BE MULTIPLIED WITH t
!-------------------------------------------------------------
REAL, INTENT(OUT) :: TS(:,:)    ! Fundamental solution
REAL, INTENT(IN)  :: DXR(:)     ! rx , ry, rz
REAL, INTENT(IN)  :: R          ! r
REAL, INTENT(IN)  :: C2,C3      ! Elastic constants
REAL       ::  Cdim         !  Cartesian dimension
INTEGER    :: NSTRES        !  No. of stress components
INTEGER    :: JJ(6), KK(6)  !  sequence of stresses in pseudo-vector
REAL       :: A,C2,C3
INTEGER    :: I,N,J,K
Cdim= UBOUND(DXR,1)
IF(CDIM == 2) THEN
   NSTRES= 3
   JJ(1:3)= (/1,2,1/)
   KK(1:3)= (/1,2,2/)
ELSE
   NSTRES= 6
   JJ= (/1,2,3,1,2,3/)
```

```
   KK= (/1,2,3,2,3,1/)
END IF
Coor_directions:&
DO I=1,Cdim
   Stress_components:&
   DO N=1,NSTRES
      J= JJ(N)
      K= KK(N)
      A= 0.
      IF(I .EQ. K) A= A + DXR(J)
      IF(J .EQ. K) A= A - DXR(I)
      IF(I .EQ. J) A= A + DXR(K)
      A= A*C3
      TS(I,N)= C2/R*(A + Cdim*DXR(I)*DXR(J)*DXR(K))
      IF(Cdim .EQ. 3) TS(I,N)= TS(I,N)/2./R
   END DO &
   Stress_components
END DO &
Coor_directions
RETURN
END SUBROUTINE SK
SUBROUTINE RK(US,DXR,R,VNORM,C3,C5,C6,C7,ny)
!---------------------------------------------------------------
!    KELVIN SOLUTION FOR STRESS COMPUTATION
!    TO BE MULTIPLIED WITH u
!---------------------------------------------------------------
REAL, INTENT(OUT) :: US(:,:)        ! Fundamental solution
REAL, INTENT(IN)  :: DXR(:)         ! rx , ry, rz
REAL, INTENT(IN)  :: R              ! r
REAL, INTENT(IN)  :: VNORM(:)       ! nx , ny , nz
REAL, INTENT(IN)  :: C3,C5,C7,ny    ! Elastic constants
REAL      ::  Cdim        !  Cartesian dimension
INTEGER   :: NSTRES       !  No. of stress components
INTEGER   :: JJ(6), KK(6) !  sequence of stresses in pseudo-vector
REAL      :: costh, B,C
Cdim= UBOUND(DXR,1)
IF(CDIM == 2) THEN
  NSTRES= 3
  JJ(1:3)= (/1,2,1/)
  KK(1:3)= (/1,2,2/)
ELSE
  NSTRES= 6
  JJ= (/1,2,3,1,2,3/)
  KK= (/1,2,3,2,3,1/)
END IF
COSTH= DOT_Product(dxr,vnorm)
Coor_directions:&
DO K=1,Cdim
  Stress_components:&
  DO N=1,NSTRES
```

```
      I= JJ(N)
      J= KK(N)
      B= 0.
      IF(I .EQ. J) B= Cdim* C3* DXR(K)
      IF(I .EQ. K) B= B + ny* DXR(J)
      IF(J .EQ. K) B= B + ny* DXR(I)
      B= COSTH * (B - C6* DXR(I) * DXR(J) * DXR(K) )
      C= DXR(J) * DXR(K) * ny
      IF(J .EQ.K) C= C + C3
      C= C* VNORM(I)
      B= B+C
      C= DXR(I) * DXR(K) * ny
      IF(I .EQ. K) C=C + C3
      C= C* VNORM(J)
      B= B+C
      C= DXR(I) * DXR(J) * Cdim* C3
      IF(I .EQ. J) C= C - C7
      C= C* VNORM(K)
      US(K,N)= (B + C) * C5/R/R
      IF(Cdim .EQ. 3) US(K,N)= US(K,N)/2./R
  END DO &
  Stress_components
END DO &
Coor_directions
RETURN
END
```

The discretised form of equation (9.38) is written as

$$\mathbf{u}(P_a) = \sum_{e=1}^{E}\sum_{n=1}^{N} \Delta\mathbf{U}_n^e \; \mathbf{t}_n^e \; - \sum_{e=1}^{E}\sum_{n=1}^{N} \Delta\mathbf{T}_n^e \; \mathbf{u}_n^e \tag{9.48}$$

where

$$\Delta\mathbf{U}_n^e = \int_{S_e} \mathbf{U}(P_a,Q) N_n dS(Q) \quad ; \quad \Delta\mathbf{T}_n^e = \int_{S_e} \mathbf{T}(P_a,Q) N_n dS(Q) \tag{9.49}$$

The discretised form of equation (9.41) is written as

$$\boldsymbol{\sigma}(P_a) = \sum_{e=1}^{E}\sum_{n=1}^{N} \Delta\mathbf{S}_n^e \; \mathbf{t}_n^e - \sum_{e=1}^{E}\sum_{n=1}^{N} \Delta\mathbf{R}_n^e \; \mathbf{u}_n^e \tag{9.50}$$

where

$$\Delta\mathbf{S}_n^e = \int_{S_e} \mathbf{S}(P_a,Q) N_n dS(Q) \quad ; \quad \Delta\mathbf{R}_n^e = \int_{S_e} \mathbf{R}(P_a,Q) N_n dS(Q) \tag{9.51}$$

These integrals may be evaluated using Gauss Quadrature, as explained in Chapter 6. For 2-D problems they are given by

$$\Delta \mathbf{U}_n^e = \sum_{k=1}^{K} \mathbf{U}(P_a, Q(\xi_k))\, N_n(\xi_k) J(\xi_k) W_k$$
$$\Delta \mathbf{T}_n^e = \sum_{k=1}^{K} \mathbf{T}(P_a, Q(\xi_k))\, N_n(\xi_k) J(\xi_k) W_k \qquad etc. \tag{9.52}$$

For 3-D elasticity we have

$$\Delta \mathbf{U}_n^e = \sum_{m=1}^{M} \sum_{k=1}^{K} \mathbf{U}(P_a, Q(\xi_k, \eta_m))\, N_n(\xi_k, \eta_m) J(\xi_k, \eta_m) W_k W_m$$
$$\Delta \mathbf{T}_n^e = \sum_{m=1}^{M} \sum_{k=1}^{K} \mathbf{T}(P_a, Q(\xi_k, \eta_m))\, N_n(\xi_k, \eta_m) J(\xi_k, \eta_m) W_k W_m \quad etc. \tag{9.53}$$

The number of Gauss points in ξ and η direction M,K needed for accurate integration, will again depend on the proximity of P_a to the element over which the integration is carried out. For computation of displacements Kernel **T** has a singularity of $1/r$ for 2-D problems and $1/r^2$ for 3-D. The number of integration points M and K are chosen according to Table 6.1. A subdivision of the region of integration as outlined in Chapter 6 will be necessary for points that are close.

9.4 PROGRAM 9.1: POSTPROCESSOR

Program Postprocessor for computing results on the boundary and inside the domain is presented. This program is exacuted after General_purpose_BEM. It reads the INPUT file which is the same as the one read by General_Purpose_BEM and contains the basic job information and the geometry of boundary elements. The results of the boundary element computation are read from file BERESULTS, which was generated by General_purpose BEM program and contains the values of u and t at boundary points. The coordinates of internal points are supplied in file INPUT2 and the internal results are written onto file OUTPUT. The program first calculates fluxes/stresses at the nodes of specified boundary elements and then temperatures/displacements and fluxes/stresses at specified points inside the domain. In the case of symmetry conditions being applied the integration has to be carried out also over the mirrored elements. A call to Subroutine MIRROR takes care of this. For calculation of internal points, the integration is carried out separately for the computation of potentials/displacements and flow/stresses, as the Kernels have different singularities. This may not be the most efficient way and an over-integration of the first Kernels may be considered to improve the efficiency, since certain computations, like the Jacobian, for example, may only be computed once for a boundary element. Another improvement in efficiency can be made by lumping together internal points, so that only one integration loop is needed for all points requiring the same number of integration points. In this case the number of computations of the *Jacobian* can be reduced significantly. Using table 6.1 and element subdivision it will be found later that the internal points may be placed quite close to the boundary.

```
PROGRAM Post_processor
!-------------------------------------------------
!     General purpose Postprocessor
!     for computing results at boundary and interior points
!-------------------------------------------------------
USE Utility_lib;USE Elast_lib;USE Laplace_lib
USE Integration_lib
USE Postproc_lib
IMPLICIT NONE
INTEGER, ALLOCATABLE :: Inci(:)     !  Incidences (one elem.)
INTEGER, ALLOCATABLE :: Incie(:,:)  !  Incidences (all elem.)
INTEGER, ALLOCATABLE :: Ldest(:)    !  Destinations (one elem.)
REAL, ALLOCATABLE    :: Elcor(:,:)  !  Element coordinates
REAL, ALLOCATABLE    :: El_u(:,:,:)!
REAL, ALLOCATABLE    :: El_t(:,:,:)!  Results of System
REAL, ALLOCATABLE    :: El_ue(:,:)  !  Diplacements of Element
REAL, ALLOCATABLE    :: El_te(:,:)  !  Traction of Element
REAL, ALLOCATABLE    :: Disp(:)     !  Diplacement results Node
REAL, ALLOCATABLE    :: Trac(:)     !  Traction results of Node
REAL, ALLOCATABLE    :: El_trac(:)  !  Traction results Element
REAL, ALLOCATABLE    :: El_disp(:)  !  Displacement of Element
REAL, ALLOCATABLE    :: xP(:,:)     !  Node co-ordinates of BE
REAL, ALLOCATABLE    :: xPnt(:)     !  Co-ordinates of int. point
REAL, ALLOCATABLE    :: Ni(:),GCcor(:),dxr(:),Vnorm(:)
CHARACTER (LEN=80)   :: Title
REAL :: Elengx,Elenge,Rmin,Glcorx(8),Wix(8),Glcore(8),Wie(8)
REAL :: Jac
REAL :: Xsi1,Xsi2,Eta1,Eta2,RJacB,RonL
REAL, ALLOCATABLE :: Flow(:),Stress(:)!  Results for bound.Point
REAL, ALLOCATABLE :: uPnt(:),SPnt(:) !  Results for int Point
REAL, ALLOCATABLE :: TU(:,:),UU(:,:) !  Kernels for u
REAL, ALLOCATABLE :: TS(:,:),US(:,:) !  Kernels for q,s
REAL, ALLOCATABLE :: Fac(:),Fac_nod(:,:)   !   Fact. for symmetry
INTEGER :: Cdim,Node,M,N,Istat,Nodel,Nel,Ndof,Cod,Nreg
INTEGER :: Ltyp,Nodes,Maxe,Ndofe,Ndofs,Ncol,ndg,ldim
INTEGER :: nod,nd,Nstres,Nsym,Isym,nsy,IPS,Nan,Nen,Ios,dofa,dofe
INTEGER :: Mi,Ki,K,I,NDIVX,NDIVSX,NDIVE,NDIVSE,MAXDIVS
REAL    :: Con,E,ny,Fact,G,C2,C3,C5,C6,C7
REAL    :: xsi,eta,Weit,R,Rlim(2)
OPEN (UNIT=1,FILE='INPUT',FORM='FORMATTED')
OPEN (UNIT=2,FILE='OUTPUT',FORM='FORMATTED')
Call Jobin(Title,Cdim,Ndof,IPS,Nreg,Ltyp,Con,E,ny,&
           Isym,nodel,nodes,maxe)
Ndofe= nodel*ndof
ldim= Cdim-1
```

```
 Nsym= 2**Isym   !   number of symmetry loops
 ALLOCATE(xP(Cdim,Nodes))   !  Array for node coordinates
 ALLOCATE(Incie(Maxe,Nodel),Inci(Nodel),Ldest(Ndofe))
 ALLOCATE(Ni(Nodel),GCcor(Cdim),dxr(Cdim),Vnorm(Cdim))
 CALL Geomin(Nodes,Maxe,xp,Incie,Nodel,Cdim)
 !  Compute constants
 IF(Ndof == 1) THEN
  Nstres= Cdim
 ELSE
  G= E/(2.0*(1+ny))
  C2= 1/(8*Pi*(1-ny))
  C3= 1.0-2.0*ny
  C5= G/(4.0*Pi*(1-ny))
  C6= 15
  C7= 1.0-4.0*ny
  Nstres= 6
  IF(Cdim == 2) THEN
   IF(IPS == 1) THEN
! Plane Strain
      C2= 1/(4*Pi*(1-ny))
      C5= G/(2.0*Pi*(1-ny))
      C6= 8
      Nstres= 4
   ELSE
      C2= (1+ny)/(4*Pi              )
! Plane Stress
      C3= (1.0-ny)/(1.0+ny)
      C5= (1.0+ny)*G/(2.0*Pi)
      C6= 8
      C7= (1.0-3.0*ny)/(1.0+ny)
      Nstres= 4
   END IF
  END IF
 END IF
 ALLOCATE(El_u(Maxe,Nodel,ndof),El_t(Maxe,Nodel,ndof)&
 ,El_te(Nodel,ndof),El_ue(Nodel,ndof),Fac_nod(Nodel,ndof))
 ALLOCATE(El_trac(Ndofe),El_disp(Ndofe))
 CLOSE(UNIT=1)
 OPEN (UNIT=1,FILE='BERESULTS',FORM='FORMATTED')
 WRITE(2,*) ' '
 WRITE(2,*) 'Post-processed Results'
 WRITE(2,*) ' '
 Elements1:&
 DO Nel=1,Maxe
    READ(1,*) ((El_u(nel,n,m),m=1,ndof),n=1,Nodel)
    READ(1,*) ((El_t(nel,n,m),m=1,ndof),n=1,Nodel)
 END DO &
 Elements1
 ALLOCATE(Elcor(Cdim,Nodel))
 CLOSE(UNIT=1)
 OPEN (UNIT=1,FILE='INPUT2',FORM='FORMATTED')
```

```
ALLOCATE(Flow(Cdim),Stress(Nstres))
!-------------------------------------------------------------
!     Computation of boundary fluxes/stresses
!--------------------------------------------------------------
WRITE(2,*) 'Results at nodes of Boundary Elements:'
READ(1,*) Nan,Nen
IF(Nan > 0) THEN
 Element_loop: &
 DO NEL= Nan,Nen
    Inci= Incie(nel,:)
    Elcor= xp(:,Inci(:))
    Eta= -1.0
    Eta_loop: &
    DO Net=1,NETA
       Xsi= -1.0
       Xsi_loop: &
       DO Nxs=1,NXSI
          IF(Ndof == 1) THEN
           Flow= 0.0
           Call BFLOW(Flow,xsi,eta,El_u(Nel,:,:),Inci,Elcor,Con)
           WRITE(2,'(A,I5,A,F6.2,A,F6.2)') 'Element',Nel,&
           ' xsi=',xsi,' eta=',eta
           WRITE(2,'(A,2F10.3)') 'Flux: ',Flow
          ELSE
           Stress= 0.0
           Call BStress(Stress,xsi,eta,El_u(Nel,:,:)&
                        ,El_t(Nel,:,:),Inci,Elcor,E,ny,IPS)
           WRITE(2,'(A,I5,A,F6.2,A,F6.2)') 'Element',Nel,&
           ' xsi=',xsi,' eta=',eta
           WRITE(2,'(A,6F9.2)') ' Stress: ',Stress
          END IF
          Xsi= Xsi +1
        END DO Xsi_loop
        Eta=Eta+1.0
     END DO Eta_loop
  END DO Element_loop
END IF
ALLOCATE(uPnt(NDOF),SPnt(NSTRES),UU(NDOF,NDOF),TU(NDOF,NDOF))
ALLOCATE(TS(Nstres,Ndof),US(Nstres,Ndof))
ALLOCATE(xPnt(Cdim),Fac(Ndofe))
ALLOCATE(Disp(Cdim),Trac(Cdim))
WRITE(2,*)''
WRITE(2,*) 'Internal Results:'
WRITE(2,*)''
Internal_points: &
DO
   READ(1,*,IOSTAT=IOS) xPnt
   IF(IOS /= 0) EXIT
   Write(2,'(A,3F10.2)') 'Coordinates: ',xPnt
!  Temperatures/Displacements at Points inside a region
   uPnt= 0.0
```

```
Element_loop1: &
DO NEL= 1,MAXE
   Symmetry_loop1:&
   DO nsy=1,Nsym
      Inci= Incie(nel,:)
      Elcor= xp(:,Inci(:))
      IF(ldim == 2) THEN
        ELengx= Dist((Elcor(:,3)+Elcor(:,2))/2.&
        ,(Elcor(:,4)+Elcor(:,1))/2.,Cdim)  ! Lxsi
        ELenge= Dist((Elcor(:,2)+Elcor(:,1))/2.&
         ,(Elcor(:,3)+Elcor(:,4))/2.,Cdim)  ! Leta
       ELSE
         Call Elength(Elengx,Elcor,nodel,ldim)
       END IF
       Ldest= 1
       Fac= 1.0
       Fac_nod=1.0
       El_ue(:,:)=El_u(Nel,:,:)
       El_te(:,:)=El_t(Nel,:,:)
       IF(Isym > 0) THEN
         DO Nod=1,Nodel
            dofa= (nod-1)*Ndof+1
            dofe= dofa+Ndof-1
            El_trac(dofa:dofe)= El_te(Nod,:)
            EL_disp(dofa:dofe)= El_ue(Nod,:)
          END DO
          CALL Mirror(Isym,nsy,Nodes,Elcor,Fac,Inci&
                      ,Ldest,El_trac,EL_disp &
                      ,nodel,ndof,Cdim)
          DO Nod=1,Nodel
             dofa= (nod-1)*Ndof+1
             dofe= dofa+Ndof-1
             El_te(Nod,:)= El_trac(dofa:dofe)
             El_ue(Nod,:)= El_disp(dofa:dofe)
             Fac_nod(Nod,:)= Fac(dofa:dofe)
          END DO
         END IF
         Rmin= Min_dist(Elcor,xPnt,Nodel,ldim,Inci)
         Mi= Ngaus(Rmin/Elengx,Cdim-1,Rlim)
         NDIVSX= 1 ; NDIVSE= 1
         RJacB=1.0
         RonL= Rmin/Elengx
         IF(Mi == 5) THEN   ! subdivision required
           IF(RonL > 0.0) NDIVSX= INT(RLim(2)/RonL) + 1
           IF(NDIVSX > MAXDIVS) MAXDIVS= NDIVSX
           Mi=4
         END IF
         CALL Gauss_coor(Glcorx,Wix,Mi)  ! Coords/Wghts x dir
         Ki= 1 ; Wie(1)= 1.0 ; Glcore(1)= 0.0
         IF(Cdim == 3) THEN
            Ki= Ngaus(Rmin/Elenge,Cdim-1,Rlim)  !
```

```
          RonL= Rmin/Elenge
          IF(Ki == 5) THEN
            IF(RonL > 0.0) NDIVSE= INT(RLim(2)/RonL) + 1
            IF(NDIVSE > MAXDIVS) MAXDIVS= NDIVSE
            Ki=4
          END IF
          CALL Gauss_coor(Glcore,Wie,Ki) ! Coords/Weights
        END IF
  IF(NDIVSX > 1 .OR. NDIVSE>1) THEN
   RJacB= 1.0/(NDIVSX*NDIVSE)
 END IF
  Xsi1=-1.0
  Eta1=-1.0
  Subdivisions_xsi:&
  DO NDIVX=1,NDIVSX
  Xsi2= Xsi1 + 2.0/NDIVSX
    Subdivisions_eta:&
    DO NDIVE=1,NDIVSE
    Eta2= Eta1 + 2.0/NDIVSE
      Gauss_points_xsi: &
      DO m=1,Mi
      xsi= Glcorx(m)
      IF(NDIVSX > 1) xsi= 0.5*(Xsi1+Xsi2)+xsi/NDIVSX
        Gauss_points_eta: &
        DO k=1,Ki
        eta= Glcore(k)
        IF(NDIVSE > 1) eta= 0.5*(Eta1+Eta2)+eta/NDIVSE
        Weit= Wix(m)*Wie(k)
        CALL Serendip_func(Ni,xsi,eta,ldim,nodel,Inci)
        CALL Normal_Jac(Vnorm,Jac,xsi,eta,ldim&
                       ,nodel,Inci,elcor)
        Fact= Weit*Jac*RJacB
        CALL Cartesian(GCcor,Ni,ldim,elcor)
        r= Dist(GCcor,xPnt,Cdim)        !  Dist. P,Q
        dxr= (GCcor-xPnt)/r             !  rx/r , ry/r  etc
        IF(Ndof .EQ. 1) THEN
          TU= U(r,Con,Cdim)   ; UU= T(r,dxr,Vnorm,Cdim)
        ELSE
          TU= UK(dxr,r,E,ny,Cdim)
          UU= TK(dxr,r,Vnorm,ny,Cdim)
        END IF
         Node_loop1:&
         DO Node=1,Nodel
          Disp= El_ue(Node,:)* Fac_nod(Node,:)
          Trac= El_te(Node,:)* Fac_nod(Node,:)
          uPnt= uPnt + (MATMUL(TU,Trac)-&
            MATMUL(UU,Disp))* Ni(Node)* Fact
         END DO &
         Node_loop1
        END DO &
        Gauss_points_eta
```

```
            END DO &
            Gauss_points_xsi
            Eta1= Eta2
          END DO Subdivisions_eta
          Xsi1= Xsi2
      END DO Subdivisions_xsi
    END DO Symmetry_loop1
  END DO Element_loop1
  WRITE(2,'(A,3F10.3)') '          u: ',uPnt
!------------------------------------------------------------
!    Computation of Fluxes/Stresses at Points inside a region
!------------------------------------------------------------
   SPnt= 0.0
   Element_loop2: &
   DO NEL= 1,MAXE
     Symmetry_loop2: &
     DO nsy=1,Nsym
       Inci= Incie(nel,:)
       Elcor= xp(:,Inci(:))
       IF(ldim == 2) THEN
         ELengx= Dist((Elcor(:,3)+Elcor(:,2))/2.&
                  ,(Elcor(:,4)+Elcor(:,1))/2.,Cdim)   ! Lxsi
         ELenge= Dist((Elcor(:,2)+Elcor(:,1))/2.&
                  ,(Elcor(:,3)+Elcor(:,4))/2.,Cdim)   ! Leta
       ELSE
         Call Elength(Elengx,Elcor,nodel,ldim)
       END IF
       Ldest= 1
       Fac= 1.0
       El_ue(:,:)=El_u(Nel,:,:)
       El_te(:,:)=El_t(Nel,:,:)
       IF(Isym > 0) THEN
            DO Nod=1,Nodel
               dofa= (nod-1)*Ndof+1
               dofe= dofa+Ndof-1
               El_trac(dofa:dofe)= El_te(Nod,:)
               EL_disp(dofa:dofe)= El_ue(Nod,:)
            END DO
            CALL Mirror(Isym,nsy,Nodes,Elcor,Fac,Inci&
                        ,Ldest,El_trac,EL_disp &
                        ,nodel,ndof,Cdim)
            DO Nod=1,Nodel
               dofa= (nod-1)*Ndof+1
               dofe= dofa+Ndof-1
               El_te(Nod,:)= El_trac(dofa:dofe)
               El_ue(Nod,:)= El_disp(dofa:dofe)
               Fac_nod(Nod,:)= Fac(dofa:dofe)
            END DO
       End IF
       Rmin= Min_dist(Elcor,xPnt,Nodel,ldim,Inci)
       Mi= Ngaus(Rmin/Elengx,Cdim,Rlim)
```

```
    NDIVSX= 1 ; NDIVSE= 1
    RJacB=1.0
    RonL= Rmin/Elengx
    IF(Mi == 5) THEN
      IF(RonL > 0.0) NDIVSX= INT(RLim(2)/RonL) + 1
      IF(NDIVSX > MAXDIVS) MAXDIVS= NDIVSX
      Mi=4
    END IF
   CALL Gauss_coor(Glcorx,Wix,Mi)  ! Coords/Wghts x dir
   Ki= 1 ; Wie(1)= 1.0 ; Glcore(1)= 0.0
   IF(Cdim == 3) THEN
     Ki= Ngaus(Rmin/Elenge,Cdim,Rlim)
     RonL= Rmin/Elenge
     IF(Ki == 5) THEN   !  subdivide
      IF(RonL > 0.0) NDIVSE= INT(RLim(2)/RonL) + 1
      IF(NDIVSE > MAXDIVS) MAXDIVS= NDIVSE
      Ki=4
     END IF
     CALL Gauss_coor(Glcore,Wie,Ki) ! Coords/Wghts h dir
   END IF
   IF(NDIVSX > 1 .OR. NDIVSE>1) RJacB= 1.0/(NDIVSX*NDIVSE)
   Xsi1=-1.0
   Eta1=-1.0
   Subdivisions_xsi1:&
   DO NDIVX=1,NDIVSX
     Xsi2= Xsi1 + 2.0/NDIVSX
     Subdivisions_eta1: &
     DO NDIVE=1,NDIVSE
      Eta2= Eta1 + 2.0/NDIVSE
       Gauss_points_xsi2: &
       DO m=1,Mi
        xsi= Glcorx(m)
        IF(NDIVSX > 1) xsi= 0.5*(Xsi1+Xsi2)+xsi/NDIVSX
         Gauss_points_eta2: &
         DO k=1,Ki
           eta= Glcore(k)
           IF(NDIVSE > 1) eta= 0.5*(Eta1+Eta2)+xsi/NDIVSE
           Weit= Wix(m)*Wie(k)
           CALL Serendip_func(Ni,xsi,eta,ldim,nodel,Inci)
           CALL Normal_Jac(Vnorm,Jac,xsi,eta,ldim&
                          ,nodel,Inci,elcor)
           Fact= Weit*Jac*RJacB
           CALL Cartesian(GCcor,Ni,ldim,elcor)
           r= Dist(GCcor,xPnt,Cdim)        !  Dist. P,Q
           dxr= (GCcor-xPnt)/r            !  rx/r , ry/r  etc
           IF(Ndof .EQ. 1) THEN
             TS(:,1)= dU(r,dxr,Cdim)
             US(:,1)= dT(r,dxr,Vnorm,Cdim)
           ELSE
             CALL SK(TS,DXR,R,C2,C3)
             CALL RK(US,DXR,R,VNORM,C3,C5,c6,C7,ny)
```

```
                END IF
                  Node_loop2:&
                  DO Node=1,Nodel
                   Disp= El_ue(Node,:)* Fac_nod(Node,:)
                   Trac= El_te(Node,:)* Fac_nod(Node,:)
                   SPnt= SPnt + (MATMUL(TS,Trac)- &
                                 MATMUL(US,Disp))* Ni(Node)* Fact
                  END DO Node_loop2
                  END DO Gauss_points_eta2
                END DO Gauss_points_xsi2
                Eta1= Eta2
              END DO Subdivisions_eta1
              Xsi1= Xsi2
            END DO Subdivisions_xsi1
          END DO Symmetry_loop2
      END DO Element_loop2
      IF(Ndof == 1) THEN
        WRITE(2,'(A,6F10.3)') '       Flux: ',SPnt
      ELSE
        IF(CDIM == 2) THEN
         IF(IPS==1) THEN
           SPnt(4)=SPnt(3)
           SPnt(3)= ny*(SPnt(1)+SPnt(2))
         ELSE
           SPnt(4)=SPnt(3)
           SPnt(3)= 0
         END IF
        END IF
        WRITE(2,'(A,6F10.3)') '     Stress: ',SPnt
      END IF
 END DO Internal_points
 END PROGRAM Post_processor
```

9.4.1 Input specification

INPUT DATA SPECIFICATION FOR Postprocessor

1.0 Boundary results

Nan, Nen First element and last element on which boundary results are to be computed

2.0 Internal point specification loop

x, y, (z) Coordinates of internal points
Specify as many as required.

9.5 GRAPHICAL DISPLAY OF RESULTS

In an engineering application, the graphical display of the results is indispensable. The display of the vector or scalar fields can be as diagrams of variation of a quantity along a line or as contour plots. The detailed description of the graphical postprocessing is beyond the scope of this book and the reader is referred to the literature on this subject. One approach to contouring is mentioned here, because it is unique to the BEM. In the BEM we are fortunate to actually have a continuous distribution of results inside the domain which is differentiable without any restriction. To take full advantage of the increased accuracy of results as compared to the FEM one may look beyond the usual interpolation schemes used there.

The idea is to determine the contours in the domain, by using a predictor/corrector scheme. For contours that start on the boundary, the starting point (x_0, y_0) is first determined for a particular contour value f_1 of the function *f(x,y)* to be contoured by using interpolation of boundary values. Next, the directions tangential and normal to the contour are determined from the condition that the value of the result to be contoured remains constant along the contour, i.e.

$$f(x, y) = f_1 \tag{9.54}$$

Along a contour the change in *f* must be zero, i.e.

$$df = \frac{\partial f}{\partial x}dx + \frac{\partial f}{\partial y}dy = 0 \tag{9.55}$$

The vectors normal and tangential to the contour are

$$\mathbf{n} = \begin{pmatrix} \dfrac{\partial f}{\partial x} \\ \dfrac{\partial f}{\partial y} \end{pmatrix} \quad \text{and} \quad \mathbf{t} = \begin{pmatrix} \dfrac{\partial f}{\partial y} \\ -\dfrac{\partial f}{\partial x} \end{pmatrix} \tag{9.56}$$

To obtain unit vectors, the components of the vectors must divided by the length

$$t = n = \sqrt{\left(\frac{\partial f}{\partial x}\right)^2 + \left(\frac{\partial f}{\partial y}\right)^2} \tag{9.57}$$

The method of contouring works as follows: A first estimate of a point on the contour (x_1, y_1) is computed, by drawing a straight line of a specified length ΔL in the direction given by $\mathbf{t}$:

$$\mathbf{x}_1 = \mathbf{x}_0 + \Delta L \cdot \frac{\mathbf{t}}{t} \tag{9.58}$$

Next the stress is computed at the point at (x_1, y_1).

The error of the prediction determined as

$$\varepsilon = f(x_1, y_1) - f_1 \tag{9.59}$$

This error is now corrected using the direction **n** computed at the point. Further points on the contour are determined by repeated application of prediction and correction until the contours meets a boundary (see Figure 9.7) or closes. It is clear from figure 9.7 that the length of the predictor must be continuously adjusted to ensure convergence of the algorithm. In the case where a contour does not start from the boundary a search for the starting point of the contour may be carried out from the boundary. Further details can be found in Noronha et al[2].

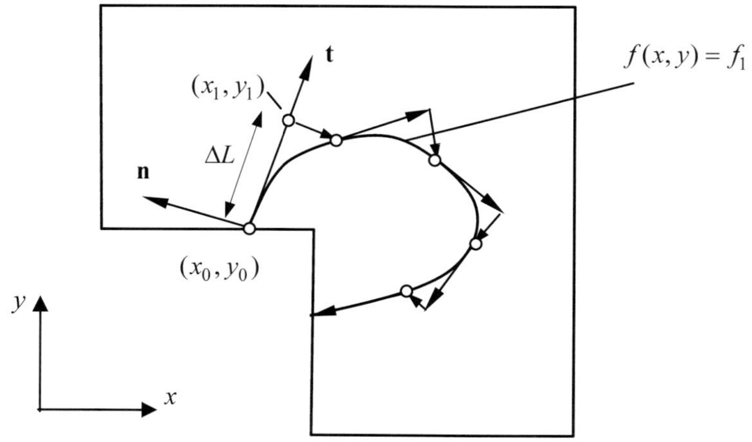

Figure 9.7 Explanation of the contouring algorithm

We show on an example in 2-D elasticity how the derivatives are determined. Taking the derivative of the stress solution we can obtain the change in the stress tensor in the x direction by

$$\frac{\partial \boldsymbol{\sigma}}{\partial x} = \int_S \frac{\partial \mathbf{S}}{\partial x}(P_a, Q)\mathbf{t}(Q)\,dS - \int_S \frac{\partial \mathbf{R}}{\partial x}(P_a, Q)\mathbf{u}(Q)\,dS \tag{9.60}$$

and in y direction by

$$\frac{\partial \boldsymbol{\sigma}}{\partial y} = \int_S \frac{\partial \mathbf{S}}{\partial y}(P_a, Q)\mathbf{t}(Q)\,dS - \int_S \frac{\partial \mathbf{R}}{\partial y}(P_a, Q)\mathbf{u}(Q)\,dS \tag{9.61}$$

where for example

$$\frac{\partial S_{ijk}}{\partial x} = \frac{2}{r} S_{ijk} r_{,x} + \frac{C_2}{r}\left[C_3 (\delta_{ki} \frac{\partial r_{,j}}{\partial x} + \delta_{kj} \frac{\partial r_{,i}}{\partial x} - \delta_{ij} \frac{\partial r_{,k}}{\partial x}) + (n+1) \frac{\partial r_{,i}}{\partial x} \frac{\partial r_{,j}}{\partial x} \frac{\partial r_{,k}}{\partial x} \right] \tag{9.62}$$

Figure 9.8 shows contour plots obtained with the new algorithm.

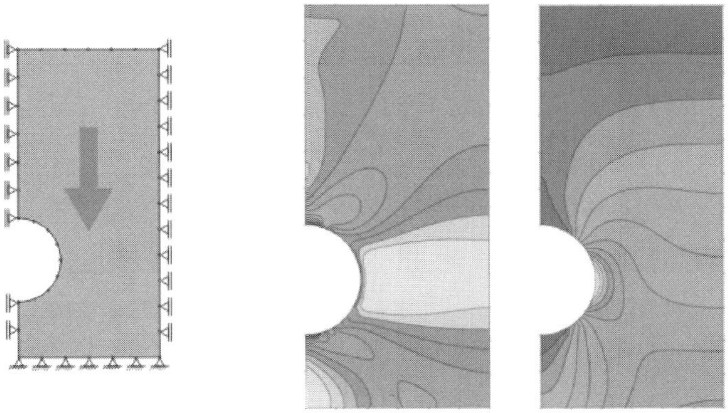

Figure 9.8 Example of a contour plot of major and minor principal stresses for a tunnel subjected to self weight.

9.6 CONCLUSIONS

In this chapter we have discussed methods for obtaining results other than values of temperature/displacement and fluxes/tractions at the nodes of boundary elements. These additional results are flows/stresses at internal points. Results exactly on the boundary elements, can be obtained by a method also known as the "stress recovery", whereby we use the shape functions of the element to determine tangential flows/stresses. The results at internal points are obtained with the fundamental solutions and are more accurate than comparable results from FEM, because they satisfy the governing differential equations exactly and – for infinite domain problems – include the effect of the infinite boundary condition. The task of computing internal results can be delegated to a postprocessor, where the user may either interactively interrogate points or define planes inside the continuum where contours are to be plotted.

It has been found that due to the high degree of singularity of the Kernel functions, care must be taken that internal points are not too close to the boundary. If the proposed numerical integration scheme is used, then there is a limiting value of R/L below which the results are in error. However, since we are able to compute the results exactly on the boundary, we may use a linear interpolation between the internal point and a point projected onto the boundary element. Finally, a method to compute very accurate contours of stresses has been presented. This scheme is based on the fact that, in contrast to the FEM, the functions that describe the variation of results are differentiable, without any loss of accuracy.

9.7 EXERCISES

Exercise 9.1

Use Program 9.1 to compute for exercise 7.1 the flow in vertical direction along a horizontal line. Compare with the theoretical solution.

Exercise 9.2

Use Program 9.1 to compute for exercise 7.3, the stress in vertical direction along a horizontal line. Compare with the theoretical solution.

Exercise 9.3

Use Program 9.1 to compute for exercise 7.4 the flow in horizontal direction along a vertical line in the middle of the rectangular region. Compare with the theoretical solution.

Exercise 9.4

Use Program 9.1 to compute for exercise 7.6 the normal and shear stress along a vertical line, as shown in Figure 9.9. Do a graphical comparison with the theoretical solution.

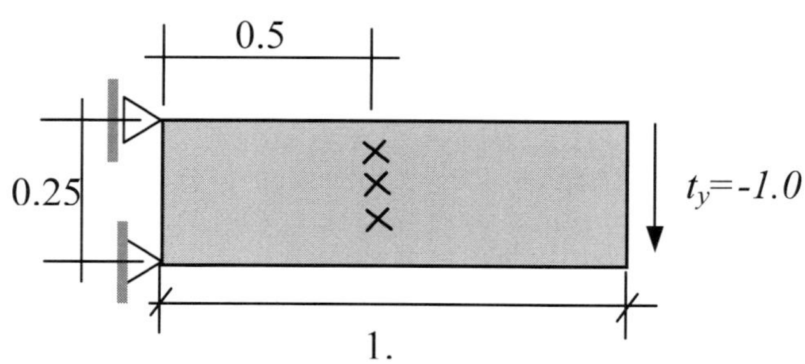

Figure 9.9 Cantilever beam with internal points

9.8 REFERENCES

1. Banerjee P.K. (1994) The Boundary Element Methods in Engineering, McGraw-HillBook Company, London.
2. Noronha M., Müller A and Pereira A.M.B. (2005) A novel pure-BEM approach for post-processing and non-linear analysis. *Proceedings McMat2005*, Joint ASME/ASCE/SES Conference on Mechanics and Materials.

10

Test Examples

Die Wahrheit wird gelebt, nicht doziert
(Truth is lived not taught)
H. Hesse

10.1 INTRODUCTION

We have now developed all the software required to perform a boundary element analysis of problems in potential flow and elasticity. The examples which we can analyse will, however, be restricted to homogeneous domains and linear material behaviour. Before we proceed further in an attempt to eliminate these restrictions, it is opportune to pause and learn, on test examples, a few things about the method especially with respect to the accuracy that can be attained. The purpose of this chapter is twofold. Firstly, the reader will learn how problems are modelled using boundary elements, with examples of simple meshes in two and three dimensions. Secondly, we will show, by comparison with theory and results from finite element meshes, the accuracy which can be obtained. We will also point out possible pitfalls, which must be avoided. As with all numerical methods, examples can be presented that favour the method and others that don't. Here we find that the BEM has difficulty dealing with cantilevers with small thickness where two opposing boundaries are close to each other. On the other hand it can deal very well with problems which involve an infinite domain. Also we will find that values at the surface are computed more accurately. This gives an indication of the range of applications where the method is superior as compared with others: those involving a large volume to surface ratio (including infinite domains) and those where the results at the boundary are important, for example stress concentration problems. In the following, several test examples will be presented ranging from the simple 2-D analysis of a cantilever beam to the 3-D analysis of a spherical excavation in an infinite continuum. In all cases we show the input file required to solve the problem with program 7.1 and 9.1 and the output obtained. The results are then analysed with respect to accuracy with different discretisations. Comparison is made with theoretical results and in some cases with finite element models.

10.2 CANTILEVER BEAM

10.2.1 Problem statement

The cantilever beam is a simple structure, which nevertheless can be used to show strengths and weaknesses of numerical methods. Here we analyse a cantilever beam with decreasing thickness and we will find that this causes some difficulties for the BEM. The problem is stated in Figure 10.1. An encastre beam is subjected to a distributed load of 10 KN at the end. The material properties are assumed to be: E= 10 000 MPa and ν=0.0. We gradually decrease the thickness *t* of the beam and observe the accuracy of results.

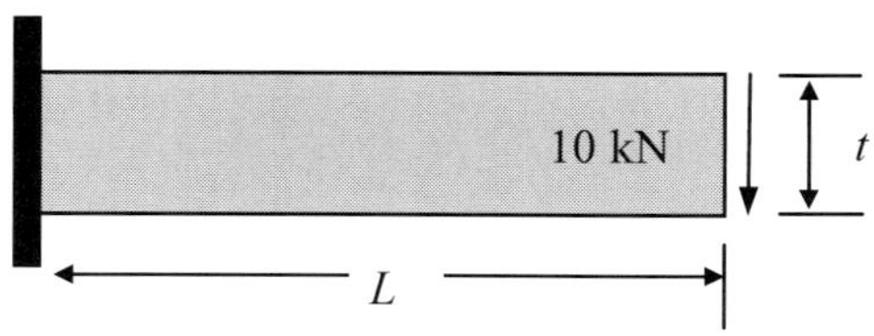

Figure 10.1 Cantilever beam: Dimensions and loading assumed

10.2.2 Boundary element discretisation and input

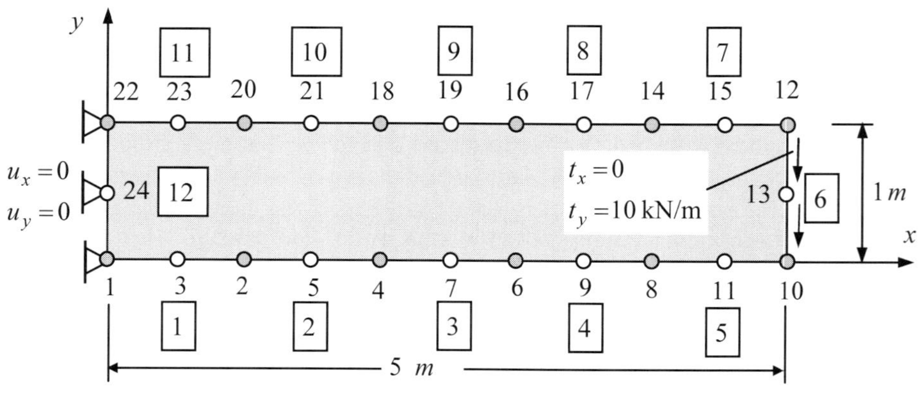

Figure 10.2 Boundary element Mesh 1 (O ...corner node, O ... mid-side node)

Figure 10.2 shows the discretisation used (12 parabolic boundary elements) and the dimension of the first mesh analysed with a ratio of t/L of 0.2. The element and node numbering as well as the boundary conditions are shown.

The input file for this problem for program 7.1 is

```
Cantilever beam
  2     ! Cdim       2-D
  2     ! Ndof        Elasticity
  2     ! ToA         Plane stress
  1     ! Nreg      finite region
  0     ! no symmetry
  2     ! Quadratic elements
 0.1000E+05 0.0000E+00       !  E,Ny
    24    !  Number of nodes
    12    !  Number of Elements
      0.000       0.000                 !   Coordinates
      1.000       0.000
      0.500       0.000
      2.000       0.000
      1.500       0.000
      3.000       0.000
      2.500       0.000
      4.000       0.000
      3.500       0.000
      5.000       0.000
      4.500       0.000
      5.000       1.000
      5.000       0.500
      4.000       1.000
      4.500       1.000
      3.000       1.000
      3.500       1.000
      2.000       1.000
      2.500       1.000
      1.000       1.000
      1.500       1.000
      0.000       1.000
      0.500       1.000
      0.000       0.500
      1      2      3        !    Incidences
      2      4      5
      4      6      7
      6      8      9
      8     10     11
     10     12     13
     12     14     15
     14     16     17
     16     18     19
     18     20     21
     20     22     23
```

```
   22    1   24
    1        !     Prescribed Dirichlet
   12   0.0 0.0 0.0 0.0 0.0 0.0
    1        !     Prescribed Neuman
    6   0.0 -10.0 0.0 -10.0 0.0 -10.0
```

10.2.3 Results

The output obtained from the program 7.1 is:

```
 Project:
 Cantilever beam
  Cartesian_dimension:           2
  Elasticity Problem
  Type of Analysis: Solid Plane Stress
  Finite Region
  No symmetry
  Quadratic Elements
  Modulus:    10000.00
  Poissons ratio:  0.0000000E+00
  Number of Nodes of System:          24
  Number of Elements of System:          12
 Node      1  Coor       0.00     0.00
 ...
 Node     24  Coor       0.00     0.50

  Incidences:
 EL      1  Inci       1    2    3
 ...
 EL     12  Inci      22    1   24

  Elements with Dirichlet BC´s:
  Element          12  Prescribed values:
   0.0000000E+00  0.0000000E+00
   0.0000000E+00  0.0000000E+00
   0.0000000E+00  0.0000000E+00

  Elements with Neuman BC´s:
  Element           6  Prescribed values:
  Node=           1  0.0000000E+00  -10.00000
  Node=           2  0.0000000E+00  -10.00000
  Node=           3  0.0000000E+00  -10.00000

  Results, Element           1
  u=      0.000      0.000     -0.027     -0.030     -0.014     -0.008
  t=    298.892      6.277      0.000      0.000      0.000      0.000
  ...
  Results, Element           6
  u=     -0.075     -0.508      0.075     -0.508      0.000     -0.508
  t=      0.000    -10.000      0.000    -10.000      0.000    -10.000
```

```
.....
Results, Element          12
u=      0.000     0.000    0.000     0.000     0.000     0.000
t=   -298.892     6.277  298.892     6.277     0.000    11.876
```

The input file for this problem for program 9.1 is

```
1 12
2.0 0.1
2.0 0.2
2.0 0.3
2.0 0.4
2.0 0.5
2.0 0.6
2.0 0.7
2.0 0.8
2.0 0.9
```

The output obtained from program 9.1 is

```
Post-processed Results
Results at Boundary Elements:
Element    1 xsi= -1.00 eta= -1.00
 Stress:   -296.90    -6.28     0.00  -298.89
Element    1 xsi=  0.00 eta= -1.00
 Stress:   -269.50     0.00     0.00     0.00
Element    1 xsi=  1.00 eta= -1.00
 Stress:   -242.10     0.00     0.00     0.00
...
Element   12 xsi= -1.00 eta= -1.00
 Stress:    298.89     0.00     0.00    -6.28
Element   12 xsi=  0.00 eta= -1.00
 Stress:      0.00     0.00     0.00   -11.88
Element   12 xsi=  1.00 eta= -1.00
 Stress:   -298.89     0.00     0.00    -6.28

 Internal Results:
Coordinates:         2.00      0.10
         u:       -0.038    -0.107
     Stress:    -144.657     0.818     0.000    -8.323
Coordinates:         2.00      0.20
         u:       -0.028    -0.107
     Stress:    -108.503     0.604     0.000   -12.658
Coordinates:         2.00      0.30
         u:       -0.019    -0.107
     Stress:     -72.370     0.421     0.000   -15.590
Coordinates:         2.00      0.40
         u:       -0.009    -0.107
```

```
        Stress:    -36.182      0.205      0.000    -17.311
   Coordinates:       2.00       0.50
             u:      0.000     -0.107
        Stress:      0.000      0.000      0.000    -17.886
   Coordinates:       2.00       0.60
             u:      0.009     -0.107
        Stress:     36.183     -0.206      0.000    -17.311
   Coordinates:       2.00       0.70
             u:      0.019     -0.107
        Stress:     72.370     -0.421      0.000    -15.590
   Coordinates:       2.00       0.80
             u:      0.028     -0.107
        Stress:    108.503     -0.603      0.000    -12.658
   Coordinates:       2.00       0.90
             u:      0.038     -0.107
        Stress:    144.655     -0.814      0.000     -8.322
```

A total of three analyses were carried out gradually reducing the value of t to 0.5 and 0.2m. The results of the analyses are summarised in Table 10.1 and compared with results obtained from the classical beam theory[1] (Bernoulli hypothesis). Compared are the maximum deflection at the free end and the bending stresses at the fixed end. Note that the maximum bending stresses are obtained directly from the analysis (these are equal to the tractions t_x at node 1 and 2 on element 12).

Table 10.1 Summary of results for cantilever beam with parabolic boundary elements

t	t/L	Mesh	Max. deflection *(mm)*		Max. stress *(MPa)*	
			Computed	Beam theory	Computed	Beam theory
1.0	0.2	1	0.508	0.500	0.299	0.300
0.5	0.1	1	3.704	4.000	1.105	1.200
0.2	0.04	1	36.74	62.50	4.38	7.50

The displacement results obtained for a ratio t/L of 0.2 include the additional effect of shear and are more accurate than the results computed by beam theory. It can be seen that results deteriorate rapidly with decreasing value of t/L and that for a ratio t/L of 0.04 the error is unacceptable. Note that for an equivalent finite element mesh, no problems arise if the thickness is reduced.

Considering that the accuracy of integration introduced with the integration scheme in Chapter 6 is quite high the reasons for this deterioration in accuracy is probably the fact that the collocation points are very close to each other in one direction and far away in the other. Since the theorem by Betti will only be satisfied at the collocation points, there seems to be a detrimental effect if these points are very unevenly spaced. Apparently a close proximity of the collocation points also causes a lack of diagonal dominance of the system of equations. Numerical experiments have shown that even if the precision of integration is further increased, errors still persist if the ratio *t/L* is decreased to a very low value.

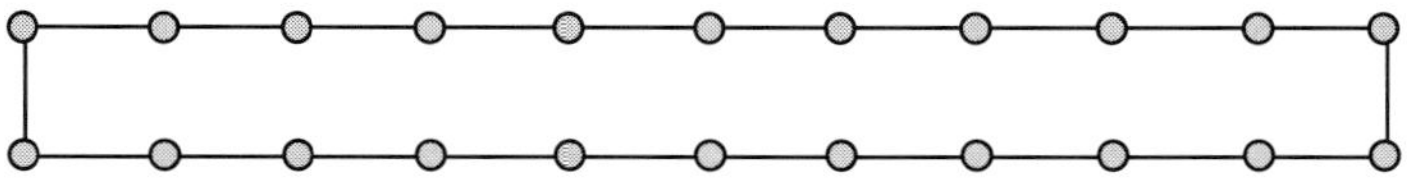

(a)

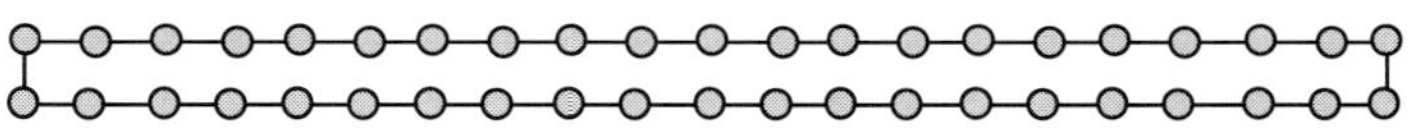

(b)

Figure 10.3 Adaptive meshes: (a) Mesh 2 and (b) Mesh 3 (only corner nodes shown)

Table 10.2 Results for refined meshes

t	*t/L*	Mesh	Max. deflection *(mm)*		Max. stress *(MPa)*	
			Computed	Beam theory	Computed	Beam theory
1.0	0.2	1	0.508	0.500	0.299	0.300
0.5	0.1	2	3.993	4.000	1.192	1.200
0.2	0.04	3	62.234	62.50	7.459	7.500

A relatively simple remedy to this problem is to increase the number of elements as the ratio *t/L* is decreased. Such adaptive meshes for *t/L*=0.1 and 0.04 are shown in Figure 10.3.As can be seen from Table 10.2 the accuracy of results is now greatly improved. The variation of normal stress and shear stress along a vertical line at a distance of 2.0 m from the fixed end is computed using Program 9.1 and plotted in Figure 10.5. The computed distribution is in good agreement with the theory for both normal and shear stresses

10.2.4 Comparison with FEM

We now make a comparison with the finite element method. The mesh shown in Fig 10.4 has the same discretisation on the boundary as the BEM mesh. If we change the thickness to length ratio we see in Table 10.3 that this has no effects on the results. When we compare the stress distributions in Figure 10.5 we can see that the FEM exactly represents the bending stress, but will only be able to approximate the parabolic distribution of the shear stress by a constant distribution.

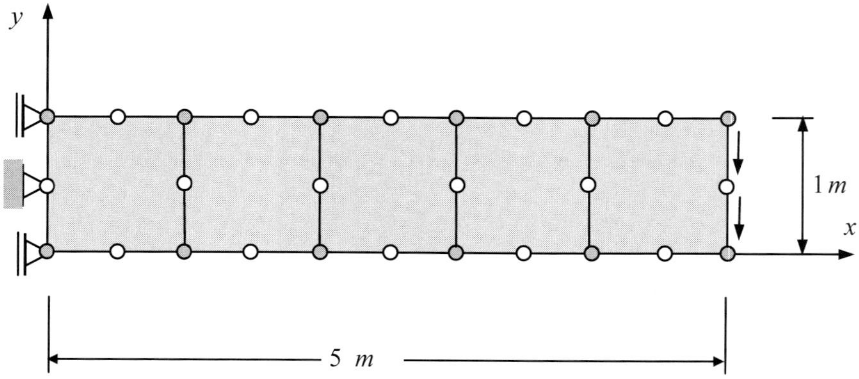

Figure 10.4 Finite element mesh

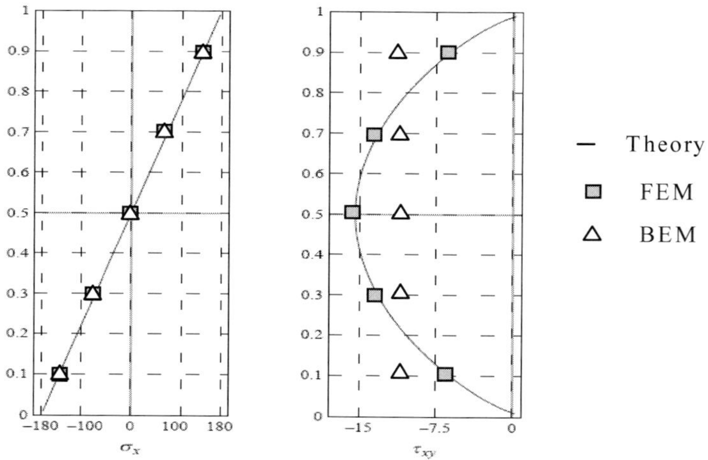

Figure 10.5 Comparison of results

Table 10.3 Results of finite element analysis

t/L	Mesh	Max. deflection *(mm)*		Max. σ_x *(MPa)*	
		Computed	Beam theory	Computed	Beam theory
0.2	1	0.515	0.500	0.300	0.300
0.1	1	4.031	4.000	1.200	1.200
0.04	1	62.58	62.50	7.500	7.500

10.2.5 Conclusions

The conclusions from this example are that there is very little difference in the discretisation effort and computing time between the FEM and the BEM for this example. We found that for slender beams the program leads to significant errors for coarse discretisations. However, there are ways in which we may improve these results, for example by using schemes other than point collocation, namely the Galerkin method mentioned briefly in Chapter 6. However this will result in significantly higher computation effort so that the BEM may loose some advantage. Another elegant and efficient way would be to include in the fundamental solution the classical beam bending theory. If this is done then there is only need to discretise the centerline of the beam thereby avoiding all the difficulties which we have experienced. A good description of this method is given by Hartmann[2].

10.3 CIRCULAR EXCAVATION IN INFINITE DOMAIN

10.3.1 Problem statement

Consider an excavation made in an infinite, homogeneous, elastic space. The elastic space is assumed to have a modulus of elasticity of 10 000 MPa, a Poisson's ratio of zero and to have been pre-stressed with a stress field of σ_x= 0.0 MPa, σ_y= -3.0 MPa, (compression) and τ_{xy}= 0.0 MPa.

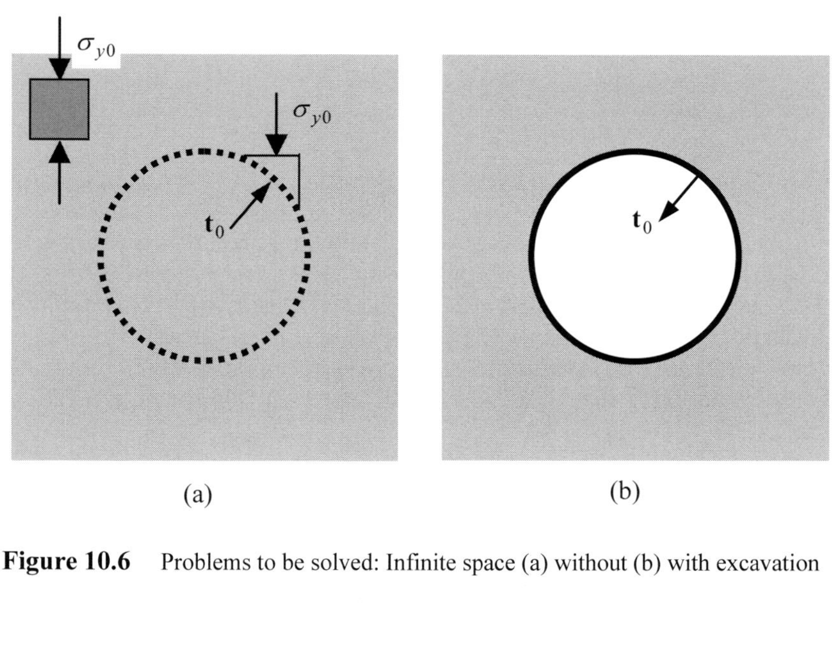

Figure 10.6 Problems to be solved: Infinite space (a) without (b) with excavation

Required are the displacements and the changed stress distribution due to excavation. To obtain this solution we actually have to solve two problems (Figure 10.6): One (trivial) one where no excavation exists and one where the supporting tractions $\mathbf{t}_0$ computed in the first step are released, i.e., applied in the opposite direction.
We can use equation (4.28) to solve the problem (a) i.e. to compute the tractions $\mathbf{t}_0$ as

$$\mathbf{t}_0 = \begin{Bmatrix} 0.0 \\ n_y \sigma_{y0} \end{Bmatrix} \tag{10.1}$$

10.3.2 Boundary element discretisation and input

To solve problem (b) we use the BEM with two planes of symmetry. For the first mesh a single parabolic element (Figure 10.7) is used. Subsequently two (Mesh 2) and four elements (Mesh 3) are used for a quarter of the boundary. The mesh is subjected to *Neuman* boundary conditions with values of $\mathbf{t}_0$ computed using (10.1) and applied as shown in Figure 10.7.

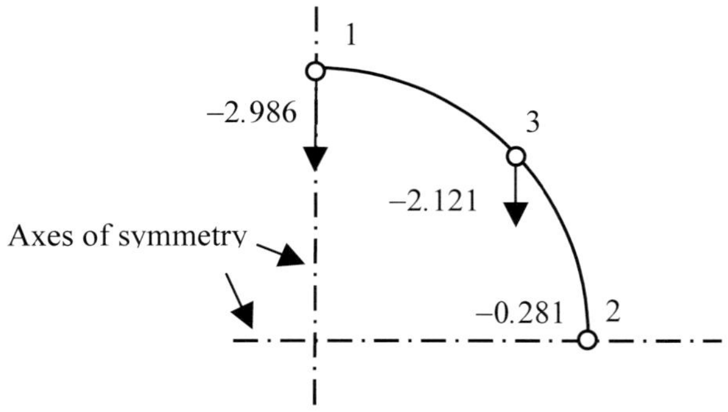

Figure 10.7 Boundary element mesh with *Neuman* boundary conditions

If the element would be able to describe an exact circle then the values of traction t_{y0} should be exactly –3.0 at node 1 and 0.0 at node 2. However, since the element can only describe a parabola, the y-component of the normal vector will not be exactly –1.0 at node 1 and not exactly 0.0 at node 2. Therefore, a small geometrical error occurs due to the coarse discretisation. Alternatively we could specify the values of traction that correspond to an exact circle (-3.0,-2.12, 0.0) The input file for program 7.1 for this problem is

```
Circular hole
2       ! 2-D
2       ! Elasticity problem
1       ! Plane strain
2       ! Finite Region
```

```
 2      ! double symmetry
 2      ! quadratic elements
0.1000E+05  0.0000E+00     ! E,Ny
    3     ! Nodes
    1     ! Elements
     0.000      1.000      ! Coordinates
     1.000      0.000
     0.707      0.707
    1    2    3     ! Incidences
    0    ! Dirichlet BC
    1    ! Neumann BC
    1    0.00000 -2.98681   0.00000 -0.28103  0.00000 -2.12132
```

The output obtained from program 7.1 is

```
Project:
 Circular hole
 Cartesian_dimension: 2
 Elasticity Problem
 Type of Analysis:  Solid Plane Strain
 Infinite Region
 Symmetry about y-z and x-z planes
 Quadratic Elements
 Modulus:   10000.00
 Poissons ratio:  0
 Number of Nodes of System:            3
 Number of Elements of System:         1
Node      1  Coor      0.00    1.00
Node      2  Coor      1.00    0.00
Node      3  Coor      0.71    0.71
 Incidences:
EL     1  Inci      1    2    3
 Elements with Dirichlet BC´s:
 Elements with Neuman BC´s:
 Element            1  Prescribed values:
  0.00  -2.986810
  0.00  -0.281030
  0.00  -2.121320
 Results, Element    1
   u=   0.00000  -0.00060   0.00029   0.00000   0.00021  -0.00041
   t=   0.00000  -2.98681   0.00000  -0.28103   0.00000  -2.12132
```

The input file for this problem for program 9.1 is

```
 1 1
 1.1   0
 1.2   0
 1.3   0
```

```
1.4  0
1.5  0
```

The output obtained from program 9.1 is

```
Post-processed Results
 Results at Boundary Elements:
Element     1 xsi= -1.00 eta= -1.00
 Stress:        2.85      3.03      0.00      -0.27
Element     1 xsi=  0.00 eta= -1.00
 Stress:       -1.52      1.47      0.00       1.52
Element     1 xsi=  1.00 eta= -1.00
 Stress:       -0.08     -5.52      0.00       0.80

 Internal Results:
Coordinates:          1.10       0.00
          u:         0.000      0.000
     Stress:        -0.647     -4.205      0.000      0.000
Coordinates:          1.20       0.00
          u:         0.000      0.000
     Stress:        -0.927     -3.103      0.000      0.000
Coordinates:          1.30       0.00
          u:         0.000      0.000
     Stress:        -1.046     -2.380      0.000      0.000
Coordinates:          1.40       0.00
          u:         0.000      0.000
     Stress:        -1.079     -1.875      0.000      0.000
Coordinates:          1.50       0.00
          u:         0.000      0.000
     Stress:        -1.067     -1.508      0.000      0.000
```

10.3.3 Results

A theoretical solution for this problem has been obtained by Kirsch[3]. In Table 10.4 the results at the boundary for the 3 meshes are compared. It can be seen that even the coarse mesh, with only one element per quarter, gives acceptable results for this problem.

Table 10.4 Results for meshes with parabolic boundary elements

Mesh	No. Elem	Max. deflection *(mm)*	Max. stress *(MPa)*	Min. stress *(MPa)*
1	1	0.60	-8,52	2.85
2	2	0.60	-8.99	2.99
3	4	0.60	-9.00	3.00
Theory		0.60	-9.00	3.00

The internal results along a horizontal line are shown in Figure 10.8. It can be seen that there is good agreement even for coarse meshes.

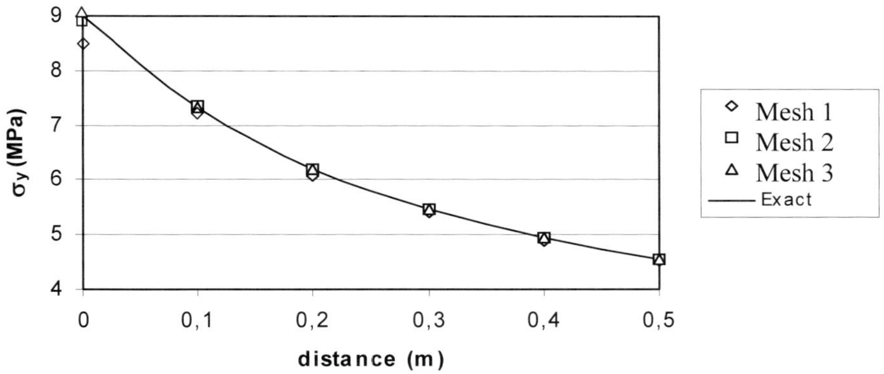

Figure 10.8 Distribution of vertical stress along a horizontal line

10.3.4 Comparison with FEM

The problem was analysed with the FEM with 3 different meshes of finite elements with quadratic shape function, as shown in Figure 10.9. Symmetry was considered by appropriate boundary conditions at the symmetry planes. This discretisation has the same variation of displacements along the excavation boundary as the boundary element mesh. Note that in the FEM we have to truncate the mesh at some distance away from the excavation. A truncation of 2 diameters away from the excavation is used here. At the truncation surface all displacements are assumed to be fixed. In order to eliminate the truncation error, coupled analyses were also made, where boundary elements were used at the edge of the FE mesh (see Chapter 16 on methods of coupling)

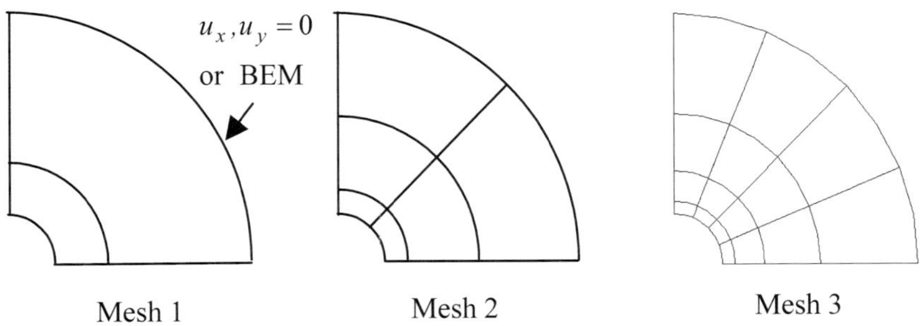

Figure 10.9 Finite element meshes used

Table 10.5 shows the results of the analysis. It can be seen that they are less accurate than the ones obtained for the BEM and that the truncation error is significant.

Table 10.5 Results for meshes with parabolic finite elements

		FEM		Coupled	
Mesh	No. Elem	u_{max} *(mm)*	σ_{max} *(MPa)*	u_{max} *(mm)*	σ_{max} *(MPa)*
1	2	0.480	-6.840	0.535	-8.48
2	6	0.494	-7.189	0.583	-9.01
3	16	0.506	-8.165	0.598	-8.97
Theory		0.600	-9.000	0.600	-9.00

10.3.5 Conclusions

In contrast to the previous example this one favours the boundary element method. We see that with the FEM we have two sources of error: one associated with the truncation of the mesh that is necessary because the method is unable to model infinite domains, the other one is that in the FEM the variation of the unknown has to be approximated by shape functions inside the continuum as well as along the boundary surface. It can be seen that without much additional effort the first error can be virtually eliminated by using the coupled method and by specifying boundary elements at the truncated boundary. This example demonstrates that the BEM is most efficient when the ratio boundary surface to volume is very small. For problems in geomechanics, where the soil/rock mass can be assumed to have infinite extent, this ratio actually approaches zero.

10.4 SQUARE EXCAVATION IN INFINITE ELASTIC SPACE

10.4.1 Problem statement

This example was chosen to demonstrate the ability of the BEM to model stress concentrations. The problem is identical to the previous one, except that the shape of the excavation is square instead of circular. The exact solution for this problem is not known but according to the theory of elasticity, a singularity of the vertical stress occurs as the corner is approached. It is known[4] that for a corner with a subtended angle of 180° (crack) the displacements tend to zero with $\sqrt{r}$ and the stresses tend to infinity with $1/\sqrt{r}$. A boundary element with quadratic variation of displacements will not be able to model this variation, so we expect some loss of accuracy for coarse meshes.

While there is obviously no point in trying to compute an infinite value of stress, the variation of the displacement can be used to compute intensity factors[5]. So the aim is rather to predict the variation of displacements accurately.

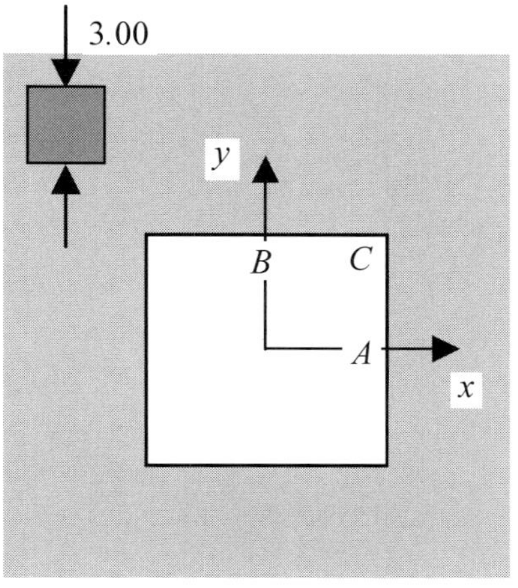

Figure 10.10 Problem statement with result points A,B,C

10.4.2 Boundary element discretisation and input

For the solution of the problem we again use the conditions of symmetry on two planes and 4 meshes, three of which are shown in Figure 10.11. The first three simply represent a uniform subdivision into 2,4 and 16 elements. Mesh 4 is a graded mesh, where the element size has been reduced near the corner.

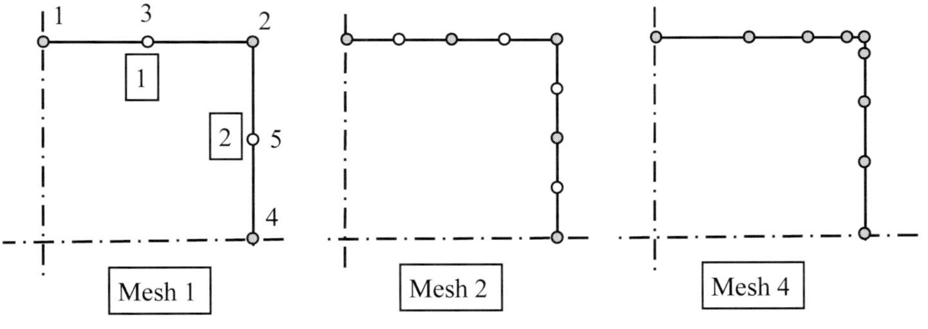

Figure 10.11 Meshes used (mid-side nodes not shown for graded mesh 4)

The input file for program 7.1 for Mesh 1 is

```
 Square excavation - Mesh 1
 2
 2
 1
 2
 2
 2
0.1000E+05
0.0000E+00
    5
    2
      0.000      1.000
      1.000      1.000
      0.500      1.000
      1.000      0.000
      1.000      0.500
    1    2    3
    2    4    5
    0
    2
    1  0.00000  -3.00000   0.00000  -3.00000  0.00000    -3.00000
    2  0.00000   0.00000   0.00000   0.00000  0.00000     0.00000
```

The output obtained from program 7.1 is

```
Project:
 Square excavation - Mesh 1
 Cartesian_dimension:                          2
 Elasticity Problem
 Type of Analysis:                             Solid Plane Strain
 Infinite Region
 Symmetry about y-z and x-z planes
 Quadratic Elements
 Modulus:                                      10000.00
 Poissons ratio:                               0
 Number of Nodes of System:                    5
 Number of Elements of System:                 2
Node      1  Coor      0.00    1.00
Node      2  Coor      1.00    1.00
Node      3  Coor      0.50    1.00
Node      4  Coor      1.00    0.00
Node      5  Coor      1.00    0.50
 Incidences:
EL     1  Inci       1    2    3
EL     2  Inci       2    4    5
 Elements with Dirichlet BC´s:
 Elements with Neuman BC´s:
 Element              1  Prescribed values:
```

```
 0.00  -3.00
 0.00  -3.00
 0.00  -3.00
Element            2  Prescribed values:
 0.00   0.00
 0.00   0.00
 0.00   0.00
Results, Element    1
  u=   0.00000  -0.00072   0.00018  -0.00031   0.00014  -0.00063
  t=   0.00000  -3.00000   0.00000  -3.00000   0.00000  -3.00000
Results, Element    2
  u=   0.00018  -0.00031   0.00020   0.00000   0.00021  -0.00012
  t=   0.00000   0.00000   0.00000   0.00000   0.00000   0.00000
```

The input file for program 9.1 for Mesh 1 is

```
1 2
1.1  0
1.2  0
1.3  0
1.4  0
1.5  0
```

The output obtained from program 9.1 is

```
Post-processed Results
 Results at Boundary Elements:
Element    1 xsi= -1.00 eta= -1.00
 Stress:      3.80     3.00      0.00      0.00
Element    1 xsi=  0.00 eta= -1.00
 Stress:      1.80     3.00      0.00      0.00
Element    1 xsi=  1.00 eta= -1.00
 Stress:     -0.20     3.00      0.00      0.00
Internal Results:
Coordinates:        1.10      0.00
          u:       0.000     0.000
     Stress:       0.132    -2.107     0.000     0.000
Coordinates:        1.20      0.00
          u:       0.000     0.000
     Stress:       0.026    -2.074     0.000     0.000
Coordinates:        1.30      0.00
          u:       0.000     0.000
     Stress:      -0.071    -1.954     0.000     0.000
Coordinates:        1.40      0.00
          u:       0.000     0.000
     Stress:      -0.161    -1.796     0.000     0.000
Coordinates:        1.50      0.00
          u:       0.000     0.000
```

```
Stress:       -0.239     -1.627      0.000      0.000
```

The boundary stresses obtained from Program 9.1 are plotted in Figure 10.12. It can be seen that the magnitude of the stress concentration depends on the fineness of the boundary element mesh near the corner and that a fine graded mesh can reasonably approximate the theoretical stress distribution at the corner.

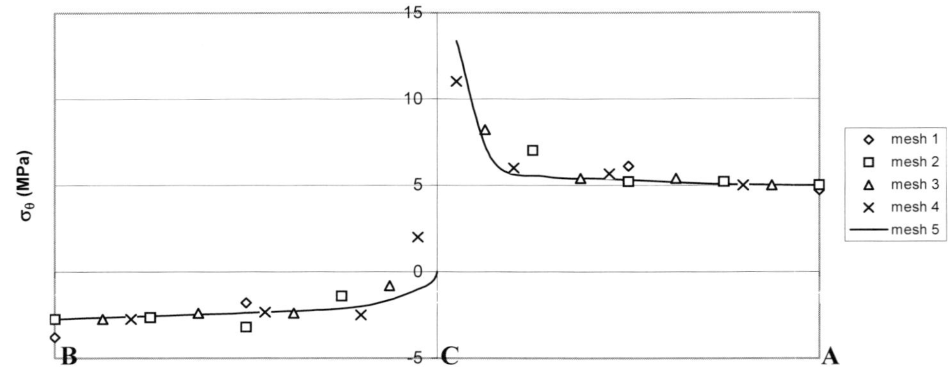

Figure 10.12 Distribution of tangential stress on Boundary BCA

10.4.3 "Quarter point" elements

A numerical trick can be used to simulate a singularity: if we move the "midside" node of an element to the "quarter point" on one side, then it will be shown that the Jacobian tends to zero as the nearest corner node is approached.

Consider the simple element in Figure 10.13, which is located along the x-axis with one point at the origin. In the derivation we transform the intrinsic coordinate ξ which ranges from -1 to +1 to $\bar{\xi}$, which ranges from 0 to 1.

Expressed in this new coordinate system the three shape functions of a quadratic element are given by

$$N_1 = 2(\bar{\xi} - 1/2)(\bar{\xi} - 1) \; ; \; N_3 = -4\bar{\xi}(\bar{\xi} - 1) \; ; \; N_2 = 2\bar{\xi}(\bar{\xi} - 1/2) \tag{10.2}$$

The coordinate x of a point with local coordinate $\bar{\xi}$ can be computed by the interpolation

$$x = \sum N_i(\bar{\xi}) x_i \tag{10.3}$$

Substituting for the coordinates of the nodes ($x_1 = 0.0, x_2 = 0.25, x_3 = 1.0$) we obtain

$$x = \bar{\xi}^2 \tag{10.4}$$

Substitution of this into (10.2) we obtain

$$N_1 = 1 + 2x - 3\sqrt{x} \quad ; \quad N_3 = -4x + 4\sqrt{x} \quad ; \quad N_2 = 2x - \sqrt{x} \tag{10.5}$$

Assuming an iso-parametric formulation the variation of the displacement u is given by

$$u_x = \sum N_i u_{xi} = u_1 + x(2u_{x1} + 2u_{x2} - 4u_{x3}) + \sqrt{x}(3u_{x1} + 4u_{x2} - u_{x3}) \tag{10.6}$$

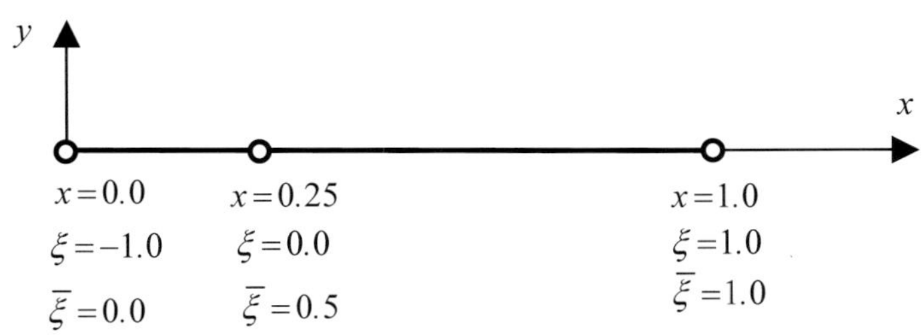

Figure 10.13 "Quarter point" boundary element

Taking point 1 as the singular point we may substitute $r=x$ and therefore the displacements tend to zero with $\sqrt{r}$ .The strains are computed by taking the derivative of the displacements and are given by

$$\varepsilon_x = \frac{\partial u_x}{\partial x} = c + \frac{d}{\sqrt{r}} \tag{10.7}$$

where c and d are constants. Since for elastic material the stresses σ_x are proportional to the strains we see that they go to infinity with $o(1/\sqrt{r})$

In Figure 10.12 is shown that with the simple expedient of moving the third node point of the element near the corner, we can obtain similar or slightly better results for the mesh 3 with the midside node moved to the "quarter point" (mesh 5) than with the graded mesh 4. Such elements have been successfully used for the computation of stress intensity factors[6].

10.4.4 Comparison with finite elements

Comparing with finite element results we find that results are influenced not only by the discretisation along the boundary, but also the subdivision inside the elastic space. In fact, any result for the stress concentration at the edge can be obtained depending on the element subdivision. A reasonably fine graded mesh is shown in Figure 10.14. It consists of 40 Elements with quadratic variation of displacements. The distribution of the tangential stress along a vertical line in Figure 10.15 however, shows that the general trend of the theoretical distribution can not be obtained.

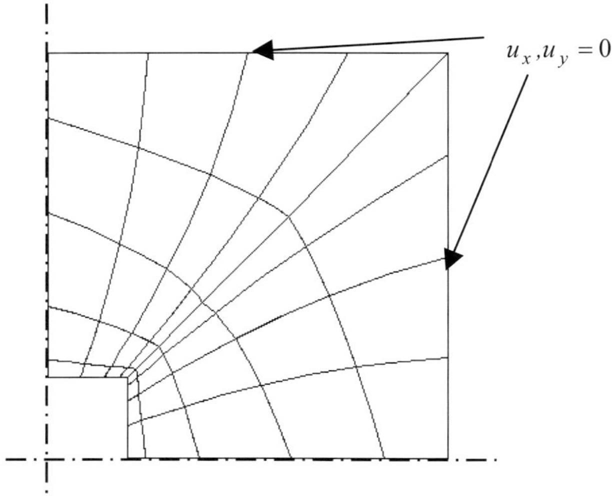

Figure 10.14 Finite Element mesh

10.4.5 Conclusions

In this example we have shown how the boundary element method deals with singularities as they sometimes arise when we have corners. These singularities are of course only theoretical, since there is no such thing as a perfect corner in nature. Also, stresses can not reach an infinite value because they will be limited by a maximum value that a material can sustain. In fracture mechanics we may compute stress intensity factors based on the variation of displacements near the crack. The BEM is well suited for the computation of such factors, but this topic is beyond the limited scope of this book and will not be discussed further. For more information the reader may consult Aliabadi[7]. We have shown in the comparison with the FEM, that since we only have to worry about approximating the variation of displacements in one direction, i.e., along the boundary, the task of finding the mesh which gives the most accurate result is simplified. In the FEM the result will depend on the approximation of the displacements inside the elastic space as well, so refinement has to be made in two directions. For a comparable discretisation on the boundary, however, we find that the BEM gives the better answer for this problem.

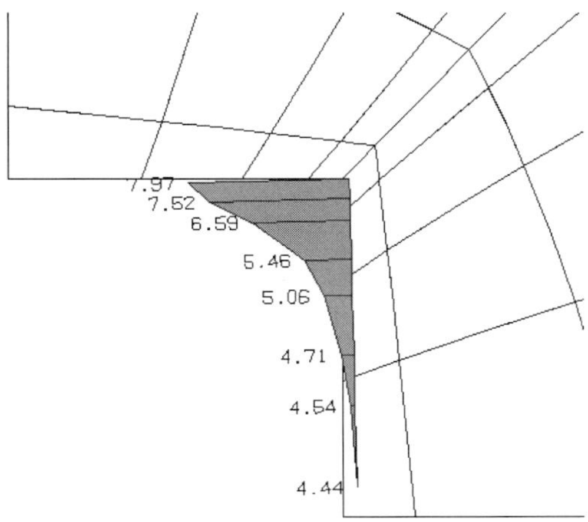

Figure 10.15 Distribution of vertical stress

10.5 SPHERICAL EXCAVATION

All problems analysed so far were two-dimensional. In order to show how more drastic savings can be made when using the BEM we show an example in 3-D.

10.5.1 Problem statement

The example is similar to the example of a circular excavation in an infinite domain except that the excavation is now spherical and the virgin stress is given by

$$\boldsymbol{\sigma}_0 = \begin{bmatrix} 0 & 0 & -1,0 & 0 & 0 & 0 \end{bmatrix}^T \tag{10.8}$$

10.5.2 Boundary element discretisation and input

Two fairly coarse discretisations are used. Both meshes consist of only 3 boundary elements, one consists of linear the other of parabolic elements. Three planes of symmetry are assumed, so only one octant of the problem had to be considered. The meshes are shown in Figure 10.17.

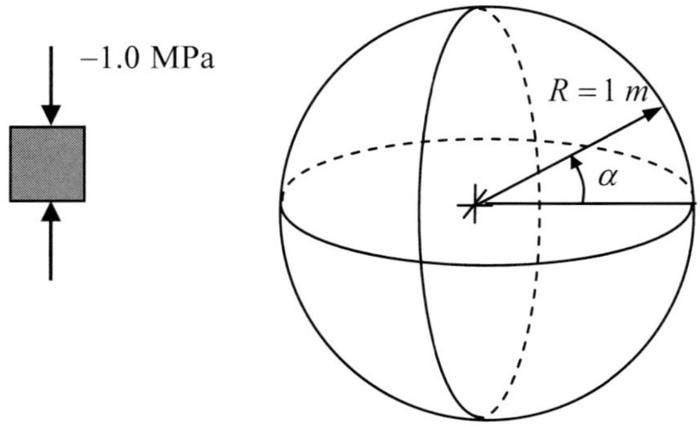

Figure 10.16 Problem statement

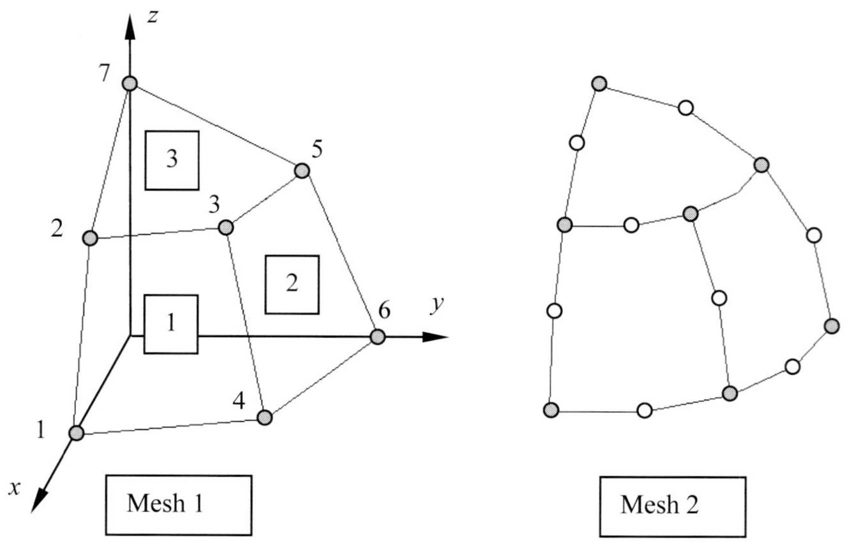

Figure 10.17 Boundary meshes used

Although the tractions $\mathbf{t}_0$ to be applied should be computed according to

$$\mathbf{t}_0 = \begin{Bmatrix} 0 \\ 0 \\ n_z \sigma_{z0} \end{Bmatrix} \tag{10.9}$$

It is easier to compute them assuming the surface to correspond exactly to a sphere. The error in doing this is expected to compensate for the error in describing the geometry by a fairly coarse discretisation. In this case the tractions are computed by

$$\mathbf{t}_0 = \begin{Bmatrix} 0 \\ 0 \\ \sigma_{z0} \cdot \sin\alpha \end{Bmatrix} \tag{10.10}$$

The input file for program 7.1 for Mesh 1 is

```
 Spherical excavation - linear elements
 3
 3
 2
 3
 1
0.1000E+04
0.0000E+00
    7
    3
    1.000      0.000      0.000
    0.707      0.000      0.707
    0.500      0.500      0.707
    0.707      0.707      0.000
    0.000      0.707      0.707
    0.000      1.000      0.000
    0.000      0.000      1.000
    1    2    3    4
    4    3    5    6
    3    2    7    5
    0
    3
    1 0. 0. 0. 0. 0. -0.707 0. 0. -0.707 0. 0. 0.
    2 0. 0. 0. 0. 0. -0.707 0. 0. -0.707 0. 0. 0.
    3 0. 0. -0.707 0. 0. -0.707 0. 0. -1.0 0. 0. -0.707
```

The output obtained from program 7.1 is

```
Project:
 Spherical excavation - linear elements

 Cartesian_dimension:             3
 Elasticity Problem
 Infinite Region
 Symmetry about all planes
 Linear Elements
 Modulus:   1000.000
```

```
 Poissons ratio:  0.0000000E+00
 Number of Nodes of System:          7
 Number of Elements of System:          3
Node     1  Coor      1.00    0.00    0.00
Node     2  Coor      0.71    0.00    0.71
Node     3  Coor      0.50    0.50    0.71
Node     4  Coor      0.71    0.71    0.00
Node     5  Coor      0.00    0.71    0.71
Node     6  Coor      0.00    1.00    0.00
Node     7  Coor      0.00    0.00    1.00

 Incidences:

EL      1  Inci       1    2    3    4
EL      2  Inci       4    3    5    6
EL      3  Inci       3    2    7    5

 Elements with Dirichlet BC´s:


 Elements with Neuman BC´s:

 Element              1  Prescribed values:
 Node=              1  0.0000000E+00  0.0000000E+00  0.0000000E+00
 Node=              2  0.0000000E+00  0.0000000E+00 -0.7070000
 Node=              3  0.0000000E+00  0.0000000E+00 -0.7070000
 Node=              4  0.0000000E+00  0.0000000E+00  0.0000000E+00
 Element              2  Prescribed values:
 Node=              1  0.0000000E+00  0.0000000E+00  0.0000000E+00
 Node=              2  0.0000000E+00  0.0000000E+00 -0.7070000
 Node=              3  0.0000000E+00  0.0000000E+00 -0.7070000
 Node=              4  0.0000000E+00  0.0000000E+00  0.0000000E+00
 Element              3  Prescribed values:
 Node=              1  0.0000000E+00  0.0000000E+00 -0.7070000
 Node=              2  0.0000000E+00  0.0000000E+00 -0.7070000
 Node=              3  0.0000000E+00  0.0000000E+00  -1.000000
 Node=              4  0.0000000E+00  0.0000000E+00 -0.7070000
 Results, Element           1
 u= 0.154E-03 0.000E+00 0.000E+00 0.744E-04 0.000E+00-0.468E-03
    0.480E-04 0.480E-04-0.467E-03 0.113E-03 0.113E-03 0.000E+00

 t=     0.000     0.000     0.000     0.000     0.000    -0.707
        0.000     0.000    -0.707     0.000     0.000     0.000

 Results, Element           2
 u= 0.113E-03 0.113E-03 0.000E+00 0.480E-04 0.480E-04-0.467E-03
    0.000E+00 0.744E-04-0.468E-03 0.000E+00 0.154E-03 0.000E+00

 t=     0.000     0.000     0.000     0.000     0.000    -0.707
        0.000     0.000    -0.707     0.000     0.000     0.000
```

```
Results, Element           3
u= 0.480E-04 0.480E-04-0.467E-03 0.744E-04 0.000E+00-0.468E-03
   0.000E+00 0.000E+00-0.723E-03 0.000E+00 0.744E-04-0.468E-03

t=     0.000      0.000     -0.707      0.000      0.000     -0.707
       0.000      0.000     -1.000      0.000      0.000     -0.707
```

The input file for Mesh 1 for program 9.1 is

```
 1 3
 1.1   0    0
 1.2   0    0
 1.3   0    0
 1.4   0    0
 1.5   0    0
 2.0   0    0
 4.0   0    0
 6.0   0    0
10.0   0    0
```

The output obtained from program 9.1 is

```
Project:
 Spherical excavation - linear elements

 Cartesian_dimension:           3
 Elasticity Problem
 Infinite Region
 Symmetry about all planes
 Linear Elements
 Modulus:   1000.000
 Poissons ratio:  0.0000000E+00
 Number of Nodes of System:           7
 Number of Elements of System:           3
Node       1  Coor       1.00     0.00     0.00
Node       2  Coor       0.71     0.00     0.71
Node       3  Coor       0.50     0.50     0.71
Node       4  Coor       0.71     0.71     0.00
Node       5  Coor       0.00     0.71     0.71
Node       6  Coor       0.00     1.00     0.00
Node       7  Coor       0.00     0.00     1.00

 Incidences:

EL       1 Inci       1    2    3    4
EL       2 Inci       4    3    5    6
EL       3 Inci       3    2    7    5
```

```
 Post-processed Results

 Results at Boundary Elements:
Element    1 xsi= -1.00 eta= -1.00
 Stress:     -0.04      0.12     -0.47     -0.08      0.08      0.17
Element    1 xsi=  1.00 eta= -1.00
 Stress:     -0.23      0.04      0.00     -0.14      0.30      0.70
Element    1 xsi= -1.00 eta=  1.00
 Stress:     -0.04      0.12     -0.47     -0.08      0.06      0.17
Element    1 xsi=  1.00 eta=  1.00
 Stress:     -0.24      0.05      0.00     -0.13      0.28      0.70
Element    2 xsi= -1.00 eta= -1.00
 Stress:      0.12     -0.04     -0.47     -0.08      0.17      0.06
Element    2 xsi=  1.00 eta= -1.00
 Stress:      0.05     -0.24      0.00     -0.13      0.70      0.28
Element    2 xsi= -1.00 eta=  1.00
 Stress:      0.12     -0.04     -0.47     -0.08      0.17      0.08
Element    2 xsi=  1.00 eta=  1.00
 Stress:      0.04     -0.23      0.00     -0.14      0.70      0.30
Element    3 xsi= -1.00 eta= -1.00
 Stress:      0.10      0.10      0.71      0.00      0.00      0.00
Element    3 xsi=  1.00 eta= -1.00
 Stress:     -0.12      0.06      0.74     -0.09      0.03      0.07
Element    3 xsi= -1.00 eta=  1.00
 Stress:      0.06     -0.12      0.74     -0.09      0.07      0.03
Element    3 xsi=  1.00 eta=  1.00
 Stress:     -0.11     -0.11      1.05     -0.21      0.14      0.14

 Internal Results:

Coordinates:        1.10      0.00      0.00
         u:  0.142E-03 0.316E-11-0.532E-10
     Stress: -0.135  0.137  -0.365   0.000   0.011   0.000
Coordinates:        1.20      0.00      0.00
         u:  0.127E-03-0.862E-12 0.780E-10
     Stress: -0.134  0.111   -0.244   0.000   0.000   0.000
Coordinates:        1.30      0.00      0.00
         u:  0.115E-03-0.209E-12 0.486E-10
     Stress: -0.121  0.091 -0.176    0.000   0.000   0.000
Coordinates:        1.40      0.00      0.00
         u:  0.103E-03-0.340E-12 0.337E-10
     Stress: -0.108 0.075   -0.132   0.000   0.000   0.000
Coordinates:        1.50      0.00      0.00
         u:  0.930E-04-0.117E-14 0.712E-11
     Stress: -0.095 0.062   -0.101   0.000   0.000   0.000
```

10.5.3 Results

According to the highlighted value in the output from 7.1, the maximum displacement for mesh 1at the top is 0.72 mm compared to a theoretical value of 0.9. To obtain the correct stress results of the excavation problem we must add the virgin stress to the values highlighted in the output from 9.1. According to the highlighted value we obtain a maximum compressive stress in the z-direction at the boundary of -1.47 MPa, compared to a theoretical value of -2.0 MPa.

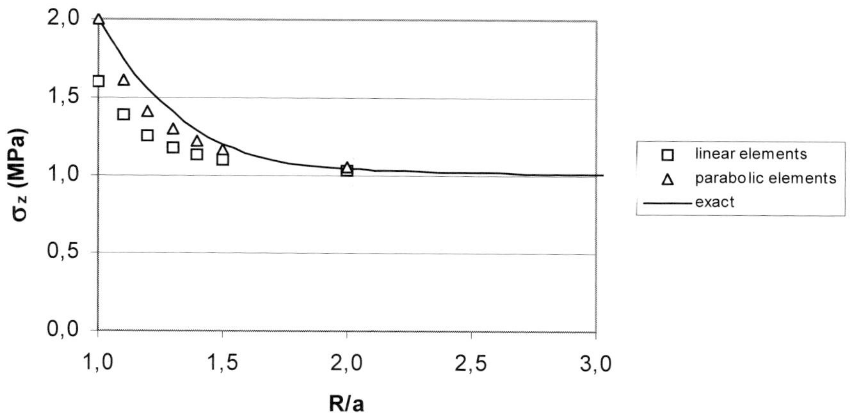

Figure 10.18 Distribution of vertical stress

The results are obviously improved for Mesh 2. For the mesh with parabolic boundary elements the maximum displacement at the crown is computed as 0.88 mm. The maximum value of stress at the meridian of the sphere is computed as 2.02 MPa compared with the theoretical solution of 2.0. The results for the internal stresses are summarized in Figure 10.18 where the vertical stress is plotted along a horizontal line originating from the meridian. It can be seen that even a relatively coarse mesh with 3 quadratic elements gives a reasonable accuracy.

10.5.4 Comparison with FEM

To be able to model this problem with the FEM we have to truncate the mesh, as with the 2-D example. A fairly coarse mesh is shown in Figure 10.19. The mesh has 135 degrees of freedom, as compared with 16 degrees of freedom for the boundary element mesh 2. The maximum displacement obtained from this analysis is 0.084 which represents a poor agreement with the theory. A much finer mesh would be required.

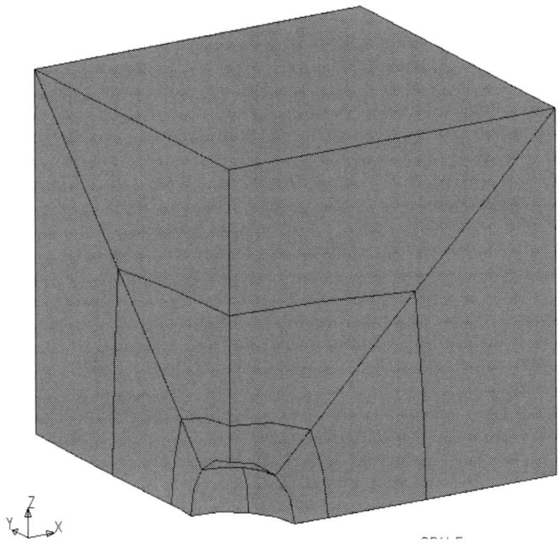

Figure 10.19 Finite element mesh

10.6 CONCLUSIONS

The purpose of this chapter was to show on a number of simple examples, how the method works and the sort of accuracy that can be expected. We have seen that some examples favour the BEM, while others do not. On the cantilever example it has been shown that care has to be taken to avoid situations where two boundary elements are too close to each other. The method based on point collocation implemented in this book shows that significant errors can be observed if surfaces are too close to each other and the mesh is coarse. On the other hand we have shown that for problems involving infinite domains, as they occur, for example in geotechnical engineering, accuracy, efficiency and user friendliness is superior to the FEM.

After reading this chapter the reader should have learned how to generate boundary element meshes and input files for Programs 7.1 Generel_purpose_BEM and 9.1 Post_processor. A good appreciation of the method, the accuracy that can be obtained and the pitfalls that should be avoided should also have been gained. The reader may now proceed to learn more about more advanced topics.

10.7 REFERENCES

1. Krätzig W.B. and Wittek U.(1995) Tragwerke 1. Springer, Berlin
2. Hartmann F. (1989) Introduction to Boundary Elements, Theory and Applications. Springer, Berlin
3. Kirsch (1898) Die Theorie der Elastizität und die Bedürfnisse der Festigkeitslehre. *Zeitschrift des Vereins deutscher Ingenieure*, **42**,797-807
4. Kanninen,M.F. and Popelar, C.H. (1985) Advanced Fracture Mechanics. Oxford Science Series15, Oxford University Press, New York.
5. Ingraffea, A.R. and Manu, C. (1980) Stress intensity factor computation in three dimensions with quarter point elements, *International Journal for Numerical Methods in Engineering*, **15**, 1427-1445.
6. Aliabadi M.H. (2002) The Boundary Element Method, Volume 2. J. Wiley.

11

Multiple regions

Imagination is more important than knowledge..
A. Einstein

11.1 INTRODUCTION

The solution procedures described so far are only applicable to homogeneous domains, as the fundamental solutions used assume that material properties do not change inside the domain being analysed. There are many instances, however, where this assumption does not hold. For example, in a soil or rock mass, the modulus of elasticity may change with depth or there might be various layers/inclusions with different properties. For some special types of heterogeneity it is possible to derive fundamental solutions, for example, if the material properties change in a simplified way (linear increase with depth). However, such fundamental solutions are often complicated and the programming effort significant[1].

In cases where we have layers or zones of different materials, however, we can develop special solution methods based on the fundamental solutions for homogeneous materials in Chapter 4. The basic idea is to consider a number of regions which are connected to each other, much like pieces of a puzzle. Each region is treated in the same way as discussed previously but can now be assigned different material properties. With this method we are able to solve *piecewise* homogeneous material problems. As we will see later, the method also allows simulating contact and cracking propagation problems.

Since at the interfaces between the regions both **t** and **u** are not known, the number of unknowns is increased and additional equations are required to solve the problem. These equations can be obtained from the conditions of equilibrium and compatibility at the region interfaces. There are two approaches which can be taken in the implementation of the method.

In the first, we modify the assembly procedure, so that a larger system of equations is now obtained including the additional unknowns at the interfaces. The second method is similar to the approach taken by the finite element method. Here we construct a "stiffness matrix", **K,** of each region, the coefficients of which are the fluxes or tractions due to unit temperatures/displacements. The matrices **K** for all regions are then assembled in the same way as with the FEM. The second method is more efficient and more amenable to implementation on parallel computers. The method may also be used for coupling boundary with finite elements, as outlined in Chapter 12. We will therefore only discuss the second method here. For the explanation of the first approach the reader is referred for appropriate text books[2,3].

11.2 STIFFNESS MATRIX ASSEMBLY

The multi-region assembly is not very efficient in cases where sequential excavation/construction (for example, in tunnelling) is to be modelled, since the coefficient matrices of all regions have to be computed and assembled every time a region is added or removed. Also, the method is not suitable for parallel processing since there the region matrices must be assembled and computed completely separately. Finally, significant efficiency gains can be made with the proposed method where only some nodes of the region are connected to other regions.

The stiffness matrix assembly, utilises a philosophy similar to that used by the finite element method. The idea is to compute a "stiffness matrix" $\mathbf{K}^N$ for each region *N*. Coefficients of $\mathbf{K}^N$ are values of **t** due to unit values of **u** at all region nodes. In potential flow problems these would correspond to fluxes due to unit temperatures while in elasticity they would be tractions due to unit displacements. To obtain the "stiffness matrix" $\mathbf{K}^N$ of a region, we simply solve the *Dirichlet* problem *M* times, where *M* is the number of degrees of freedom of the BE region nodes. For example, to get the first column of $\mathbf{K}^N$, we apply a unit value of temperature or of displacement in x-direction, as shown in Figure 11.1 while setting all other node values to zero.

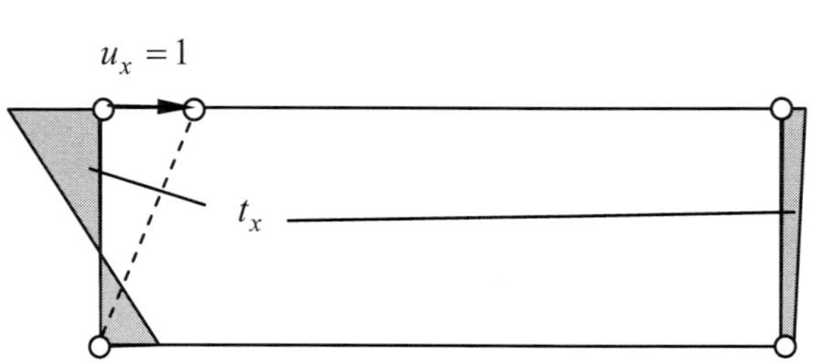

Figure 11.1 Example of computation of "stiffness coefficients": Cantilever beam subjected to a unit displacement $u_x = 1$ showing the traction distribution obtained from Program 7.1

For computation of *Dirichlet* problems we use equation (7.3), with a modified right hand side

$$[\Delta U]\{t\}_1 = [\Delta T]\{u\}_1 \tag{11.1}$$

Here $[\Delta T], [\Delta U]$ are the assembled coefficient matrices, $\{t\}_1$ is the first column of the stiffness matrix $\mathbf{K}^{\mathbf{M}}$ and $\{u\}_1$ is a vector with a unit value in the first row ,i.e

$$\{u\}_1 = \begin{Bmatrix} 1 \\ 0 \\ 0 \\ \vdots \end{Bmatrix} \tag{11.2}$$

If we perform the multiplication of $[\Delta T]\{u\}_1$ it can be easily seen that the right hand side of equation (11.1) is simply the first column of matrix $[\Delta T]$. The computation of the region "stiffness matrix" is therefore basically a solution of $[\Delta U]\{t\}_i = \{F\}_i$, with *N* right hand sides $\{F\}_i$, where each right hand side corresponds to a column in $[\Delta T]$. Each solution vector $\{t\}_i$ represents a column in $\mathbf{K}$, i.e.,

$$\mathbf{K}^N = \begin{bmatrix} \{t\}_1 & \{t\}_2 & \cdots \end{bmatrix} \tag{11.3}$$

For each region *(N)* we have the following relationship between $\{\mathbf{t}\}$ and $\{\mathbf{u}\}$:

$$\{t\}^N = \mathbf{K}^N \{u\}^N \tag{11.4}$$

To compute, for example, the problem of heat flow past an isolator, which is not impermeable but has conductivity different to the infinite domain, we specify two regions, an infinite and a finite one, as shown in figure 11.2. Note that the outward normals of the two regions point in directions opposite to each other (Figure 11.3). First we compute matrices $\mathbf{K}^I$ and $\mathbf{K}^{II}$ for each region separately and then we assemble the regions using the conditions for flow balance and uniqueness of temperature in the case of potential problems and equilibrium and compatibility in the case of elasticity. These conditions are written as

$$\{t\} = \{t\}^I + \{t\}^{II} \quad ; \quad \{u\}^I = \{u\}^{II} \tag{11.5}$$

The assembled system of equations for the example in Figure 11.2 is simply:

$$\{t\} = \left(\mathbf{K}^{\mathrm{I}} + \mathbf{K}^{\mathrm{II}}\right)\{u\} \tag{11.6}$$

which can be solved for $\{u\}$ if $\{t\}$ is known.

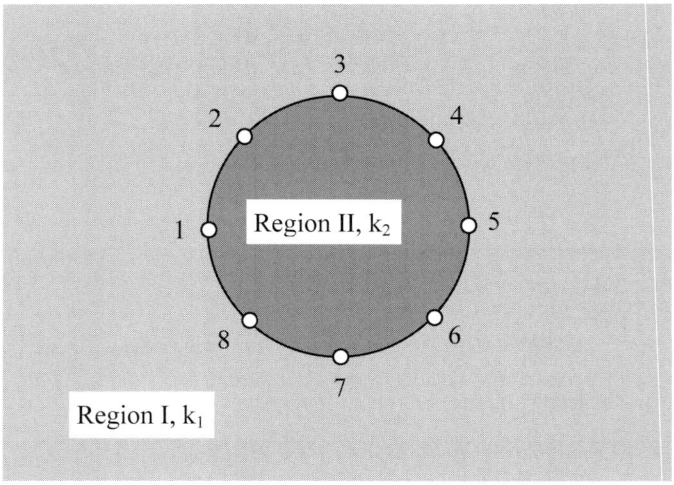

Figure 11.2 Example of a multi-region analysis: inclusion with different conductivity in an infinite domain

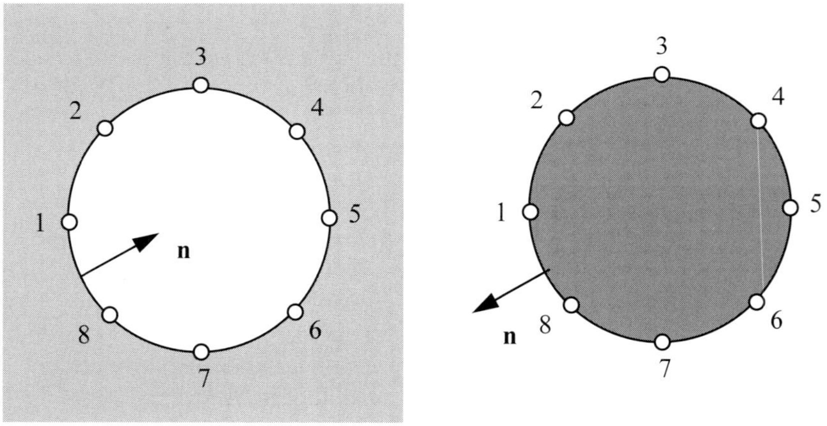

Region I (infinite) Region II (finite)

Figure 11.3 The two regions of the problem

11.2.1 Partially coupled problems

In many cases we have problems where not all nodes of the regions are connected (these are known as partially coupled problems). Consider for example the modified heat flow problem in Figure 11.4 where an additional circular impermeable isolator is specified on the right hand side.

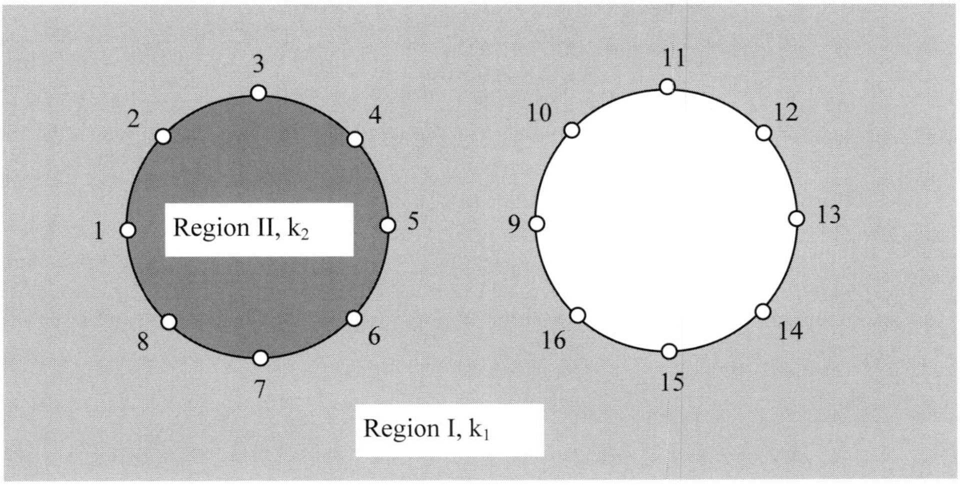

Figure 11.4 Problem with a circular inclusion and an isolator

Here only some of the nodes of region I are connected to region II. It is obviously more efficient to consider in the calculation of the stiffness matrix only the interface nodes, i.e. only of those nodes that are connected to a region. It is therefore proposed that we modify our procedure in such a way that we first solve the problem with zero values of **u** at the interface between region I and II and then solve the problem where unit values of **u** are applied at each node in turn.

For partially coupled problems we therefore have to solve the following types of problems (this is explained on a heat flow problem but can be extended to elasticity problems by replacing t with **t** and u with **u**):

1. **Solution of system with "fixed" interface nodes**

The first one is where boundary conditions are applied at the nodes which are not connected to other regions (free nodes) and *Dirichlet* boundary conditions with zero prescribed values are applied at the nodes which are connected to other regions (coupled nodes). For each region we can write the following system of equations:

$$[B]^N \begin{Bmatrix} \{t\}_{c0}^N \\ \{x\}_{f0}^N \end{Bmatrix} = \{F\}_0^N \tag{11.7}$$

where $[B]^N$ is the assembled left hand side and $\{F\}_o^N$ contains the right hand side due to given boundary conditions for region *N*. Vector $\{t\}_{co}^N$ contains the heat flow at the coupled nodes and vector $\{x\}_{fo}^N$ either temperatures or heat flow at the free nodes of region *N*, depending on the boundary conditions prescribed (*Dirichlet* or *Neumann*).

2. **Solution of system with unit values applied at the interface nodes**

The second problem to be solved for each region is to obtain the solution due to *Dirichlet* boundary condition of unit value applied at each of the interface nodes in turn and zero prescribed values at the free nodes. The equations to be solved are

$$[B]^N \begin{Bmatrix} \{t\}_{cn}^N \\ \{x\}_{fn}^N \end{Bmatrix} = \{F\}_n \qquad n = 1, 2 \ldots N_c \tag{11.8}$$

where $\{F\}_n^N$ is the right hand side computed for a unit value of *u* at node *n*. The vector $\{t\}_{cn}^N$ contains the heat flow at the coupled nodes and $\{x\}_{fn}^N$ the temperature or heat flow at the free nodes, for the case of unit *Dirichlet* boundary conditions at node *n*. N_c equations are obtained where N_c is the number of interface nodes in the case of the potential problem (in the case of elasticity problems it refers to the number of interface degrees of freedom). Note that the left hand side of the system of equations, $[B]^N$, is the same for the first and second problem and that $\{F\}_n$ simply corresponds to the nth column of $[\Delta T]$.

After the solution of the first two problems $\{t\}_c^N$ and $\{x\}_f^N$ can be expressed in terms of $\{u\}_c^N$ by:

$$\begin{Bmatrix} \{t\}_c^N \\ \{x\}_f^N \end{Bmatrix} = \begin{Bmatrix} \{t\}_{c0}^N \\ \{x\}_{f0}^N \end{Bmatrix} + \begin{bmatrix} \mathbf{K}^N \\ \mathbf{A}^N \end{bmatrix} \{u\}_c^N \tag{11.9}$$

where $\{u\}_c^N$ contains the temperatures at the interface nodes of region *N* and the matrices $\mathbf{K}^N$ and $\mathbf{A}^N$ are defined by:

$$\mathbf{K}^N = \begin{bmatrix} \{t\}_{c1} & \cdots & \{t\}_{cN_c} \end{bmatrix}^N \quad ; \quad \mathbf{A}^N = \begin{bmatrix} \{x\}_{c1} & \cdots & \{x\}_{cN_c} \end{bmatrix}^N \tag{11.10}$$

3. Assembly of regions, calculation of interface unknowns

After all the region stiffness matrices $\mathbf{K}^N$ have been computed they are assembled to a system of equations which can be solved for the unknown $\{u\}_c$.

For the assembly we use conditions of heat balance and uniqueness of temperature or equilibrium and compatibility as discussed previously. This results in the following system of equations

$$[\mathbf{K}]\{u\}_c = \{F\} \tag{11.11}$$

where $[\mathbf{K}]$ is the assembled "stiffness matrix" of the interface nodes and $\{F\}$ is the assembled right hand side. This system is solved for the unknown $\{u\}_c$ at the nodes of all interfaces of the problem.

4. Calculation of unknowns at the free nodes of region *N*

After the interface unknown have been determined the values of t at the interface ($\{t\}_c^N$) and the value of u or t at the free nodes ($\{x\}_f^N$) are determined for each region by the application of

$$\begin{Bmatrix} \{t\}_c^N \\ \{x\}_f^N \end{Bmatrix} = \begin{Bmatrix} \{t\}_{c0}^N \\ \{x\}_{f0}^N \end{Bmatrix} + \begin{bmatrix} \mathbf{K}^N \\ \mathbf{A}^N \end{bmatrix} \{u\}_c^N \tag{11.12}$$

Note that $\{u\}_c^N$ is obtained by gathering values from the vector of unknown at all the interfaces $\{u\}_c$.

11.2.2 Example

The procedure is explained in more detail on a simple example in potential flow. Consider the example in Figure 11.5 which contains two homogeneous regions. *Dirichlet* boundary conditions with prescribed zero values are applied on the left side and *Neuman* BC's on the right side as shown. All other boundaries are assumed to have *Neuman* BC with zero prescribed values. The interface only involves nodes 2 and 3 and therefore only 2 interface unknowns exist.

For an efficient implementation it will be necessary to renumber the nodes for each region, i.e. introduce a separate local numbering for each region. This will not only allow each region to be treated completely independently but also save storage space, because nodes not on the interface will belong to one region only.

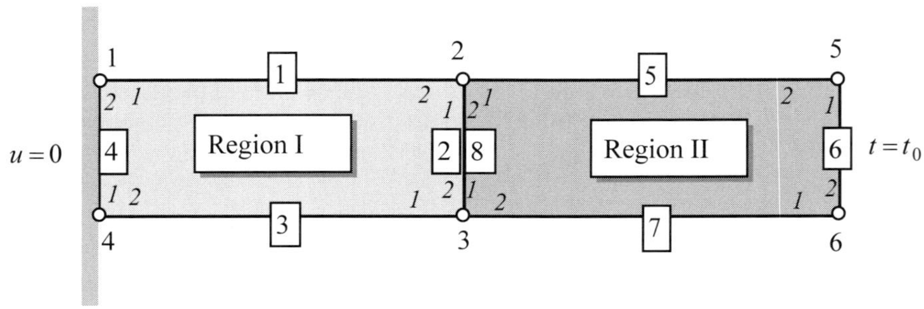

Figure 11.5 Example for stiffness assembly, partially coupled problem with global node numbering; local (element) numbering shown in italics.

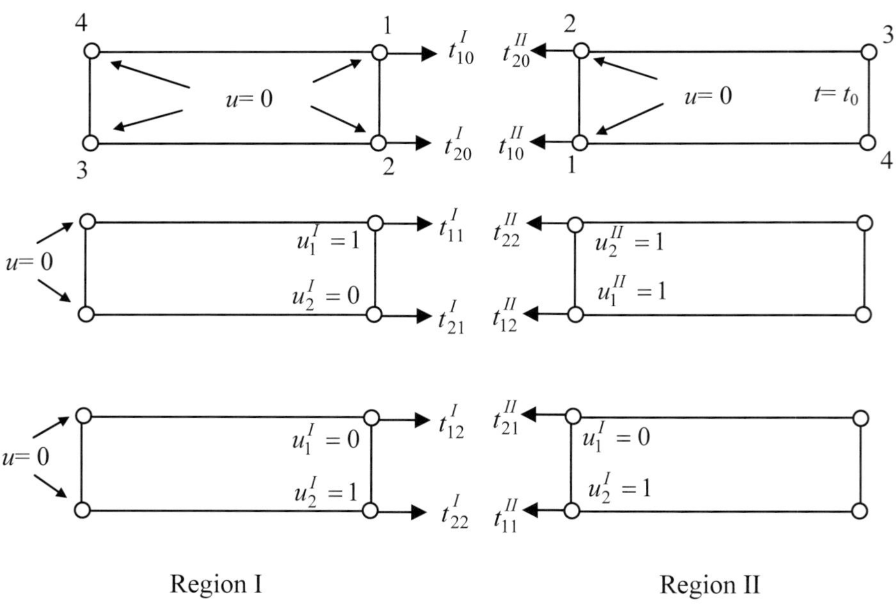

Figure 11.6 The different problems to be solved for regions I and II (potential problem)

On the top of Figure 11.6 we show the local (region) numbering that is adapted for region I and II. The sequence in which the nodes of the region are numbered is such that the interface nodes are numbered first. We also depict in the same figure the problems which have to be solved for obtaining vector $\{t\}_{co}$ and the two rows of matrix **K** and **A**. It is obvious that for the first problem to be solved for region I, where for all nodes u=0, $\{t\}_{co}$ will also be zero. Following the procedure in chapter 7 and referring to the element numbering of Figure 11.5 we obtain the following integral equations for the second and third problem for region *I*.

$$
\begin{aligned}
u_2 = 1 \quad &\Rightarrow \Delta U_{1i}^{2}\, t_{11}^{I} + \Delta U_{2i}^{2}\, t_{21}^{I} + \Delta U_{1i}^{4}\, t_{31}^{I} + \Delta U_{2i}^{4}\, t_{41}^{I} = \left(\Delta T_{2i}^{1} + \Delta T_{1i}^{2}\right)\ 1 \\
u_3 = 1 \quad &\Rightarrow \Delta U_{1i}^{2}\, t_{12}^{2} + \Delta U_{2i}^{2}\, t_{22}^{I} + \Delta U_{1i}^{4}\, t_{32}^{I} + \Delta U_{2i}^{4}\, t_{42}^{I} = \left(\Delta T_{1i}^{3} + \Delta T_{2i}^{2}\right)\ 1
\end{aligned}
\tag{11.13}
$$

for i=*1,2,3,4*. In Equation 11.13, two subscripts have been introduced for t: the first subscript refers to the node number where t is computed and the second to the node number where the unit value of u is applied. The roman superscript refers to the region number. The notation for ΔU and ΔT is the same as defined in Chapter 7, i.e. the first subscript defines the node number and the second the collocation point number; the superscript refers to the boundary element number (in square areas in Figure 11.5).

This gives the following system of equations with two right hand sides

$$
\begin{bmatrix}
\Delta U_{11}^{2} & \Delta U_{21}^{2} & \Delta U_{11}^{4} & \Delta U_{21}^{4} \\
\Delta U_{12}^{2} & \Delta U_{22}^{2} & \Delta U_{12}^{4} & \Delta U_{22}^{4} \\
\Delta U_{13}^{2} & \Delta U_{23}^{2} & \Delta U_{13}^{4} & \Delta U_{23}^{4} \\
\Delta U_{14}^{2} & \Delta U_{24}^{1} & \Delta U_{14}^{4} & \Delta U_{24}^{4}
\end{bmatrix}
\left\{\begin{array}{c:c}
t_{11}^{I} & t_{12}^{I} \\
t_{21}^{I} & t_{22}^{I} \\
t_{31}^{I} & t_{32}^{I} \\
t_{41}^{I} & t_{42}^{I}
\end{array}\right\}
=
\left\{\begin{array}{c:c}
\Delta T_{21}^{1} + \Delta T_{11}^{2} & \Delta T_{21}^{2} + \Delta T_{11}^{3} \\
\Delta T_{22}^{1} + \Delta T_{12}^{2} & \Delta T_{22}^{1} + \Delta T_{12}^{2} \\
\Delta T_{23}^{1} + \Delta T_{13}^{2} & \Delta T_{23}^{1} + \Delta T_{13}^{2} \\
\Delta T_{24}^{1} + \Delta T_{14}^{2} & \Delta T_{24}^{1} + \Delta T_{14}^{2}
\end{array}\right\}
\tag{11.14}
$$

After solving the system of equations we obtain

$$
\{\mathbf{t}\}_c^{\mathrm{I}} = \mathbf{K}^{\mathrm{I}}\{\mathbf{u}\}_c^{\mathrm{I}} \quad ; \quad \{\mathbf{x}\}_f^{\mathrm{I}} = \mathbf{A}^{\mathrm{I}}\{\mathbf{u}\}_c^{\mathrm{I}}
\tag{11.15}
$$

where

$$
\begin{aligned}
&\{t\}_c^{\mathrm{I}} = \begin{Bmatrix} t_1^{I} \\ t_2 \end{Bmatrix} \quad ; \quad \mathbf{K}^{I} = \begin{bmatrix} t_{11}^{I} & t_{12}^{I} \\ t_{21}^{I} & t_{22}^{I} \end{bmatrix} \quad ; \quad \{u\}_c^{\mathrm{I}} = \begin{Bmatrix} u_1^{I} \\ u_2^{I} \end{Bmatrix} \\
&\{x\}_f^{\mathrm{I}} = \begin{Bmatrix} t_3^{I} \\ t_4^{I} \end{Bmatrix} \quad ; \quad \mathbf{A}^{I} = \begin{bmatrix} t_{31}^{I} & t_{32}^{I} \\ t_{41}^{I} & t_{42}^{I} \end{bmatrix}
\end{aligned}
\tag{11.16}
$$

where $\{t\}_c$ and $\{u\}_c$ refer to the values of *t* and *u* at the interface.

For region II we have for the case of zero *u* at the interface nodes

$$\begin{aligned} &\Delta U_{2i}^{8}\, t_{20}^{II} + \Delta U_{1i}^{8}\, t_{10}^{II} - \left(\Delta T_{1i}^{6} + \Delta T_{2i}^{5}\right) u_{30}^{II} \\ &-\left(\Delta T_{1i}^{6} + \Delta T_{2i}^{7}\right) u_{40}^{II} = -\left(\Delta U_{1i}^{6} + \Delta U_{2i}^{6}\right)\, t_0 \end{aligned} \tag{11.17}$$

This gives

$$\begin{bmatrix} \Delta U_{11}^{8} & \Delta U_{21}^{8} & -\left(\Delta T_{11}^{6} + \Delta T_{21}^{5}\right) & -\left(\Delta T_{11}^{6} + \Delta T_{21}^{7}\right) \\ \Delta U_{12}^{8} & \Delta U_{22}^{8} & -\left(\Delta T_{12}^{6} + \Delta T_{22}^{5}\right) & -\left(\Delta T_{12}^{6} + \Delta T_{22}^{7}\right) \\ \Delta U_{13}^{8} & \Delta U_{23}^{8} & -\left(\Delta T_{13}^{6} + \Delta T_{23}^{5}\right) & -\left(\Delta T_{13}^{6} + \Delta T_{23}^{7}\right) \\ \Delta U_{14}^{8} & \Delta U_{24}^{8} & -\left(\Delta T_{14}^{6} + \Delta T_{24}^{5}\right) & -\left(\Delta T_{14}^{6} + \Delta T_{24}^{7}\right) \end{bmatrix} \begin{Bmatrix} t_{10}^{II} \\ t_{20}^{II} \\ u_{30}^{II} \\ u_{40}^{II} \end{Bmatrix} = \begin{Bmatrix} -\left(\Delta U_{11}^{6} + \Delta U_{21}^{6}\right) t_0 \\ -\left(\Delta U_{12}^{6} + \Delta U_{22}^{6}\right) t_0 \\ -\left(\Delta U_{13}^{6} + \Delta U_{23}^{6}\right) t_0 \\ -\left(\Delta U_{14}^{6} + \Delta U_{24}^{6}\right) t_0 \end{Bmatrix} \tag{11.18}$$

which can be solved for the values at the interface and free nodes

$$\{t\}_{co}^{\mathrm{I}} = \begin{Bmatrix} t_{10}^{II} \\ t_{40}^{II} \end{Bmatrix} \quad ; \quad \{x\}_{f0}^{\mathrm{I}} = \begin{Bmatrix} u_{20}^{II} \\ u_{30}^{II} \end{Bmatrix} \tag{11.19}$$

In our notation $\{t\}_{co}^{\mathrm{I}}$ refers to the values of *t* at the interface for the case where *u*=0 at the interface. $\{x\}_{co}^{\mathrm{I}}$ refers to the values of *u* at the free nodes (where *Neumann* boundary conditions have been applied).

For $u_2 = 1$ and $u_3 = 1$ we obtain

$$\begin{aligned} &\Delta U_{1i}^{8}\, t_{12}^{II} + \Delta U_{2i}^{8}\, t_{22}^{II} - \left(\Delta T_{1i}^{6} + \Delta T_{2i}^{5}\right) u_{32}^{II} - \left(\Delta T_{1i}^{6} + \Delta T_{2i}^{7}\right) u_{42}^{II} = \left(\Delta T_{1i}^{5} + \Delta T_{2i}^{8}\right)\, 1 \\ &\Delta U_{2i}^{8}\, t_{11}^{II} + \Delta U_{2i}^{8}\, t_{21}^{II} - \left(\Delta T_{1i}^{6} + \Delta T_{2i}^{5}\right) u_{31}^{II} - \left(\Delta T_{1i}^{6} + \Delta T_{2i}^{7}\right) u_{41}^{II} = \left(\Delta T_{2i}^{7} + \Delta T_{1i}^{8}\right)\, 1 \end{aligned} \tag{11.20}$$

The solutions can be written as

$$\{t\}_c^{\mathrm{II}} = \{t\}_{c0}^{\mathrm{II}} + \mathbf{K}^{\mathrm{II}} \{u\}_c^{\mathrm{II}} \quad ; \quad \{x\}_f^{\mathrm{II}} = \{x\}_{f0}^{\mathrm{II}} + \mathbf{A}^{\mathrm{II}} \{u\}_c^{\mathrm{II}} \tag{11.21}$$

where

$$\{t\}_c^{\mathrm{II}} = \begin{Bmatrix} t_1^{II} \\ t_2^{II} \end{Bmatrix} \quad ; \quad \mathbf{K}^{II} = \begin{bmatrix} t_{11}^{II} & t_{12}^{II} \\ t_{21}^{II} & t_{22}^{II} \end{bmatrix} \quad ; \quad \{t\}_{co}^{\mathrm{II}} = \begin{Bmatrix} t_{10}^{II} \\ t_{20}^{II} \end{Bmatrix} \quad ; \quad \{u\}_c^{\mathrm{I}} = \begin{Bmatrix} u_1^{II} \\ u_2^{II} \end{Bmatrix}$$

$$\{x\}_c^{\mathrm{II}} = \begin{Bmatrix} u_3^{II} \\ u_4^{II} \end{Bmatrix} \quad ; \quad \mathbf{A}^{II} = \begin{bmatrix} u_{31}^{II} & u_{32}^{II} \\ u_{41}^{II} & u_{42}^{II} \end{bmatrix} \quad ; \quad \{x\}_{co}^{\mathrm{II}} = \begin{Bmatrix} u_{30}^{II} \\ u_{40}^{II} \end{Bmatrix} \tag{11.22}$$

The equations of compatibility or preservation of heat at the interface can be written as

$$\begin{Bmatrix} t_{1c}^{I} \\ t_{2c}^{I} \end{Bmatrix} + \begin{Bmatrix} t_{2c}^{II} \\ t_{1c}^{II} \end{Bmatrix} = 0 \quad ; \quad \begin{Bmatrix} u_{1c}^{I} \\ u_{2c}^{I} \end{Bmatrix} = \begin{Bmatrix} u_{2c}^{II} \\ u_{1c}^{II} \end{Bmatrix} = \begin{Bmatrix} u_2 \\ u_3 \end{Bmatrix} \tag{11.23}$$

Substituting (11.16) and (11.22) into (11.23) we obtain

$$\mathbf{K}\ \{u\}_c + \{t\} = 0 \tag{11.24}$$

where

$$\mathbf{K} = \begin{bmatrix} t_{11}^{I} + t_{22}^{II} & t_{12}^{I} + t_{21}^{II} \\ t_{22}^{I} + t_{11}^{II} & t_{21}^{I} + t_{12}^{II} \end{bmatrix} \quad ; \quad \{u\}_c = \begin{Bmatrix} u_2 \\ u_3 \end{Bmatrix} \quad ; \quad \{t\} = \begin{Bmatrix} t_{10}^{I} \\ t_{20}^{I} \end{Bmatrix} \tag{11.25}$$

This system can be solved for the interface unknowns. The calculation of the other unknowns is done separately for each region. For region I we have

$$\{t\}_c^{\mathrm{I}} = \mathbf{K}^{\mathrm{I}}\{u\}_c^{\mathrm{I}} \quad ; \quad \{x\}_f^{\mathrm{I}} = \mathbf{A}^{\mathrm{I}}\{u\}_c^{\mathrm{I}} \tag{11.26}$$

Whereas for region II

$$\{t\}_c^{\mathrm{II}} = \{t\}_{c0}^{\mathrm{II}} + \mathbf{K}^{\mathrm{II}}\{u\}_c^{\mathrm{II}} \quad ; \quad \{x\}_f^{\mathrm{II}} = \{x\}_{f0}^{\mathrm{II}} + \mathbf{A}^{\mathrm{II}}\{u\}_c^{\mathrm{II}} \tag{11.27}$$

If we consider the equivalent elasticity problem of a cantilever beam, we see (Figure 11.7) then for region *II* the problem where the interface displacements are fixed gives the tractions at the interface corresponding to a shortened cantilever beam. If u_x=1 is applied only a rigid body motion results and therefore no resulting tractions at the interface occur. The application of u_y=1 however will result in shear tractions at the interface.

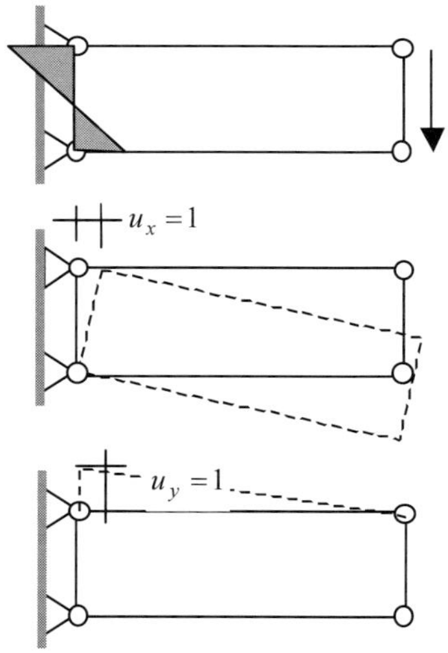

Figure 11.7 Effect of application of *Dirichlet* boundary conditions on region II of cantilever beam (elasticity problem)

11.3 COMPUTER IMPLEMENTATION

We now consider the computer implementation of the stiffness matrix assembly method. We divide this into two tasks. First we develop a SUBROUTINE Stiffness_BEM for the calculation of matrix **K**. If the problem is not fully coupled, then this subroutine will also determine the matrix **A** and the solutions for zero values of **u** at the interface. Secondly we develop a program General_purpose_BEM_Multi.

For an efficient implementation (where zero entries in the matrices are avoided) we must consider 3 different numbering systems, each one is related to the global numbering system as shown in Figure 11.5:

1. **Element numbering.** This is the sequence in which the nodes to which an element connects are entered in the element incidence vector. In the example in Figure 11.5 we have only two element nodes (*1,2*). Table 11.1 has two main columns: One termed "in global numbering" which shows the node numbers as they appear in Figure 11.5 and the other termed "in region numbering" as they appear on the top of Figure 11.6.

2. **Region numbering.** This numbering is used for computing the "stiffness matrix" of a region. For this the element node numbers are specified in "region numbering".

Table 11.2 depicts the "region incidences" i.e. the sequence of node numbers of a region.

3. **Interface numbering.** This is basically the sequence in which the interface nodes are entered in the interface incidence vector. For the example problem the interface incidences are given in Table 11.3. This sequence is determined in such a way that the first node of the first interface element will start the sequence. Note that the interface incidences are simply the first two values of the region incidence vector.

For problems involving more than one unknown per node the incidence vectors have to be expanded to *Destination* vectors as explained in Chapter 7.

Table 11.1 Incidences of boundary elements in global and local numbering

Boundary Element number	in global numbering		in region numbering	
	1.	*2.*	*1.*	*2.*
1	1	2	4	1
2	2	3	1	2
3	3	4	2	3
4	4	1	3	4
5	2	5	2	3
6	5	6	3	4
7	6	3	4	1
8	3	2	1	2

Table 11.2 Region incidences

	1.	*2.*	*3.*	*4.*
Region *I*	2	3	4	1
Region *II*	3	2	5	6

Table 11.3 Interface incidences

	1.	*2.*
Region *I*	2	3
Region *II*	3	2

11.3.1 Subroutine Stiffness_BEM

The tasks for the Stiffness_BEM are essentially the same as for the General_Purpose_BEM, except that the boundary conditions that are considered are expanded. We add a new boundary code,2 , which is used to mark nodes at the interface.

The input parameters for **SUBROUTINE Stiffness_BEM** are the incidence vectors of the boundary elements, which describe the boundary of the region, the coordinates of the nodes and the Boundary conditions. Note that the vector of incidences as well as the coordinates has to be in the local (region) numbering. **SUBROUTINE AssemblySTIFF** is basically the same as **SUBROUTINE Assembly**, except that a boundary code 2 for interface conditions has been added. Boundary code 2 is treated the same as code 1 (*Dirichlet*) except that columns of $[\Delta \mathbf{T}]^e$ are assembled into the array RhsM (multiple right hand sides). **SUBROUTINE Solve** is modified into **Solve_Multi**, which can handle both single (Rhs) and multiple (RhsM) right hand sides.

The output parameter of the SUBROUTINE is stiffness matrix **K** and for partially coupled problems in addition matrix **A** as well as $\{\mathbf{t}\}_c$. The rows and columns of these matrices will be numbered in a local (interface) numbering. The values of $\{\mathbf{u}\}_{f0}$ and $\{\mathbf{t}\}_{c0}$ are stored in the array El_res which contains the element results. They can be added at element level.

We show below the library module Stiffness_lib which contains all the necessary declarations and subroutines for the computation of the stiffness matrix. The symmetry option has been left out in the implementation shown to simplify the coding.

```
MODULE Stiffness_lib
USE Utility_lib ; USE Integration_lib ; USE Geometry_lib
IMPLICIT NONE
INTEGER :: Cdim    !   Cartesian dimension
INTEGER :: Ndof    !   No. of degeres of freedom per node
INTEGER :: Nodel   !   No. of nodes per element
INTEGER :: Ndofe   !   D.o.F´s / Elem
REAL    :: C1,C2   !   material constants
INTEGER, ALLOCATABLE :: Bcode(:,:)
REAL, ALLOCATABLE :: Elres_u(:,:),Elres_t(:,:) !  El. results
CONTAINS
SUBROUTINE Stiffnes_BEM(Nreg,maxe,xP,incie,Ncode,Ndofc,KBE,A,TC)
!--------------------------------------------
!    Computes the stiffness matrix of a boundary element region
!    no symmetry implemented
!--------------------------------------------
INTEGER,INTENT(IN)  :: Nreg          !  Region code
INTEGER,INTENT(IN)  :: maxe          !  Number of boundary elements
REAL, INTENT(IN)    :: xP(:,:)       !  Array of node coordinates
INTEGER, INTENT(IN):: Incie(:,:)    !  Array of incidences
INTEGER, INTENT(IN):: Ncode(:)      !  Global restraint code
INTEGER, INTENT(IN):: Ndofc         !  No of interface D.o.F.
```

```
REAL(KIND=8), INTENT(OUT)  :: KBE(:,:) !  Stiffness matrix
REAL(KIND=8), INTENT(OUT)  :: A(:,:) ! u due to uc=1
REAL(KIND=8), INTENT(OUT)  :: TC( :) ! t due to uc=0
INTEGER, ALLOCATABLE :: Ldeste(:,:)!  Element destinations
REAL(KIND=8), ALLOCATABLE :: dUe(:,:),dTe(:,:),Diag(:,:)
REAL(KIND=8), ALLOCATABLE :: Lhs(:,:)
REAL(KIND=8), ALLOCATABLE :: Rhs(:),RhsM(:,:) ! right hand sides
REAL(KIND=8), ALLOCATABLE :: u1(:),u2(:,:)      ! results
REAL, ALLOCATABLE :: Elcor(:,:)
REAL    :: v3(3),v1(3),v2(3)
INTEGER :: Nodes,Dof,k,l,nel
INTEGER :: n,m,Ndofs,Pos,i,j,nd
Nodes= UBOUND(xP,2)  !   total number of nodes of region
Ndofs= Nodes*Ndof    !  Total degrees of freedom of region
ALLOCATE(Ldeste(maxe,Ndofe)) ! Elem. destination vector
!-------------------------------------------
!     Determine Element destination vector
!----------------------------------------------
Elements:&
DO Nel=1,Maxe
 k=0
 DO n=1,Nodel
   DO m=1,Ndof
     k=k+1
     IF(Ndof > 1) THEN
       Ldeste(Nel,k)= ((Incie(Nel,n)-1)*Ndof + m)
     ELSE
       Ldeste(Nel,k)= Incie(Nel,n)
     END IF
   END DO
 END DO
END DO &
Elements
ALLOCATE(dTe(Ndofs,Ndofe),dUe(Ndofs,Ndofe))
ALLOCATE(Diag(Ndofs,Ndof))
ALLOCATE(Lhs(Ndofs,Ndofs),Rhs(Ndofs),RhsM(Ndofs,Ndofs))
ALLOCATE(u1(Ndofs),u2(Ndofs,Ndofs))
ALLOCATE(Elcor(Cdim,Nodel))
!-------------------------------------------------------------
!  Compute and assemble element coefficient matrices
!-------------------------------------------------------------
Lhs= 0
Rhs= 0
Elements_1:&
DO Nel=1,Maxe
  Elcor(:,:)= xP(:,Incie(nel,:))!    gather element coords
  IF(Cdim == 2) THEN
    IF(Ndof == 1) THEN
      CALL Integ2P(Elcor,Incie(nel,:),Nodel,Nodes,xP,C1,dUe,dTe)
    ELSE
      CALL Integ2E(Elcor,Incie(nel,:),Nodel,Nodes&
```

```
                ,xP,C1,C2,dUe,dTe)
    END IF
   ELSE
    CALL Integ3(Elcor,Incie(nel,:),Nodel,Nodes,xP,Ndof &
               ,C1,C2,dUe,dTe)
   END IF
   CALL AssemblyMR(Nel,Lhs,Rhs,RhsM,DTe,Due&
                 ,Ldeste(nel,:),Ncode,Diag)
 END DO &
 Elements_1
 !--------------------------------------------------------------
 ! Add azimuthal integral for infinite regions
 !--------------------------------------------------------------
 IF(Nreg == 2) THEN
   DO m=1, Nodes
    DO n=1, Ndof
     k=Ndof*(m-1)+n
     Diag(k,n) = Diag(k,n) + 1.0
    END DO
   END DO
 END IF
 !---------------------------------------------------------------
 ! Add Diagonal coefficients
 !---------------------------------------------------------------
 Nodes_global: &
 DO m=1,Nodes
   Degrees_of_Freedoms_node: &
   DO n=1,Ndof
    DoF = (m-1)*Ndof + n    !  global degree of freedom no.
    k = (m-1)*Ndof + 1      !  address in coeff. matrix (row)
    l = k + Ndof - 1        !  address in coeff. matrix (column)
    IF (NCode(DoF) == 1) THEN  ! Dirichlet - Add Diagonal to Rhs
     CALL Get_local_DoF(Maxe,Dof,Ldeste,Nel,Pos)
     Rhs(k:l) = Rhs(k:l) - Diag(k:l,n)*Elres_u(Nel,Pos)
    ELSE                       ! Neuman - Add Diagonal to Lhs
     Lhs(k:l,Dof)= Lhs(k:l,Dof) + Diag(k:l,n)
    END IF
   END DO &
   Degrees_of_Freedoms_node
 END DO &
 Nodes_global
 !    Solve problem
 CALL Solve_Multi(Lhs,Rhs,RhsM,u1,u2)
 !-------------------------------------
 ! Gather element results due to
 ! "fixed" interface nodes
 !-------------------------------------
 Elements2: &
 DO nel=1,maxe
   D_o_F1:  &
   DO nd=1,Ndofe
```

```
    IF(NCode(Ldeste(nel,nd)) == 0) THEN
      Elres_u(nel,nd) =  u1(Ldeste(nel,nd))
    ELSE IF(NCode(Ldeste(nel,nd)) == 1) THEN
      Elres_t(nel,nd) =  u1(Ldeste(nel,nd))
    END IF
  END DO &
  D_o_F1
END DO &
Elements2
!-----------------------------------
!   Gather stiffness matrix KBE and matrix A
!-----------------------------------
Interface_DoFs: &
DO N=1,Ndofc
 KBE(N,:)= u2(N,:)
 TC(N)= u1(N)
END DO &
Interface_DoFs
Free_DoFs: &
DO N=Ndofc+1,Ndofs
  A(N,:)= u2(N,:)
END DO &
Free_DoFs
DEALLOCATE (Ldeste,dUe,dTe,Diag,Lhs,Rhs,RhsM,u1,u2,Elcor)
RETURN
END SUBROUTINE Stiffnes_BEM

SUBROUTINE Solve_Multi(Lhs,Rhs,RhsM,u,uM)
!--------------------------------------------
!    Solution of system of equations
!    by Gauss Elimination
!    for multple right hand sides
!--------------------------------------------
REAL(KIND=8) ::   Lhs(:,:)   !   Equation Left hand side
REAL(KIND=8) ::   Rhs(:)     !   Equation right hand side 1
REAL(KIND=8) ::   RhsM(:,:)  !   Equation right hand sides 2
REAL(KIND=8) ::   u(:)       !   Unknowns 1
REAL(KIND=8) ::   uM(:,:)    !   Unknowns 2
REAL(KIND=8) ::   FAC
INTEGER  M,Nrhs            !    Size of system
INTEGER  i,n,nr
M= UBOUND(RhsM,1) ; Nrhs= UBOUND(RhsM,2)
!  Reduction
Equation_n: &
DO n=1,M-1
   IF(ABS(Lhs(n,n)) < 1.0E-10) THEN
     CALL Error_Message('Singular Matrix')
   END IF
   Equation_i: &
   DO i=n+1,M
     FAC= Lhs(i,n)/Lhs(n,n)
```

```
     Lhs(i,n+1:M)= Lhs(i,n+1:M) - Lhs(n,n+1:M)*FAC
      Rhs(i)= Rhs(i) - Rhs(n)*FAC
      RhsM(i,:)= RhsM(i,:) - RhsM(n,:)*FAC
    END DO &
    Equation_i
  END DO &
  Equation_n
  !     Backsubstitution
  Unknown_1: &

  DO n= M,1,-1
   u(n)= -1.0/Lhs(n,n)*(SUM(Lhs(n , n+1:M)*u(n+1:M)) - Rhs(n))
  END DO &
  Unknown_1
  Load_case: &
  DO Nr=1,Nrhs
    Unknown_2: &
    DO n= M,1,-1
     uM(n,nr)= -1.0/Lhs(n,n)*(SUM(Lhs(n , n+1:M)*uM(n+1:M , nr))&
                                 - RhsM(n,nr))
    END DO &
    Unknown_2
  END DO &
  Load_case
  RETURN
  END SUBROUTINE Solve_Multi

  SUBROUTINE AssemblyMR(Nel,Lhs,Rhs,RhsM,DTe,DUe,Ldest,Ncode,Diag)
  !---------------------------------------------
  !  Assembles Element contributions DTe , DUe
  !  into global matrix Lhs, vector Rhs
  !  and matrix RhsM
  !---------------------------------------------
  INTEGER,INTENT(IN)          :: NEL
  REAL(KIND=8)                :: Lhs(:,:) !  Eq.left hand side
  REAL(KIND=8)                :: Rhs(:)   !  Right hand side
  REAL(KIND=8)                :: RhsM(:,:) ! Matrix of r. h. s.
  REAL(KIND=8), INTENT(IN)::  DTe(:,:),DUe(:,:)   !  Element arrays
  INTEGER , INTENT(IN)        :: LDest(:) ! Element destination vector
  INTEGER , INTENT(IN)        :: NCode(:) ! Boundary code (global)
  REAL(KIND=8) :: Diag(:,:) ! Diagonal coeff of DT
  INTEGER :: n,Ncol,m,k,l
  DoF_per_Element:&
  DO m=1,Ndofe
   Ncol=Ldest(m)        !   Column number
   IF(BCode(nel,m) == 0) THEN     !   Neumann BC
     Rhs(:) = Rhs(:) + DUe(:,m)*Elres_t(nel,m)
!     The assembly of dTe depends on the global BC
    IF (NCode(Ldest(m)) == 0) THEN
      Lhs(:,Ncol)=  Lhs(:,Ncol) + DTe(:,m)
    ELSE
```

```
      Rhs(:) = Rhs(:) - DTe(:,m) * Elres_u(nel,m)
    END IF
  ELSE IF(BCode(nel,m) == 1) THEN    !   Dirichlet BC
    Lhs(:,Ncol) = Lhs(:,Ncol) - DUe(:,m)
    Rhs(:)= Rhs(:) - DTe(:,m) * Elres_u(nel,m)
  ELSE IF(BCode(nel,m) == 2) THEN    !   Interface
    Lhs(:,Ncol) = Lhs(:,Ncol) - DUe(:,m)
    RhsM(:,Ncol)= RhsM(:,Ncol) - DTe(:,m)
  END IF
 END DO &
 DoF_per_Element
 !      Sum of off-diagonal coefficients
 DO n=1,Nodel
   DO k=1,Ndof
     l=(n-1)*Ndof+k
     Diag(:,k)= Diag(:,k) - DTe(:,l)
   END DO
 END DO
 RETURN
 END SUBROUTINE AssemblyMR
 END MODULE Stiffness_lib
```

11.4 PROGRAM 11.1: GENERAL PURPOSE PROGRAM, DIRECT METHOD, MULTIPLE REGIONS

Using the library for stiffness matrix computation we now develop a general purpose program for the analysis of multi-region problems. The input to the program is the same as for one region, except that we must now specify additional information about the regions. A region is specified by a list of elements that describe its boundary, a region code that indicates if the region is finite or infinite and the symmetry code. In order to simplify the code, however symmetry will not be considered here and therefore the symmetry code must be set to zero.
The various tasks to be carried out are

1. **Detect interface elements, number interface nodes/degrees of freedom**

   The first task of the program will be to determine which elements belong to an interface between regions and to establish a local interface numbering. Interface elements can be detected by the fact that two boundary elements connect to the exactly same nodes, although not in the same sequence, since the outward normals will be different. The number of interface degrees of freedom will determine the size of matrices **K** and **A**.

2. **For each region**

   a. Establish local (region) numbering for element incidences

For the treatment of the individual regions we have to renumber the nodes/degrees of freedom for each region into a local (region) numbering system, as explained previously. The incidence and destination vectors of boundary elements, as well as coordinate vector, are modified accordingly.

b. Determine **K** and **A** and results due to "fixed" interface nodes

The next task is to determine matrix **K**. At the same time we assemble it into the global system of equations using the interface destination vector. For partially coupled problems, we calculate and store, at the same time, the results for the elements due to zero values of $\{\mathbf{u}\}_c$ at the interface. These values are stored in the element result vectors Elres_u and Elres_t. Matrix **A** and the vector $\{\mathbf{t}\}_c$ are also determined and stored.

3. Solve global system of equations

The global system of equations is solved for the interface unknowns $\{\mathbf{u}\}_c$

4. For each region determine $\{\mathbf{t}\}_c$ and $\{\mathbf{u}\}_f$

Using equation (11.27) the values for the fluxes/tractions at the interface and (for partially coupled problems) the temperatures/displacements at the free nodes are determined and added to the values already stored in Elres_u and Elres_t. Note that before Equation (11.27) can be used the interface unknowns have to be gathered from the interface vector using the relationship between interface and region numbering in Table 11.3.

```
PROGRAM General_purpose_MRBEM
!-----------------------------------------------
!     General purpose BEM program
!     for solving elasticity and potential problems
!     with multiple regions
!--------------------------------------------------
USE Utility_lib; USE Elast_lib; USE Laplace_lib
USE Integration_lib; USE Stiffness_lib
IMPLICIT NONE
INTEGER, ALLOCATABLE :: NCode(:,:) ! Element BC´s
INTEGER, ALLOCATABLE :: Ldest_KBE(:) ! Interface destinations
INTEGER, ALLOCATABLE :: TypeR(:)  ! Type of BE-regions
REAL, ALLOCATABLE :: Elcor(:,:)   ! Element coordinates
REAL, ALLOCATABLE :: xP(:,:) ! Node co-ordinates
REAL, ALLOCATABLE :: Elres_u(:,:) ! Element results
REAL, ALLOCATABLE :: Elres_t(:,:) ! Element results
REAL(KIND=8), ALLOCATABLE :: KBE(:,:,:) ! Region stiffness
REAL(KIND=8), ALLOCATABLE :: A(:,:,:)   ! Results due to ui=1
REAL(KIND=8), ALLOCATABLE :: Lhs(:,:),Rhs(:) ! global matrices
```

```
  REAL(KIND=8), ALLOCATABLE :: uc(:) !   interface unknown
  REAL(KIND=8), ALLOCATABLE :: ucr(:)! interface unknown(region)
  REAL(KIND=8), ALLOCATABLE :: tc(:)   ! interface tractions
  REAL(KIND=8), ALLOCATABLE :: xf(:)   ! free unknown
  REAL(KIND=8), ALLOCATABLE :: tcxf(:) ! unknowns of region
  REAL, ALLOCATABLE   :: XpR(:,:)  ! Region node coordinates
  REAL, ALLOCATABLE   :: ConR(:)   ! Conductivity of regions
  REAL, ALLOCATABLE   :: ER(:)     ! Youngs modulus of regions
  REAL, ALLOCATABLE   :: nyR(:)    ! Poissons ratio of regions
  REAL                :: E,ny,Con
  INTEGER,ALLOCATABLE:: InciR(:,:)! Incidences (region)
  INTEGER,ALLOCATABLE:: Incie(:,:)! Incidences (global)
  INTEGER,ALLOCATABLE:: IncieR(:,:) ! Incidences (local)
  INTEGER,ALLOCATABLE:: ListC(:)   ! List of interface nodes
  INTEGER,ALLOCATABLE:: ListEC(:,:) ! List of interface Elem.
  INTEGER,ALLOCATABLE:: ListEF(:,:) ! List of free Elem.
  INTEGER,ALLOCATABLE:: LdestR(:,:) ! Destinations(local
numbering)
  INTEGER,ALLOCATABLE:: Nbel(:)     ! Number of BE per region
  INTEGER,ALLOCATABLE:: NbelC(:)   ! Number of Interf. Elem./reg.
  INTEGER,ALLOCATABLE:: NbelF(:)   !Number of free elem./region
  INTEGER,ALLOCATABLE:: Bcode(:,:)  ! BC for all elements
  INTEGER,ALLOCATABLE:: Ldeste(:,:) ! Destinations (global)
  INTEGER,ALLOCATABLE:: LdesteR(:,:)!Destinations (local)
  INTEGER,ALLOCATABLE:: NodeR(:)   ! No. of nodes of Region
  INTEGER,ALLOCATABLE:: NodeC(:)   ! No. of nodes on Interface
  INTEGER,ALLOCATABLE:: ListR(:,:)  ! List of Elements/region
  INTEGER,ALLOCATABLE:: Ndest(:,:)
  INTEGER   :: Cdim    ! Cartesian dimension
  INTEGER   :: Nodes   ! No. of nodes of System
  INTEGER   :: Nodel   ! No. of nodes per element
  INTEGER   :: Ndofe   ! D.o.F´s of Element
  INTEGER   :: Ndof    ! No. of degrees of freedom per node
  INTEGER   :: Ndofs   ! D.o.F´s of System
  INTEGER   :: NdofR   ! Number of D.o.F. of region
  INTEGER   :: NdofC   ! Number of interface D.o.F. of region
  INTEGER   :: NdofF   ! Number D.o.F. of free nodes of region
  INTEGER   :: NodeF   ! Number of free Nodes of region
  INTEGER   :: NodesC  ! Total number of interface nodes
  INTEGER   :: NdofsC  ! Total number of interface D.o.F.
  INTEGER   :: Toa     ! Type of analysis (plane strain/stress)
  INTEGER   :: Nregs   ! Number of regions
  INTEGER   :: Ltyp    ! Element type(linear = 1, quadratic = 2)
  INTEGER   :: Isym    ! Symmetry code
  INTEGER   :: Maxe    ! Number of Elements of System
  INTEGER   :: nr,nb,ne,ne1,nel
  INTEGER   :: n,node,is,nc,no,ro,co
  INTEGER   :: k,m,nd,nrow,ncln,DoF_KBE,DoF
  CHARACTER(LEN=80) :: Title
  !-----------------------------------------------------
  !   Read job information
```

```
!----------------------------------------------------------
OPEN (UNIT=1,FILE='INPUT',FORM='FORMATTED') !  Input
OPEN (UNIT=2,FILE='OUTPUT',FORM='FORMATTED')!  Output
Call JobinMR(Title,Cdim,Ndof,Toa,Ltyp,Isym,nodel,nodes,maxe)
Ndofs= Nodes * Ndof            ! D.O.F's of System
Ndofe= Nodel * Ndof            ! D.O.F's of Element
Isym= 0  !   no symmetry considered here
ALLOCATE(Ndest(Nodes,Ndof))
Ndest= 0
READ(1,*)Nregs   !   read number of regions
ALLOCATE(TypeR(Nregs),Nbel(Nregs),ListR(Nregs,Maxe))
IF(Ndof == 1)THEN
 ALLOCATE(ConR(Nregs))
ELSE
 ALLOCATE(ER(Nregs),nyR(Nregs))
END IF
CALL Reg_Info(Nregs,ToA,Ndof,TypeR,ConR,ER,nyR,Nbel,ListR)
ALLOCATE(xP(Cdim,Nodes))  !  Array for node coordinates
ALLOCATE(Incie(Maxe,Nodel)) !  Array for incidences
CALL Geomin(Nodes,Maxe,xp,Incie,Nodel,Cdim)
ALLOCATE(BCode(Maxe,Ndofe))
ALLOCATE(Elres_u(Maxe,Ndofe),Elres_t(Maxe,Ndofe))
CALL BCinput(Elres_u,Elres_t,Bcode,nodel,ndofe,ndof)
!-----------------------------------------
!     Determine Element destination vector for assembly
!-----------------------------------------
ALLOCATE(Ldeste(Maxe,Ndofe))
Elements_of_region2:&
DO Nel=1,Maxe
 k=0
 DO n=1,Nodel
   DO m=1,Ndof
     k=k+1
     IF(Ndof > 1) THEN
       Ldeste(Nel,k)= ((Incie(Nel,n)-1)*Ndof + m)
     ELSE
       Ldeste(Nel,k)= Incie(Nel,n)
     END IF
   END DO
 END DO
END DO &
Elements_of_region2
!-------------------------------------------
!    Detect interface elements,
!    assign interface boundary conditions
!    Determine number of interface nodes
!-------------------------------------------
ALLOCATE(ListC(Nodes))
NodesC=0
ListC=0
Elements_loop: &
```

```
DO ne=1,Maxe
 Elements_loop1: &
 DO nel=ne+1,Maxe
   IF(Match(Incie(nel,:),Incie(ne,:))) THEN
     BCode(ne,:)= 2 ; BCode(nel,:)= 2   !  assign interface BC
     Element_nodes: &
     DO n=1,nodel
       Node= Incie(ne,n)
       is= 0
       Interface_nodes: &
       DO nc=1,NodesC
         IF(Node == ListC(nc)) is= 1
       END DO &
       Interface_nodes
       IF(is == 0) THEN
         NodesC= NodesC + 1
         ListC(NodesC)= Node
       END IF
     END DO &
     Element_nodes
     EXIT
   END IF
 END DO &
 Elements_loop1
END DO &
Elements_loop
NdofsC= NodesC* Ndof
ALLOCATE(InciR(Nregs,Nodes),IncieR(Maxe,Nodel))
ALLOCATE(KBE(Nregs,NdofsC,NdofsC),A(Nregs,Ndofs,Ndofs))
ALLOCATE(Lhs(NdofsC,NdofsC),Rhs(NdofsC),uc(NdofsC),tc(NdofsC))
ALLOCATE(NodeR(Nregs),NodeC(Nregs))
ALLOCATE(ListEC(Nregs,maxe))
ALLOCATE(ListEF(Nregs,maxe))
ALLOCATE(LdesteR(Maxe,Ndofe))
ALLOCATE(Ldest_KBE(Ndofs))
ALLOCATE(NCode(Nregs,Ndofs))
ALLOCATE(LdestR(Nregs,Ndofs))
ALLOCATE(NbelC(Nregs))
ALLOCATE(NbelF(Nregs))
LdesteR= 0
Ncode= 0
NbelF= 0
NbelC= 0
!-------------------------------------------
!    Assign local (region) numbering
!    and incidences of BE in local numbering
!-------------------------------------------
ListEC= 0
ListEF= 0
DoF_KBE= 0
Regions_loop_1: &
```

```
DO nr=1,Nregs
 node= 0
 Elements_of_region: &
 DO nb=1,Nbel(nr)
   ne= ListR(nr,nb)
   Interface_elements: &
   IF(Bcode(ne,1) == 2) THEN
     NbelC(nr)= NbelC(nr) + 1
     ListEC(nr,NbelC(nr))= ne
     Nodes_of_Elem: &
     DO n=1,Nodel
!     check if node has allready been entered
      is=0
      DO no=1,node
        IF(InciR(nr,no) == Incie(ne,n)) THEN
          is= 1
          EXIT
        END IF
      END DO
      IF(is == 0) THEN
        node=node+1
        InciR(nr,node)= Incie(ne,n)
        IncieR(ne,n)= node
      ELSE
        IncieR(ne,n)= no
      END IF
     END DO &
     Nodes_of_Elem
   END IF &
   Interface_elements
 END DO &
 Elements_of_region
 NodeC(nr)= Node      ! No of interface nodes of Region nr
 NdofC= NodeC(nr)*Ndof  ! D.o.F. at interface of Region nr
 Elements_of_region1: &
 DO nb=1,Nbel(nr)
   ne= ListR(nr,nb)
   Free_elements: &
   IF(Bcode(ne,1) /= 2) THEN
     NbelF(nr)= NbelF(nr) + 1
     ListEF(nr,NbelF(nr))= ne
     Nodes_of_Elem1: &
     DO n=1,Nodel
       is=0
     DO no=1,node
       IF(InciR(nr,no) == Incie(ne,n)) THEN
         is= 1
         EXIT
       END IF
     END DO
     IF(is == 0) THEN
```

```
        node=node+1
        InciR(nr,node)= Incie(ne,n)
        IncieR(ne,n)= node
      ELSE
        IncieR(ne,n)= no
      END IF
     END DO &
     Nodes_of_Elem1
   END IF &
   Free_elements
  END DO &
  Elements_of_region1
  NodeR(nr)= node                    !   number of nodes per region
  !-------------------------------------------
  !     Determine Local Element destination vector
  !-------------------------------------------
  Elements:&
  DO Nel=1,Nbel(nr)
   k=0
   ne= ListR(nr,Nel)
   DO n=1,Nodel
     DO m=1,Ndof
      k=k+1
      IF(Ndof > 1) THEN
        LdesteR(ne,k)= ((IncieR(ne,n)-1)*Ndof + m)
      ELSE
        LdesteR(ne,k)= IncieR(ne,n)
      END IF
     END DO
   END DO
  END DO &
  Elements
  !-------------------------------------------
  !     Determine Local Node destination vector
  !-------------------------------------------
  n= 0
  DO no=1, NodeR(nr)
   DO m=1, Ndof
     n= n + 1
     LdestR(nr,n)= (InciR(nr,no)-1) * Ndof + m
   END DO
  END DO
!-------------------------------------------
!     Determine global Boundary code vector for assembly
!-------------------------------------------
  NdofR= NodeR(nr)*Ndof ! Total degrees of freedom of region
  DoF_o_System: &
  DO nd=1,NdofR
   DO Nel=1,Nbel(nr)
     ne=ListR(nr,Nel)
   DO m=1,Ndofe
```

```
IF (nd == LdesteR(ne,m) .and. NCode(nr,nd) == 0) THEN
      NCode(nr,nd)= NCode(nr,nd)+BCode(ne,m)
     END IF
    END DO
   END DO
 END DO &
 DoF_o_System
END DO &
Regions_loop_1
Regions_loop_2: &
DO nr=1,Nregs
 !-----------------------------------
 !  allocate coordinates in local(region) numbering
 !-----------------------------------
 ALLOCATE(XpR(Cdim,NodeR(nr)))
 Region_nodes: &
 DO Node=1,NodeR(nr)
   XpR(:,Node)= Xp(:,InciR(nr,node))
 END DO &
 Region_nodes
 !-------------------------------------------------------------
 !   Determine interface destination vector for region assembly
 !-------------------------------------------------------------
 No_o_Interfaceelements:&
 DO n=1, NbelC(nr)
   ne= ListEC(nr,n)
   DoF_o_Element:&
   DO m=1, Ndofe
     DoF= Ldeste(ne,m)
     IF(Ldest_KBE(DoF) == 0)THEN
       DoF_KBE= DoF_KBE + 1
       Ldest_KBE(DoF)= DoF_KBE
     END IF
   END DO &
    DoF_o_Element
 END DO &
No_o_Interfaceelements
 NdofR= NodeR(nr)*Ndof ! Total degrees of freedom of region
 NdofC= NodeC(nr)*Ndof ! D.o.F. of interface of Region nr
 E=ER(nr)
 ny=nyR(nr)
 CALL Stiffness_BEM(nr,XpR,Nodel,Ndof,Ndofe&
 ,NodeR,Ncode(nr,:),NdofR,NdofC,KBE(nr,:,:)&
 ,A(nr,:,:),tc,Cdim,Elres_u,Elres_t,IncieR&
 ,LdesteR,Nbel,ListR,TypeR,Bcode,Con,E,ny,Ndest,Isym)
 DO ro=1,NdofC
   DoF= LdestR(nr,ro)
   Nrow= Ldest_KBE(DoF)
   Rhs(Nrow)= Rhs(Nrow) + tc(ro)
   DO co=1, NdofC
     DoF= LdestR(nr,co)
```

```
     Ncln= Ldest_KBE(DoF)
     Lhs(Nrow,Ncln)= Lhs(Nrow,Ncln) - KBE(nr,ro,co)
   END DO
  END DO
  DEALLOCATE (XPR)
 END DO &
 Regions_loop_2
 DEALLOCATE(tc)
 !-----------------------------
 !   solve for interface unknown
 !-----------------------------
 CALL Solve(Lhs,Rhs,uc)
 !----------------------------
 !  compute and add effect of interface displ.
 !----------------------------
 Regions_loop_3: &
 DO nr=1,Nregs
 !  gather region interface displacements
   NdofC= NodeC(nr)*Ndof
  ALLOCATE(ucr(NdofC))
   Interface_dof: &
  DO n=1,NdofC
   DoF= LdestR(nr,n)
   ucr(n)= uc(Ldest_KBE(DoF))
  END DO &
  Interface_dof
  !-------------------------------------------------------------
  ! Store Interfacedisplacements into Elres_u
  !-------------------------------------------------------------
  Interface_DoF1:&
  DO nd=1, NdofC
   DO n=1, Nbel(nr)
     ne=ListR(nr,n)
     DO m=1,Ndofe
      IF(nd == LdesteR(ne,m))THEN
       Elres_u(ne,m)= Elres_u(ne,m) + ucr(nd)
      END IF
     END DO
   END DO
  END DO &
   Interface_DoF1
 !   effects of interface displacement in local (region)
numbering
  NdofR= NodeR(nr)*Ndof
  NdofF= (NodeR(nr) - NodeC(nr))*Ndof  !   d.o.F , free nodes
  ALLOCATE(tc(NdofC),xf(NdofF),tcxf(NdofR))
  tc= 0.0;  xf= 0.0;  tcxf= 0.0
  tc= Matmul(KBE(nr,1:NdofC,1:NdofC),ucr)
  xf= Matmul(A(nr,1:NdofF,1:NdofC),ucr)
 tcxf(1:NdofC)= tc
 tcxf(NdofC+1:NdofR)= xf
```

```
 !----------------------------------------------------------------
 ! Store Interface tractions into Elres_t
 !----------------------------------------------------------------
 DO nd=1, NdofC
   DO n=1, NbelC(nr)
     ne=ListEC(nr,n)
     DO m=1, Ndofe
       IF(nd == LdesteR(ne,m))THEN
         Elres_t(ne,m)= Elres_t(ne,m)  + tcxf(nd)
       END IF
     END DO
   END DO
 END DO
 !------------------------------------------------------------
 ! Store Results of free nodes into Elres_u or Elres_t
 !------------------------------------------------------------
DO nd=NdofC+1, NdofR
   DO n=1, NbelF(nr)
     ne=ListEF(nr,n)
     DO m=1, Ndofe
       IF(nd == LdesteR(ne,m))THEN
         IF(Ncode(nr,nd) == 0)THEN
           Elres_u(ne,m)= Elres_u(ne,m)  + tcxf(nd)
         ELSE IF(Bcode(ne,m) == 1)THEN
           Elres_t(ne,m)= Elres_t(ne,m)  + tcxf(nd)
         END IF
       END IF
     END DO
   END DO
 END DO
 DEALLOCATE(tc,xf,tcxf,ucr)
END DO &
Regions_loop_3
!--------------------------
!    Print out results
!--------------------------
CLOSE(UNIT=2)
OPEN(UNIT=2,FILE= 'BERESULTS',FORM='FORMATTED')
Elements_all:&
DO nel=1,Maxe
 WRITE(2,*)  ' Results, Element ',nel
 WRITE(2,*)  'u=' , (Elres_u(nel,m), m=1,Ndofe)
 WRITE(2,*)  't=' , (Elres_t(nel,m), m=1,Ndofe)
END DO &
Elements_all
END PROGRAM General_purpose_MRBEM
```

11.4.1 User's manual

The input data to be supplied in the data file INPUT are described below. Free field input is used, that is, numbers are separated by blanks. However, all numbers including zero entries must be specified.

The input is divided into three parts. First, general information about the problem is read in. Next, the mesh geometry is specified. The problem may consist of linear and quadratic elements, as shown in Figure 11.8. The sequence in which node numbers have to be entered when specifying the incidences is also shown. Note that this order determines the direction of the outward normal, which has to point away from the material. For 3-D elements, if node numbers are entered in an anticlockwise sense the outward normal points towards the viewer. Finally, information about regions has to be specified. For each region we must input the number of boundary elements that describe the region, the region code (finite or infinite) and the material properties.

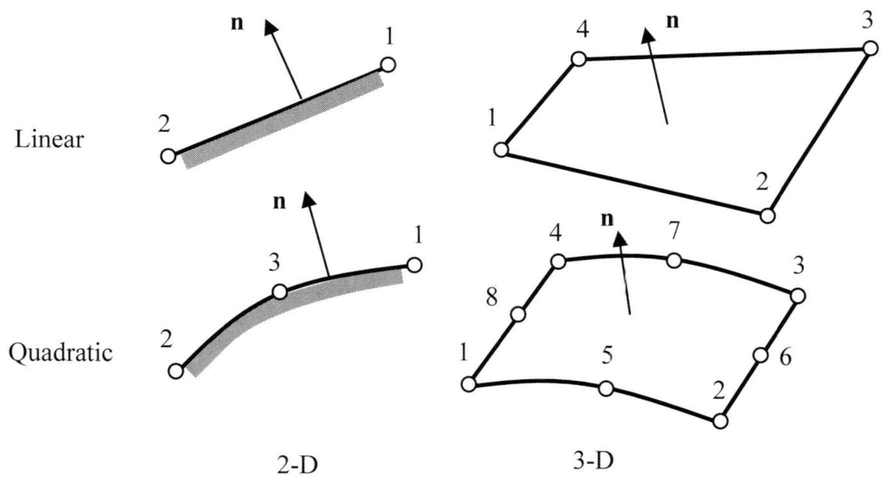

Figure 11.8 Element library

INPUT DATA SPECIFICATION FOR General_purpose-BEM program

1.0	Title specification	
	TITLE	Project title (max 60 characters)
2.0	Cartesian dimension of problem	
	Cdim	Cartesian dimension
	2= two-dimensional problem	
	3= three-dimensional problem	

3.0	Problem type specification	
	Ndof	Degree of freedom per node 1 = potential problem 2,3 = elasticity problem
4.0	Element type specification	
	Ltyp	Element type 1= linear 2= quadratic
5.0	Node specification	
	Nodes	Number of nodes
6.0	Element specification	
	Maxe	Number of elements
7.0	Region specification	
	Nregs	Number of regions

For Nregs regions DO

8.0	Region specification	
	TypeR	Type of region (1=finite, 2=infinite)
9.0	Symmetrycode	
	Isym	Symmetry code (must be set to zero)
10.0	Material properties	
	C1,C2	Material properties **C**1= k (conductivity) for Ndof=1 = E (Modulus of elasticity) for Ndof=2 **C**2= Poisson´s ratio for Ndof=1
11.0	Number of elements	
	Nbel	Number of boundary elements/region
12.0	List of elements	
	ListR(1:Nbel)	List of elements belonging to region

END DO for each region

13.0	Loop over nodes	
	x,y,(z)	Node coordinates
14.0	Loop over all elements	
	Inci (1:Element nodes)	Global node numbers of element nodes
15.0	*Dirichlet* boundary conditions	
	NE_u	Number of elements with *Dirichlet* BC
16.0	Prescribed values for *Dirichlet* BC for **NE_u** elements	
	Nel, Elres_u(1 : Element D.o.F.)	Specification of boundary condition **Nel** = Elem. number to be assigned BC **Elres_u** = Prescribed values for all degrees of freedom of element: all d.o.F first node; all d.o.F second node etc.
17.0	*Neuman* boundary conditions	

NE_t	Number of elements with *Neuman* BC Only specify for non-zero prescribed values.
18.0 Prescribed values for *Neuman* BC for **NE_t** elements	
Nel, Elres_t(1 : Element D.o.F.)	Specification of boundary condition **Nel** = Elem. number to be assigned BC **Elres_t** = Prescribed values for all degrees of freedom of element: all d.o.F first node; all d.o.F second node etc.

11.4.2 Sample problem

The example problem is the same as the cantilever in Chapter 10, except that two regions are specified instead of one. The mesh is shown in Figure 11.9.

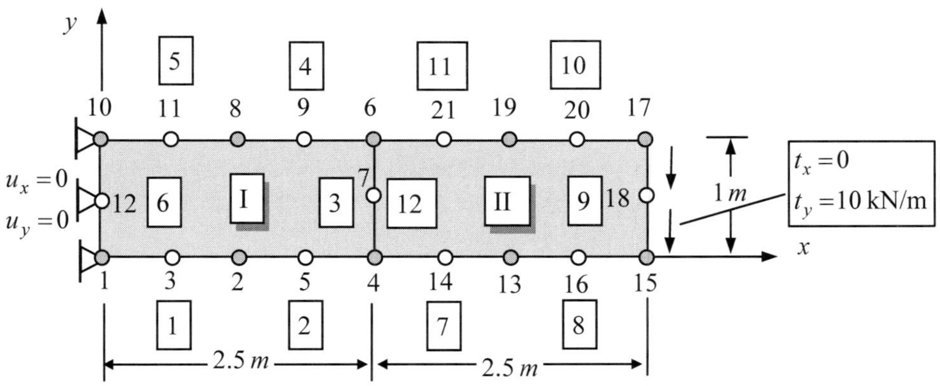

Figure 11.9 Cantilever beam multi-region mesh

The input file for this problem is

```
Cantilever beam multi-region    !  Title
 2                   !  Cartesian dimension
 2                   !  Elasticity problem
 1                   !  T.o.A.= plane strain
 2                   !  Parabolic elements
   21                !  Nodes
   12                !  Elements
    2                !  Number of regions
 1                   !  Region 1: Type of region= Finite
 0                   !  No symmetry
```

```
 0.1000E+05 0.0000E+00    ! E , ν
    6                 !  Number of elements
    1 2 3 4 5 6       !  List of elements
  1                   !  Region 2: Type of region= Finite
  0                   !  No symmetry
 0.1000E+05 0.0000E+00    ! E , ν
    6                 !  Number of elements
  7 8 9 10 11 12      !  List of elements
       0.000      0.000    !    Node coordinates
       1.250      0.000
       0.625      0.000
       2.500      0.000
       1.875      0.000
       2.500      1.000
       2.500      0.500
       1.250      1.000
       1.875      1.000
       0.000      1.000
       0.625      1.000
       0.000      0.500
       3.750      0.000
       3.125      0.000
       5.000      0.000
       4.375      0.000
       5.000      1.000
       5.000      0.500
       3.750      1.000
       4.375      1.000
       3.125      1.000
     1     2     3  !   Element incidences
     2     4     5
     4     6     7
     6     8     9
     8    10    11
    10     1    12
     4    13    14
    13    15    16
    15    17    18
    17    19    20
    19     6    21
     6     4     7
     1
  6 0.0 0.0 0.0 0.0 0.0 0.0    !  Dirichlet BC
   1
 9 0.0 -10.0 0.0 -10.0 0.0 -10.0    ! Neumann BC
```

The output from program 11.1 is as follows

```
  Project:
   Cantilever beam multi-region
```

```
  Cartesian_dimension:          2
  Elasticity Problem
  Type of Analysis: Solid Plane Strain
  Quadratic Elements
 Number of Nodes of System:            21
 Number of Elements of System:            12
 Region          1
 Finite region
 No symmetry
 Youngs modulus:    10000.00
 Poissons ratio:   0.0000000E+00
 List of boundary elements:
 1           2           3           4           5           6
 Region          2
 Finite region
 No symmetry
 Youngs modulus:    10000.00
 Poissons ratio:   0.0000000E+00
 List of boundary elements:
 7           8           9           10          11          12
Node       1  Coor        0.00      0.00
Node       2  Coor        1.25      0.00
Node       3  Coor        0.63      0.00
Node       4  Coor        2.50      0.00
Node       5  Coor        1.88      0.00
Node       6  Coor        2.50      1.00
Node       7  Coor        2.50      0.50
Node       8  Coor        1.25      1.00
Node       9  Coor        1.88      1.00
Node      10  Coor        0.00      1.00
Node      11  Coor        0.63      1.00
Node      12  Coor        0.00      0.50
Node      13  Coor        3.75      0.00
Node      14  Coor        3.13      0.00
Node      15  Coor        5.00      0.00
Node      16  Coor        4.38      0.00
Node      17  Coor        5.00      1.00
Node      18  Coor        5.00      0.50
Node      19  Coor        3.75      1.00
Node      20  Coor        4.38      1.00
Node      21  Coor        3.13      1.00

 Incidences:
EL      1  Inci       1     2     3
EL      2  Inci       2     4     5
EL      3  Inci       4     6     7
EL      4  Inci       6     8     9
EL      5  Inci       8    10    11
EL      6  Inci      10     1    12
EL      7  Inci       4    13    14
EL      8  Inci      13    15    16
```

```
EL      9  Inci      15    17    18
EL     10  Inci      17    19    20
EL     11  Inci      19     6    21
EL     12  Inci       6     4     7

Elements with Dirichlet BC´s:

 Element               6  Prescribed values:
  0.0000000E+00  0.0000000E+00
  0.0000000E+00  0.0000000E+00
  0.0000000E+00  0.0000000E+00

 Elements with Neuman BC´s:

 Element               9  Prescribed values:
  0.0000000E+00  -10.00000
  0.0000000E+00  -10.00000
  0.0000000E+00  -10.00000
 .
 .
 Results, Element      9
 u= -7.3849E-02 -0.5012
     7.3849E-02 -0.5012
     1.0140E-09 -0.5011
 .
 .
Results, Element      12
 u=  5.5386E-02 -0.1582
    -5.5386E-02 -0.1582
     1.0792E-09 -0.1582
 t=  -147.7       5.933
      147.7       5.933
    -1.2626E-06  12.060
```

It can be seen that the maximum displacement is 0.5012, as compared with the theoretical value of 0.500 and that the multi-region method does not result in any loss of accuracy.

11.5 CONCLUSIONS

In this chapter we have extended the capabilities of the programs, so that problems with piecewise non-homogeneous material properties can be handled. The "stiffness matrix assembly" approach taken is quite different from the methods published in text books and uses some ideas of the finite element method. There are several advantages: since each region can be treated completely separately the method is well suited to parallel processing because each processor could be assigned to the computation of the stiffness matrix of one region. Furthermore, with this method it is possible to model sequential excavation and construction as is required, for example for tunnelling[4]. By choosing to implement the method we have also laid the groundwork for the coupling with the finite element method so that there is not much more theory to discuss in Chapter 16. The multi-region method extends the capability of the BEM not only to handle non-homogeneous domains but also, as will be demonstrated later, can be applied to contact and crack propagation problems[5].

11.6 EXERCISES

Exercise 11.1

Use program 11.1 to analyse the cantilever beam in Figure 11.10 consisting of two materials. Assume $\nu = 0$ and different ratios of E_1 / E_2 (1.0, 2.0, 5.0). Compute the internal stresses for each region using program 9.1 and plot along a vertical line.

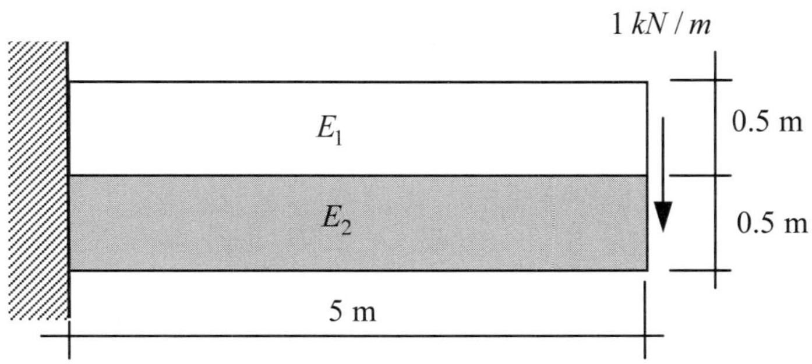

Figure 11.10 Dimensions of cantilever beam

Exercise 11.2

Use program 11.1 to analyse the circular excavation in an inhomogeneous prestressed ground ($\sigma_{vertical} = -1.0$, $\sigma_{horiz.} = 0$) shown in Figure 11.11 (see also Exercise 7.3). Assume $\nu = 0$ and $E_2 / E_1 = 0.5$. Use different values of distance a (2, 5, 10 m). Determine the effect on the maximum displacements.

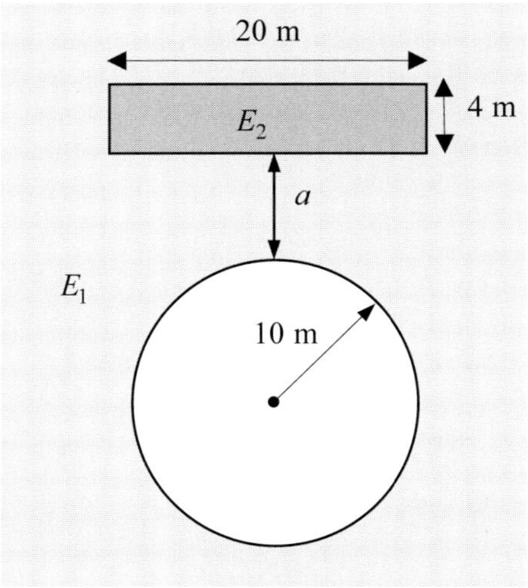

Figure 11.11 Description of example for exercise 11.2

11.7 REFERENCES

1. H-Y Kuo and T. Chen(2005) Steady and transient Green's functions for anisotropic conduction in an exponentially graded solid. *International Journal of Solids and Structures* , 42(3-4): 1111-1128
2. Banerjee P.K. (1994) The Boundary Element Methods in Engineering. McGraw-Hill Book Company, London.
3. Butterfield,R and Tomlin,G.R. (1972) Integral techniques for solving zoned anisotropic continuum problems. *Int. Conf. Variational Methods in Engineering*, Southampton University: 9/31-9/53
4. Beer G. and Dünser Ch. Boundary element analysis of problems in tunnelling, John Booker Memorial Symposium, (J.Carter ed). AA. Balkema, Rotterdam.
5. Beer G. (1993) An efficient numerical method for modelling initiation and propagation of cracks along a material interface. *Int. J. Numer. Methods Eng.* 36 (21): 3579-3594.

12

Dealing with corners and changing geometry

He who goes beneath the surface
does this at his own risk
O. Wilde

12.1 INTRODUCTION

The multi-region method outlined in the previous chapter works well if interfaces between regions are smooth, i.e. where interface points have a unique tangent. If the boundary is not smooth but has corners and edges, i.e. the outward normals are different at elements adjacent to the node, then the normal flow or normal tractions are also different on each side. Such a case would arise, for example, if the shape of the inclusion in the example of the previous chapter is square (Figure 12.1) instead of circular. In this case, two values of normal flow or two sets of traction vectors would have to be computed at the corner node instead of one. However, the integral equations allow the computation at a node of only one value of t for potential problems and one vector of **t** for elasticity problems.

In some applications of numerical simulation one has to deal with geometries that change during the analysis. For example the analysis of tunnel construction involves a changing excavation surface as the tunnel is constructed.

This chapter deals in some detail with the treatment of corners in the boundary element method and the efficient analysis of problems with changing geometries.

12.2 CORNERS AND EDGES

A number of schemes for dealing with the problem of sharp corners, where the solution for **t** is not unique have been proposed in the past.

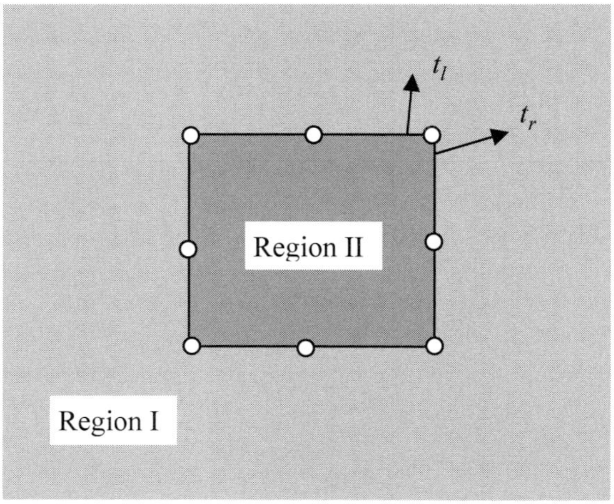

Figure 12.1 Example of a multi-region problem with corners

The following are some methods that have been suggested:

- Numerically round off the corner by using an average outward normal, i.e., an average of all normal vectors of elements connecting to the node. This is not really correct, as the geometry of the element should be rounded off too.

- The unknown values of **t** are computed by extrapolation from the nodes adjacent to the corner node[1]. This method is not difficult to implement but its accuracy would greatly depend on the size of the boundary elements adjacent to the corner.

- Use of auxiliary equations[2,6] based on stress symmetry and on the differential equation of equilibrium to compute extra values of **t**. Another method to solve the problem is based on derivations of potentials or displacements on each side of the corner to get the appropriate number of equations for solving multi valued flow or tractions[3].

- Use discontinuous elements[4] introduced in section 3.7.2. Here **t** is actually not computed right at the corner but slightly inside the element. Therefore two sets of **t** may be computed at each side of the corner.

The approaches which add auxiliary equations to the system of equation have been implemented and tested[6], and it has been found that with a careful implementation they do improve the results at corners. However, auxiliary equations are somehow artificial and often the results depend on the fineness of the mesh. Especially for problems with changing geometries/boundary conditions, where the loading of the current analysis step depends on the previous steps, these methods do not work correctly and give erroneous results. It was found that for a multi-region analysis only discontinuous elements give satisfactory results and guarantee equilibrium at interface elements adjacent to corner and edge nodes. Therefore only discontinuous elements are discussed in more detail here.

12.2.1 Discontinuous elements

The method implemented here uses a continuous discretisation of the geometry and a discontinuous interpolation as explained in chapter 3. By this method the collocation points, where unknowns are determined, are placed inside elements, whereas the geometry nodes remain the same. Two programs are developed to demonstrate discontinuous elements, prog71_discont (single region program) and prog111_discont (multiple region program). Because we use continuous discretisation of the geometry the same input files can be used as for the programs prog71 and prog111.

12.2.2 Numerical integration for one-dimensional elements

In this section the numerical integration in 2D is explained for the case of elasticity. The approach for potential problems is similar. The integrals which have to be evaluated over a discontinuous element, shown in Figure 3.22, are for elasticity problems

$$\Delta\mathbf{U}_{ni}^{e} = \int_{-1}^{1} N_n(\xi)\,\mathbf{U}(P_i,\xi)\,J(\xi)\,d\xi, \quad \Delta\mathbf{T}_{ni}^{e} = \int_{-1}^{1} N_n(\xi)\,\mathbf{T}(P_i,\xi)\,J(\xi)\,d\xi \qquad (12.1)$$

where $N_n(\xi)$ are linear or quadratic discontinuous shape functions. The Jacobian $J(\xi)$ is evaluated with continuous shape functions and with the coordinates of the element nodes.

When point P_i is not one of the element nodes, both integrals can be evaluated by Gauss Quadrature and the integrals in Equation (12.1) can be replaced by a sum where the number of integration points M is a function of the proximity of P_i to the integration region as explained in Chapter 6

$$\begin{aligned} \Delta\mathbf{T}_{ni}^{e} &\approx \sum_{m=1}^{M} N_n(\xi_m)\,\mathbf{T}(P_i,\xi_m)\,J(\xi_m)\,W_m \\ \Delta\mathbf{U}_{ni}^{e} &\approx \sum_{m=1}^{M} N_n(\xi_m)\,\mathbf{U}(P_i,\xi_m)\,J(\xi_m)\,W_m \end{aligned} \qquad (12.2)$$

We now investigate the other possibilities for the location of P_i.

P_i is located at one of the discontinuous nodes inside the element and the shape function value at this node is zero

Since $N_i \in O(r)$ the singularity of the kernel shape function product cancels out if $Q(\xi)$ approaches P_i. The integral of the kernel shape function product remains regular, but the fact, that P_i is located inside the element and the integrand is discontinuous at P_i, requires a splitting of the integration region into two sub regions.

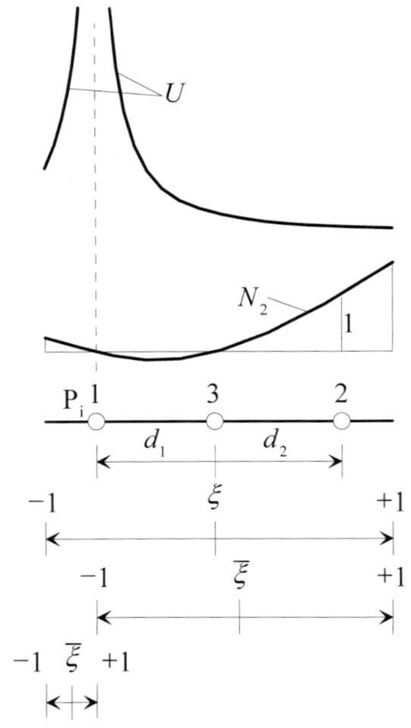

Figure 12.2 Integration when P_i is located at local node 1 and the shape function value is zero

The integrals of equation (12.1) have to be evaluated separately and added for the 2 sub regions

$$\Delta \mathbf{U}^e_{ni} = \sum_{r=1}^{2} \int_{-1}^{1} N_n(\xi)\,\mathbf{U}(P_i,\xi)\,J(\xi)\frac{d\xi}{d\bar{\xi}}d\bar{\xi}, \quad \Delta \mathbf{T}^e_{ni} = \sum_{r=1}^{2} \int_{-1}^{1} N_n(\xi)\,\mathbf{T}(P_i,\xi)\,J(\xi)\frac{d\xi}{d\bar{\xi}}d\bar{\xi} \quad (12.3)$$

The local coordinate system is changed from ξ to $\bar{\xi}$ for the left and the right sub region in the following way:

***P_i* located at node 1:**

Left sub region: $\xi = -\frac{d_1+1}{2} + \bar{\xi}\left(\frac{1-d_1}{2}\right)$ $\qquad \frac{d\xi}{d\bar{\xi}} = \frac{1-d_1}{2}$

Right sub region: $\xi = -\frac{d_1-1}{2} + \bar{\xi}\left(\frac{1+d_1}{2}\right)$ $\qquad \frac{d\xi}{d\bar{\xi}} = \frac{1+d_1}{2}$

***P_i* located at node 2:**

Left sub region: $\xi = \frac{d_2-1}{2} + \bar{\xi}\left(\frac{1+d_2}{2}\right)$ $\qquad \frac{d\xi}{d\bar{\xi}} = \frac{1+d_2}{2}$

Right sub region: $\xi = \frac{d_2+1}{2} + \bar{\xi}\left(\frac{1-d_2}{2}\right)$ $\qquad \frac{d\xi}{d\bar{\xi}} = \frac{1-d_2}{2}$

***P_i* located at node 3:**

Left sub region: $\xi = \frac{1}{2}\left(\bar{\xi}-1\right)$ $\qquad \frac{d\xi}{d\bar{\xi}} = \frac{1}{2}$

Right sub region: $\xi = \frac{1}{2}\left(\bar{\xi}+1\right)$ $\qquad \frac{d\xi}{d\bar{\xi}} = \frac{1}{2}$

***P_i* is located at one of the discontinuous nodes inside the element and the shape function value at that node is not zero**

In this case the kernel shape function product $\Delta\mathbf{U}_{\mathrm{ni}}^{\mathrm{e}}$ is weakly singular and $\Delta\mathbf{T}_{\mathrm{ni}}^{\mathrm{e}}$ is strongly singular. $\Delta\mathbf{T}_{\mathrm{ni}}^{\mathrm{e}}$ is evaluated with the rigid body motion approach explained in chapter 6. The diagonal terms of the displacement fundamental solution consist of a logarithmic function $\ln\left(\frac{1}{r}\right)$. For the integration of this function a modified Gauss Quadrature, *Gauss-Laguerre*[5] is used. The off-diagonal terms consist of non- singular terms. Thus, these Kernel shape function products can be integrated regularly and in the same way as explained above.

The logarithmic function $\ln\left(\frac{1}{r}\right)$ is integrated in the following form

$$\int_0^1 f(\bar{\xi})\ln\left(\frac{1}{\bar{\xi}}\right)d\bar{\xi} \approx \sum_{m=1}^{M} W_m f(\bar{\xi}_m) \tag{12.4}$$

As can be seen in equation (12.4) the integration interval is from 0 to 1. This interval has to be applied for the sub regions as shown in Figure 12.3.

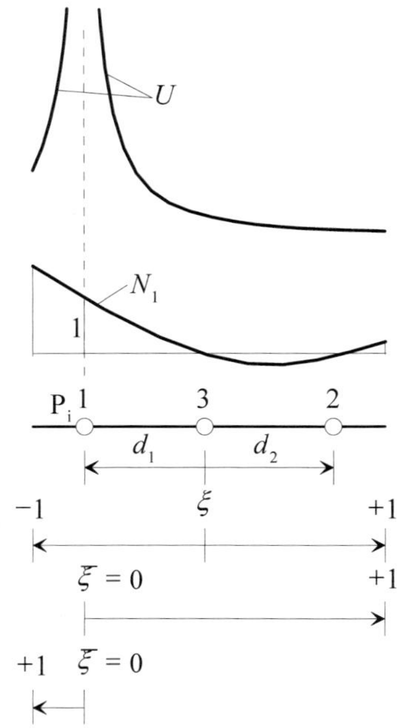

Figure 12.3 Integration when P_i is located at local node 1 and the shape function value is one

The following transformations from intrinsic coordinate ξ to $\bar{\xi}$ have to be applied depending on the location of the collocation node inside the element.

***P**$_i$* **located at node 1:**

Left sub region: $\xi = -d_1 - \bar{\xi}(1-d_1)$ $\qquad \dfrac{d\xi}{d\bar{\xi}} = -(1-d_1)$

Right sub region: $\xi = -d_1 + \bar{\xi}(1+d_1)$ $\qquad \dfrac{d\xi}{d\bar{\xi}} = 1+d_1$

***P**$_i$* **located at node 2:**

Left sub region: $\xi = d_2 - \bar{\xi}(1+d_2)$ $\qquad \dfrac{d\xi}{d\bar{\xi}} = -(1+d_2)$

Right sub region: $\xi = d_2 + \bar{\xi}(1-d_2)$ $\qquad \dfrac{d\xi}{d\bar{\xi}} = 1-d_2$

***P**$_i$* **located at node 3**

Left sub region: $\xi = \bar{\xi}$ $\qquad \dfrac{d\xi}{d\bar{\xi}} = 1$

Right sub region: $\xi = \bar{\xi}$ $\qquad \dfrac{d\xi}{d\bar{\xi}} = 1$

The weakly singular part of the integral of Kernel ΔU_{xx}^{e} is for example

$$\Delta U_{xx}^{e} = C \cdot C_1 \sum_{r=1}^{2} \int_{-1}^{1} N_n(\xi) \ln\left(\frac{1}{r}\right)(P_i,\xi)\, J(\xi) \frac{d\xi}{d\bar{\xi}} d\bar{\xi} \tag{12.5}$$

For straight elements the radius r can be expressed in the following way

$$r = a \cdot \xi \quad \textit{where} \quad a = \frac{L}{2} \frac{d\xi}{d\bar{\xi}} \tag{12.6}$$

The factor a is a constant and depends on $d\xi/d\bar{\xi}$ and the element length L. Because of this, Equation (12.5) can be split into two parts. The first part can be integrated by the Quadrature of *Gauss-Laguerre* and the second one can be evaluated regularly

$$\Delta U_{xx}^{e} = C \cdot C_1 \sum_{r=1}^{2} \left(\int_{0}^{1} N_n(\xi) \ln\left(\frac{1}{\xi}\right) J(\xi) \frac{d\xi}{d\bar{\xi}} d\bar{\xi} + \int_{0}^{1} N_n(\xi) \ln\left(\frac{1}{a}\right) J(\xi) \frac{d\xi}{d\bar{\xi}} d\bar{\xi} \right) \tag{12.7}$$

The integration limits of the second integral in equation (12.7) are from 0 to 1 and when using normal *Gauss* Quadrature, the integral has to be transformed to the limits -1 to $+1$ as follows

$$\int_{0}^{1} N_n(\xi) \ln\left(\frac{1}{a}\right) J(\xi) \frac{d\xi}{d\bar{\xi}} d\bar{\xi} = \int_{-1}^{1} N_n(\xi) \ln\left(\frac{1}{a}\right)(P_i,\xi)\, J(\xi) \frac{d\xi}{d\bar{\xi}} \frac{d\bar{\xi}}{d\tilde{\xi}} d\tilde{\xi} \tag{12.8}$$

The transformation from $\bar{\xi}$ to $\tilde{\xi}$ is independent of the location of the collocation point and the sub region. The transformation has the following form:

$$\bar{\xi} = \frac{\tilde{\xi} + 1}{2} \quad \text{and} \quad \frac{d\bar{\xi}}{d\tilde{\xi}} = \frac{1}{2}$$

12.2.3 Numerical implementation

Two dimensional discontinuous elements (linear and quadratic) are implemented in the single region program prog71_discont and in the multi region program prog111_discont.

In the Module **Integration_lib** a new subroutine called **Integ2E_Disc** is implemented which manages the integration. This subroutine itself calculates the regular integrals (when the collocation point is not located at the element). Inside this routine two subroutines are called (**NonsingularIntegration2D** and **SingularIntegration2D**) which are responsible for the integration when the collocation point is located at the element. The subroutine **NonsingularIntegration2D** performs the integration for the case where the shape function value is zero at the collocation point and the subroutine **SingularIntegration2D** is calculating the weakly singular behaviour of the kernel $\mathbf{U}$. An additional subroutine **ShapefunctionDisc** has been added to the module **Geometry_lib** which calculates the discontinuous shape functions needed in the subroutines of integration.

```
MODULE Integration_lib

SUBROUTINE Integ2E_Disc(Elcor,Inci,Nodel,Ncol,xP,E&
,ny,dUe,dTe,Ndest,Isym)
!--------------------------------------------------
!    Computes  [ dT] e and [ dU] e for 2-D elasticity problems
!    by numerical integration for discontinuous elements
!--------------------------------------------------
IMPLICIT NONE
REAL, INTENT(IN)     :: Elcor(:,:)      !   Element coordinates
INTEGER, INTENT(IN)  :: Ndest(:,:)      !   Node destinations
INTEGER, INTENT(IN)  :: Inci(:)         !   Element Incidences
INTEGER, INTENT(IN)  :: Nodel           !   No. of Element Nodes
INTEGER , INTENT(IN) :: Ncol            !   Number of points Pi
INTEGER , INTENT(IN) :: Isym
REAL, INTENT(IN)    :: E,ny            !   Elastic constants
REAL, INTENT(IN)     :: xP(:,:)     !   Array with coll. coords.
REAL(KIND=8), INTENT(OUT)    :: dUe(:,:),dTe(:,:)
REAL :: epsi= 1.0E-4
REAL :: Eleng,Rmin,RonL,Glcor(8),Wi(8),Ni(Nodel),Vnorm(2)&
,GCcor(2)
REAL :: Jac,dxr(2),UP(2,2),TP(2,2), xsi, eta, r, dxdxb,Pi,C,C1
REAL :: d1,d2
INTEGER :: i,j,k,m,n,Mi,nr,ldim,cdim,iD,nD,Nreg
d1=0.8 ; d2=0.8     ! offsets for discont. nodes
Pi=3.14159265359
C=(1.0+ny)/(4*Pi*E*(1.0-ny))
ldim= 1
cdim=ldim+1
CALL Elength(Eleng,Elcor,nodel,ldim)
dUe= 0.0
dTe= 0.0
Colloc_points: DO i=1,Ncol
  Rmin= Min_dist1(Elcor,xP(:,i),Nodel,inci,ELeng,Eleng,ldim)
  RonL= Rmin/Eleng                        !  R/L
!    Integration off-diagonal coeff.  -> normal Gauss Quadrature
  Mi= Ngaus(RonL,1)
```

```
   Mi=8
   Call Gauss_coor(Glcor,Wi,Mi)   ! Assign coords/Weights
   Gauss_points: DO m=1,Mi
     xsi= Glcor(m)
     CALL Normal_Jac(Vnorm,Jac,xsi,eta,ldim,nodel,Inci,elcor)
     CALL Serendip_func(Ni,xsi,eta,ldim,nodel,Inci)
     CALL Cartesian(GCcor,Ni,ldim,elcor)
     r= Dist(GCcor,xP(:,i),cdim)                  !  Dist. P,Q
     dxr= (GCcor-xP(:,i))/r                !  rx/r , ry/r
     CALL ShapefunctionDisc(Ni,xsi,eta,ldim,nodel,Inci)
     UP= UK(dxr,r,E,ny,Cdim) ; TP= TK(dxr,r,Vnorm,ny,Cdim)
     Node_points: DO n=1,Nodel
     Direction_P: DO j=1,2
       IF(Isym == 0)THEN
         iD= 2*(i-1) + j
       ELSE
         iD= Ndest(i,j)                    !  line number in array
       END IF
       IF (id == 0) CYCLE
       Direction_Q: DO k= 1,2
         nD= 2*(n-1) + k                   !  column number in array
         IF(HasEntry(inci, i) ==  .FALSE. ) THEN
! i is not an element node
           dUe(iD,nD)= dUe(iD,nD) + Ni(n)*UP(j,k)*Jac*Wi(m)
           dTe(iD,nD)= dTe(iD,nD) + Ni(n)*TP(j,k)*Jac*Wi(m)
         END IF
       END DO Direction_Q
     END DO Direction_P
    END DO Node_points
  END DO Gauss_points
END DO Colloc_points
CALL NonSingularIntegration2D(Elcor,Inci,Nodel,xP&
                                 ,E,ny,dUe,dTe,Ndest,Isym,d1,d2)
CALL SingularIntegration2D(Elcor,Inci,Nodel,xP&
                                 ,E,ny,dUe,dTe,Ndest,Isym,d1,d2)
RETURN
END SUBROUTINE Integ2E_Disc

SUBROUTINE NonSingularIntegration2D(Elcor,Inci,Nodel,xP&
,E,ny,dUe,dTe,Ndest,Isym, d1, d2)
!-------------------------------------------------
! Computes  nonsingular Integrals for the element
! when the collocation node coincides with an element node
! and the shape function is zero
!------------------------------------------------
IMPLICIT NONE
REAL, INTENT(IN)      :: Elcor(:,:)      !   Element coordinates
INTEGER, INTENT(IN) :: Ndest(:,:)      !   Node destination vector
INTEGER, INTENT(IN) :: Inci(:)         !   Element Incidences
INTEGER, INTENT(IN) :: Nodel           !   No. of Element Nodes
INTEGER , INTENT(IN):: Isym
```

```
REAL, INTENT(IN)    :: E,ny          !   Elastic constants
REAL, INTENT(IN)    :: xP(:,:)       ! Coordinates of disc. nodes
REAL(KIND=8), INTENT(INOUT)    :: dUe(:,:),dTe(:,:)
REAL             :: epsi= 1.0E-4
REAL             :: Glcor(8),Wi(8),Ni(Nodel),Vnorm(2),GCcor(2)
REAL             :: Jac,dxr(2),UP(2,2),TP(2,2), xsi, eta, r, dxdxb
REAL, INTENT(IN)    :: d1, d2
INTEGER  :: m,n,Mi,ldim,cdim,iD,nD,nreg, ns, i, j, k
ldim= 1
cdim=ldim+1
Element_nodes:DO n=1,Nodel
  i=Inci(n)          ! Collocation node
  Shape_function: DO ns=1, Nodel
   IF(n == ns) CYCLE   ! only if N is zero
   Region_Loop: DO nreg=1, 2
     Mi= 8
     Call Gauss_coor(Glcor,Wi,Mi)
     Gauss_points: DO m=1,Mi
       SELECT CASE (n)
         CASE (1)                              ! Node1
          IF(nreg==1)THEN                       ! right
           xsi= (1.-d2)/2. + Glcor(m) *  (1.+d2)/2.
           dxdxb= (1.+d2)/2.
          ELSE IF(nreg==2)THEN                  ! left
           xsi= (-1.-d2)/2. + Glcor(m) *  (1.-d2)/2
           dxdxb= (1.-d2)/2
          END IF
         CASE (2)                              ! Node2
           IF(nreg==1)THEN                      ! right
             xsi= (1.+d1)/2. + Glcor(m)*(1.-d1)/2
             dxdxb= (1.-d1)/2
           ELSE IF(nreg==2)THEN                 ! left
             xsi= (-1.+d1)/2. + Glcor(m)*(1.+d1)/2
             dxdxb= (1.+d1)/2
           END IF
         CASE (3)                              ! Node3
           IF(nreg==1)THEN                      ! right
             xsi= 0.5 + Glcor(m) * 0.5
             dxdxb= 0.5
           ELSE IF(nreg==2)THEN                 ! left
             xsi= -0.5 + Glcor(m) * 0.5
             dxdxb= 0.5
           END IF
         CASE DEFAULT
       END SELECT
       CALL Normal_Jac(Vnorm,Jac,xsi,eta,ldim,nodel,Inci,elcor)
       CALL Serendip_func(Ni,xsi,eta,ldim,nodel,Inci)
       CALL Cartesian(GCcor,Ni,ldim,elcor)
       r= Dist(GCcor,xP(:,i),cdim)               !  Dist. P,Q
       dxr= (GCcor-xP(:,i))/r           !  rx/r , ry/r
       CALL ShapefunctionDisc(Ni,xsi,eta,ldim,nodel,Inci)
```

```
      UP= UK(dxr,r,E,ny,Cdim) ; TP= TK(dxr,r,Vnorm,ny,Cdim)
      Direction1:DO j=1,2
        IF(Isym == 0)THEN
          iD= 2*(i-1) + j
        ELSE
          iD= Ndest(i,j)                   !  line number in array
        END IF
        IF (id == 0) CYCLE
        Direction2: DO k= 1,2
          nD= 2*(ns-1) + k                      !  column number in array
          dUe(iD,nD)= dUe(iD,nD) + Ni(ns)*UP(j,k)*Jac*dxdxb*Wi(m)
          dTe(iD,nD)= dTe(iD,nD) + Ni(ns)*TP(j,k)*Jac*dxdxb*Wi(m)
        END DO Direction2
      END DO Direction1
     END DO Gauss_points
    END DO Region_Loop
  END DO Shape_function
END DO Element_nodes
END SUBROUTINE NonSingularIntegration2D

SUBROUTINE SingularIntegration2D(Elcor,Inci,Nodel,xP,E&
,ny,dUe,dTe,Ndest,Isym, d1, d2)
!-----------------------------------------------------
! Computes  nonsingular Integrals for the element
! when the collocation node coincide with an element node
! and the shape function is not zero
!-----------------------------------------------------
IMPLICIT NONE
REAL, INTENT(IN)     :: Elcor(:,:)       !   Element coordinates
INTEGER, INTENT(IN) :: Ndest(:,:)        !   Node destination vector
INTEGER, INTENT(IN) :: Inci(:)           !   Element Incidences
INTEGER, INTENT(IN) :: Nodel             !   No. of Element Nodes
INTEGER , INTENT(IN):: Isym
REAL, INTENT(IN)    :: E,ny             !   Elastic constants
REAL, INTENT(IN)    :: xP(:,:)              ! Coordinates of
discontinuous nodes
REAL(KIND=8), INTENT(INOUT)    :: dUe(:,:),dTe(:,:)
REAL   :: epsi= 1.0E-4
REAL   :: Glcor(12),Wi(12),Ni(Nodel),Vnorm(2),GCcor(2)
REAL   :: Jac,dxr(2),UP(2,2),TP(2,2), xsi, eta, r, dxdxb, dxbdxp&
, C, C1, Ellength, Radius
REAL           :: f1, f2
REAL, INTENT(IN)    :: d1, d2
INTEGER           :: m,n,Mi,ldim,cdim,iD,nD,nreg, ns, i, j, k
ldim= 1                              ! Element dimension
cdim=ldim+1
Pi=3.14159265359
C=(1.0+ny)/(4*Pi*E*(1.0-ny))
!-----------------------------------------------------
! Integration of off-diagonal coefficients
! for U kernel at singular point
```

```
! -> not singular -> normal gauss quadrature
!---------------------------------------------------
Element_nodes:DO n=1,Nodel
  i=Inci(n)          ! Collocation node
  Shape_function: DO ns=1, Nodel
  ! Node at element -> shape function in integral term
    IF(n /= ns) CYCLE
   Region_Loop: DO nreg=1, 2
    Mi= 8
     Call Gauss_coor(Glcor,Wi,Mi)
     Gauss_points: DO m=1,Mi
       SELECT CASE (n)
        CASE (1)                          ! Node1
         IF(nreg==1)THEN                     ! right
          xsi= (1.-d2)/2. + Glcor(m) *  (1.+d2)/2.
          dxdxb= (1.+d2)/2.
         ELSE IF(nreg==2)THEN                ! left
          xsi= (-1.-d2)/2. + Glcor(m) *  (1.-d2)/2
           dxdxb= (1.-d2)/2
         END IF
        CASE (2)                          ! Node2
         IF(nreg==1)THEN                     ! right
          xsi= (1.+d1)/2. + Glcor(m)*(1.-d1)/2
          dxdxb= (1.-d1)/2
         ELSE IF(nreg==2)THEN                ! left
          xsi= (-1.+d1)/2. + Glcor(m)*(1.+d1)/2
          dxdxb= (1.+d1)/2
         END IF
        CASE (3)                          ! Node3
         IF(nreg==1)THEN                     ! right
          xsi= 0.5 + Glcor(m) * 0.5
          dxdxb= 0.5
         ELSE IF(nreg==2)THEN                ! left
          xsi= -0.5 + Glcor(m) * 0.5
          dxdxb= 0.5
         END IF
        CASE DEFAULT
       END SELECT
       CALL Normal_Jac(Vnorm,Jac,xsi,eta,ldim,nodel,Inci,elcor)
       CALL Serendip_func(Ni,xsi,eta,ldim,nodel,Inci)
       CALL Cartesian(GCcor,Ni,ldim,elcor)
       r= Dist(GCcor,xP(:,i),cdim)             !  Dist. P,Q
       dxr= (GCcor-xP(:,i))/r          !  rx/r , ry/r
       CALL ShapefunctionDisc(Ni,xsi,eta,ldim,nodel,Inci)
        Direction1:DO j=1,2
         IF(Isym == 0)THEN
          iD= 2*(i-1) + j
         ELSE
            iD= Ndest(i,j)
         END IF
         IF (id == 0) CYCLE
```

```
        Direction2:  DO k= 1,2
          nD= 2*(ns-1) + k
          dUe(iD,nD)= dUe(iD,nD)&
                               + Ni(n)*C*dxr(j)*dxr(k)*Jac*dxdxb*Wi(m)
        END DO Direction2
       END DO Direction1
      END DO Gauss_points
    END DO Region_Loop
  END DO Shape_function
END DO Element_nodes
!-----------------------------------------------------
! Integration of diagonal coefficients
! for U kernel at singular point
! -> singular ->  gauss laguerre quadrature
!-----------------------------------------------------
C= C*(3.0-4.0*ny)
CALL Elength(Ellength,Elcor,Nodel,ldim)
Element_nodes_1: DO n=1,Nodel
  i=Inci(n)          ! Collocation node
  Shape_function_1: DO ns=1, Nodel
    IF(n .NE. ns) CYCLE
   Region_Loop_1: DO nreg=1, 2
     Mi= 12
      Call Gauss_Laguerre_coor(Glcor,Wi,Mi)
      Gauss_points_1: DO m=1,Mi
       SELECT CASE (n)
        CASE (1)                          ! Node1
         IF(nreg==1)THEN                   ! left
          xsi= -d1 - (1. - d1 ) * Glcor(m)
          dxdxb= 1.- d1
         ELSE IF(nreg==2)THEN              ! right
          xsi= -d1 + (1. + d1 ) * Glcor(m)
          dxdxb= ( 1. + d1 )
         END IF
        CASE (2)                          ! Node2
         IF(nreg==1)THEN                   ! left
          xsi= d2 - (1. + d2 ) * Glcor(m)
          dxdxb= 1. + d2
         ELSE IF(nreg==2)THEN              ! right
          xsi= d2 + (1. - d2 ) * Glcor(m)
          dxdxb=  1. - d2
         END IF
        CASE (3)                          ! Node3
         IF(nreg==1)THEN                   ! left
          xsi=  - Glcor(m)
          dxdxb= 1.
         ELSE IF(nreg==2)THEN              ! right
          xsi= Glcor(m)
          dxdxb=  1.
         END IF
        CASE DEFAULT
```

```
      END SELECT
      Call Normal_Jac(Vnorm,Jac,xsi,eta,ldim,nodel,Inci,elcor)
        CALL ShapefunctionDisc(Ni,xsi,eta,ldim,nodel,Inci)
        Direction1_1: DO j=1,2
          IF(Isym == 0)THEN
            iD= 2*(i-1) + j
          ELSE
            iD= Ndest(i,j)                   !  line number in array
          END IF
          IF (id == 0) CYCLE
          nD= 2*(ns-1) + j                    !  column number in array
          dUe(iD,nD)= dUe(iD,nD) + Ni(n)*C*Jac*dxdxb*Wi(m)
        END DO Direction1_1
        END DO Gauss_points_1
        Mi= 8
        Call Gauss_coor(Glcor,Wi,Mi)
        Gauss_points2:  DO m=1,Mi
        SELECT CASE (n)
        CASE (1)                                ! Node1
          IF(nreg==1)THEN                         ! right
            xsi= -d2 + (1. + d2 ) *  (1+Glcor(m))/2
            dxdxb= (1.+ d2 )
            dxbdxp= 0.5
          ELSE IF(nreg==2)THEN                    ! left
            xsi= -d2 - (1. - d2 ) *  (1-Glcor(m))/2
            dxdxb= ( d2 - 1. )
            dxbdxp= -0.5
          END IF
        CASE (2)                                ! Node2
          IF(nreg==1)THEN                         ! right
            xsi= d1 + (1. - d1 ) *   (1+Glcor(m))/2
            dxdxb= 1. - d1
            dxbdxp= 0.5
          ELSE IF(nreg==2)THEN                    ! left
            xsi= d1 - (1. + d1 ) * (1-Glcor(m))/2
            dxdxb= - (1.+d1)
            dxbdxp= -0.5
          END IF
        CASE (3)                                ! Node3
          IF(nreg==1)THEN                         ! right
            xsi= (1+Glcor(m))/2
            dxdxb= 1.
            dxbdxp= 0.5
           ELSE IF(nreg==2)THEN                   ! left
             xsi= - (1-Glcor(m))/2
             dxdxb= - 1.
             dxbdxp= -0.5
          END IF
          CASE DEFAULT
          END SELECT
          Radius= (Ellength/2 ) * ABS(dxdxb)
```

```
            C1= LOG( 1/Radius )*C
            CALL Normal_Jac(Vnorm,Jac,xsi,eta,ldim,nodel,Inci,elcor)
            CALL ShapefunctionDisc(Ni,xsi,eta,ldim,nodel,Inci)
             Direction1_2: DO j=1,2
              IF(Isym == 0)THEN
                iD= 2*(i-1) + j
              ELSE
                iD= Ndest(i,j)
              END IF
              IF(iD == 0)CYCLE
              nD= 2*(ns-1) + j
              dUe(iD,nD)= dUe(iD,nD) &
                          + Ni(n)*C1* Jac* dxdxb*  dxbdxp* Wi(m)
             END DO Direction1_2
            END DO Gauss_points2
          END DO Region_Loop_1
        END DO Shape_function_1
     END DO Element_nodes_1
END SUBROUTINE SingularIntegration2D
```

For the discontinuous version of both programs (prog71_discont and prog111_discont) the input files of the continuous versions can be used. At the beginning of each program, the coordinates of the collocation nodes are calculated from the node coordinates of the element and given values of d_1 and d_2. From the input file given incidences are transformed to the discontinuous version, too, from which the degrees of freedom are specified. There were no other major changes necessary.

12.2.4 Test Example – Single Region

To test the implementation the following example a cantilever beam (same example as used in chapter 10) shown in Figure 12.4 is considered. The example is designed to show that the discontinuous elements give good results even for the case where no corners are present.

The diagram in Figure 12.5 shows the vertical displacements along the cantilever beam. The results for the discontinuous meshes shown in Figure 12.5 are extrapolated from the discontinuous points to the element nodes according to the interpolation functions and are compared with discontinuous elements for two different meshes. The first mesh consists of 3 quadratic elements along the length and the second has 5 elements along the length. For both one element for the height is used.

As is shown in Figure 12.5 the vertical displacements at the length side of the beam are the same and agree very well with the analytical solution, except for the coarse mesh with continuous elements. For the discontinuous mesh with 3 elements and for the continuous mesh with 5 elements on the length side the same accuracy of results are obtained. These two meshes are comparable because they have exact the same number of degrees of freedom.

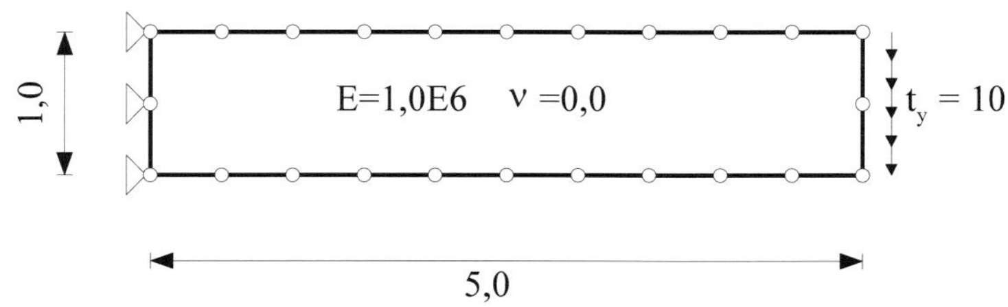

Figure 12.4 Cantilever beam

The expected error for the discontinuous displacement at common nodes of adjacent elements nodes is less then 0.1%.

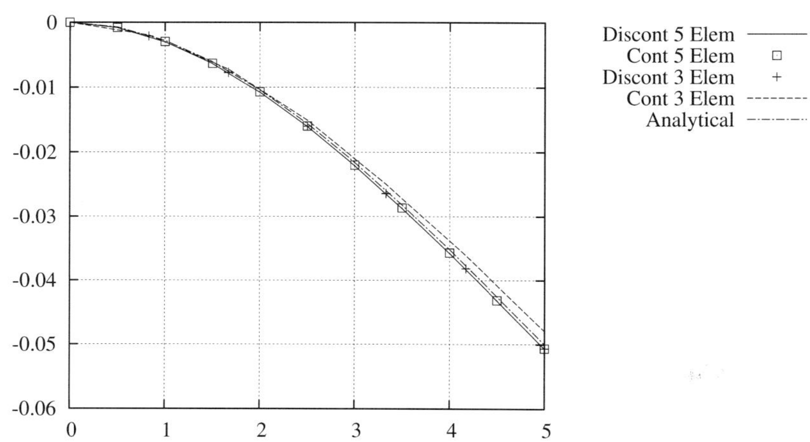

Figure 12.5 Vertical displacements

12.2.5 Test Example – Multiple Regions

This example is a cube with a distributed boundary load of 10 KN/m^2 on the top of the cube. The geometry is shown in Figure 12.6 and the material parameters for all regions are E=1000kN/m^2, ν=0. For the purpose of demonstrating the corner problem the cube is subdivided into four regions. Region 1 and 2 is discretised with 8 linear elements. Region 3 and 4 consists of 6 linear elements. The points *B* and *D* of regions 3 and 4 are corner nodes. These points are located at the interface between regions and therefore need special attention. The calculation is done two times, first with the program prog111 which uses continuous elements and then with the program prog111_discont, the discontinuous version of the multi-region program. If we compare the tractions at interface elements in Figures 12.7, 12.8 with 12.9 at the interface between regions we see that the value, that should be constant, fluctuates widely if continuous elements are used.

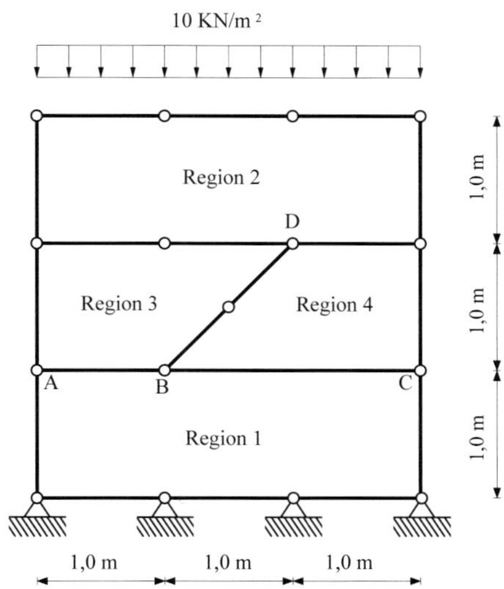

Figure 12.6 Vertical Displacements

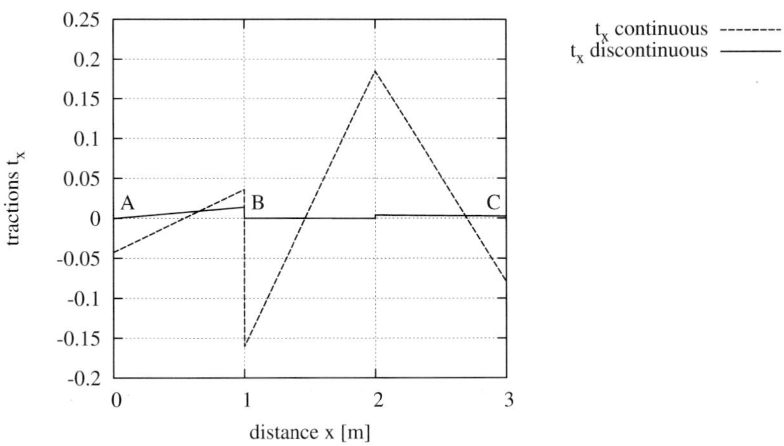

Figure 12.7 Tractions t_x at the boundary of regions 3 and 4 along the line $\overline{ABC}$

If discontinuous elements are used the tractions, which are now evaluated at points slightly inside, show no fluctuation and only a small jump which is due to coarseness of the mesh. Indeed the diagram in Figure 12.7 indicates a gross violation of equilibrium conditions if continuous elements are used because for $\nu = 0$ the tractions should be equal to zero, everywhere.

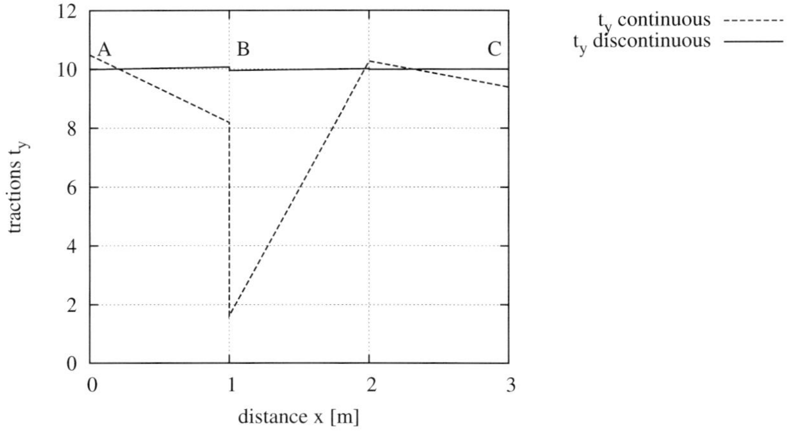

Figure 12.8 Tractions t_y at the regions 3 and 4 along the line $\overline{ABC}$

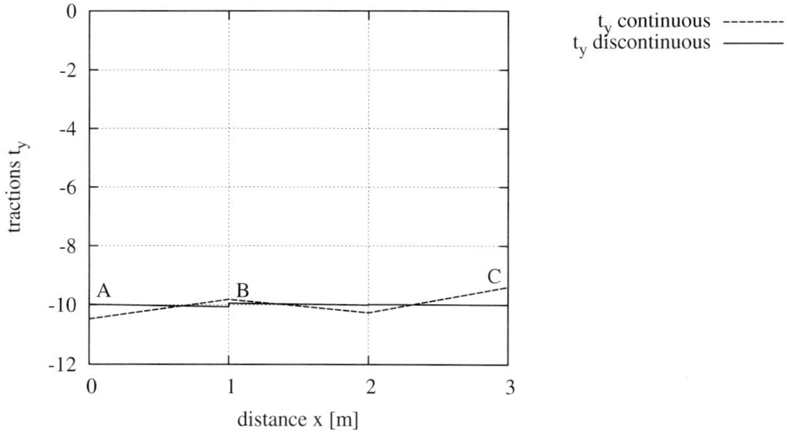

Figure 12.9 Tractions t_y at the regions 1 along the line $\overline{ABC}$

12.3 DEALING WITH CHANGING GEOMETRY

In this chapter we turn our attention to problems where the geometry is changing throughout the analysis process. Due to the change of the geometry, boundary conditions may also change. An example is the modelling of a tunnel excavation process[6]. Here the domain is assumed to be of infinite or semi infinite extent and only the boundary of the tunnel has to be meshed by elements.

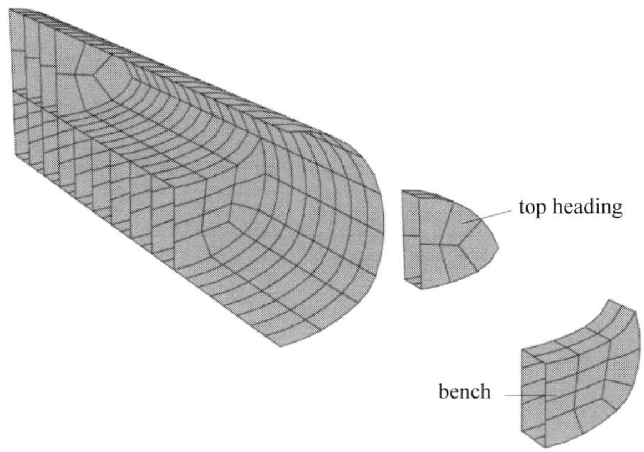

Figure 12.10 Example for a staged excavation process in 3D (only half of the mesh shown)

As shown in Figure 12.10 the multiple region BEM[7] is used to model the excavation. In tunnelling with the New Austrian Tunnelling Method, excavation advances in steps of several meters, either by excavating the full cross section or parts of it. In the example shown in Figure 12.10 a two stage excavation (top heading and bench) is shown. Figure 12.11 illustrates how excavation is modelled with a multi-region BEM.

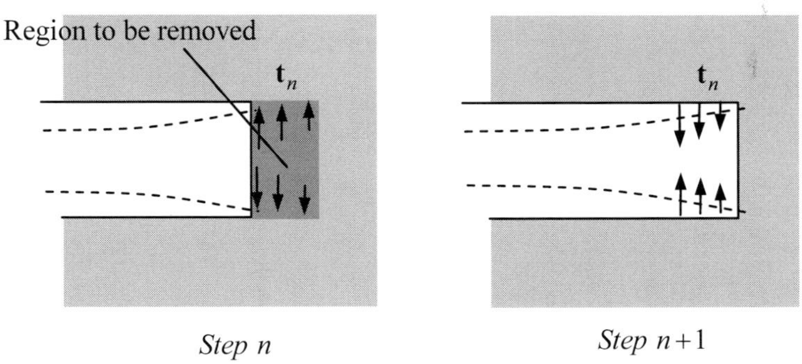

Figure 12.11 The steps in modelling excavation

The volumes of material to be excavated are discretised by boundary elements and represent boundary element regions in a multi-region analysis. According to the multi-region algorithm explained in the previous chapter, stiffness matrices are calculated for each region separately. Each excavation step is simulated by the deactivation of a region.

When a region is deactivated then the tractions at the interfaces of the removed region have to be applied to the mesh in order to restore equilibrium conditions. We can observe that boundary conditions for the boundary elements of the region representing the fully excavated tunnel change from *Interface* to *Neumann* condition.

The implementation of the activation and deactivation process in a computer code is not a trivial task and the detailed discussion related to the architectural design of software is outside the scope of this book. However, we will point out the drastic effects that corners and edges can have on the results for problems of changing boundary conditions if not properly addressed. In the following we restrict ourselves to two-dimensional problems.

12.3.1 Example

In Figure 12.12 a staged excavation of 10 steps is shown. We assume an excavation in 2D under plane strain conditions and this means excavation with infinite extend out of plane. This of course is not a real tunnel excavation, but serves well to explain the method. The mesh consists of 10 regions for top heading and bench. All these finite regions are embedded in an infinite region, which represent the infinite extent of the continuum.

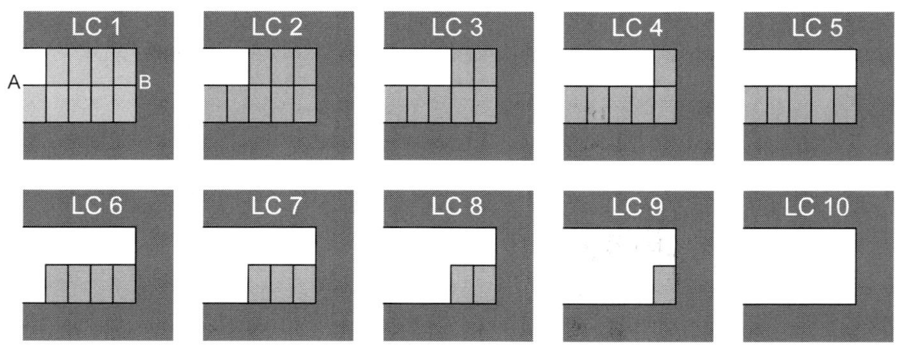

Figure 12.12 Example for a staged excavation process in 2D

The excavation process is modelled by the de-activation of regions that represent excavated material. First 5 top heading regions are excavated successively and then 5 regions at bench. The sequence of excavation is shown in Figure 12.12. The material parameters are E= 5000 *MN/m2* and ν=0. The virgin stress field is given as follows:

$$\sigma_{x0} = -5,0 \; MN/m^2 \quad \sigma_{y0} = -5,0 \; MN/m^2 \quad \tau_{xy0} = 0,0 \; MN/m^2 .$$

When regions are removed some elements will change boundary conditions from *Interface* to *Neumann*. The loading for *Neumann* elements is calculated from the stresses calculated at previous load cases. For the first stage the virgin stresses are applied.

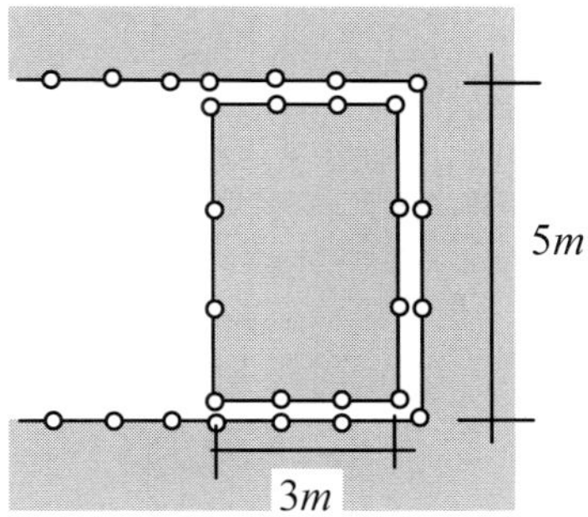

Figure 12.13 Discretisation of regions (only corner nodes shown)

The discretisation of the regions is shown in Figure 12.13. For a finite region 3 quadratic elements are used on all sides. The discretisation of the infinite region matches the mesh of the finite regions.

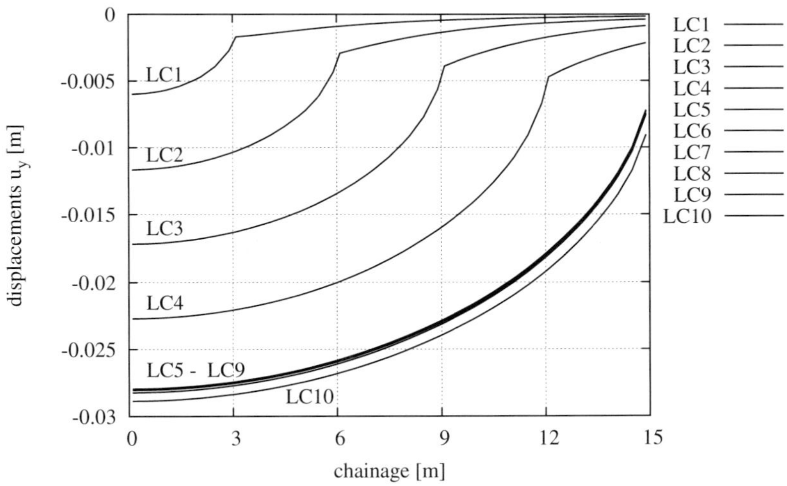

Figure 12.14 Vertical displacements for LC1 to LC10

In Figure 12.14 the vertical displacements at the top of the excavation (crown) is shown for all load cases for the sequential calculation using discontinuous elements. To verify these results an analysis was also performed for the case of the excavation made in one step (single region problem) for the selected load cases 4 and 7. Because this is a linear problem the sequential excavation and the one step excavation results should be the same.

The geometry of these single region meshes is shown in Figure 12.15. Only the boundary of the excavated part is discretised and the excavation is done in one single step. As the boundary conditions for all elements are of *Neumann* type there is no corner problem involved for both geometries. Thus, these calculations are performed with continuous elements.

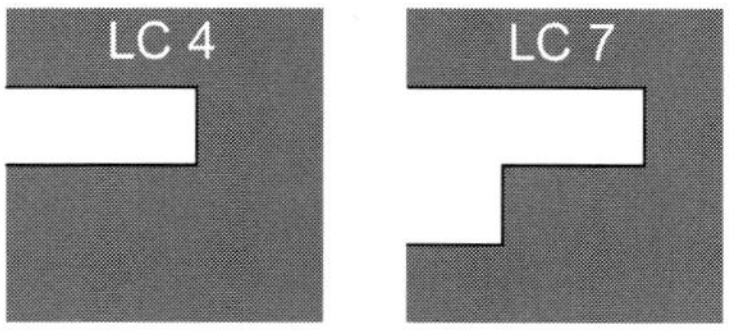

Figure 12.15 Single region meshes for LC4 and LC7

The vertical displacements at the crown are shown in Figure 12.16 for the multi region calculation with discontinuous elements and the single region calculation with continuous elements. As can be seen the results are in excellent agreement.

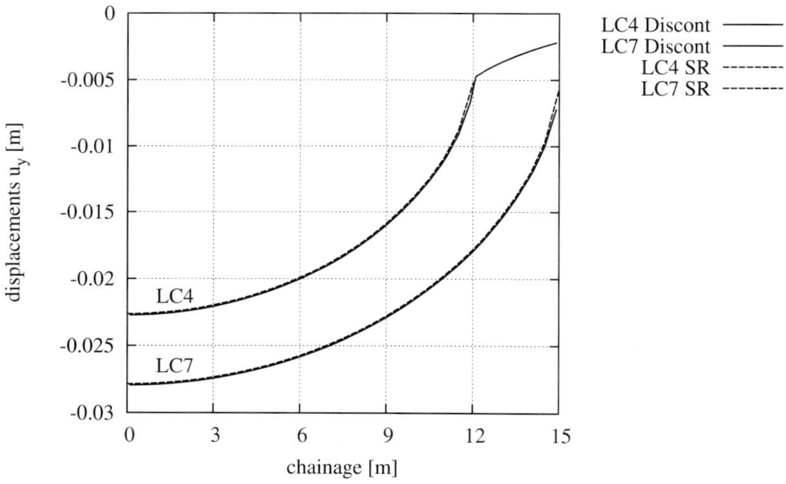

Figure 12.16 Vertical displacements for LC4 and LC7

In the following the effect of the corner problem is pointed out. For the load cases LC1 to LC5 the calculations are done twice, first with continuous elements and second with discontinuous elements, both with the sequential multi-region algorithm. In Figure 12.17 the vertical displacements at the line $\overline{AB}$ (indicated in Figure 12.12) for the LC1 to LC5 are compared. As can be seen the results for continuous elements contain a large error and the errors accumulate from each load case to the other.

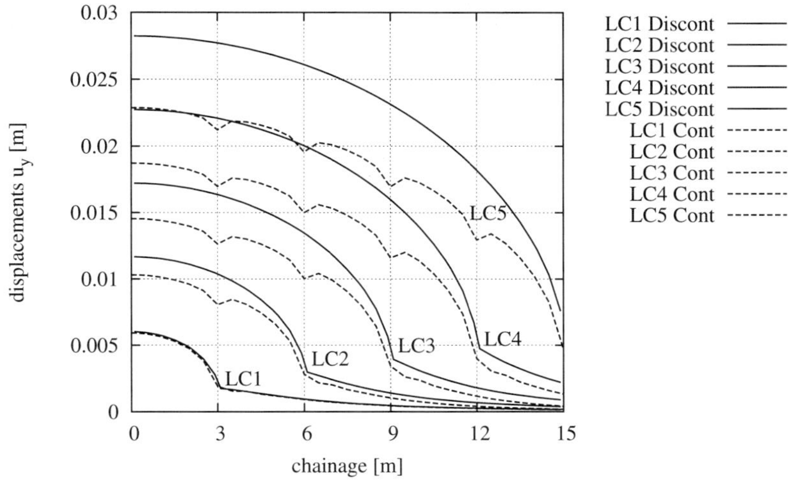

Figure 12.17 Vertical displacements for LC1 to LC5 for the calculation with continuous and discontinuous elements

The reason for these errors is the erroneous calculation of tractions at corner nodes for continuous elements. In the sequential algorithm the tractions computed at a previous step is applied as loading of the following calculation step. Because of this fact the results are getting worse from step to step.

12.4 ALTERNATIVE STRATEGY

The strategy for modelling excavation problems is expensive, especially for 3-D problems, since the total number of interface degrees of freedom can become quite large if many excavation stages are considered. An alternative strategy, involving only one region, is explained for the same example as before and for load cases 1-5. The idea is to calculate (by the post-processing procedure explained in Chapter 9) after an analysis the stress distribution along a line that represents the boundary of the next excavation step (Figure 12.18). However, at the sharp corners A and B the stress is theoretically infinite and can not be determined by post-processing. To overcome this problem it is suggested to evaluate the stress very close to the edge. We propose that the location is specified by an intrinsic coordinate of value $\xi = -0,90$ of the element that will model the new excavation surface. The final stress distribution for this step is obtained by extrapolation using a similar procedure as for the discontinuous elements (Figure 12.18 right). Note that this distance is chosen quite arbitrary and the choice will affect the final results. After the computation we compute the tractions that will be applied at the next excavation step as

$$\mathbf{t} = \boldsymbol{\sigma} \cdot \mathbf{n} \tag{12.9}$$

Note that the resulting traction to be applied at the new excavation surface for load case 4 is the sum of tractions obtained by internal stress evaluation for load cases 1 to 3 plus the tractions due to the virgin stress field. For the analysis of the next load case, the mesh of the single infinite region representing the excavated tunnel surface is changed by removing the face elements and adding a row of elements representing the next stage of excavation.

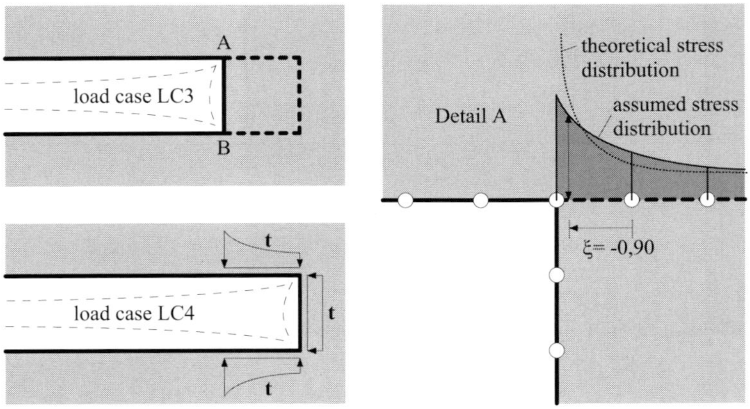

Figure 12.18 Vertical displacements at tunnel crown

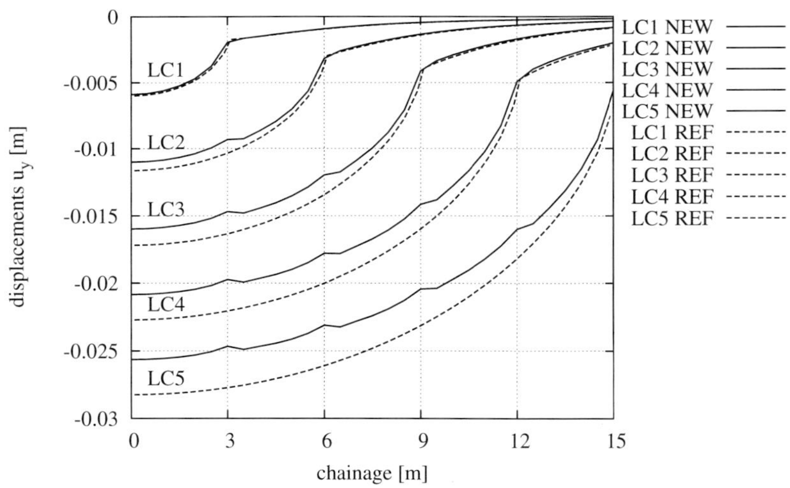

Figure 12.19 Vertical displacements at tunnel crown

The results of vertical displacements along the crown of the tunnel are shown in Figure 12.19 for load cases 1 to 5. These results are compared with the reference solution. There is some difference and this can be attributed to approximation made for the stress distribution near the corners. It seems that the resultant excavation force is not accurately computed and this error accumulates load case after load case. Obviously some improvements are possible by adjusting the stress distribution so the resultant excavation force is closer to the actual one.

12.5 CONCLUSIONS

The correct treatment of corners and edges is of great importance for some applications, in particular for applications where the boundary conditions as well as the geometry are changing during the calculation process. It was found out, that from all possibilities to improve the results at corner nodes discontinuous elements give the best results. Of course additional degrees of freedom are introduced by this method. For simplicity all elements have been treated as discontinuous here. This increases the size of the equation system drastically, especially in 3D. It is much more efficient to use discontinuous nodes only where they are needed, i.e. only at corner and edge nodes where the traction is discontinuous. The manner in which the interpolation functions are presented in chapter 3 makes possible a mixture of discontinuous and continuous functions in one element. When dealing with changing geometries as in sequential excavation problems the multi-region analysis with discontinuous elements gives good results. However, the effort can be quite considerable especially for 3-D applications because with each excavation stage modelled the number of regions and hence the interface degrees of freedom increase. An alternative method that involves only one region seems attractive but the accuracy still has to be improved.

12.6 REFERENCES

1. Beer G. and Watson J.O. (1995) Introduction to Finite and Boundary Element Methods for Engineers. J. Wiley.
2. Gao X.W. and Davies T. (2001) Boundary element programming in mechanics. Cambridge University Press, London.
3. Sladek V. and Sladek J. (1991) Why use double nodes in BEM? *Engineering Analysis with Boundary Elements* **8**: 109-112.
4. Aliabadi M. H. (2002) The Boundary Element Method (Volume 2). J. Wiley.
5. Stroud, A.H. and Secrest, D. (1966) Gaussian Quadrature Formulas. Prentice-Hall, Englewood Cliffs, New Jersey.
6. Duenser C. (2007) Simulation of sequential tunnel excavation with the Boundary Element Method. Monographic Series TU Graz,Austria.
7. Duenser C., Beer G. (2001) Boundary element analysis of sequential tunnel advance. *Proceedings of the ISRM regional symposium,* Eurock: 475-480.

13

Body Forces

Gravitation is not responsible
for people falling in love
J. Keppler

13.1 INTRODUCTION

The advantages of the boundary element method over the FEM that no elements are required inside the domain, also has some disadvantages: loading may only be applied at the boundary, but not inside the domain. A number of problems exist where applying loading inside the domain is necessary, for example

- where sources (of heat or water) or forces have to be considered inside the domain
- where self weight or centrifugal forces have to be considered
- where initial strains are applied inside the domain, for example when material is subjected to swelling.

In addition, as we will see later, for the analysis of domains exhibiting nonlinear material behaviour, for which we cannot find fundamental solutions, the problem can be considered as one where initial stresses are generated inside the domain.

In this chapter we will discuss methods which allow us to consider such loads commonly known as *body forces*. Here we will distinguish between those which are constant, such as for example, self weight and those which vary inside the domain. We will find that we can deal with constant body forces in a fairly straightforward way since the volume integrals which occur can be transformed into surface integrals. In the case where they are not constant, however, the only way to deal with volume integration is by providing additional volume discretisation.

We will start this chapter by revisiting Betti's theorem as derived for integral equations but now we will consider the additional effect of body forces.

13.2 GRAVITY

First we deal with gravity forces, for example those generated by self weight. If the material is homogeneous then these forces per unit volume are constant inside the domain. We expand Betti`s theorem used in Chapter 5, to derive the integral equations taking into account the effect of body forces.

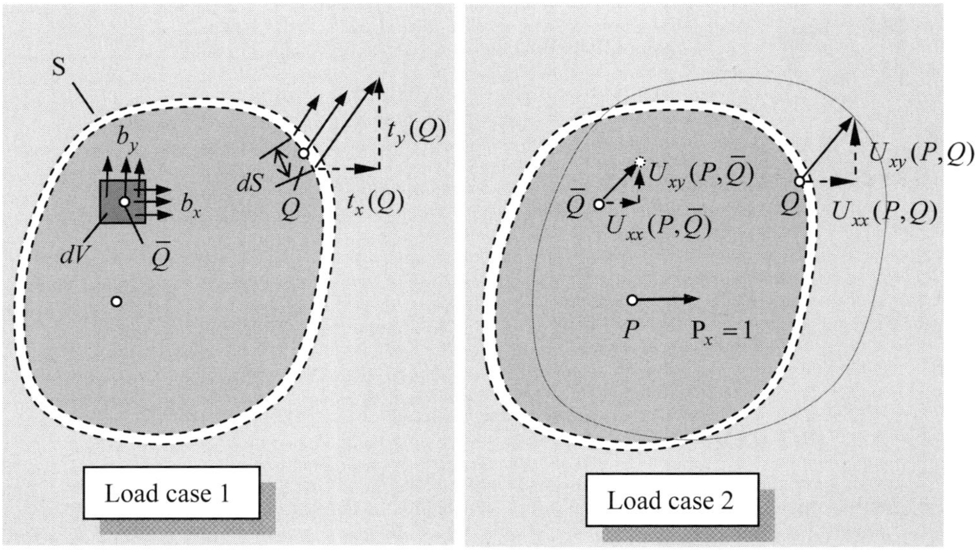

Figure 13.1 Application of Betti´s theorem including the effect of body forces

As shown in Figure 13.1 for 2-D problems, the forces of load case 1 consist of boundary tractions **t** (components t_x and t_y) and of body forces **b** (components b_x, b_y) defined as forces per unit volume.

The work done by the loads of load case 1 times the displacements of load case 2, W_{12} is computed by

$$W_{12} = \int_S (t_x(Q)\,U_{xx}(P,Q) + t_y(Q)\,U_{xy}(P,Q)\,)dS(Q) + \int_V (b_x(\bar{Q})U_{xx}(P,\bar{Q}) + b_y(\bar{Q})U_{xy}(P,\bar{Q})dV \tag{13.1}$$

The work done by the displacements of load case 1 times the forces of load case 2, W_{21} is the same as explained in Chapter 5

$$W_{21} = \int_S \left[u_x(Q)T_{xx}(P,Q) + u_y(Q)T_{xy}(P,Q)\right] dS + 1\,u_x(P) \tag{13.2}$$

The integral equations, including the body force effect can be written as:

$$\mathbf{u}(P)=\int_S \mathbf{U}(P,Q)\mathbf{t}(Q)dS-\int_S \mathbf{T}(P,Q)\mathbf{u}(Q)dS+\int_V \mathbf{U}(P,\bar{Q})\mathbf{b}(\bar{Q})dV \tag{13.3}$$

where the last integral in equation (13.3) is a volume integral. It can be shown[1] that for body forces which are constant over volume V, this integral can be transformed into a surface integral

$$\int_V \mathbf{U}(P,Q)\mathbf{b}(Q)dV=\int_S \mathbf{G}dS \tag{13.4}$$

where for 2-D and 3-D problems

$$\mathbf{G}=\begin{bmatrix}G_x\\G_y\end{bmatrix}\quad;\quad \mathbf{G}=\begin{bmatrix}G_x\\G_y\\G_z\end{bmatrix} \tag{13.5}$$

For 3-D problems the coefficients of **G** may be computed from[1]

$$G_i=\frac{1}{8\pi\,\mathrm{G}}\left(b_i\cos\theta-\frac{1}{2(1-\nu)}n_i\cos\psi\right) \tag{13.6}$$

where x,y,z may be substituted for i, G is the shear modulus, $\cos\theta$ has been defined previously in Chapter 4 and

$$\cos\psi=\mathbf{b}\bullet\frac{1}{r}\mathbf{r} \tag{13.7}$$

Vectors **n** and **r** are the normal vector and the position vector, as defined in Chapter 4. For plane strain problems we have[1] :

$$G_i=\frac{1}{8\pi G}\left(2\ln\frac{1}{r}-1\right)\left(b_i\cos\theta-\frac{1}{2(1-\nu)}n_i\cos\psi\right) \tag{13.8}$$

The discretised form of equation (13.3) can be written as

$$\mathbf{cu}(P_i)+\sum_{e=1}^{E}\sum_{n=1}^{N}\Delta\mathbf{T}_{ni}^{e}\,\mathbf{u}_n^e=\sum_{e=1}^{E}\sum_{n=1}^{N}\Delta\mathbf{U}_{ni}^{e}\,\mathbf{t}_n^e+\sum_{e=1}^{E}\Delta\mathbf{G}_i^e \tag{13.9}$$

where

$$\Delta \mathbf{G}_i^e = \int_{S_e} \mathbf{G}(P_i, Q)\, dS(Q) \tag{13.10}$$

For the three-dimensional case, no singularity occurs as P approaches Q and, therefore, the minimum integration order with which we are able to accurately compute the surface area of the element can be used. The analysis of problems with constant body forces proceeds the same way as before, except that an additional right hand side term is assembled. The final system of equations will be.

$$[\mathbf{T}]\{\mathbf{u}\} = \{\mathbf{F}\} + \{\mathbf{F}\}_b \tag{13.11}$$

where the components of $\mathbf{F}_b$ for the *i-th* collocation point are

$$\mathbf{F}_{ib} = \sum_{e=1}^{E} \Delta \mathbf{G}_i^e \tag{13.12}$$

13.2.1 Post-processing

When computing internal results the effect of body forces has to be included. For calculation of displacements

$$\mathbf{u}(P_a) = \sum_{e=1}^{E}\sum_{n=1}^{N} \Delta \mathbf{U}_{na}^e\, \mathbf{t}_n^e - \sum_{e=1}^{E}\sum_{n=1}^{N} \Delta \mathbf{T}_{na}^e\, \mathbf{u}_n^e + \sum_{e=1}^{E} \Delta \mathbf{G}_a^e \tag{13.13}$$

and for computation of stresses

$$\boldsymbol{\sigma}(P_a) = \sum_{e=1}^{E}\sum_{n=1}^{N} \Delta \mathbf{S}_{na}^e\, \mathbf{t}_n^e - \sum_{e=1}^{E}\sum_{n=1}^{N} \Delta \mathbf{R}_{na}^e\, \mathbf{u}_n^e + \sum_{e=1}^{E} \Delta \hat{\mathbf{S}}_a^e \tag{13.14}$$

where

$$\Delta \hat{\mathbf{S}}_a^e = \int_{S_e} \hat{\mathbf{S}}(P_a, Q)\, dS(Q) \tag{13.15}$$

Matrix $\mathbf{S}$ is obtained by differentiating (13.6) or (13.8) and multiplying with the constitutive matrix $\mathbf{D}$.

$$\hat{\mathbf{S}} = \begin{Bmatrix} S_{xx} \\ S_{yy} \\ S_{xy} \end{Bmatrix} \text{ for } 2-\text{D} \text{ and } \hat{\mathbf{S}} = \begin{Bmatrix} S_{xx} \\ S_{yy} \\ S_{zz} \\ S_{xy} \\ S_{yz} \\ S_{xz} \end{Bmatrix} \text{ for } 3-\text{D} \tag{13.16}$$

For 3-D problems we have[1]

$$\hat{S}_{ij} = \frac{1}{8\pi\, r} \begin{Bmatrix} \cos\theta(b_i r_{,j} + b_j r_{,i}) + \dfrac{1}{1-\nu} \nu \delta_{ij} \left(\cos\theta \cos\psi - \cos\phi\right) \\ -\dfrac{1}{2}\left[\cos\psi(n_i r_{,j} + n_j r_{,i}) + (1-2\nu)(b_i n_j + b_j n_i)\right] \end{Bmatrix} \tag{13.17}$$

where $\cos\theta$ and $\cos\psi$ have been defined previously, δ_{ij} is the Kronecker delta defined in Chapter 4 and

$$cos\,\phi = \mathbf{b} \bullet \mathbf{n} \tag{13.18}$$

For plane strain problems we have[1]

$$\hat{S}_{ij} = \frac{1}{8\pi} \begin{Bmatrix} 2\cos\theta(b_i r_{,j} + b_j r_{,i}) \\ + \dfrac{1}{1-\nu} \left[\nu \delta_{ij} \begin{pmatrix} 2\cos\theta\cos\psi - \cos\phi \\ +(1 - 2\ln\dfrac{1}{r})\cos\phi \end{pmatrix} - \cos\psi(n_i r_{,j} + n_j r_{,i}) \right] \\ -\dfrac{1}{2}\left[\cos\psi(n_i r_{,j} + n_j r_{,i}) + (1-2\nu)(b_i n_j + b_j n_i)\right] \end{Bmatrix} \tag{13.19}$$

Two subroutines **Grav_dis** and **Grav_stress**, which compute matrices **G** and $\hat{\mathbf{S}}$ needed for the gravity load case are added to the library **Elasticity.lib** . The subroutines can be used to compute the element contributions for assembly of the right hand side.

```
SUBROUTINE Grav_dis(GK,dxr,r,Vnor,b,G,ny)
!--------------------------------------------
!   FUNDAMENTAL SOLUTION FOR Displacements
!   Gravity Loads(Kelvin solution)
!--------------------------------------------
IMPLICIT NONE
REAL              :: GK(:)       !   Fundamental solution
REAL,INTENT(IN)   :: dxr(:)      !   rx/r etc.
REAL,INTENT(IN)   :: r           !
REAL,INTENT(IN)   :: Vnor(:)     !   normal vector
REAL,INTENT(IN)   :: b (:)       !   gravity force vector
REAL,INTENT(IN)   :: G           !   Shear modulus
REAL,INTENT(IN)   :: ny          !   Poisson's ratio
INTEGER           :: Cdim        !   Cartesian dimension
REAL              :: c1,c2,costh,Cospsi  !   Temps
C1= 1.0/(8*Pi*G)
C2=1.0/(2.0*(1.0-ny))
Costh= DOT_PRODUCT(Vnor ,DXR)Cospsi= DOT_PRODUCT(b,DXR)
 IF(Cdim == 2) THEN
 C1= C1*(2.0*LOG(1.0/r)-1.0)
 GK= C1*(b*costh - C2*Vnor*cospsi)
ELSE
 GK= C1*(b*costh - C2*Vnor*cospsi)
END IF
RETURN
END

SUBROUTINE Grav_stress(SK,dxr,r,Vnor,b,G,ny)
!--------------------------------------------
!   FUNDAMENTAL SOLUTION FOR Stresses
!   Gravity Loads(Kelvin solution)
!--------------------------------------------
IMPLICIT NONE
REAL              :: SK(:)        !   Kernel
REAL,INTENT(IN)   :: dxr(:)       !   rx/r etc.
REAL,INTENT(IN)   :: r            !
REAL,INTENT(IN)   :: Vnor(:)      !   normal vector
REAL,INTENT(IN)   :: b (:)        !   body force vector
REAL,INTENT(IN)   :: G            !   Shear modulus
REAL,INTENT(IN)   :: ny           !   Poisson's ratio
INTEGER           :: Cdim         !   Cartesian dimension
INTEGER           :: II(6),JJ(6)  !   Order of stress components
REAL              :: c,c1,c2,c3,c4,costh,Cospsi,Cosphi   !   Temps
C2=1.0/(1.0-ny)
C3= 1-2.0*ny
Costh= DOT_PRODUCT(Vnor ,DXR)
Cospsi= DOT_PRODUCT(b,DXR)
Cosphi= DOT_PRODUCT(b,Vnor)
IF(Cdim == 2) THEN    !   Two-dimensional solution
 C1= 1.0/(8*Pi)
```

```
  C4= 1.0 - 2.0*LOG(1.0/r)
  II(1:3)= (/1,2,1)
  JJ(1:3)= (/1,2,2)
  Stress_components: &
  DO N=1,3
   I= II(N) ; J= JJ(N)
   IF(I == J) THEN
    C= ny*(2.0*costh*cospsi-cosphi)+(1.0-2.0*LOG(1/r))cosphi
   ELSE
    C= 0.0
   END IF
   SK(N)= C2*(C - cospsi*(Vnor(I)*dxr(J)+ Vnor(J)*dxr(I)) &
        - 0.5 *(cospsi*(Vnor(I)*dxr(J)+ Vnor(J)*dxr(I)) &
                  + C3*(b(I)*Vnor(J)   + b(J)*Vnor(I)))
  END DO
  Stress_components
  SK= C1*SK
ELSE                    !   Three-dimensional solution
   II= (/1,2,3,1,2,3)
   JJ= (/1,2,3,2,3,1)
   C1= 1.0/(8*Pi*r)
   Stress_components1: &
   DO N=1,6
     I= II(N) ; J= JJ(N)
     C=0.
    IF(I == J) THEN
     C= c2*ny*(costh*cospsi-cosphi)
    ELSE
     C= 0.0
    END IF
    SK(N)= 2.0*costh*(b(I)dxr(J)+ b(J)dxr(I))+ C &
         -0.5 *(cospsi*(Vnor(I)*dxr(J)+ Vnor(J)*dxr(I))&
                 + C3*(b(I)*Vnor(J)   + b(J)*Vnor(I)))
  END DO &
  Stress_components1
  SK= C1*SK
END IF
RETURN
END
```

13.3 INTERNAL CONCENTRATED FORCES

It is sometimes necessary to apply concentrated forces inside the domain. An example of this is the simulation of a pre-stressed rock bolt in tunnelling, where a concentrated force is generated inside the domain. According to figure 13.2, additional work is done by a concentrated force **F** acting at point $\bar{Q}$.

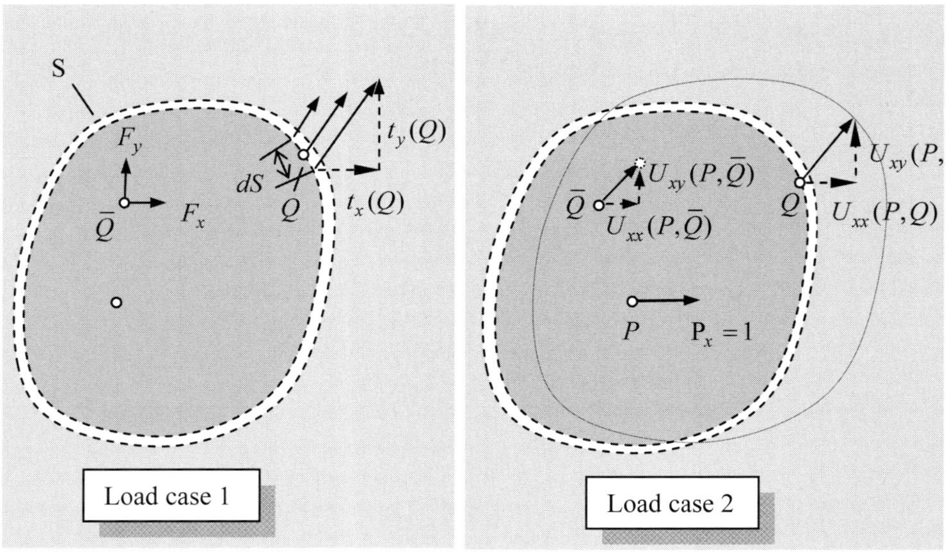

Figure 13.2 Application of Betti´s theorem including the effect of internal concentrated forces

The work done by the loads of load case 1, times the displacements of load case 2, W_{12} is computed by

$$W_{12} = \int_S (t_x(Q)\,U_{xx}(P,Q) + t_y(Q)\,U_{xy}(P,Q))\,dS(Q) + F_x U_{xx}(P,\bar{Q}) + F_y(\bar{Q}) U_{xy}(P,\bar{Q}) \tag{13.20}$$

The work done by the displacements of load case 1 times the forces of load case 2, W_{21} is the same as explained in Chapter 5. Using Betti's theorem the following integral equation is obtained, which includes the effect of the concentrated laod:

$$\mathbf{u}(P) = \int_S \mathbf{U}(P,Q)\mathbf{t}(Q)\,dS - \int_S \mathbf{T}(P,Q)\mathbf{u}(Q)\,dS + \mathbf{U}(P,\bar{Q})\mathbf{F}(\bar{Q}) \tag{13.21}$$

where

$$\mathbf{F} = \begin{pmatrix} F_x(\bar{Q}) \\ F_y(\bar{Q}) \end{pmatrix} \tag{13.22}$$

The discretised form can be written as

$$\mathbf{c}\mathbf{u}(P_i) + \sum_{e=1}^{E}\sum_{n=1}^{N} \Delta\mathbf{T}_{ni}^{e}\,\mathbf{u}_n^e = \sum_{e=1}^{E}\sum_{n=1}^{N} \Delta\mathbf{U}_{ni}^{e}\,\mathbf{t}_n^e + \mathbf{F}\mathbf{U}(P_i,\bar{Q}) \tag{13.23}$$

The final system of equations will be.

$$[\mathbf{T}]\{\mathbf{u}\} = \{\mathbf{F}\} + \{\mathbf{F}\}_P \tag{13.24}$$

where the components of $\mathbf{F}_P$ for the *i-th* collocation point are

$$\mathbf{F}_{iP} = \mathbf{FU}(P_i, \bar{Q}) \tag{13.25}$$

13.3.1 Post-processing

When computing internal results, the effect of the internal force has to be included. For calculation of displacements

$$\mathbf{u}(P_a) = \sum_{e=1}^{E}\sum_{n=1}^{N} \Delta\mathbf{U}_{na}^{e}\,\mathbf{t}_n^{e} - \sum_{e=1}^{E}\sum_{n=1}^{N} \Delta\mathbf{T}_{na}^{e}\,\mathbf{u}_n^{e} + \mathbf{F}\cdot\mathbf{U}(P_a, \bar{Q}) \tag{13.26}$$

whereas for computation of stresses we have

$$\boldsymbol{\sigma}(P_a) = \sum_{e=1}^{E}\sum_{n=1}^{N} \Delta\mathbf{S}_{na}^{e}\,\mathbf{t}_n^{e} - \sum_{e=1}^{E}\sum_{n=1}^{N} \Delta\mathbf{R}_{na}^{e}\,\mathbf{u}_n^{e} + \mathbf{F}\cdot\mathbf{S}(P_a, \bar{Q}) \tag{13.27}$$

13.4 INTERNAL DISTRIBUTED LINE FORCES

We now consider the effect of distributed line forces that may be shear forces acting in the rock mass due to a rock bolt. According to figure 13.3, additional work is done by a distributed force **f** acting along a line.

The work done by the loads of load case 1 times the displacements of load case 2, W_{12} is computed by

$$\begin{aligned} W_{12} = &\int_S (t_x(Q)\,U_{xx}(P,Q) + t_y(Q)\,U_{xy}(P,Q))\,dS(Q) \\ &+ \int_{\bar{S}} \left(f_x U_{xx}(P,\bar{Q}) + f_y(\bar{Q}) U_{xy}(P,\bar{Q}\right) d\bar{S} \end{aligned} \tag{13.28}$$

where the last integral is over the line on which the distributed force acts. The work done by the displacements of load case 1 times the forces of load case 2, W_{21} is the same as explained in Chapter 5.

The integral equations including the body force effect can be written as:

$$\mathbf{u}(P) = \int_S \mathbf{U}(P,Q)\,\mathbf{t}(Q)\,dS - \int_S \mathbf{T}(P,Q)\,\mathbf{u}(Q)\,dS + \int_{\bar{S}} \mathbf{U}(P,\bar{Q})\,\mathbf{f}(\bar{Q}) \tag{13.29}$$

where

$$\mathbf{f} = \begin{pmatrix} f_x(\bar{Q}) \\ f_y(\bar{Q}) \end{pmatrix} \tag{13.30}$$

The discretised form can be written as

$$\mathbf{cu}(P_i) + \sum_{e=1}^{E}\sum_{n=1}^{N} \Delta\mathbf{T}_{ni}^{e}\,\mathbf{u}_n^e = \sum_{e=1}^{E}\sum_{n=1}^{N} \Delta\mathbf{U}_{ni}^{e}\,\mathbf{t}_n^e + \int_{\bar{S}} \mathbf{f}\mathbf{U}(P_i,\bar{Q})dS(\bar{Q}) \tag{13.31}$$

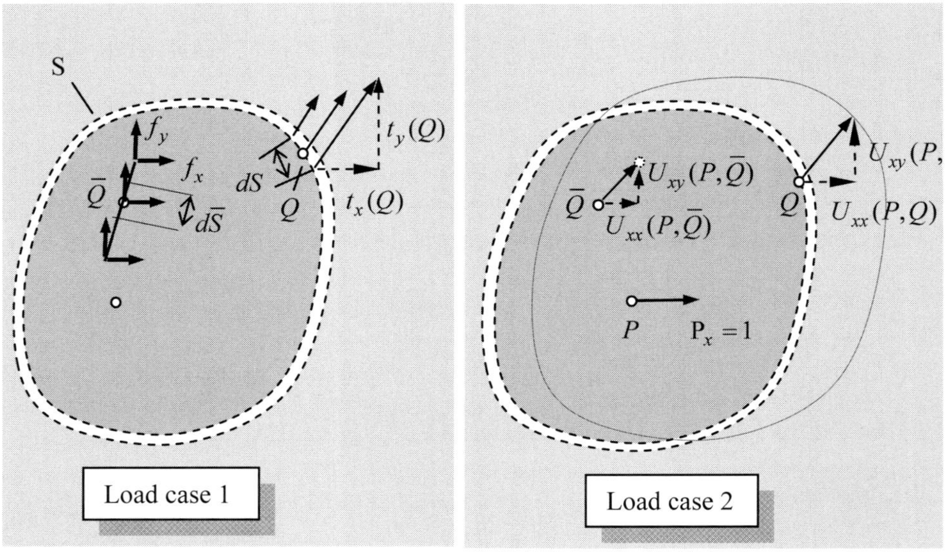

Figure 13.3 Application of Betti´s theorem including the effect of internal distributed forces

To evaluate the last line integral we propose to use internal cells. The cells are actually exactly like the 1-D boundary elements introduced in Chapter 3 but are used for the integration only. If the variation of **f** along the line is linear or quadratic then only one linear or quadratic cell element is required for the integration. Using the interpolation as discussed in Chapter 3

$$\mathbf{f} = \sum_{n=1}^{2(3)} N_n(\xi)\,\mathbf{f}_n \tag{13.32}$$

where $\mathbf{f}_n$ are the nodal values of **f**, we obtain

$$\int_{\bar{S}} \mathbf{f}\mathbf{U}(P_i,\bar{Q})dS(\bar{Q}) = \sum_{n=1}^{2(3)} \mathbf{f}_n \Delta\mathbf{U}_{ni}^{e} \tag{13.33}$$

where

$$\Delta \mathbf{U}_{ni}^{e} = \int_{\bar{S}} N_n \cdot \mathbf{U}(P_i, \bar{Q}) \cdot d\bar{S} \tag{13.34}$$

These integrals may be evaluated using Gauss integration. The final system of equations will be

$$[\mathbf{T}]\{\mathbf{u}\} = \{\mathbf{F}\} + \{\mathbf{F}\}_p \tag{13.35}$$

where the components of $\mathbf{F}_p$ for the *i-th* collocation point are

$$\mathbf{F}_{ip} = \sum_{n=1}^{2(3)} \mathbf{f}_n \Delta \mathbf{U}_{ni}^{e} \tag{13.36}$$

13.4.1 Post-processing

When computing internal results the effect of the internal force has to be included. For calculation of displacements at point P_a

$$\mathbf{u}(P_a) = \sum_{e=1}^{E}\sum_{n=1}^{N} \Delta \mathbf{U}_{na}^{e}\, \mathbf{t}_n^{e} - \sum_{e=1}^{E}\sum_{n=1}^{N} \Delta \mathbf{T}_{na}^{e}\, \mathbf{u}_n^{e} + \sum_{n=1}^{2(3)} \mathbf{f}_n \Delta \mathbf{U}_{na}^{e} \tag{13.37}$$

whereas for computation of stresses we have

$$\boldsymbol{\sigma}(P_a) = \sum_{e=1}^{E}\sum_{n=1}^{N} \Delta \mathbf{S}_n^{e}\, \mathbf{t}_n^{e} - \sum_{e=1}^{E}\sum_{n=1}^{N} \Delta \mathbf{R}_n^{e}\, \mathbf{u}_n^{e} + \sum_{n=1}^{2(3)} \mathbf{f}_n \Delta \mathbf{S}_{na}^{e} \tag{13.38}$$

where

$$\Delta \mathbf{S}_{na}^{e} = \int_{\bar{S}} N_n \cdot \mathbf{S}(P_a, \bar{Q}) \cdot d\bar{S} \tag{13.39}$$

13.5 INITIAL STRAINS

There are problems where strains are generated inside domains that are not associated with loading by forces. Examples are thermal strains generated by a temperature increase and strains due to swelling of soil. Invariably these strains will not be constant over the whole domain. Therefore, it will no longer be possible to transform the volume integrals into surface integrals. If we assume that the solid is subjected to a non-uniform volumetric strain (caused for example by a temperature increase) given by

$$\boldsymbol{\varepsilon}_0 = \begin{pmatrix} \varepsilon_{x0} \\ \varepsilon_{y0} \end{pmatrix} \tag{13.40}$$

additional work will be done.

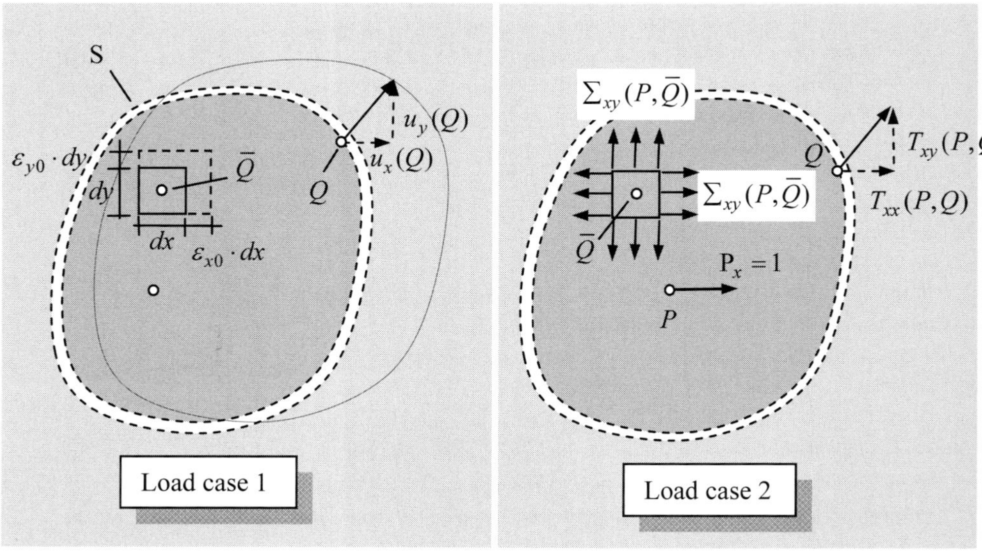

Figure 13.4 Application of the Betti theorem including the effect of initial strains

Referring to figure 13.4, the work done by the displacements/strains of load case 1 times the forces/stresses of load case 2 is given by:

$$\begin{aligned} W_{12} &= \int_S (u_x(Q)\, T_{xx}(P,Q) + u_y(Q) T_{xy}(P,Q))\, dS(Q) \\ &+ \iint \varepsilon_{x0} dx \cdot \Sigma_{xx}(P,\bar{Q})\, dy + \varepsilon_{y0} dy \cdot \Sigma_{yx}(P,\bar{Q})\, dx \end{aligned} \tag{13.41}$$

where $\Sigma_{xx}(P,\bar{Q})$ and $\Sigma_{yx}(P,\bar{Q})$ are the stresses at $\bar{Q}$ due to a unit force in x direction at P. Here we assume that only volumetric initial strains are present, even though it is obvious that shear strains could easily be included. The work done by the displacements of load case 2 times the forces/stresses of load case 1 is the same as for the case where no initial strains are applied.

Applying Betti's theorem we obtain

$$\mathbf{u}(P) = \int_S \mathbf{U}(P,Q)\mathbf{t}(Q)dS - \int_S \mathbf{T}(P,Q)\mathbf{u}(Q)dS + \int_V \mathbf{\Sigma}(P,\overline{Q})\boldsymbol{\varepsilon}_0(\overline{Q})dV \qquad (13.42)$$

where

$$\mathbf{\Sigma} = \begin{bmatrix} \Sigma_{xx} & \Sigma_{xy} \\ \Sigma_{yx} & \Sigma_{yy} \end{bmatrix} \qquad (13.43)$$

For 3-D problems where strain ε_{z0} is also present, matrix $\mathbf{\Sigma}$ is expanded to

$$\mathbf{\Sigma} = \begin{bmatrix} \Sigma_{xx} & \Sigma_{xy} & \Sigma_{xz} \\ \Sigma_{yx} & \Sigma_{yy} & \Sigma_{yz} \\ \Sigma_{zx} & \Sigma_{zy} & \Sigma_{zz} \end{bmatrix} \qquad (13.44)$$

The fundamental solution is given by[2]

$$\Sigma_{ik} = \frac{-C_2}{r^n}\left(C_3(2\delta_{ik}r_{,i} - r_{,k}) + C_4 r_{,i}^2 r_{,k}\right) \qquad (13.45)$$

where x,y,z may be substituted for i,k as usual. The values for the constants are given in Table 12.1

Table 12.1 Constants for fundamental solution for initial strains

	Plane strain	Plane stress	3-D
n	1	1	2
C_2	$1/4\pi(1-\nu)$	$(1+\nu)/4\pi$	$1/8\pi(1-\nu)$
C_3	$1-2\nu$	$(1-\nu)/(1+\nu)$	$1-2\nu$
C_4	2	2	3

A FUNCTION for computing Matrix $\mathbf{\Sigma}$ is written and added to the Elasticity_lib. FUNCTION SigmaK returns an array of dimension 2x2 or 3x3 with fundamental solutions for normal stresses.

```
FUNCTION SigmaK(dxr,r,E,ny,Cdim)
!--------------------------------------------
!   FUNDAMENTAL SOLUTION FOR Normal Stresses
!   isotropic material (Kelvin solution)
!--------------------------------------------
REAL,INTENT(IN)    :: dxr(:)            ! rx/r etc.
REAL,INTENT(IN)    :: r                 ! r
REAL,INTENT(IN)    :: E                 ! Young's modulus
REAL,INTENT(IN)    :: ny                ! Poisson's ratio
INTEGER,INTENT(IN) :: Cdim              ! Cartesian dimension
REAL               :: SigmaK(Cdim,Cdim) ! Returns array CdimxCdim
INTEGER            :: n,i,j
REAL               :: G,c,c2,c3,c4      ! Temps
G= E/(2.0*(1+ny))
SELECT CASE (Cdim)
  CASE (2)          !    Plane strain solution
    n= 1
    c2= 1.0/(4.0*Pi*(1.0-ny))
    c3= 1.0-2.0*ny
    c4= 2.0
  CASE(3)           !     Three-dimensional solution
    n= 2
    c2= 1.0/(8.0*Pi*(1.0-ny))
    c3= 1.0-2.0*ny
    c4= 3.0
  CASE DEFAULT
END SELECT
Direction_Pi: &
DO i=1,Cdim
  Direction_Sigma: &
  DO j=1,Cdim
    IF(i == j) THEN
      SigmaK(i,i)= -c2/r**n*(c3*dxr(i)+c4*dxr(i)**3)
    ELSE
      SigmaK(i,j)= -c2/r**n*(-c3*dxr(i) + c4*dxr(i)**2*dxr(j))
    END IF
END DO &
Direction_Sigma
END DO &
Direction_Pi
RETURN
END FUNCTION SigmaK
```

The discretised form can be written as

$$\mathbf{c}\mathbf{u}\left(P_i\right)+\sum_{e=1}^{E}\sum_{n=1}^{N}\Delta\mathbf{T}_{ni}^{e}\,\mathbf{u}_{n}^{e}=\sum_{e=1}^{E}\sum_{n=1}^{N}\Delta\mathbf{U}_{ni}^{e}\,\mathbf{t}_{n}^{e}+\int_{V}\mathbf{\Sigma}\left(P_i,\bar{Q}\right)\boldsymbol{\varepsilon}_{0}\left(\bar{Q}\right)dV \tag{13.46}$$

We propose to evaluate the volume integral numerically with the Gauss Quadrature method. To apply this method, however, the volume where initial strains are specified needs to be discretised, i.e., subdivided into cells. We use two-dimensional cells for the discretisation of 2-D problems and three-dimensional cells for 3-D problems. The cells have already been introduced in Chapter 3. For the interpolation of the strains inside an element we have for plane problems with either linear (N=4) or quadratic (N=8) shape functions N_n

$$\boldsymbol{\varepsilon}_0(\xi,\eta)=\sum_{n=1}^{N} N_n(\xi,\eta)\,\boldsymbol{\varepsilon}_{0n}^{e} \tag{13.47}$$

The last integral in Eq. (13.46) is replaced by a sum of integrals over cells

$$\int_V \boldsymbol{\Sigma}\left(P_i,\bar{Q}\right)\boldsymbol{\varepsilon}_0\left(\bar{Q}\right)dV=\sum_{c=1}^{N_c}\sum_{n=1}^{N}\Delta\boldsymbol{\Sigma}_{\mathrm{ni}}^{\mathrm{c}}\;\boldsymbol{\varepsilon}_{0n} \tag{13.48}$$

where

$$\Delta\boldsymbol{\Sigma}_{\mathrm{ni}}^{\mathrm{c}}=\int_{V_c} N_n\cdot\boldsymbol{\Sigma}(P_i,\bar{Q})\cdot dV \tag{13.49}$$

The final system of equations will be.

$$[\mathbf{T}]\{\mathbf{u}\}=\{\mathbf{F}\}+\{\mathbf{F}\}_{\varepsilon} \tag{13.50}$$

This means that the presence of initial strains will result in an additional right hand side $\{\mathbf{F}\}_{\varepsilon}$ where the components of $\mathbf{F}_{\varepsilon}$ for the *i-th* collocation point are

$$\mathbf{F}_{i\varepsilon}=\sum_{c=1}^{N_c}\sum_{n=1}^{N}\boldsymbol{\Sigma}_{\mathrm{ni}}^{\mathrm{c}}\;\boldsymbol{\varepsilon}_{0n} \tag{13.51}$$

13.5.1 Post-processing

In post-processing the effect of the initial strains has to be included. For calculation of displacements at point P_a we use Eq. (13.42)

$$\mathbf{u}\left(P_a\right)=\int_S \mathbf{U}\left(P_a,Q\right)\mathbf{t}\left(Q\right)dS-\int_S \mathbf{T}\left(P_a,Q\right)\mathbf{u}\left(Q\right)dS+\int_V \boldsymbol{\Sigma}\left(P_a,\bar{Q}\right)\boldsymbol{\varepsilon}_0\left(\bar{Q}\right)dV \tag{13.52}$$

For obtaining strains and stresses we have to take the derivative of the displacement

$$\begin{aligned}\mathbf{u}_{,j}\left(P_a\right)=&\int_S \mathbf{U}_{,j}\left(P_a,Q\right)\mathbf{t}\left(Q\right)dS-\int_S \mathbf{T}_{,j}\left(P_a,Q\right)\mathbf{u}\left(Q\right)dS\\&+\int_V \boldsymbol{\Sigma}_{,j}\left(P_a,\bar{Q}\right)\boldsymbol{\varepsilon}_0\left(\bar{Q}\right)dV\end{aligned} \tag{13.53}$$

From this equation the strains and stresses may be computed using Equations (4.34) and (4.45). If strains or stresses are evaluated inside a region that is subjected to initial strains then the integrand in last integral in (13.53) approaches infinity as $\bar{Q}$ approaches P_a and special care has to be taken in evaluating this integral. If we assume a small volume of exclusion around P_a, then we can split the integral into two parts (Figure 13.5).

$$\int_V \boldsymbol{\Sigma}_{,j}\left(P_a,\bar{Q}\right)\boldsymbol{\varepsilon}_0\left(\bar{Q}\right) dV = \int_{V-V_\varepsilon} \boldsymbol{\Sigma}_{,j}\left(P_a,\bar{Q}\right)\boldsymbol{\varepsilon}_0\left(\bar{Q}\right) dV + \int_{V_\varepsilon} \boldsymbol{\Sigma}_{,j}\left(P_a,\bar{Q}\right)\boldsymbol{\varepsilon}_0\left(\bar{Q}\right) dV \quad (13.54)$$

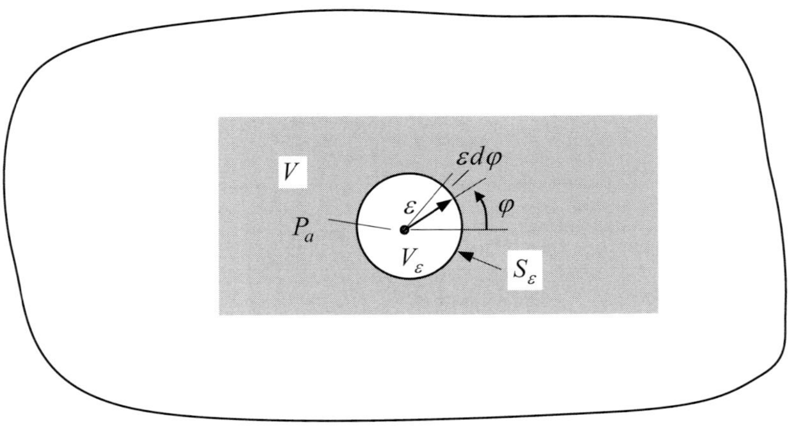

Figure 13.5 Region of exclusion for the computation of volume integral

The first integral only exists as a Cauchy principal value and we will discuss its evaluation in Chapter 15. The second integral can be evaluated analytically. If we assume that $\varepsilon \rightarrow 0$, then we can assume $\boldsymbol{\varepsilon}_0$ to be constant (i.e. $\boldsymbol{\varepsilon}_0(\bar{Q}) = \boldsymbol{\varepsilon}_0(P_a)$) and we have

$$\int_{V_\varepsilon} \boldsymbol{\Sigma}_{,j}\left(P_a,\bar{Q}\right)\boldsymbol{\varepsilon}_0\left(\bar{Q}\right) dV = \boldsymbol{\varepsilon}_0\left(P_a\right)\cdot \int_{V_\varepsilon} \boldsymbol{\Sigma}_{,j}\left(P_a,\bar{Q}\right) dV \quad (13.55)$$

The volume integral can be transformed into an integral over the surface of the exclusion S_ε

$$\int_{V_\varepsilon} \boldsymbol{\Sigma}_{,j}\left(P_a,\bar{Q}\right) dV = \int_{S_\varepsilon} \boldsymbol{\Sigma}\left(P_a,\bar{Q}\right)\cdot \mathbf{n}\ dV \quad (13.56)$$

We explain the analytical integration on an example in plane strain.

The two components of the integral in (13.55) are:

$$\begin{Bmatrix} \varepsilon_{x0}\left(P_a\right)\cdot \int\limits_{S_\varepsilon} \Sigma_{xx}\left(P_a,\bar{Q}\right)\cdot n_x \; dS + \varepsilon_{x0}\left(P_a\right)\cdot \int\limits_{S_\varepsilon} \Sigma_{yx}\left(P_a,\bar{Q}\right)\cdot n_y \; dS \\ \varepsilon_{y0}\left(P_a\right)\cdot \int\limits_{S_\varepsilon} \Sigma_{xy}\left(P_a,\bar{Q}\right)\cdot n_x \; dS + \varepsilon_{y0}\left(P_a\right)\cdot \int\limits_{S_\varepsilon} \Sigma_{yy}\left(P_a,\bar{Q}\right)\cdot n_y \; dS \end{Bmatrix} \tag{13.57}$$

According to Figure 13.5 we have $r=\varepsilon$; $\mathbf{n}=\mathbf{r}$; $r_{,x}=\cos\varphi$; $r_{,y}=\sin\varphi$. Therefore we have for example:

$$\int\limits_{S_\varepsilon} \Sigma_{xx}\left(P_a,\bar{Q}\right)\cdot n_x \; dS = \int\limits_0^{2\pi} \frac{-C_2}{\varepsilon}\left(\sin\varphi + C_4 \sin\varphi^3\right)\cdot \sin\varphi \cdot \varepsilon d\varphi = \frac{1-5\nu}{15-\nu} \tag{13.58}$$

Applying Equations (4.34) and (4.45) with (13.53) we obtain for the stress

$$\boldsymbol{\sigma}\left(P_a\right) = \sum_{e=1}^{E}\sum_{n=1}^{N} \Delta\mathbf{S}_n^e \; \mathbf{t}_n^e - \sum_{e=1}^{E}\sum_{n=1}^{N} \Delta\mathbf{R}_n^e \; \mathbf{u}_n^e + \sum_{c=1}^{N_c}\sum_{n=1}^{N} \Delta\hat{\boldsymbol{\Sigma}}_n^c \boldsymbol{\varepsilon}_{0n} + \mathbf{H}\cdot\boldsymbol{\varepsilon}_0(P_a) \tag{13.59}$$

where

$$\Delta\hat{\boldsymbol{\Sigma}}_{\mathrm{ni}}^{\mathrm{c}} = \int\limits_{V_c} N_n \cdot \hat{\boldsymbol{\Sigma}}(P_a,\bar{Q})\cdot dV \tag{13.60}$$

and

$$\hat{\boldsymbol{\Sigma}} = \begin{bmatrix} \hat{\Sigma}_{\mathrm{xxx}} & \hat{\Sigma}_{\mathrm{xxy}} \\ \hat{\Sigma}_{yy\mathrm{x}} & \hat{\Sigma}_{yyy} \\ \hat{\Sigma}_{\mathrm{x}y\mathrm{x}} & \hat{\Sigma}_{\mathrm{x}yy} \end{bmatrix} \text{ for } 2-\mathrm{D} \text{ and } \hat{\boldsymbol{\Sigma}} = \begin{bmatrix} \hat{\Sigma}_{\mathrm{xxx}} & \hat{\Sigma}_{\mathrm{xx}y} & \hat{\Sigma}_{\mathrm{xx}z} \\ \hat{\Sigma}_{yy\mathrm{x}} & \hat{\Sigma}_{yyy} & \hat{\Sigma}_{yyz} \\ \hat{\Sigma}_{zz\mathrm{x}} & \hat{\Sigma}_{zzy} & \hat{\Sigma}_{zzz} \\ \hat{\Sigma}_{\mathrm{x}y\mathrm{x}} & \hat{\Sigma}_{\mathrm{x}yy} & \hat{\Sigma}_{\mathrm{x}yz} \\ \hat{\Sigma}_{yz\mathrm{x}} & \hat{\Sigma}_{yzy} & \hat{\Sigma}_{yzz} \\ \hat{\Sigma}_{\mathrm{x}z\mathrm{x}} & \hat{\Sigma}_{\mathrm{x}zy} & \hat{\Sigma}_{\mathrm{x}zz} \end{bmatrix} \text{ for } 3-\mathrm{D} \tag{13.61}$$

The coefficients of $\hat{\boldsymbol{\Sigma}}$ are given by[2]

$$\begin{aligned} \hat{\Sigma}_{ijk}\left(P,\bar{Q}\right) = \frac{C_7}{r^{n+1}}\Big\{ & C_3 2\cdot\delta_{ik}\delta_{jk} - C_{14}\delta_{ij} \\ & + (n+1)\nu\left(\delta_{ik}r_{,j}r_{,k} + \delta_{jk}r_{,i}r_{,k} + \delta_{ik}r_{,j}r_{,k} + \delta_{jk}r_{,i}r_{,k}\right) \\ & + (n+1)\left[C_3 r_{,i}r_{,j} + C_{15}\delta_{ij}{r_{,k}}^2 - C_6 r_{,i}r_{,j}{r_{,k}}^2\right]\Big\} \end{aligned} \tag{13.62}$$

and those of **H**

$$H_{ijk} = C_{16}\left[C_{12} 2\delta_{ik}\delta_{jk} + C_{17}\delta_{ij}\right] \tag{13.63}$$

where the constants are given in Table 12.2.

Table 12.2 Constants for fundamental solutions

	Plane strain	Plane stress	3-D
n	1	1	2
C_7	1-4 ν	(1-3 ν)/(1+ ν)	1-4 ν
C_3	1-2 ν	(1- ν)/(1+ ν)	1-2 ν
C_{14}	1	(1-3 ν)/(1+ ν)	1-4 ν
C_{15}	1	(1- ν)/(1+ ν)	1-2 ν
C_6	4	4	5
C_{16}	G/(4(1-ν)	G(1+ ν)/4	G/(15(1-ν)
C_{12}	1	1	7-5ν
C_{17}	1-4ν	1	2+10ν

13.6 INITIAL STRESSES

The last type of body forces considered here are initial stresses. As will be seen later in the chapter on plasticity, these may correspond to the plastic stresses generated when a point goes into the plastic range. The capability to deal with initial stresses, therefore, will be important for the application of the BEM to nonlinear material response, in particular plasticity. The consideration of initial stresses follows a similar line as the consideration of initial strains. We start again with the theorem of Betti, but instead of initial strains we consider stresses applied inside the domain. If they exist, it is obvious that additional work will be done.

Referring to Figure 13.6 the work done by the tractions/stresses of load case 1 times the displacement/strains of load case 2 is given by:

$$\begin{aligned} W_{12} &= \int_S (t_x(Q)\,U_{xx}(P,Q) + t_y(Q)\,U_{xy}(P,Q))\,dS(Q) \\ &+ \iint \sigma_{x0} dy \cdot E_{xxx}\left(P,\bar{Q}\right) dx + \sigma_{y0} dx \cdot E_{xyy}\left(P,\bar{Q}\right) dy + \tau_{xy0} dx \cdot E_{xxy}\left(P,\bar{Q}\right) dy \end{aligned} \tag{13.64}$$

where E_{xxx} , E_{xyy} , E_{xxy} are the fundamental solutions for strains at point $\bar{Q}$ due to a unit x-force at P.

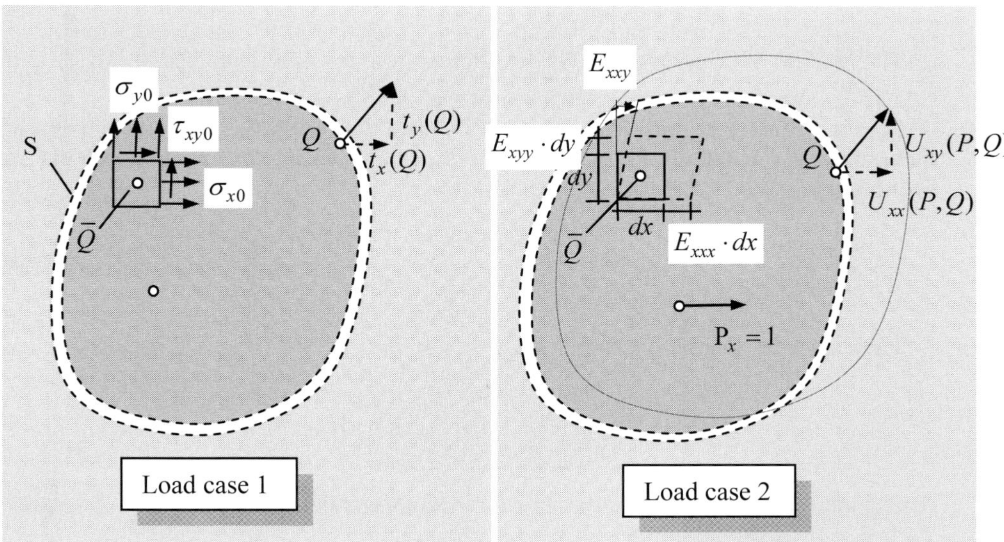

Figure 13.6 Application of the theorem of Betti including the effect of initial strains

The work done by the forces/stresses of load case 2 times the displacements of load case one is the same as for the case without initial stresses.
After applying the theorem by Betti we obtain

$$\mathbf{u}(P) = \int_S \mathbf{U}(P,Q)\mathbf{t}(Q)\,dS - \int_S \mathbf{T}(P,Q)\mathbf{u}(Q)\,dS + \int_V \mathbf{E}(P,\bar{Q})\boldsymbol{\sigma}_0(\bar{Q})\,dV \qquad (13.65)$$

The detailed implementation of initial stresses is discussed in Chapter 15, dealing with non-linear problems.

13.7 NUMERICAL INTEGRATION OVER CELLS

The integrals over the cells are evaluated numerically using Gauss Quadrature. For the line integrals in (13.34) we use linear cells. Changing from x,y coordinates to intrinsic coordinate ξ we get

$$\Delta\mathbf{U}^e_{ni} = \int_{\bar{S}} N_n \cdot \mathbf{U}(P_i,\bar{Q}) \cdot d\bar{S} = \int_{-1}^{1} N_n(\xi) \cdot \mathbf{U}(P_i,\bar{Q}) \cdot J(\xi) \cdot d\xi \qquad (13.66)$$

The numerical integration is given by

$$\Delta\mathbf{U}^{e}_{ni} = \sum_{m=1}^{M} N_n(\xi_m)\cdot \mathbf{U}(P_i,\xi_m)\cdot J(\bar{Q}(\xi_m))\cdot W_m \tag{13.67}$$

The number of Gauss points M is determined from the minimum distance of P_i to the cell, as explained in Chapter 6, J is the *Jacobian* and W_m are weight factors.

For the volume integrals occurring in (13.49) we use plane cells for 2-D problems and three-dimensional cells for 3-D problems. For plane problems the expression in intrinsic coordinates is

$$\Delta\boldsymbol{\Sigma}^{\mathrm{c}}_{\mathrm{ni}} = \int_{V_c} N_n \cdot \boldsymbol{\Sigma}(P_i,\bar{Q})\cdot dV = \int_{-1}^{+1}\int_{-1}^{+1} N_n(\xi,\eta)\cdot \boldsymbol{\Sigma}(P_i,\bar{Q}(\xi,\eta))\cdot J(\xi,\eta)\cdot d\xi\cdot d\eta \tag{13.68}$$

The numerical integration formula is

$$\Delta\boldsymbol{\Sigma}^{\mathrm{c}}_{\mathrm{ni}} = \sum_{m=1}^{M}\sum_{k=1}^{K} N_n(\xi_m,\eta_k)\cdot \boldsymbol{\Sigma}(P_i,\bar{Q}(\xi_m,\eta_k))\cdot J(\xi_m,\eta_k)\cdot W_m\cdot W_k \tag{13.69}$$

where the number of integration points in ξ,η directions M and K are determined according to the proximity of P_i to the cell. For three-dimensional problems we have

$$\Delta\boldsymbol{\Sigma}^{\mathrm{c}}_{\mathrm{ni}} = \int_{-1}^{+1}\int_{-1}^{+1}\int_{-1}^{+1} N_n(\xi,\eta,\zeta)\cdot \boldsymbol{\Sigma}(P_i,\bar{Q}(\xi,\eta,\zeta))\cdot J(\xi,\eta,\zeta)\cdot d\xi\cdot d\eta\cdot d\zeta \tag{13.70}$$

The numerical integration formula is

$$\Delta\boldsymbol{\Sigma}^{\mathrm{c}}_{\mathrm{ni}} = \sum_{l=1}^{L}\sum_{m=1}^{M}\sum_{k=1}^{K} N_n(\xi_m,\eta_k,\zeta_l)\cdot \boldsymbol{\Sigma}(P_i,(\xi_m,\eta_k,\zeta_l))\cdot J(\xi_m,\eta_k,\zeta_l)\cdot W_m\cdot W_k\cdot W_l \tag{13.71}$$

13.8 IMPLEMENTATION

For the implementation into general purpose program 7.1 we insert a call to a subroutine **Body_force** that adds the influence of body forces. Here we will only show one example of an implementation of body force: the effect of initial volumetric strains defined at a given cell mesh for a plane elasticity problem. Although the implementation will be general and allow for a variation of initial strains inside the domain, the input will be restricted, for purposes of simplicity, to a constant strain for all cells. Following the explanations given in this Chapter, the reader may however have no great difficulty in programming the other body force effects and extend this to 3-D applications. Subroutine **Body_force** will read first the information about the number of cell nodes, cells and the initial strain to be applied. Then the coordinates of the cell nodes and the incidences of the cells are read in. Next the additional right hand side is computed. According to Eq. (13.51) the right hand side is given by:

$$\mathbf{F}_{i\varepsilon} = \sum_{c=1}^{N_c}\sum_{n=1}^{N} \Delta\Sigma_{\mathrm{ni}}^{\mathrm{c}}\ \boldsymbol{\varepsilon}_{0n} =$$

$$\sum_{c=1}^{N_c}\sum_{n=1}^{N}\sum_{m=1}^{M}\sum_{k=1}^{K} N_n(\xi_m,\eta_k)\cdot\Sigma(P_i,\bar{Q}(\xi_m,\eta_k))\cdot J(\xi_m,\eta_k)\cdot W_m\cdot W_k\cdot\boldsymbol{\varepsilon}_{0n} \tag{13.72}$$

The implementation therefore involves 5 Do loops: One over collocation points i, number of cells N_c, over cell nodes N and over the Gauss points m and k. In the innermost DO loop there is a matrix vector product. Note that in the Subroutine it is assumed that the cells are sufficiently far away from the collocation points so that a constant number of Gauss points is sufficient. A listing in Subroutine Body force is given below.

```
SUBROUTINE Body_force(F,CDim,xP,NCol,Isym,E,ny)
!---------------------------------------------------------------
!  Adds contribution of body force terms (initial strain)
!  to the right hand side vector F
!  This implementation is only for linear cells and plane
!  problems
!---------------------------------------------------------------
USE Utility_Lib; USE Elast_lib; USE Integration_lib
IMPLICIT NONE
INTEGER , INTENT(IN)          :: CDim
REAL , INTENT(IN)             :: E
REAL , INTENT(IN)             :: ny
INTEGER , INTENT(IN)          :: NCol
INTEGER , INTENT(IN)          :: Isym
REAL , INTENT  (IN)           :: xP(Cdim,Ncol)
REAL(KIND=8), INTENT(INOUT) :: F(CDim*Ncol)    !  right hand side
INTEGER, ALLOCATABLE :: InciC(:,:)   !  Cell Incidences
INTEGER, ALLOCATABLE :: Inci(:)
REAL, ALLOCATABLE      :: xPC(:,:)     !  Cell Node co-ordinates
REAL, ALLOCATABLE      :: Ni(:),Elcor(:,:)
INTEGER    :: NodesC,Ncells,NodelC,ldimC,IOS
INTEGER    :: m,n,k,Node,Nc,i,ii,jj,Mi,Ki,iD
REAL       :: Eps0(2),SigK(Cdim,Cdim),GCcor(3)
REAL       :: Glcorx(8),Glcore(8),Wix(8),Wie(8),Vnorm(3)
REAL       :: Jac,Weit,xsi,eta,r,dxr(Cdim)
IF(Cdim > 2) RETURN   ! This coding is for plane problems only
IF(ISym > 0) RETURN   ! Symmetry not considered here
```

```
READ(1,*,IOSTAT=IOS) NodesC
IF(IOS /= 0) RETURN    !   No body force effects
Write(2,*) 'Number of cell nodes=',NodesC
READ(1,*) Ncells
Write(2,*) 'Number of cells=',Ncells
READ(1,*) Eps0
Write(2,*) 'Eps0=',Eps0
NodelC=4          !  only linear elements considered
ldimC= 2          !  plane cells
ALLOCATE(xPC(Cdim,NodesC))    !  Array for node coordinates
ALLOCATE(InciC(Ncells,NodelC),Inci(NodelC))
ALLOCATE(Ni(nodelC),ELCOR(Cdim+1,nodelC))
!--------------------------------------------------------
!  Read Cell Node Co-ordinates
!--------------------------------------------------------
DO Node=1,NodesC
 READ(1,*) (xPC(M,Node),M=1,Cdim)
 WRITE(2,'(A5,I5,A8,3F8.2)') 'Node ',Node,&
          '  Coor   ',(xPC(M,Node),M=1,Cdim)
END DO
!--------------------------------------------------------
!  Read Cell Incidences
!--------------------------------------------------------
WRITE(2,*)''
WRITE(2,*)'Incidences: '
WRITE(2,*)''
DO Nc=1,Ncells
  READ(1,*) (InciC(Nc,n),n=1,NodelC)
  WRITE(2,'(A3,I5,A8,4I5)')'EL ',Nc,'  Inci  ',InciC(Nc,:)
END DO
!---------------------------------------------------------
!   compute contribution to right hand side
!--------------------------------------------------------
Colloc_points: DO i=1,Ncol
 Cells: DO nc=1,Ncells
  Mi=4 ;  Ki=4  !  cell is sufficiently far away from Pi
  Call Gauss_coor(Glcorx,Wix,Mi)
  Call Gauss_coor(Glcore,Wie,Ki)
  Elcor(1:2,:)= xPC(1:2,InciC(nc,:))
  Elcor(3,:)= 0.0 ! we are using 2-D boundary element as a cell
  Inci=InciC(nc,:)
  Gauss_points_xsi: DO m=1,Mi
   xsi= Glcorx(m)
   Gauss_points_eta: DO k=1,Ki
    eta= Glcore(k)
    CALL Serendip_func(Ni,xsi,eta,ldimC,nodelC,Inci)
    Call Normal_Jac(Vnorm,Jac,xsi,eta,ldimC,nodelC,Inci,elcor)
    Weit= Wix(m)*Wie(k)*Jac
    CALL Cartesian(GCcor,Ni,ldimC,elcor)
    r= Dist(GCcor(1:2),xP(:,i),Cdim)
    dxr= (GCcor(1:2)-xP(1:2,i))/r
```

```
     SigK= SigmaK(dxr,r,E,ny,Cdim)
     Direction_P: DO ii=1,Cdim
      iD= Cdim*(i-1) + ii      !  Position in F array
      Direction_Q: DO jj=1,Cdim
        Node_points: DO n=1,NodelC
          F(iD)= F(iD) + Ni(n)*SigK(ii,jj)*Weit*Eps0(jj)
        END DO Node_points
      END DO Direction_Q
     END DO Direction_P
    END DO Gauss_points_eta
  END DO Gauss_points_xsi
END DO Cells
END DO Colloc_points
DEALLOCATE(xPC)
DEALLOCATE(InciC)
DEALLOCATE(Ni,Elcor)
RETURN
END Subroutine Body_force
```

In Program General_purpose_BEM we have to insert a call to Body_force as shown

```
....
END DO &
Elements_1
CALL Body_force(F,CDim,xP,Nodes,Isym,E,ny)
!-------------------------------------------------------------
!  Add azimuthal integral for infinite regions
!-------------------------------------------------------------
...
```

13.8.1 Input data specification for Body_force

The input data for the cell mesh and the specification of the initial strain is supplied via file INPUT. If no body force effects are present **SUBROUTINE Body_force** returns immediately. The geometry of the cell and the sequence of the input of node numbers is shown in Figure 13.7.

INPUT DATA SPECIFICATION FOR Body_force

1.0	Node specification	
	Nodes	Number of cell nodes
2.0	Element specification	
	Ncells	Number of cells
3.0	Initial strain specification	
	Eps0(1:2)	Initial strains $\varepsilon_{x0}, \varepsilon_{y0}$

4.0 Loop over nodes
x, y, (z) Node coordinates
5.0 Loop over all cells
Inci (1:4) Incidences of cells

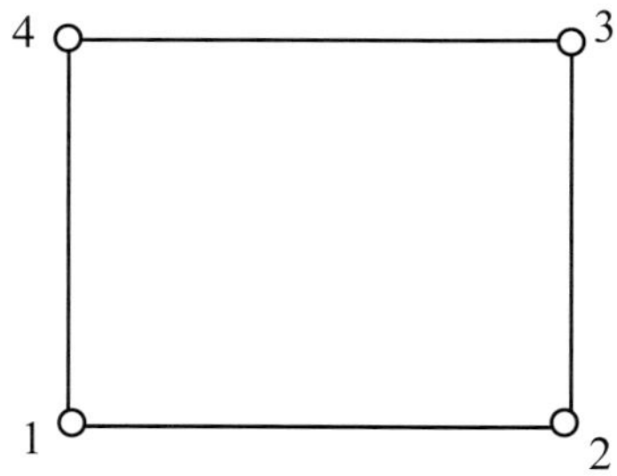

Figure 13.7 Linear plane cell

13.9 SAMPLE INPUT FILE AND RESULTS

To test the Subroutine a small example in included here. It is an example of a circular opening in an infinite domain, where part of the domain is subjected to swelling. The effect of the swelling on the deformations at the boundary of the hole is required. The material properties for the domain are assumed to be E= 1000.0 and ν= 0.0. The swelling zone is assumed to be subjected to an initial strain of 0.1 in the vertical direction. The mesh with quadratic boundary elements and linear cells is shown in Figure 13.8. Note that the numbering for the cells is completely separated from the numbering of the boundary elements so we can start with number 1.

The file INPUT is

```
Circular hole with swelling
 2     ! Cdim
 2     ! Ndof
 1     ! Toa Plane strain
 2     ! Nreg   infinite eregion
 0     ! ISym  no symetry
 2     ! Ltyp quads
0.1000E+05  0.0000E+00
    8    ! Nodes
    4    ! Elements
    1.000     1.000
    0.707     0.707
    1.000     0.000
    0.707    -0.707
    0.000    -1.000
```

```
   -0.707    -0.707
   -1.000     0.000
   -0.707     0.707
    1    3    2
    3    5    4
    5    7    6
    7    1    8
    0
    0
    6   ! Cell Nodes
    2   ! Cells
0.0 0.1  !   vertical strain
   -1.000     2.000
    0.000     2.000
    1.000     2.000
    1.000     2.500
    0.000     2.500
   -1.000     2.500
    1    2    5    6
    2    3    4    5
```

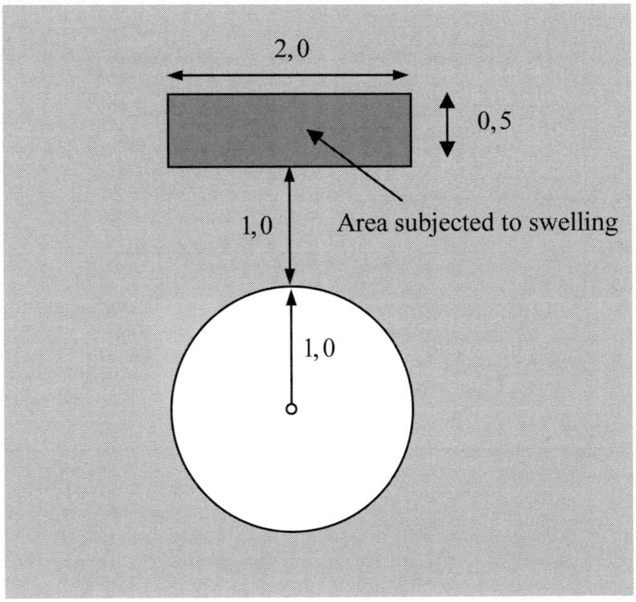

Figure 13.8 Problem specification

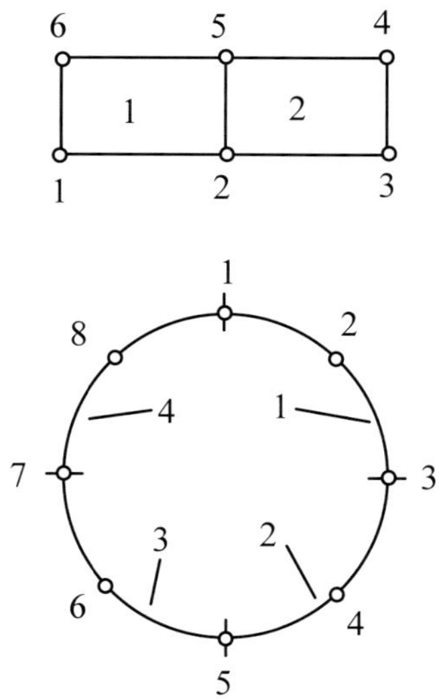

Figure 13.9 Mesh used

The output obtained from the program is

```
Project:
 Circular hole with swelling
 Cartesian_dimension:           2
 Elasticity Problem
 Type of Analysis: Solid Plane Strain
 Infinite Region
 No symmetry
 Quadratic Elements
 Modulus:    10000.00
 Poissons ratio:  0.0000000E+00
 Number of Nodes of System:           8
 Number of Elements of System:           4
Node      1  Coor      1.00     1.00
Node      2  Coor      0.71     0.71
Node      3  Coor      1.00     0.00
Node      4  Coor      0.71    -0.71
Node      5  Coor      0.00    -1.00
Node      6  Coor     -0.71    -0.71
Node      7  Coor     -1.00     0.00
Node      8  Coor     -0.71     0.71
 Incidences:
EL     1  Inci       1    3    2
EL     2  Inci       3    5    4
EL     3  Inci       5    7    6
EL     4  Inci       7    1    8
```

```
 Elements with Dirichlet BC´s:
 Elements with Neuman BC´s:
 Number of cell nodes=          6
 Number of cells=          2
 Eps0=  0.0000000E+00  0.1000000
Node      1  Coor      -1.00     2.00
Node      2  Coor       0.00     2.00
Node      3  Coor       1.00     2.00
Node      4  Coor       1.00     2.50
Node      5  Coor       0.00     2.50
Node      6  Coor      -1.00     2.50
 Incidences:
EL      1  Inci       1    2    5    6
EL      2  Inci       2    3    4    5
Results, Element           1
 u=-0.471E-02-0.245E-01-0.171E-02-0.195E-01-0.323E-02-0.257E-01
 t=      0.000     0.000     0.000     0.000     0.000     0.000
 Results, Element          2
 u=-0.171E-02-0.195E-01 0.612E-03-0.120E-01-0.397E-03-0.142E-01
 t=      0.000     0.000     0.000     0.000     0.000     0.000
 Results, Element          3
 u= 0.612E-03-0.120E-01-0.798E-03-0.156E-01 0.802E-03-0.123E-01
 t=      0.000     0.000     0.000     0.000     0.000     0.000
 Results, Element          4
 u=-0.798E-03-0.156E-01-0.471E-02-0.245E-01-0.104E-02-0.251E-01
 t=      0.000     0.000     0.000     0.000     0.000     0.000
```

13.10 CONCLUSIONS

In this chapter we have dealt with the treatment of various types of effects occurring inside the domain, where no boundary elements exist. We have loosely called these effects body forces, even though some of these, for example the initial strains were not forces at all. With the capability to deal with these effects the range of application of the BEM has been expanded and we have also laid the foundations for the chapter that deals with plasticity. In the solution of material non-linear problems we can visualise the redistribution of stresses due to plasticity, as body forces which are generated once plasticity occurs.

The effect of body forces is essentially an additional right hand side that is generated in the system of equations. If the body forces are constant, then this term can be computed as a surface integral using numerical integration. Otherwise, a mesh of cells has to be created in order to enable a volume integration to be carried out. Those critical of the BEM might suggest that the main attraction of the method, surface instead of volume discretisation, would be lost. However this is not the case. Cells are only needed where internal loading occurs and there are no additional degrees of freedom associated with the nodes of a cell mesh. This is demonstrated by the example shown where cells were only provided in the swelling zone. An equivalent finite element discretisation would have to cover the entire domain, including the part where no swelling occurs. Also note that the number of unknowns at the nodes of the boundary elements was not increased by the fact that cells were required to compute the right hand side of the system of equations.

The implementation of body forces requires additional fundamental solutions which have been added to the library. It is obvious that volumetric loading effects also occur in potential problems but the discussion of these in more detail is beyond the scope of this book.

We have only shown one example of implementation: the treatment of initial strains as they may occur in problems where part of the domain, is subjected to swelling. However with the knowledge of programming gained in previous chapters and the explanation of the theory in this chapter the reader should be able to perform the implementation of the other types of body forces into program General_purpose_BEM.

13.11 EXERCISES

Exercise 13.1

Write a Subroutine similar to **SUBROUTINE Body_force** that computes the right hand side for gravity as discussed in section 13.2. Implement this into program **General_Purpose_BEM** and test on an example of a cube subjected to self weight.

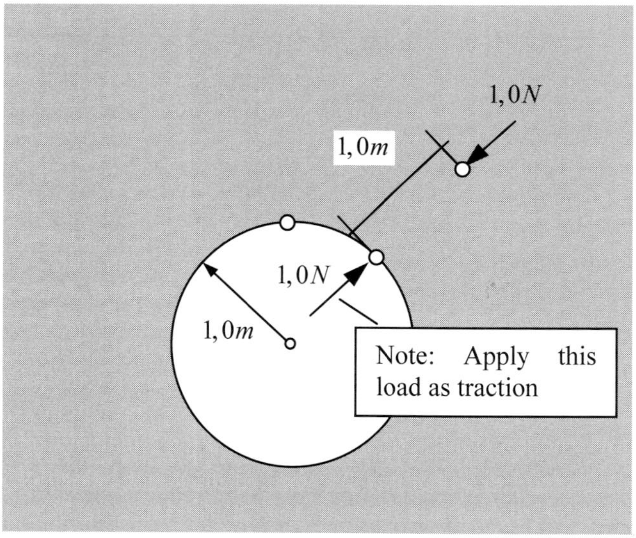

Figure 13.10 Test example for exercise 13.1

Exercise 13.1
Write a Subroutine similar to SUBROUTINE Body_force that computes the right hand side for concentrated forces in the domain. Test on the example in Figure 13.10 of a hole in an infinite domain subjected to a pre-stressing force of a rock bolt.

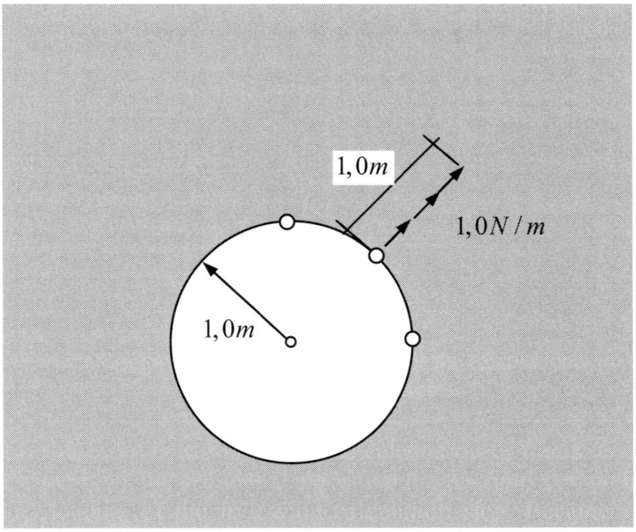

Figure 13.11 Test example for Exercise 13.2

Exercise 13.2
Write a Subroutine similar to SUBROUTINE Body_force that computes the right hand side for distributed line forces in the domain. Test on the example in Figure 13.10 of a hole in an infinite domain subjected to a distributed force (for example of a grouted rock bolt).

13.12 REFERENCES

1. Brebbia A., Telles J.C.F and Wrobel, L.C. (1984) Boundary Element Techniques. Springer Verlag, NewYork
2. Banerjee P.K. The Boundary Element Methods in Engineering, McGraw-Hill, London.

14

Dynamic Analysis

Do not worry about your difficulties in Mathematics,
I can assure you mine are still greater.
Albert Einstein

14.1 INTRODUCTION

So far we have discussed problems that are independent of time and neglected inertial effects. It is therefore appropriate to extend the discussion of the theory and implementation to problems in dynamics. We have already seen that the Boundary Element Method has distinct advantages over the Finite Element Method for static problems involving infinite domains. This advantage is even more pronounced for dynamic problems since we will see that the fundamental solutions used in the BEM implicitly fulfil the radiation conditions. It is known that the FEM which requires the truncation of the mesh has the problem that waves may be reflected at truncation boundaries. In this Chapter we will only give an overview of the implementation of dynamic problems. For more details the reader is referred to relevant publications[1,2,3].

In dynamics we distinguish between cases where the field variables are dependent on time in a harmonic or general way. A field variable at a point Q which depends on time in a harmonic way can be expressed as

$$u(Q,t) = a(Q)(\sin \omega\, t + i \cos \omega\, t) \tag{14.1}$$

where $i = \sqrt{-1}$, $a(Q)$ is the amplitude at point Q and ω is the frequency. The use of complex numbers allows compact representation of sinusoidal and cosine effects. Equation (14.1) can also be written as

$$u(Q,t) = a(Q)e^{-i\omega t} \tag{14.2}$$

The solution for these types of problems can be carried out in the "frequency domain" reducing the problem to the determination of the time-independent variable $a(Q)$ at frequency ω.

If the field variables are not harmonic then the solution has to proceed in the "time domain" also known as a transient problem.

A "frequency domain" analysis may be motivated by:

- Analysing the response due to steady state excitations of the type $p(Q',t) = p_0(Q') \cdot e^{-i\omega t}$ assuming that a steady state has been reached, i.e. that the influence of transient effects has become negligible.
- Transformation of a problem involving non-harmonic excitation from the time domain into the Laplace domain. We can also use a Fourier transform of the type

$$u(Q,t) = 1/\sqrt{2\pi} \int_{\mathbb{R}} \hat{a}(Q,\omega) \cdot e^{-i\omega t} d\mathbb{R} \tag{14.3}$$

For each value of ω the transformed unknown $\hat{a}(Q,\omega)$ can be calculated for discrete values of ω. The advantage of this is that a greater range of fundamental solutions is available in the frequency domain.

In this Chapter we will first start with the simplest problem, namely the scalar wave equation in the frequency domain and then proceed to the time domain. Next, general dynamic problems are discussed. This follows the philosophy of the book where potential problems with one degree of freedom where introduced first for static problems.

14.2 SCALAR WAVE EQUATION, FREQUENCY DOMAIN

The scalar wave equation governs many physical phenomena for example the transverse motions of membranes or the propagation of pressure in an acoustic fluid. For a time harmonic problem the governing differential equation is given by

$$\frac{\partial^2 u}{\partial x^2} + \frac{\partial^2 u}{\partial y^2} + \frac{\partial^2 u}{\partial z^2} + k^2 u + F = 0 \tag{14.4}$$

or

$$\Delta u(Q,\omega) + k^2 u(Q,\omega) + F(Q,\omega) = 0 \tag{14.5}$$

where in the case of propagation in a fluid u is the pressure amplitude and $k = \omega / c$, also known as the wave number. c denotes the wave velocity which is a material constant and

F is the body source distribution. The Equation (14.4) is also known as the *Helmholz* equation.

14.2.1 Fundamental solutions

The fundamental solution for the pressure amplitude $U(P,Q,\omega)$ is for a point source of unit amplitude located at point P, i.e. the solution of:

$$\Delta U + k^2 U + \delta(P-Q) = 0 \tag{14.6}$$

where $\delta(P-Q)$ is the Dirac delta function. The Dirac delta function is defined as

$$\begin{aligned} &\delta(P-Q) = 0 \quad \text{when} \quad P \neq Q \\ &\int_\Omega \delta(P-Q)\,d\Omega = 1 \end{aligned} \tag{14.7}$$

The fundamental solution for the pressure at Q is given by

$$U(P,Q,\omega) = \frac{e^{ikr}}{4\pi r} \tag{14.8}$$

where r is the distance between P and Q.

Note that for the static case this reverts to $U(P,Q) = 1/4\pi r$. The derivatives of the pressure are given by

$$U_{,j} = -\frac{r_{,j}}{4\pi r^2}(1-ikr)\cdot e^{ikr} \tag{14.9}$$

where x,y,z may be substituted for j and $r_{,j}$ has been defined in Chapter 4. The derivative of U in the direction $\mathbf{n}$, here referred to as T, is given by

$$T = U_{,x}\cdot n_x + U_{,y}\cdot n_y + U_{,z}\cdot n_z = -\frac{\cos\theta}{4\pi r^2}(1-ikr)\cdot e^{ikr} \tag{14.10}$$

where $\cos\theta$ has been defined in Chapter 4. If we integrate the three-dimensional solution over $-\infty \leq z \leq \infty$ we obtain the solution for the plane problem as

$$U = \frac{i}{4}H_0^1(kr) \quad ; \quad U_{,j} = \frac{ik}{4}H_1^1(kr) \tag{14.11}$$

where H_0^1 and H_1^1 are Hankel functions of the first kind and order 0 and 1, respectively.

The Hankel functions can be expanded to[4]

$$H_0^1(x) = 1 + \frac{2i}{\pi} \cdot (\gamma + \ln\frac{x}{2}) + \ldots$$
$$H_1^1(x) = \frac{2i}{\pi x} + \frac{ix}{\pi} \cdot (\gamma - \frac{1}{2} + \ln\frac{x}{2}) + \frac{x}{2} + \ldots \qquad (14.12)$$

where $\gamma = 0.57721$ (Euler constant) and here only the first terms of the expansion are shown. A plot of the fundamental solution for U is shown in Figure 14.1.

14.2.2 Boundary integral equations

As with the static system the reciprocal theorem of Betti can be used to obtain the integral equation. In the absence of body sources (distributed over the domain) we obtain

$$\hat{c}u(P) = \int_S U(P,Q,\omega) \cdot q(Q,\omega) dS - \int_S T(P,Q,\omega) \cdot u(Q,\omega) dS \qquad (14.13)$$

where the pressure gradient is defined as $q = \frac{\partial u}{\partial n}$ and $\hat{c}$ depends on the boundary shape.

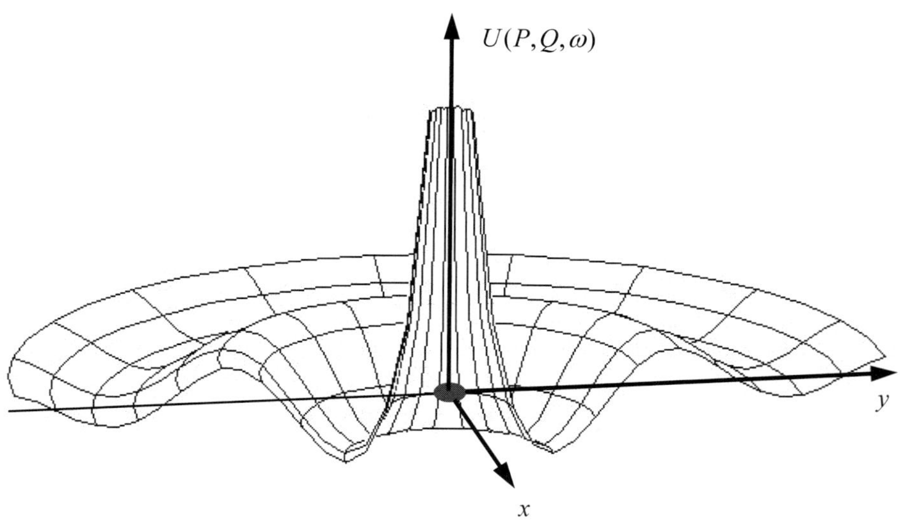

Figure 14.1 Plot of the fundamental solution U for real part and $k = \omega / c = 4$

14.2.3 Numerical implementation

In a well posed boundary value problem either u or q is specified on the boundary. Furthermore the integral equation (14.13) must be satisfied for any source point P. If we ensure the satisfaction at a discrete number of points P_i then we can get as many equations that are necessary to compute all the unknowns (point collocation).

The solution of (14.13) can be achieved by discretisation of the boundary of the problem as for the static case. We introduce the interpolations

$$u(Q,\omega) = \sum_{j=1}^{J} N_j \cdot u_j^e(\omega) \quad ; \quad q(Q,\omega) = \sum_{j=1}^{J} N_j \cdot q_j^e(\omega) \tag{14.14}$$

where J is the number of element nodes, N_j are shape functions introduced in Chapter 3 and u_j^e, q_j^e are nodal values of u and q at element e.

The discretised form of equation (14.13) is

$$\hat{c}u\left(P_i\right) + \sum_{e=1}^{E}\sum_{j=1}^{J} \Delta T_{ji}^e \, u_j^e = \sum_{e=1}^{E}\sum_{j=1}^{J} \Delta U_{ji}^e \, q_j^e \tag{14.15}$$

where E is the number of elements and

$$\Delta U_{ji}^e = \int_{S_e} N_j \, U\left(P_i, Q, \omega\right) dS\left(Q\right) \quad , \quad \Delta T_{ji}^e = \int_{S_e} N_j \, T\left(P_i, Q, \omega\right) dS\left(Q\right) \tag{14.16}$$

The integration over the boundary elements can be carried out using Gauss Quadrature as for the static case. However all variables must be declared COMPLEX type in the program. An example of programming the fundamental solution is given below. The complex function `Hankel0` may be obtained from mathematical libraries or may be programmed using the approximation given here (however, note that only a very limited number of terms are considered).

```
COMPLEX FUNCTION UW(r,k,Cdim)
!------------------------------
!   Fundamental solution for scalar wave equation
!   Pressure
!-----------------------------
USE Hankel
REAL,INTENT(IN)       ::  r   !   Distance between P and Q
REAL,INTENT(IN)       ::  k   !   wave number
INTEGER,INTENT(IN)    :: Cdim    !   Cartesian dimension (2-D,3-D)
REAL                  ::  Pi
COMPLEX :: C0,Hankel0
Pi= 3.141592654
SELECT CASE (CDIM)
   CASE (2)               !  Two-dimensional solution
```

```
      C0= CMPLX(0,0.25)
      UW= C0*Hankel0(k* r)
    CASE (3)              !  Three-dimensional solution
      C0= CMPLX(0,k* r)
      UW= 1/(4.0* k* r* Pi)*EXP(C0)
    CASE DEFAULT
      UW=0.0
      WRITE (11,*)'Cdim not equal 2 or 3 in Function UW(...)'
  END SELECT
  END FUNCTION UW
```

14.3 SCALAR WAVE EQUATION, TIME DOMAIN

For the case where $u=u(Q,t)$ is not harmonic but transient, the scalar wave equation is given by

$$\Delta u - \frac{1}{c^2}\ddot{u} + F = 0 \tag{14.17}$$

where overdots mean differentiation with respect to time t, c is the wave velocity and F is a specified body source. The assumption is of an isotropic and homogeneous medium. For a well posed problem we must have initial and boundary conditions. The initial conditions at time 0 are specified as

$$u(Q,0) = u_0(Q) \quad ; \quad \dot{u}(Q,0) = v_0(Q) \tag{14.18}$$

The condition $u_0(Q) = v_0(Q) = 0$ is termed *initial rest* or *quiescent past*,

14.3.1 Fundamental solutions

A fundamental solution of the differential equation can be found by assuming an impulsive point source at P applied at time $t=\tau$, in an infinite domain. Therefore we seek the solution of

$$\Delta U - \frac{1}{c^2}\ddot{U} + \delta(P-Q)\delta(t-\tau) = 0 \tag{14.19}$$

where a Dirac Delta function has been introduced for the time and space. The Dirac Delta function for the space has been discussed previously; the one for the time is defined as

$$\begin{aligned} &\delta(t-\tau) = 0 \quad \text{when} \quad t \neq \tau \\ &\int_{-\infty}^{\infty} \delta(t-\tau)d\tau = 1 \end{aligned} \tag{14.20}$$

The 3-D fundamental solution for the pressure U at point Q at time t due to an impulsive source at point P at time τ is given by

$$U(P,\tau,Q,t)=\frac{1}{4\pi r}\delta(t-\tau-r/c) \tag{14.21}$$

The gradient of the pressure, T, in direction $\mathbf{n}$ is given by

$$T(P,\tau,Q,t)=\frac{\cos\theta}{4\pi r}\left\{-\frac{1}{r}\delta(t-\tau-r/c)+\frac{1}{c}\dot{\delta}(t-\tau-r/c)\right\} \tag{14.22}$$

.

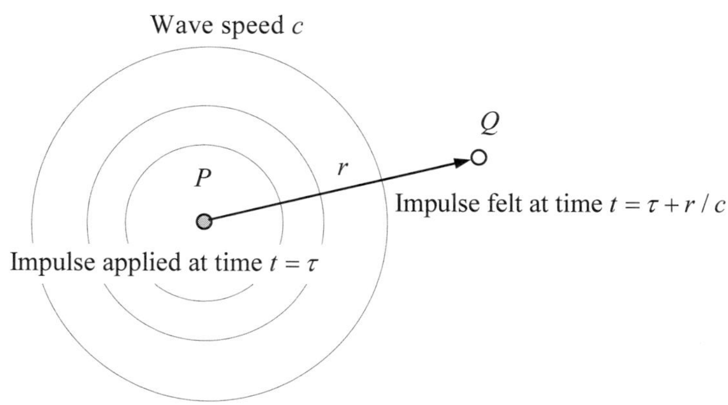

Figure 14.2 Diagram explaining causality

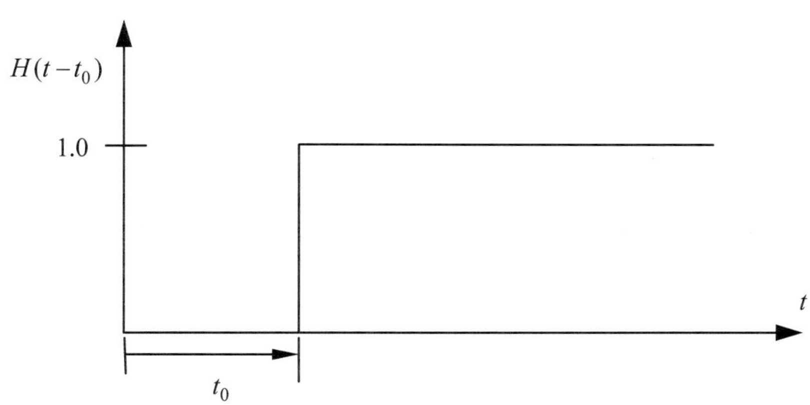

Figure 14.3 Heaviside function

The interpretation of the fundamental solution U is as follows (Figure 14.2): If an impulse is applied at P then the wave takes some time to reach point Q (this is also known as *causality*). This time depends on the speed of the wave c and the distance (r) of the point Q from P, and is computed as r/c. Therefore point Q will only feel the impulse after a time lag of r/c.

Integration of (14.21) in the z-direction gives the plane solution as

$$U(P,\tau,Q,t) = \frac{c}{2\pi} \cdot \frac{1}{c^2(t-\tau)^2 - r^2} H(t-\tau-r/c) \tag{14.23}$$

and

$$T(P,\tau,Q,t) = \frac{\cos\theta}{2\pi} \frac{rc}{\left(c^2(t-\tau)^2 - r^2\right)^{3/2}} \cdot H\left(t-\tau-r/c\right) \tag{14.24}$$

where H is the Heaviside function defined as (Figure 14.3):

$$\begin{aligned} H(t-t_0) &= 0 \quad \text{for} \quad t < t_0 \\ H(t-t_0) &= 1 \quad \text{for} \quad t \geq t_0 \end{aligned} \tag{14.25}$$

14.3.2 Boundary integral equations

Before we proceed with deriving the integral equations the concept of the convolution integral, which is commonly used in structural dynamics[5], is explained on a system with one degree of freedom. The system is subjected to a transient load $P(t)$ and we want to determine the response of the system due to this loading, $u(t)$. We divide the transient loading $P(t)$ into a sequence of impulses of magnitude $P(t)d\tau$ (Figure 14.4a). The response of the system to one such impulse can be written as

$$du(t) = P(\tau)d\tau \cdot h(t-\tau) \tag{14.26}$$

where $h(t-\tau)$ is the response due to a unit value of $P(t)$ (Figure 14.5).

If we integrate all the impulses over time t, the response due to the given loading $P(t)$ is obtained as (Figure 14.4b)

$$u(t) = \int_0^t P(\tau) \cdot h(t-\tau)d\tau \tag{14.27}$$

This is also known as the *Duhamel* integral equation.

The *time convolution*[*] *integral* can also be written as

$$u(t) = P(\tau) * h(t - \tau) \tag{14.28}$$

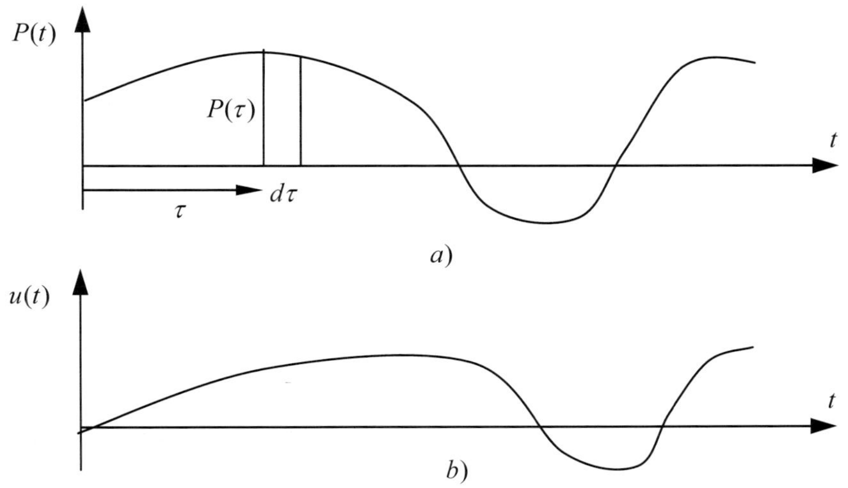

Figure 14.4 a) transient load and b) response of the system

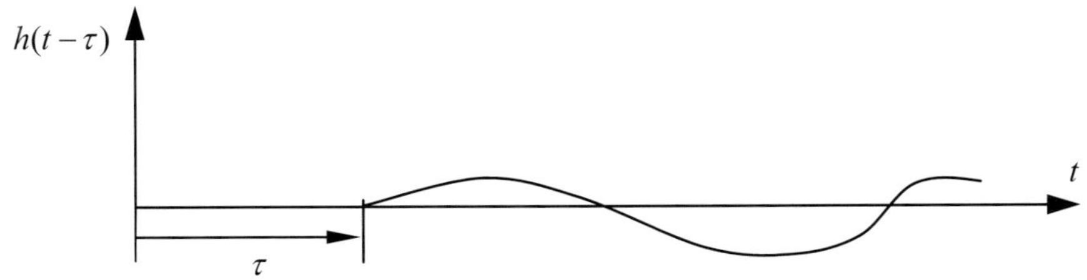

Figure 14.5 Response of system to a unit impulse applied at time τ

[*] The word *convolution* is a term that has been coined by mathematicians. The term is akin to the word "folding" which is actually the term used in German (*Faltung*). It refers to the time integration of a product.

The integral can be approximated numerically using the *Convolution Quadrature Method* (CQM) first introduced by Lubich[6]

$$P(\tau) * h(t-\tau) \approx \sum_{k=0}^{n} w_{n-k}\left(\hat{h}(s), \Delta t\right) P(k\Delta t) \tag{14.29}$$

where Δt is a time step so that $t = n\Delta t$, $\hat{h}(s)$ is the Laplace transform[7] of h and w_n are the convolution quadrature weights.

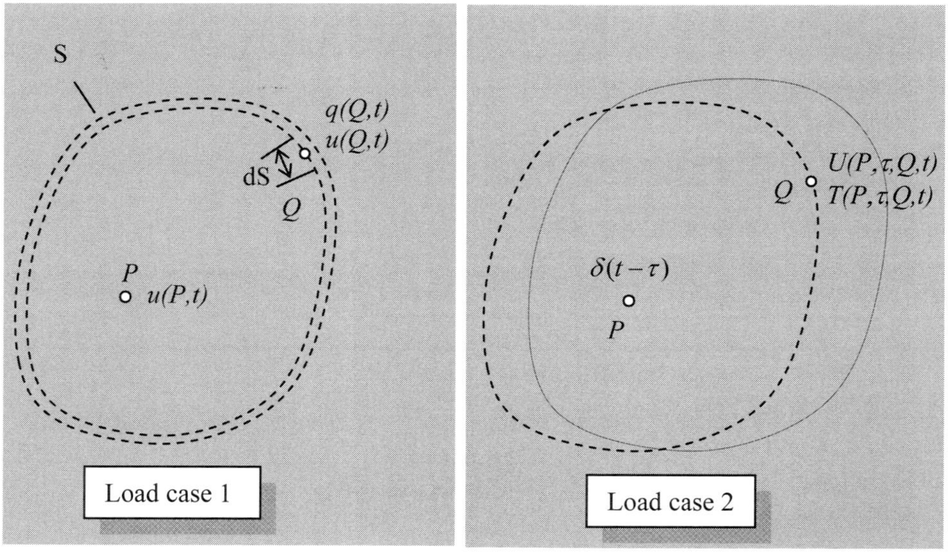

Figure 14.6 Load cases considered for the derivation of the integral equation

The reciprocal theorem in dynamics specifies a relationship between two dynamic states. It is an extension of the reciprocal theorem by Betti (a rigorous proof is given by Wheeler[8]). We apply the reciprocal theorem to the scalar wave problem and two distinct dynamical "load" cases. The first load case is the one, the solution of which we want to obtain at a time instant t. The second load case is the case where a unit impulse is applied at time τ (Figure 14.6). The reciprocal theorem gives

$$\int_{\tau=0}^{t} \delta(t-\tau) u(P,\tau) d\tau + \int_{\tau=0}^{t} \int_{S} T(P,\tau,Q,t) u(Q,\tau) \cdot dS d\tau = \int_{\tau=0}^{t} \int_{S} U(P,\tau,Q,t) q(Q,\tau) \cdot dS d\tau \tag{14.30}$$

or rearranging and introducing the above notation for the time integrals

$$u(P,t) = \int_S [U(P,\tau,Q,t) * q(Q,\tau) - T(P,\tau,Q,t) * u(Q,\tau)] \cdot dS \tag{14.31}$$

where

$$\begin{aligned} U(P,\tau,Q,t) * q(Q,t) &= \int_{\tau=0}^{t} U(P,\tau,Q,t) q(Q,\tau) d\tau \\ T(P,\tau,Q,t) * u(Q,t) &= \int_{\tau=0}^{t} T(P,\tau,Q,t) u(Q,\tau) d\tau \end{aligned} \tag{14.32}$$

Taking the limiting value as P approaches the boundary

$$\hat{c}u(P,t) = \int_S [U(P,\tau,Q,t) * q(Q,\tau) - T(P,\tau,Q,t) * u(Q,\tau)] \cdot dS \tag{14.33}$$

where $\hat{c}$ is the jump term arising from taking the limit as Q approaches the boundary.

14.3.3 Numerical implementation

For the solution of the integral equation we have to discretise the problem in space as for the static case. Since the numerical integration using the CQM previously introduced is quite complex because it involves a Laplace transform we propose an alternative approach.

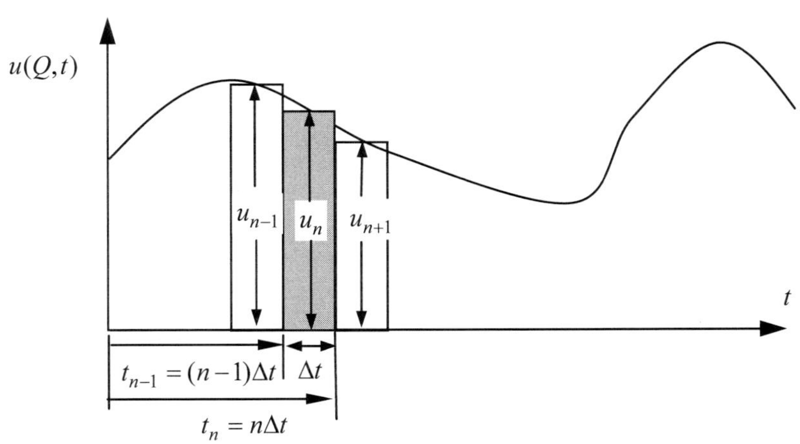

Figure 14.7 Discretisation in time with a constant shape function

We discretise the total time into arbitrary small steps of size Δt, then we have

$$u(Q,t)=\sum_{n=1}^{N}\tilde{N}_n(t)\cdot u_n(Q) \quad ; \quad q(Q,t)=\sum_{n=1}^{N}\tilde{N}_n(t)\cdot q_n(Q) \tag{14.34}$$

where $\tilde{N}_n(t)$ are shape functions in time and u_n and q_n are the pressure and pressure gradient at time step n (at time $t_n = n\Delta t$). If we assume the variation of u and q to be constant within one time step Δt, then the convolution integrals may be evaluated analytically. In this case the shape functions are

$$\tilde{N}_n(t)=H\left(t-t_{n-1}\right)-H(t-t_n) \tag{14.35}$$

where H is the Heaviside function. The time interpolation is shown in Figure 14.7.

Substituting (14.34) into (14.33) we obtain the integral equation discretised in time and written for the time t_N (time step N):

$$\hat{c}u_N(P)=\int_S [U(P,\tau,Q,t_N)*q(Q,\tau)-T(P,\tau,Q,t_N)*u(Q,\tau)]\cdot dS \tag{14.36}$$

The convolution integrals are approximated by

$$U(P,\tau,Q,t_N)*q(Q,\tau)\approx\sum_{n=1}^{N}q_n(Q)\cdot\Delta U_{Nn} \tag{14.37}$$

and

$$T(P,\tau,Q,t_N)*u(Q,\tau)\approx\sum_{n=1}^{N}u_n(Q)\cdot\Delta T_{Nn} \tag{14.38}$$

where

$$\Delta U_{Nn}=\int_{t_{n-1}}^{t_n}U(P,\tau,Q,t_N)d\tau \quad ; \quad \Delta T_{Nn}=\int_{t_{n-1}}^{t_n}T(P,\tau,Q,t_N)d\tau \tag{14.39}$$

This means that only the fundamental solutions are inside the integrals and these may be integrated analytically[3].

The time discretised integral equation now becomes

$$\hat{c}u_N(P)=\int_S\sum_{n=1}^{N}\Delta U_{Nn}\cdot q_n(Q)\cdot dS(Q)-\int_S\sum_{n=1}^{N}\Delta T_{Nn}\cdot u_n(Q)\cdot dS(Q) \tag{14.40}$$

or taking the sum outside the integral

$$\hat{c}u_N(P) = \sum_{n=1}^{N} \int_S \Delta U_{Nn} \cdot q_n(Q) \cdot dS(Q) - \sum_{n=1}^{N} \int_S \Delta T_{Nn} \cdot u_n(Q) \cdot dS(Q) \qquad (14.41)$$

For each time step N we get an integral equation. In a well posed boundary value problem either u or q is specified on the boundary and the values of u and q are known at the beginning of the analysis (t=0). Furthermore the integral equation (11.41) must be satisfied for any source point P. If we ensure the satisfaction at a discrete number of points P_i then we can get for each time step N as many equations that are necessary to compute the unknowns. Similar to static problems we specify the points P_i to be the node points of the boundary element mesh (point collocation). To solve the integral equation we introduce the discretisation in space of Chapter 3:

$$u_n(Q) = \sum_{j=1}^{J} N_j \cdot u_{nj}^e \quad ; \quad q_n(Q) = \sum_{j=1}^{J} N_j \cdot q_{nj}^e \qquad (14.42)$$

where u_n, q_n are pressure and pressure gradients at Q; u_{nj}^e, q_{nj}^e refer to values of u and q at node j of element e at time step n and N_j are shape functions. Substitution of (14.42) into (14.41) gives

$$\hat{c}u_N(P_i) = \sum_{n=1}^{N} \sum_{e=1}^{E} \sum_{j=1}^{J} \Delta U_{ijNn}^e \cdot q_{nj}^e - \sum_{n=1}^{N} \sum_{e=1}^{E} \sum_{j=1}^{J} \Delta T_{ijNn}^e \cdot u_{nj}^e \qquad (14.43)$$

where

$$\Delta U_{ijNn}^e = \int_{S_e} \Delta U_{Nn}(P_i) \cdot N_j \cdot J \cdot dS \qquad (14.44)$$

and

$$\Delta T_{ijNn}^e = \int_{S_e} \Delta T_{Nn}(P_i) \cdot N_j \cdot J \cdot dS \qquad (14.45)$$

J is the Jacobian and E is the number of Elements.

If we define vectors $\{u\}_n$ and $\{q\}_n$ to contain all nodal values of pressure and pressure gradient at the nodes at time increment N we can rewrite Equation (14.43) in matrix form

$$\sum_{n=1}^{N} [T]_n \{u\}_n = \sum_{n=1}^{N} [U]_n \{q\}_n \qquad (14.46)$$

If we solve for time step N, the results for the previous time steps are known and can be put to the right hand side:

$$[T]_N\{u\}_N=[U]_N\{q\}_N-\sum_{n=1}^{N-1}[T]_n\{u\}_n+\sum_{n=1}^{N-1}[U]_n\{q\}_n \tag{14.47}$$

or

$$[T]_N\{u\}_N=[U]_N\{q\}_N+\{F\} \tag{14.48}$$

where the vector $\{F\}$ contains the effect of the time history. The coefficients of $\{F\}$ are

$$F_i=\sum_{n=1}^{N-1}\sum_{e=1}^{E}\sum_{j=1}^{J}\Delta U_{ijNn}^{e}\cdot q_{nj}-\sum_{n=1}^{N-1}\sum_{e=1}^{E}\sum_{j=1}^{J}\Delta T_{ijNn}^{e}\cdot u_{nj} \tag{14.49}$$

14.4 ELASTODYNAMICS

We now turn our attention to general problems in elasticity. The differential equation for dynamics in the frequency domain can be written in matrix form as:

$$(\lambda+G)\cdot\nabla\nabla\mathbf{u}+G\cdot\nabla\mathbf{u}+\rho\cdot\mathbf{b}=\rho\omega^2\mathbf{u} \tag{14.50}$$

where $\mathbf{b}$ is a body force vector

$$\mathbf{u}=\begin{pmatrix}u_x\\u_y\\u_z\end{pmatrix} \quad ; \quad \nabla\nabla=\begin{pmatrix}\dfrac{\partial^2}{\partial x^2} & \dfrac{\partial^2}{\partial x\partial y} & \dfrac{\partial^2}{\partial x\partial z}\\ \dfrac{\partial^2}{\partial y\partial x} & \dfrac{\partial^2}{\partial y^2} & \dfrac{\partial^2}{\partial y\partial z}\\ \dfrac{\partial^2}{\partial z\partial x} & \dfrac{\partial^2}{\partial z\partial y} & \dfrac{\partial^2}{\partial z^2}\end{pmatrix} \tag{14.51}$$

and

$$\nabla=\begin{pmatrix}\dfrac{\partial^2}{\partial x^2}+\dfrac{\partial^2}{\partial y^2}+\dfrac{\partial^2}{\partial z^2} & 0 & 0\\ 0 & \dfrac{\partial^2}{\partial x^2}+\dfrac{\partial^2}{\partial y^2}+\dfrac{\partial^2}{\partial z^2} & 0\\ 0 & 0 & \dfrac{\partial^2}{\partial x^2}+\dfrac{\partial^2}{\partial y^2}+\dfrac{\partial^2}{\partial z^2}\end{pmatrix} \tag{14.52}$$

ρ is the mass density and G, λ are elastic constants introduced in Chapter 4 and ω is the frequency.

The differential equation for dynamics in the time domain can be written in matrix form as:

$$(\lambda + G) \cdot \nabla\nabla\mathbf{u} + G \cdot \nabla\mathbf{u} + \rho \cdot \mathbf{b} = \rho\ddot{\mathbf{u}} \tag{14.53}$$

where the acceleration vector is defined as

$$\ddot{\mathbf{u}} = \begin{pmatrix} \ddot{u}_x \\ \ddot{u}_y \\ \ddot{u}_z \end{pmatrix} \tag{14.54}$$

Equation (14.51) can be re-written in terms of pressure and shear velocities, c_1, c_2

$$(c_1^2 - c_2^2) \cdot \nabla\nabla\mathbf{u} + c_2^2 \cdot \nabla\mathbf{u} + \mathbf{b} = \ddot{\mathbf{u}} \tag{14.55}$$

where $c_1^2 = (\lambda + 2G)/\rho$, $c_2^2 = G/\rho$.

14.4.1 Fundamental solutions

Fundamental solutions are obtained for a concentrated impulse applied at P at time τ, i.e. for the case of a body force of

$$b_j = \delta(P - Q) \cdot \delta(t - \tau) \tag{14.56}$$

where δ is the Dirac Delta function introduced earlier.

For 3-D problems the fundamental solution for the displacement is given by:

$$U_{ij}(P,\tau,Q,t) = \frac{1}{4\pi\rho r} \begin{bmatrix} \dfrac{r_{,i} \cdot r_{,j}}{c_1^2} \cdot \delta(t - \dfrac{r}{c_1}) - \dfrac{1}{c_2^2}\left(\delta_{ij} - r_{,i} \cdot r_{,j}\right) \cdot \delta(t - \dfrac{r}{c_2}) \\ + \left(3r_{,i} \cdot r_{,j} - \delta_{ij}\right) \displaystyle\int_{1/c_1}^{1/c_2} \delta(t - \lambda r) \cdot \lambda \cdot d\lambda \end{bmatrix} \tag{14.57}$$

14.4.2 Boundary integral equations

The integral equation is obtained in a similar way as for the scalar wave equation except that vectors **u** and **t** are used for the displacements and tractions.

The integral equation is given by

$$\hat{c}\mathbf{u}(P,t) = \int_S [\mathbf{U}(P,\tau,Q,t) * \mathbf{t}(Q,\tau) - \mathbf{T}(P,\tau,Q,t) * \mathbf{u}(Q,\tau)] \cdot dS \tag{14.58}$$

where **U** and **T** are matrices containing the fundamental solutions.

14.4.3 Numerical implementation

For the solution of the integral equation we discretise the problem in time as well as in space as for the scalar wave equation. If we discretise the total time into equal (arbitrary small) steps of size Δt then we have

$$\mathbf{u}(Q,t) = \sum_{n=1}^{N} \tilde{N}_n(t) \cdot \mathbf{u}_n(Q) \quad ; \quad \mathbf{t}(Q,t) = \sum_{n=1}^{N} \tilde{N}_n(t) \cdot \mathbf{t}_n(Q) \tag{14.59}$$

Following the steps for the scalar problem and assuming a constant shape function we obtain the discretised integral equation for time step N as

$$\hat{c}\mathbf{u}_N(P) = \sum_{n=1}^{N} \int_S \Delta\mathbf{U}_{Nn} \cdot \mathbf{t}_n(Q) \cdot dS(Q) - \sum_{n=1}^{N} \int_S \Delta\mathbf{T}_{Nn} \cdot \mathbf{u}_n(Q) \cdot dS(Q) \tag{14.60}$$

where

$$\Delta\mathbf{U}_{Nn} = \int_{t_{n-1}}^{t_n} \mathbf{U}(P,\tau,Q,t_N) d\tau \quad ; \quad \Delta\mathbf{T}_{Nn} = \int_{t_{n-1}}^{t_n} \mathbf{T}(P,\tau,Q,t_N) d\tau \tag{14.61}$$

Introducing the space discretisation

$$\mathbf{u}_n(Q) = \sum_{j=1}^{J} N_j \cdot \mathbf{u}_{nj} \quad ; \quad \mathbf{t}_n(Q) = \sum_{j=1}^{J} N_j \cdot \mathbf{t}_{nj} \tag{14.62}$$

where $\mathbf{u}_n, \mathbf{t}_n$ are displacements and tractions at Q, $\mathbf{u}_{nj}^e, \mathbf{t}_{nj}^e$ refer to values of **u** and **t** at node j of element e at time step n and N_j are shape functions. Substitution of (14.62) into (14.60) gives

$$\hat{c}\mathbf{u}_N(P_i) = \sum_{n=1}^{N} \sum_{e=1}^{E} \sum_{j=1}^{J} \Delta\mathbf{U}_{ijNn}^e \cdot \mathbf{t}_{nj}^e - \sum_{n=1}^{N} \sum_{e=1}^{E} \sum_{j=1}^{J} \Delta\mathbf{T}_{ijNn}^e \cdot \mathbf{u}_{nj}^e \tag{14.63}$$

where

$$\Delta\mathbf{U}_{ijNn}^e = \int_{S_e} \Delta\mathbf{U}_{Nn}(P_i) \cdot N_j \cdot |J| \cdot dS \tag{14.64}$$

and

$$\Delta\mathbf{T}^{e}_{ijNn} = \int_{S_e} \Delta\mathbf{T}_{Nn}(P_i)\cdot N_j \cdot |J| \cdot dS \tag{14.65}$$

where $|J|$ is the Jacobian.

If we define vectors $\{\mathbf{u}\}_n$ and $\{\mathbf{t}\}_n$ to contain all nodal values of displacements and tractions at the nodes at time increment N we have

$$\sum_{n=1}^{N}[\mathbf{T}]_n\{\mathbf{u}\}_n = \sum_{n=1}^{N}[\mathbf{U}]_n\{\mathbf{t}\}_n \tag{14.66}$$

or

$$[\mathbf{T}]_N\{\mathbf{u}\}_N = [\mathbf{U}]_N\{\mathbf{t}\}_N + \{\mathbf{F}\} \tag{14.67}$$

where the vector $\{\mathbf{F}\}$ contains the effect of the time history:

$$\mathbf{F}_i = \sum_{n=1}^{N-1}\sum_{e=1}^{E}\sum_{j=1}^{J}\Delta\mathbf{U}^{e}_{ijNn}\cdot\mathbf{t}_{nj} + \sum_{n=1}^{N-1}\sum_{e=1}^{E}\sum_{j=1}^{J}\Delta\mathbf{T}^{e}_{ijNn}\cdot\mathbf{u}_{nj} \tag{14.68}$$

14.5 MULTIPLE REGIONS

The approach used for the dynamic analysis with multiple regions is very similar to the one introduced for statics in Chapter 11. The difference is that instead of applying unit *Dirichlet* boundary conditions at the interface between regions we apply unit impulses. We only consider a fully coupled problem to simplify the explanation that we present here. The details of a partially coupled analysis are given by Pereira et al.[9]

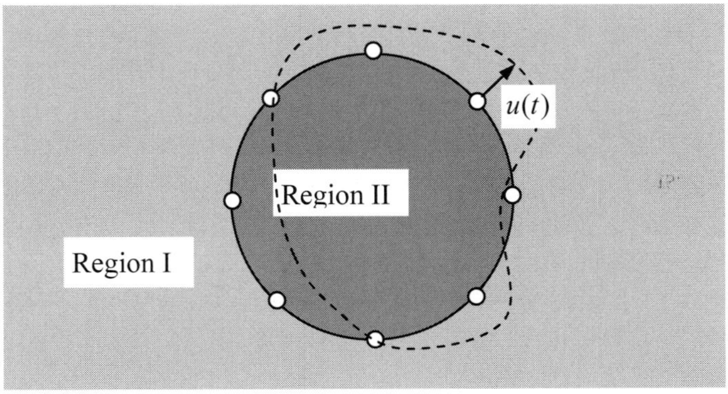

Figure 14.8 Example for explaining the analysis of multiple regions

Consider the problem of an inclusion (with different properties) in an infinite domain in Figure 14.8. We separate the regions and show the displacements and tractions. Between the regions the conditions of equilibrium and compatibility must be satisfied

$$\{\mathbf{t}\}_0 + \{\mathbf{t}\}^I + \{\mathbf{t}\}^{II} = 0 \quad ; \quad \{\mathbf{u}\}^I = \{\mathbf{u}\}^{II} \tag{14.69}$$

where $\{\mathbf{t}\}^I, \{\mathbf{t}\}^{II}$ are interface tractions for region *I* and *II* and $\{\mathbf{t}\}_0$ are applied tractions. $\{\mathbf{u}\}^I, \{\mathbf{u}\}^{II}$ are the interface displacements. We attempt to derive a relationship between the tractions and the displacements at the interface between each region.

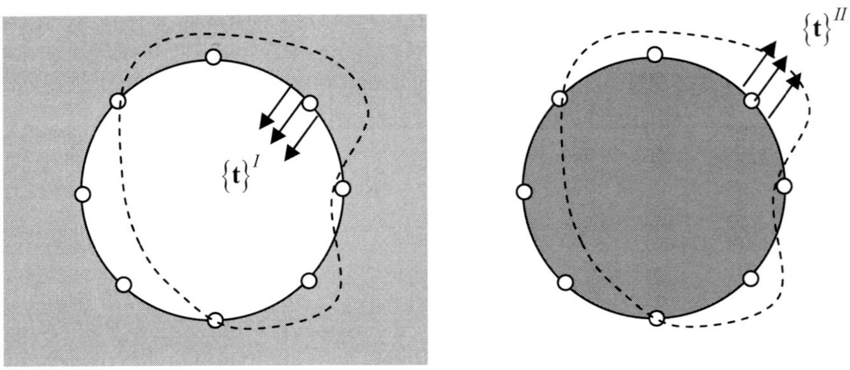

Figure 14.9 Separated regions

For this we consider each region separately and apply a (transient) unit displacement at each node while keeping the other displacements zero. We use the concept of the *Duhamel* integral introduced earlier to obtain the transient tractions due to transient unit displacements. If we do this then we obtain the following relationship between tractions and displacements for region *i*

$$\{\mathbf{t}(t)\}^i = \int_0^t \tilde{\mathbf{K}}(t,\tau)^i \{\mathbf{u}(t)\}^i \, d\tau \tag{14.70}$$

where $\tilde{\mathbf{K}}(t,\tau)^i$ is a unit displacement impulse response matrix whose coefficients represent the transient traction components due to an impulsive unit displacement $\delta(t-\tau)$ applied at time τ. Matrix $\tilde{\mathbf{K}}(t,\tau)^i$ can be computed in the Laplace domain using the CQM introduced above. This is discussed in detail by Pereira[10].

To solve the fully coupled problem the time may be divided into *n* time steps Δt. Then Equation (14.70) may be written for time step *n* as

$$\{\mathbf{t}(n\Delta t)\}^i = \sum_{m=0}^{n} \left[\bar{\mathbf{K}}^i((n-m)\Delta t)\{\mathbf{u}(m\Delta t)\}^i + \{\mathbf{t}(m\Delta t)\}_0 \right] \tag{14.71}$$

where $\bar{\mathbf{K}}^i(n\Delta t)$ is a "dynamic stiffness matrix" of region i similar to the one obtained in Chapter 11. Introducing the compatibility and equilibrium equations (14.69) we obtain the equations for the solution of interface displacements at time $n\Delta t$ [9]

$$\left(\bar{\mathbf{K}}^I(0)+\bar{\mathbf{K}}^{II}(0)\right)\{\mathbf{u}(n\Delta t)\} = \\ -\{\mathbf{t}(n\Delta t)\}_0 - \sum_{m=0}^{n-1}\left[\left(\bar{\mathbf{K}}^I((n-m)\Delta t)+\bar{\mathbf{K}}^{II}((n-m)\Delta t)\right)\{\mathbf{u}(m\Delta t)\}+\{\mathbf{t}(m\Delta t)\}_0\right] \tag{14.72}$$

14.6 EXAMPLES

Here we show two examples involving multiple regions. The first is meant to ascertain the accuracy of the method, the second to show a practical application.

14.6.1 Test example

A standard benchmark example commonly used to validate transient dynamic formulations is the wave propagation in a rod, as shown in Figure 14.10. The material properties of the rod are $E = 2.1\text{x}10^{11}$ N/m^2, $\nu = 0$ and $\rho = 7850$ kg/m^3 (steel). The road is divided into two regions. A Heaviside compression load of magnitude 1 kN/m^2 is applied on the free end of the rod.

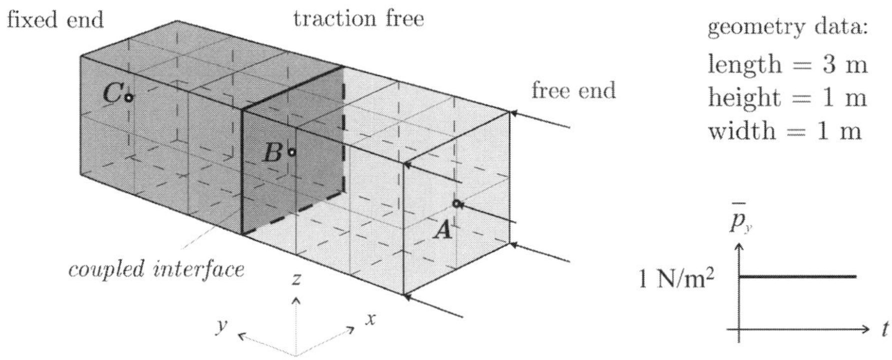

Figure 14.10 Step function excitation of a free-fixed steel rod

In the following, all results are normalized by their corresponding static values, i.e., the displacements by $u_s = 1.4218\text{x}10^{-11}$m and the tractions by $t_s = 1$ kN/m^2, respectively. The displacements at points **A** and **B** (free end and coupled interface) and the traction in longitudinal direction at the fixed end are plotted versus time in Figure 14.11 and Figure 14.12, respectively. These results are obtained for different time step sizes. Taking as reference a parameter $\beta = c\Delta t\,/\,r$, where r is the element length, it is possible to identify a range of values that depend on the time step size where the results are satisfactory i.e., stable and accurate. It can be observed, that the results are in good agreement with the analytic solution and with the numerical results for single region problem published for example by Schanz[11]. Excellent agreement with the analytic solution is obtained for the time step $\beta = 0.25$, however the results for $\beta = 0.10$ are unstable. The larger time steps (e.g., $\beta = 1.50$) tend to smooth the results due to larger numerical damping and introduce some phase shift. Nevertheless, the results for all time step sizes inside the interval $0.20 < \beta < 1.50$ are satisfactory.

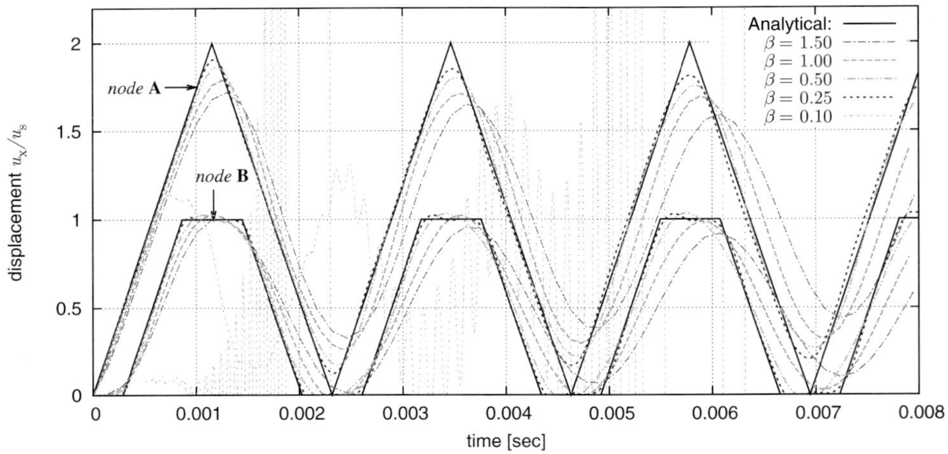

Figure 14.11 Longitudinal normalized displacements at nodes **A** and **B**.

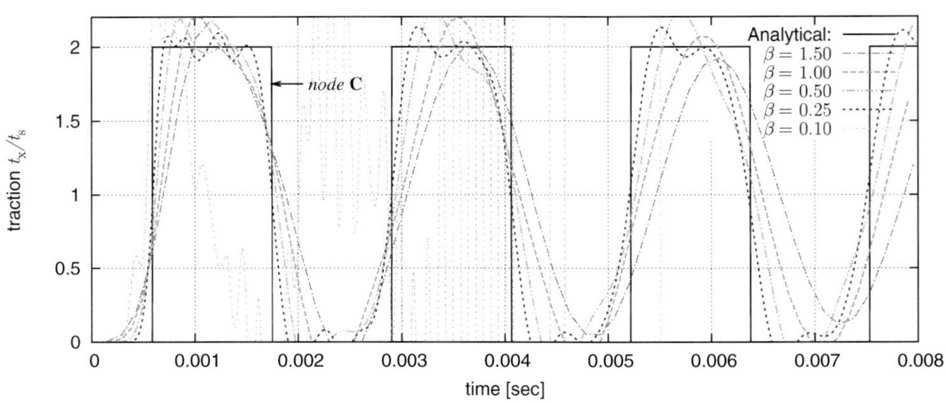

Figure 14.12 Longitudinal normalized tractions at the fixed end (node **C**).

14.6.2 Practical application

This is a practical application in tunnelling. The tunnel depicted in Figure 14.13 is located in a piecewise heterogeneous rock mass with two different properties. The loading is a suddenly applied point load of magnitude F at the tunnel face.

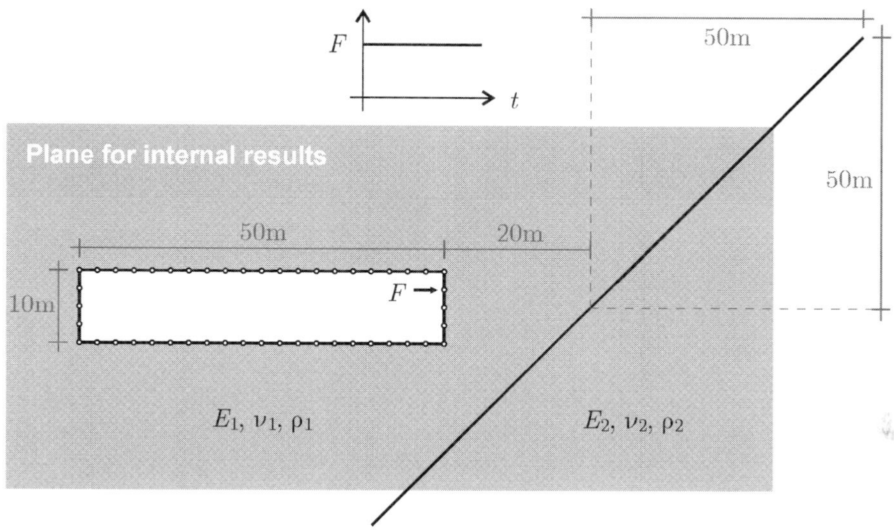

Figure 14.13 Problem statement

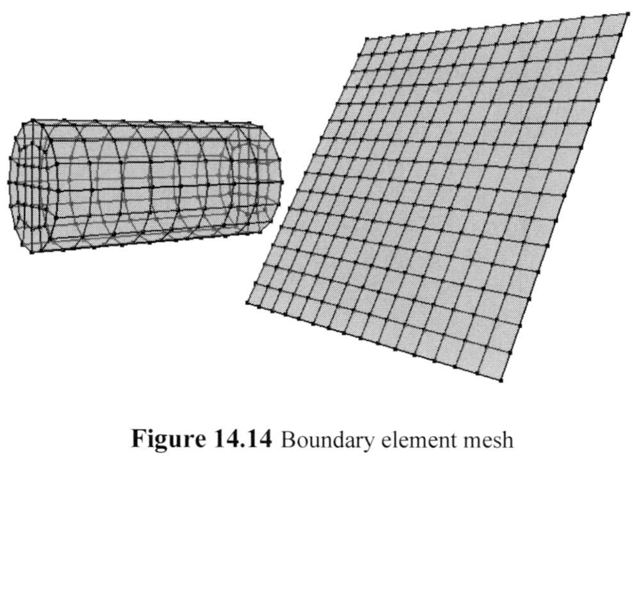

Figure 14.14 Boundary element mesh

The boundary element mesh consists of 2 regions and linear boundary elements, as shown in Figure 14.14. Results of the analysis are shown in Figure 14.15, for two different time steps and values of ratios of Young's modulus.

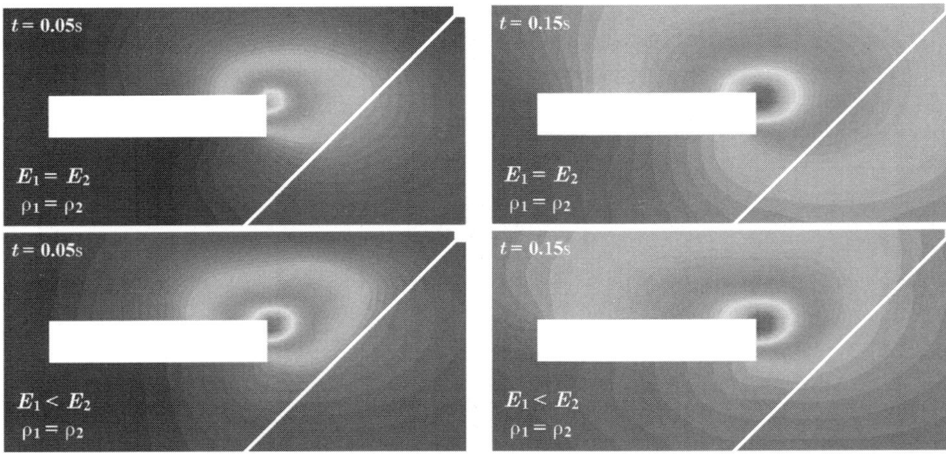

Figure 14.15 Contours of absolute displacement for two different ratios of modulus and times

14.7 REFERENCES

1. Banerjee P.K. (1994) The Boundary Element Methods in Engineering, McGraw Hill Book Company, London.
2. Manolis G.D. and Beskos D.E. (1988) Boundary Element Methods in Elastodynamics. Unwin-Hyman, London.
3. Dominguez, J. (1993) Boundary Elements in Dynamics. Computational Mechanics Publications, Southampton.
4. Bonnet, M. (1995) Boundary Integral Equation Methods for Solids and Fluids, J.Wiley.
5. Chopra A.K. (2007) Dynamics of Structures. Pearson Prentice Hall.
6. Lubich C. (1988) Convolution quadrature and discretized operational calculus I. *Numerische Mathematik*. **52**: 129-145.
7. Kreyszig, E. (1999) Advanced Engineering Mathematics. J. Wiley
8. Wheeler L.T and Sternberg E. (1968) Some theorems in classical elastodynamics. *Arch. Rational Mech. Anal.* **31**:51-90.
9. Pereira A. (2006) A Duhamel integral approach based on BEM to 3-D elasto-dynamic multi-region problems. IABEM, Verlag der TU Graz, Austria, 71-74.
10. Pereira A. (2008) PhD thesis, Graz University of Technology, Austria.
11. Schanz M. (2001) Wave propagation in viscoelastic and poroelastic continua. Springer, Berlin

15

Nonlinear Problems

παντα ρει
(Everything flows)
Aristotle

15.1 INTRODUCTION

So far we have discussed problems where there is a linear relationship between applied loading and displacement, or between applied flow and temperature/potential. The system of equations

$$[\mathbf{T}]\{\mathbf{u}\}=\{\mathbf{F}\} \tag{15.1}$$

corresponds to a linear analysis, if $\{\mathbf{u}\}$ is a linear function of $\{\mathbf{F}\}$.

The linearity of (15.1) is only guaranteed if certain assumptions are made when deriving the system of equations. These assumptions are:

1. The relationships between flux and temperature/potential or stresses and strains are linear
2. Matrix $[\mathbf{T}]$ is not affected by changes in geometry of the boundary that occurs during loading
3. Boundary conditions do not change during loading

Indeed, we have implicitly relied on these assumptions to be true in all our previous derivations of the theory.

An example where the first assumption is violated is elasto- or visco-plastic material behaviour (this is generally referred to as material nonlinear behaviour). The second one is violated if displacements significantly change the boundary shape (large displacement problems). Finally, the third no longer holds true for contact problems, where either the

Dirichlet boundary or the interface conditions between regions change during loading, thereby affecting the assembly of $[\mathbf{T}]$. An example of an elastic sphere on a rigid surface, shown in Figure 15.1. After deformation two nodes, indicated by dark circles may change from *Neuman* to *Dirichlet* boundary condition.

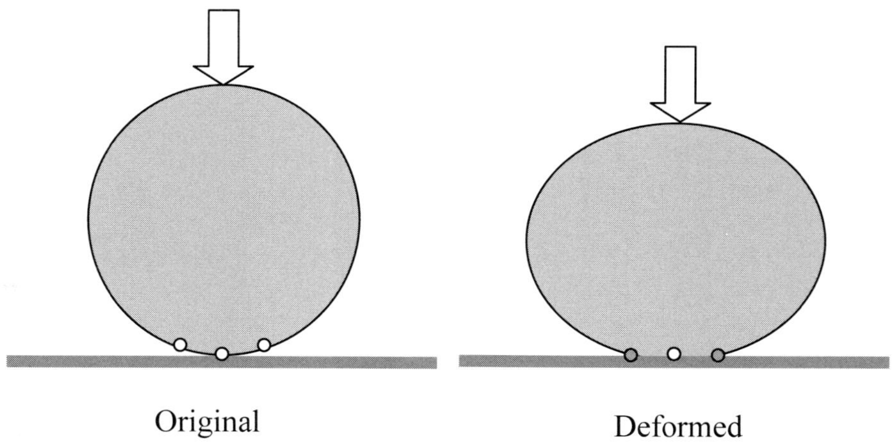

Figure 15.1 Example of nonlinear analysis: contact problem

If one of the above-mentioned assumptions are not satisfied, then the relationship between $\{\mathbf{u}\}$ and $\{\mathbf{F}\}$ will become nonlinear. In a nonlinear analysis matrix $[\mathbf{T}]$ becomes itself a function of the unknown vector $\{\mathbf{u}\}$. It is therefore not possible to solve the system of equations directly.

In this chapter we shall discuss solution methods for nonlinear problems starting with the general solution process. We will then discuss two different types of nonlinear behaviour, plasticity and contact problems. We shall see that solution methods for these types of problems are very similar to the ones employed by the finite element method. We will also find that the BEM is well suited to deal with contact problems because boundary tractions are used as primary unknown.

15.2 GENERAL SOLUTION PROCEDURE

The method proposed is to first find a solution with the assumption that the conditions for linearity are satisfied, i.e. we solve

$$[\mathbf{T}]_0\{\mathbf{x}\}_0 - \{\mathbf{F}\} = 0 \tag{15.2}$$

where $[\mathbf{T}]_0$ is the "linear" coefficient matrix. With solution vector $\{\mathbf{x}\}_0$ (which contains either displacements or tractions depending on boundary conditions) a check is then made to see whether all linearity assumptions have been satisfied, for example, we may check if the internal stresses (computed by post-processing) violate any yield condition, or if boundary conditions have changed because of deformations. If any one of these "linearity" conditions has not been satisfied this means that matrix $[\mathbf{T}]$ has changed during loading, i.e, instead of equation (15.2) we have

$$[\mathbf{T}]_1\{\mathbf{x}\}_0-\{\mathbf{F}\}=\{\mathbf{R}\}_1 \tag{15.3}$$

Here $[\mathbf{T}]_1$ is the changed matrix, also referred to as "tangent" matrix, and $\{\mathbf{R}\}_1$ is a residual vector. Therefore the solution has to be corrected.

We compute the first correction to $\{\mathbf{x}\}$, $\{\dot{\mathbf{x}}\}$ as

$$[\mathbf{T}]_1\{\dot{\mathbf{x}}\}_1=\{\mathbf{R}\}_1 \tag{15.4}$$

where the overdot means increment and proceed with these corrections until the residual vector $\{\mathbf{R}\}$ approaches zero.

Final displacements/tractions are obtained by summing all corrections:

$$\{\mathbf{x}\}=\{\mathbf{x}\}_0+\{\dot{\mathbf{x}}\}_1+\cdots+\{\dot{\mathbf{x}}\}_N \tag{15.5}$$

where N is the number of iterations to achieve convergence. The solution is assumed to have converged if the norm of the current residual vector is much smaller than the first residual vector, i.e., when

$$\frac{\|\mathbf{R}_N\|}{\|\mathbf{R}_1\|}\leq Tol \tag{15.6}$$

where Tol is a specified tolerance.

Alternative to the system of equations (15.4) we may use the "linear" matrix throughout the iteration, that is, equation (15.4) is modified to

$$[\mathbf{T}]_0\{\dot{\mathbf{x}}\}_1=\{\mathbf{R}\}_1 \tag{15.7}$$

This will obviously result in slower convergence but will save us computing a new left hand side and a new solution of the system of equations, only a re-solution with a new right hand side is required. This will be the approach that we will consider here.

15.3 PLASTICITY

There are two ways in which nonlinear material behaviour may be considered: elasto-plasticity and visco-plasticity[1]. Regardless of the method used, the aim is to obtain initial strains or stresses. Using the procedures outlined in Chapter 13 residuals $\{\mathbf{R}\}$ may be computed directly from initial stresses.

15.3.1 Elasto-plasticity

In the theory of elasto-plasticity we define a yield function $F(\boldsymbol{\sigma}, C_1, C_2....) = 0$ which specifies a limiting value of stress $\boldsymbol{\sigma}$ (C1, C2, etc., are plastic material parameters). Stress states can only be such that F is negative (elastic states) or zero (plastic states). Positive values of F are not allowed. Here we restrict the discussion to materials that exhibit no hardening, although it is clear that the numerical procedures are applicable to hardening materials also.

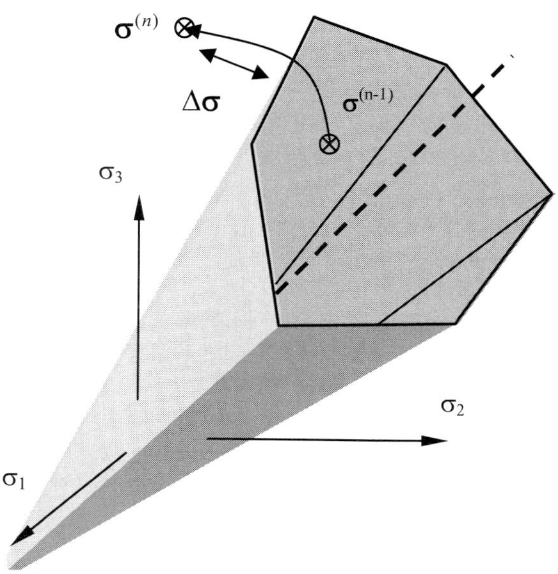

Figure 15.2 Mohr-Coulomb yield surface showing elastic and inadmissible stress state

A popular yield function for soil and rock material is the Mohr-Coulomb[2] condition which can be expressed as a surface in principal stress space by

$$F(\boldsymbol{\sigma}) = \frac{\sigma_1 + \sigma_3}{2}\sin\phi - \frac{\sigma_1 - \sigma_3}{2} - c\cos\phi = 0 \tag{15.8}$$

where σ_1 and σ_3 are maximum and minimum principal stresses, c is cohesion and ϕ the angle of friction. The yield function is plotted as a surface in the principal stress space in Figure 15.2. We assume that the loading is applied in increments. After the solution it may occur that stresses that were in an elastic state at a previous load increment *n-1* (i.e. $F < 0$) change to an inadmissible state ($F > 0$) at the current increment *n* (Figure 15.2).

Therefore, this stress state has to be corrected back to the yield surface. To do this we have to isolate the plastic and elastic components. If the state of stress is such that $F(\boldsymbol{\sigma}) < 0$, then theory of elasticity governs the relationship between stress and strain, i.e. (see Chapter 4)

$$\boldsymbol{\sigma} = \mathbf{D}\boldsymbol{\varepsilon} \tag{15.9}$$

For stress states where $F(\boldsymbol{\sigma}) = 0$, elastic strains $\boldsymbol{\varepsilon}_e$ as well as plastic strains $\boldsymbol{\varepsilon}_p$ may be present, i.e. the total strain , $\boldsymbol{\varepsilon}$, consists of two parts[3]

$$\boldsymbol{\varepsilon} = \boldsymbol{\varepsilon}_e + \boldsymbol{\varepsilon}_p \tag{15.10}$$

For this case the stress-strain law can only be written incrementally as

$$d\boldsymbol{\sigma} = \mathbf{D}_{ep} d\boldsymbol{\varepsilon} \tag{15.11}$$

where $\mathbf{D}_{ep}$ is the elasto-plastic constitutive matrix and $d\boldsymbol{\varepsilon}$ is the total strain increment.

To determine $\mathbf{D}_{ep}$ we must determine the plastic strain increment. This is

$$d\boldsymbol{\varepsilon}_p = \lambda \frac{\partial Q}{\partial \boldsymbol{\sigma}} \tag{15.12}$$

where Q is a *flow function* whose definition is similar to F. If $Q \equiv F$ then this is known as *associated flow rule*. On the yield surface (F=0) we can write for the stress increment

$$d\boldsymbol{\sigma} = \mathbf{D} \cdot d\boldsymbol{\varepsilon}_e = \mathbf{D} \cdot \left(d\boldsymbol{\varepsilon} - d\boldsymbol{\varepsilon}_p \right) = \mathbf{D} \cdot \left(d\boldsymbol{\varepsilon} - \lambda \cdot \frac{\partial Q}{\partial \boldsymbol{\sigma}} \right) \tag{15.13}$$

The stress increment can therefore be split into two parts (one elastic and one plastic)

$$d\boldsymbol{\sigma} = d\boldsymbol{\sigma}_e - d\boldsymbol{\sigma}_p \tag{15.14}$$

The condition that $F > 0$ is not allowed means that for any increment $d\boldsymbol{\sigma}$ the change in F must be zero, i.e.

$$\frac{\partial F}{\partial \boldsymbol{\sigma}} \cdot d\boldsymbol{\sigma} = 0 \tag{15.15}$$

Substitution of (15.13) into (15.15) gives after some algebra

$$\lambda = \frac{1}{\beta}\frac{\partial F}{\partial \boldsymbol{\sigma}}\mathbf{D}d\boldsymbol{\varepsilon} \qquad \text{where} \qquad \beta = \left(\frac{\partial F}{\partial \boldsymbol{\sigma}}\right)^T \mathbf{D}\frac{\partial Q}{\partial \boldsymbol{\sigma}} \tag{15.16}$$

Substituting of (15.16) into (15.13) gives for the plastic stress increment

$$d\boldsymbol{\sigma}_p = \mathbf{D}_p \cdot d\boldsymbol{\varepsilon} \tag{15.17}$$

where

$$\mathbf{D}_p = \mathbf{D} - \mathbf{D}_{ep} = \frac{1}{\beta}\mathbf{D}\frac{\partial Q}{\partial \boldsymbol{\sigma}}\left(\frac{\partial F}{\partial \boldsymbol{\sigma}}\right)^T \mathbf{D} \tag{15.18}$$

This relationship only holds true if the stress state is actually on the yield surface (F=0). In the case where during a load increment a point goes from an elastic to a plastic state and violates the yield condition in the process (i.e. F>0) then the plastic stress increment has to be related to the plastic strain increment rather than the total strain increment.

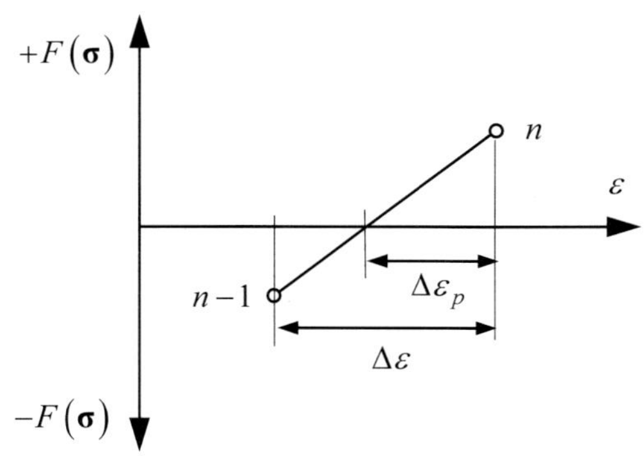

Figure 15.3 Determination of plastic part of the strain increment

Using a simple linear approximation the plastic strain increment is computed by (Figure 15.3):

$$\Delta\boldsymbol{\varepsilon}_p = r\Delta\boldsymbol{\varepsilon} \tag{15.19}$$

where Δ has been substituted for d to indicate that the increments are no longer infinitesimally small and

$$r = \frac{F(\boldsymbol{\sigma}^{(n)})}{F(\boldsymbol{\sigma}^{(n)}) - F(\boldsymbol{\sigma}^{(n-1)})} \tag{15.20}$$

Table 15.1 gives an overview of the factor r for different situations of the stress state at the beginning (*n-1*) and the end of the load increment (*n*).

Table 15.1 Values of r for various cases

Case	*n-1*	*n*	*r*
1	$F<0$	$F<0$	0.0
2	$F<0$	$F>0$	Eq. (15.20)
3	$F=0$	$F>0$	1.0
4	$F=0$	$F<0$	0.0

After a load increment the stresses have been wrongly computed if they are outside the yield surface (F>0). They should have to be computed according to

$$\Delta\boldsymbol{\sigma} = \Delta\boldsymbol{\sigma}_e - \Delta\boldsymbol{\sigma}_p \tag{15.21}$$

where

$$\Delta\boldsymbol{\sigma}_p = \mathbf{D}_p \Delta\boldsymbol{\varepsilon}_p = \mathbf{D}_p r \Delta\boldsymbol{\varepsilon} \tag{15.22}$$

Therefore the stresses have to be corrected by

$$\boldsymbol{\sigma}_0 = \Delta\boldsymbol{\sigma}_p \tag{15.23}$$

This stress can be assumed as an "initial stress" generated in the domain. Equation (15.20) is only approximate since a linear variation has been used. Therefore, when checking the stress state after the correction applied it will not lie exactly on the yield surface. The discussion of so called "return algorithms" to ensure this are beyond the scope of this text and the reader is referred to the relevant literature on this subject[4].

15.3.2 Visco-plasticity

The concept of visco-plasticity allows F($\boldsymbol{\sigma}$) to be greater than zero[5]. A positive yield function simply means that the stress state has a higher plastic potential. The stresses are then allowed to creep back to a lower plastic potential (Figure 15.4) and eventually to the yield surface. This takes into consideration the fact that the material requires time to "react" to changes in stress and also allows the consideration of creep behaviour.

The strain rate, at which "creeping" takes place, is assumed to be proportional to the plastic potential. That is

$$\frac{\partial}{\partial t}\boldsymbol{\varepsilon}_P = \frac{1}{\eta}\Phi(F)\frac{\partial Q}{\partial \boldsymbol{\sigma}} \tag{15.24}$$

where

$$\Phi(F)=0 \quad ; \quad for \quad F<0$$
$$\Phi(F)=F \quad ; \quad for \quad F>0 \tag{15.25}$$

In the above equations η is a material parameter describing its time dependent behaviour (viscosity).

A visco-plastic analysis proceeds in time steps and a visco-plastic strain increment is computed at each time step by:

$$\Delta\boldsymbol{\varepsilon}_P = \frac{\partial \boldsymbol{\varepsilon}_P}{\partial t}\Delta t \tag{15.26}$$

where Δt is a time increment. The initial stresses for the computation of the residual are

$$\boldsymbol{\sigma}_0 = \Delta\boldsymbol{\sigma}_P = \mathbf{D}\Delta\boldsymbol{\varepsilon}_P \tag{15.27}$$

The time increment Δt cannot be chosen freely but has to satisfy certain stability conditions to prevent oscillations[5].

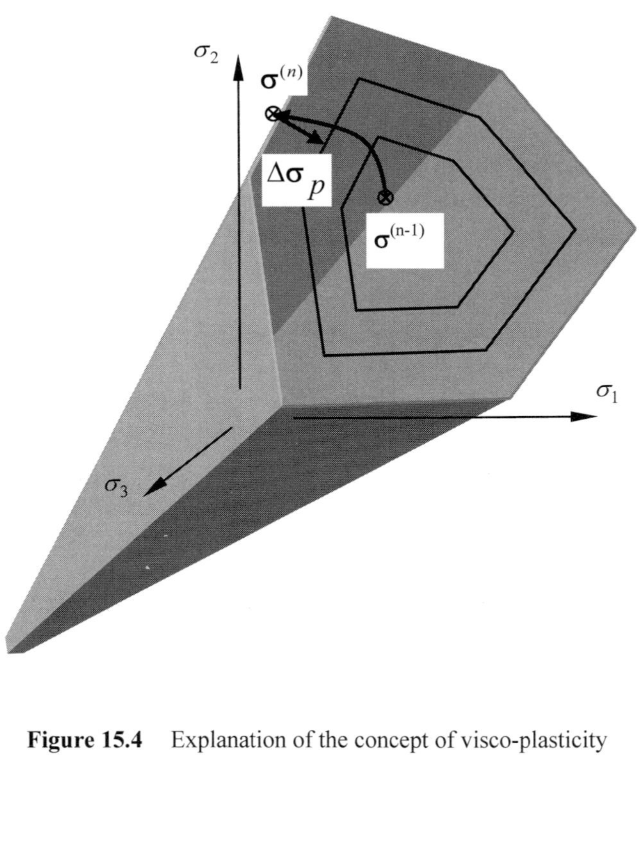

Figure 15.4 Explanation of the concept of visco-plasticity

15.3.3 Method of solution

For the solution of problems in plasticity we use a similar method as in Finite Elements known as the "initial stress" method. In this method we compute initial stresses as outlined in the previous section and apply this as loading. For this we have to amend the discretisation of the problem. In addition to surface elements we require the specification of volume cells in the parts of the domain that are likely to yield, for the integration of initial stresses. These volume cells have been discussed in Chapter 3. Figure 15.5 shows examples of discretisations for a cantilever beam and a circular hole in an infinite domain. The discretisations actually look almost like finite element meshes and it could be argued that one might as well use finite elements for this problem.

However, there are subtle differences:

- There is no requirement of continuity, i.e. elements do not need to connect to each other as finite elements need.
- There are no additional unknown associated with the mesh of volume cells. Therefore the system of equations does not increase in size.
- The representation of stress is still more accurate than with the FEM.
- The mesh of cells only needs to cover zones where plastic behaviour is expected.

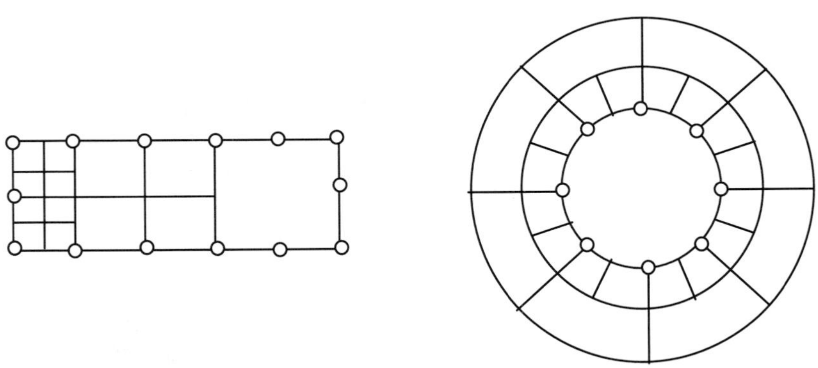

Figure 15.5 Volume cells for the example of a cantilever beam and a circular hole

The iterative process is described in the structure chart in Fig 15.6. First we may divide the total applied load into increments to optimise the number of iterations. Then we solve for the unknown displacements/tractions with the applied loading. With the boundary results we compute the stresses at each cell node and check the yield condition. If $F>0$ is detected then the "initial stress" is computed as explained previously. The residual vector $\{\mathbf{R}\}$ is computed as will be explained later and a new solution $\{\dot{\mathbf{x}}\}_m$ computed and accumulated. The iterations proceed until the norm of the residual vanishes.

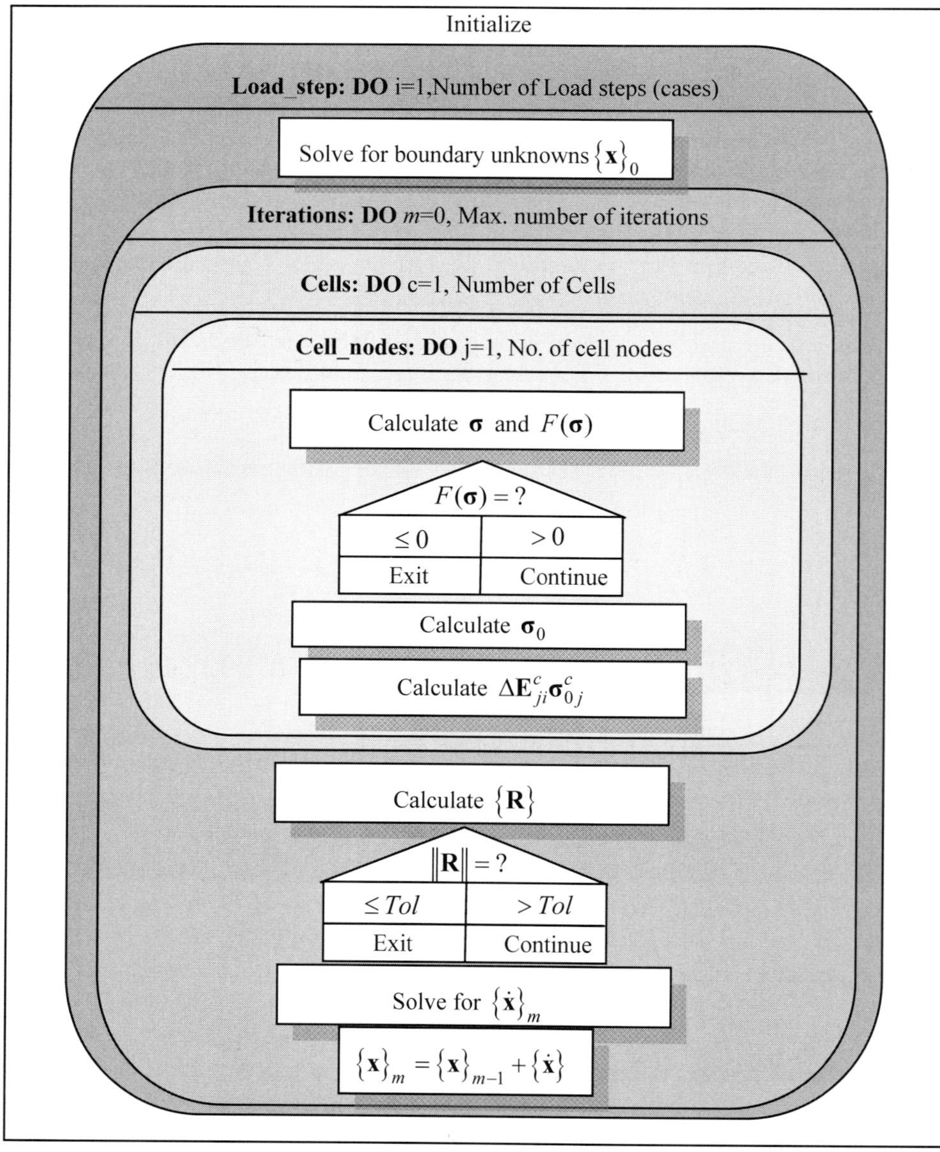

Figure 15.6 Structure chart for plasticity

15.3.4 Calculation of residual {R}

After the initial (linear) analysis the system of equations that has to be solved is

$$[\mathbf{T}]\{\dot{\mathbf{x}}\} = \{\mathbf{R}\} \tag{15.28}$$

where the components of the residual vector are given by

$$\{\mathbf{R}\} = \begin{pmatrix} \mathbf{R}_1 \\ \mathbf{R}_2 \\ \vdots \end{pmatrix} \tag{15.29}$$

and

$$\mathbf{R}_i = \sum_{c=1}^{C}\sum_{n=1}^{N} \Delta\mathbf{E}_{ni}^{c}\dot{\boldsymbol{\sigma}}_{0n}^{c} \tag{15.30}$$

where C is the number of Cells, N is the number of cell nodes and $\dot{\boldsymbol{\sigma}}_{0n}^{c}$ is the "initial stress" increment computed at node n of cell c.

The evaluation of integrals $\Delta\mathbf{E}_{ni}^{c}$ is similar to the evaluation of $\Delta\boldsymbol{\Sigma}_{ni}^{c}$, that has been discussed in section 13.7. For plane problems the expression for $\Delta\mathbf{E}_{ni}^{c}$ in intrinsic coordinates is

$$\Delta\mathbf{E}_{ni}^{c} = \int_{V_c} N_n\, \mathbf{E}\left(P_i, \bar{Q}\right) dV = \int_{-1}^{+1}\int_{-1}^{+1} N_n(\xi,\eta)\mathbf{E}\left(P_i, \bar{Q}(\xi,\eta)\right) J(\xi,\eta)\,\mathrm{d}\xi\,\mathrm{d}\eta \tag{15.31}$$

Using Gauss Quadrature the formula can be replaced by

$$\Delta\mathbf{E}_{ni}^{c} \approx \sum_{m=1}^{M}\sum_{k=1}^{K} N_n(\xi_m,\eta_k)\mathbf{E}(P_i, \bar{Q}(\xi_m,\eta_k))J(\xi_m,\eta_k)\, W_m W_k \tag{15.32}$$

where M and K are the number of integration points in ξ and η directions, respectively.

For 3D problems we have

$$\Delta\mathbf{E}_{ni}^{c} = \int_{-1}^{+1}\int_{-1}^{+1}\int_{-1}^{+1} N_n(\xi,\eta,\zeta)\mathbf{E}\left(P_i, \bar{Q}(\xi,\eta,\zeta)\right) J(\xi,\eta,\zeta)\,\mathrm{d}\xi\,\mathrm{d}\eta\,\mathrm{d}\zeta \tag{15.33}$$

and the integration formula is

$$\Delta\mathbf{E}_{ni}^{c} \approx \sum_{l=1}^{L}\sum_{m=1}^{M}\sum_{k=1}^{K} N_n(\xi_m,\eta_k,\zeta_l)\mathbf{E}(P_i, \bar{Q}(\xi_m,\eta_k,\zeta_l))J(\xi_m,\eta_k,\zeta_l)\, W_m W_k W_l \tag{15.34}$$

with *L, M* and *K* being the number of integration points in ξ , η and ζ directions. The matrix **E** is given by

$$\mathbf{E} = \begin{pmatrix} E_{xxx} & \cdots & E_{xxz} \\ \vdots & \ddots & \vdots \\ E_{zxx} & \cdots & E_{zxz} \end{pmatrix} \tag{15.35}$$

with the coefficients

$$E_{ijk}\left(P,\bar{Q}\right) = \frac{-C}{r^n}\left[C_3\left(r_{,k}\delta_{ij} + r_{,j}\delta_{ik}\right) - r_{,i}\delta_{jk} + C_4 r_{,i} r_{,j} r_{,k}\right] \tag{15.36}$$

where *x,y,z* may be substituted for *i,j,k* and the constants are given in Table 15.2

Table 15.2 Constants for fundamental solution **E**

	Plane strain	Plane stress	3-D
n	1	1	2
C	1/8πG(1−ν)	(1+ν)/8πG	1/16πG(1−ν)
C_3	1-2 ν	(1- ν)/(1+ ν)	1-2 ν
C_4	2	2	3

The above formulae are valid for the case where none of the cell nodes is the collocation point. The special case where one of the cell nodes coincides with a collocation point, P_i, the kernel $\Delta\mathbf{E}^c_{ni}$ tends to infinity with $o(1/r)$ for 2-D problems and $o(1/r^2)$ for 3-D problems. To evaluate the volume integral for this case we subdivide a cell into sub cells, as shown in Figure 15.7.

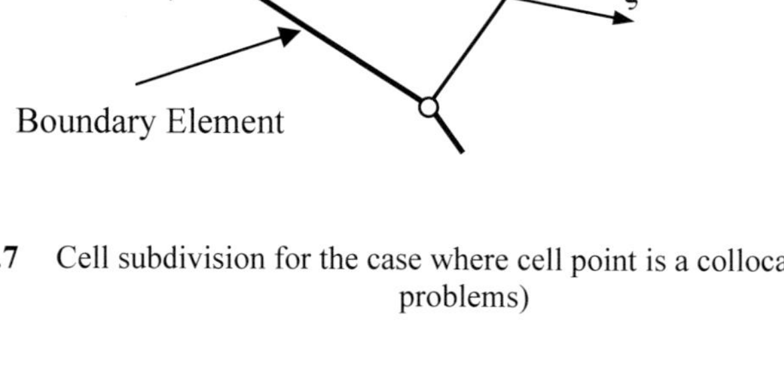

Figure 15.7 Cell subdivision for the case where cell point is a collocation point (plane problems)

For 2-D problems the subdivision is carried out in exactly the same way as for the evaluation of the boundary integrals for 3-D problems, i.e., the square domain is mapped into triangular domains where the apex of the triangle is located at P_i (see section 6.3.7). Equation (15.31) is rewritten as

$$\Delta\mathbf{E}^{c}_{ni} = \sum_{m=1}^{sc} \int_{-1}^{1}\int_{-1}^{1} \mathbf{E}(P_i, \bar{Q}(\xi,\eta)) N_n J(\xi,\eta) \frac{\partial(\xi,\eta)}{\partial(\bar{\xi},\bar{\eta})} d\bar{\xi}\, d\bar{\eta} \approx$$
$$\sum_{m=1}^{sc}\sum_{j=1}^{J}\sum_{k=1}^{K} \mathbf{E}(P_i, \bar{Q}(\bar{\xi}_j,\bar{\eta}_k)) N_n(\bar{\xi}_j,\bar{\eta}_k) J(\bar{\xi}_j,\bar{\eta}_k) \bar{J}(\bar{\xi}_j,\bar{\eta}_k) W_j W_k \tag{15.37}$$

where sc is the number of sub-cells, which is equal to 2 if the collocation point P_i is at a cell corner node, or 3 if it is a middle node of the cell. The computation of the Jacobian $\bar{J}$ of the transformation from sub element coordinates $\bar{\xi},\bar{\eta}$ to intrinsic coordinates ξ,η is explained in 6.3.7. Since the Jacobian of this transformation tends to zero with $o(r)$ as point P_i is approached, the singularity is cancelled out.

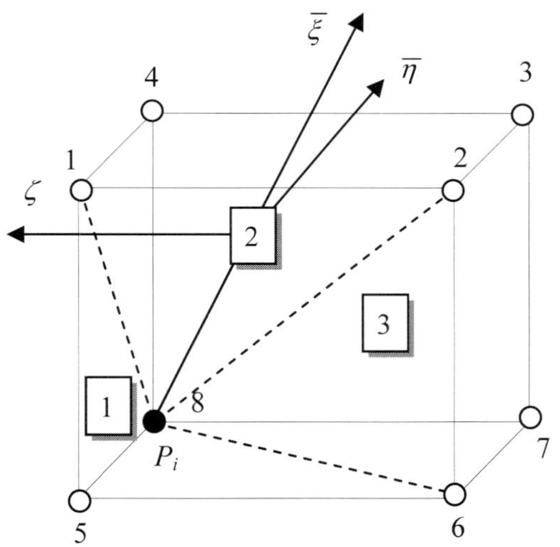

Figure 15.8 Subdivision method for computing singular volume integrals (3-D problems).

For three-dimensional problems, if one of the nodes of the cell is a collocation point, a subdivision, analogous to the 2-D case, into tetrahedral sub-cells with locally defined co-ordinate, as shown in Figure 15.8, is used. The integral over the cell is expressed as

$$\sum_{m=1}^{sc} \int_{-1}^{1}\int_{-1}^{1}\int_{-1}^{1} \mathbf{E}(P_i, Q(\xi,\eta,\zeta)) N_n J(\xi,\eta,\zeta) \frac{\partial(\xi,\eta,\zeta)}{\partial(\bar{\xi},\bar{\eta},\bar{\zeta})} d\bar{\xi}\, d\bar{\eta}\, d\bar{\zeta} \approx$$
$$\sum_{m=1}^{sc}\sum_{j=1}^{J}\sum_{k=1}^{K}\sum_{l=1}^{L} \mathbf{E}(P_i, Q(\bar{\xi}_j,\bar{\eta}_k,\bar{\zeta}_l)) N_n(\bar{\xi}_j,\bar{\eta}_k,\bar{\zeta}_l) J(\bar{\xi}_j,\bar{\eta}_k,\bar{\zeta}_l) \bar{J} W_j W_k W_l \tag{15.38}$$

where *sc* is the number of sub-cells which equals to 3 for collocation point at a corner node, and 4 for collocation point at a mid-node. $\overline{J}$ is the Jacobian of the transformation from ξ, η, ς to $\overline{\xi}, \overline{\eta}, \overline{\varsigma}$ coordinates.

This transformation is given by

$$\xi = \sum_{n=1}^{5} \overline{N}_n(\overline{\xi},\overline{\eta},\overline{\zeta})\xi_{l(n)} \; ; \; \eta = \sum_{n=1}^{5} \overline{N}_n(\overline{\xi},\overline{\eta},\overline{\zeta})\eta_{l(n)} \; ; \; \varsigma = \sum_{n=1}^{5} \overline{N}_n(\overline{\xi},\overline{\eta},\overline{\zeta})\zeta_{l(n)} \qquad (15.39)$$

Where *l(n)* is an array that indicates the local number of node *l*. For the sub-cell 2 in Figure 15.8 for example $l(n) = (4,1,2,3,8)$. More details can be found in [6].

The shape functions are defined as

$$\begin{aligned}
\overline{N}_1 &= \frac{1}{8}(1+\overline{\xi})(1+\overline{\eta})(1+\overline{\zeta}) \quad ; \quad \overline{N}_2 = \frac{1}{8}(1+\overline{\xi})(1-\overline{\eta})(1+\overline{\zeta}) \\
\overline{N}_3 &= \frac{1}{8}(1+\overline{\xi})(1-\overline{\eta})(1-\overline{\zeta}) \quad ; \quad \overline{N}_4 = \frac{1}{8}(1+\overline{\xi})(1+\overline{\eta})(1-\overline{\zeta}) \\
\overline{N}_5 &= \frac{1}{8}(1-\overline{\xi})
\end{aligned} \qquad (15.40)$$

The Jacobian is defined as

$$\overline{J} = \begin{vmatrix} \frac{\partial \xi}{\partial \overline{\xi}} & \frac{\partial \eta}{\partial \overline{\xi}} & \frac{\partial \zeta}{\partial \overline{\xi}} \\ \frac{\partial \xi}{\partial \overline{\eta}} & \frac{\partial \eta}{\partial \overline{\eta}} & \frac{\partial \zeta}{\partial \overline{\eta}} \\ \frac{\partial \xi}{\partial \overline{\zeta}} & \frac{\partial \eta}{\partial \overline{\zeta}} & \frac{\partial \zeta}{\partial \overline{\zeta}} \end{vmatrix} \qquad (15.41)$$

where

$$\frac{\partial \xi}{\partial \overline{\xi}} = \sum_{n=1}^{5} \frac{\partial \overline{N}_n}{\partial \overline{\xi}}(\overline{\xi},\overline{\eta},\overline{\zeta})\xi_{l(n)} \qquad \text{etc.} \qquad (15.42)$$

The Jacobian tends to zero with $o(r^2)$ thereby cancelling out the singularity. Having computed the residual $\{\mathbf{R}\}$ due to an initial stress state $\{\dot{\boldsymbol{\sigma}}\}_0$ we solve the problem for the boundary unknowns $\{\dot{\mathbf{x}}\}$. The next step is to compute stress increments at the cell and boundary nodes.

15.3.5 Computation of stresses at cell nodes

For computation of the internal stress results the following equation is used[7,8]

$$\dot{\boldsymbol{\sigma}}(P_a) = \int_S \mathbf{S}(P_a,Q)\dot{\mathbf{t}}(Q)\,dS - \int_S \mathbf{R}(P_a,Q)\dot{\mathbf{u}}(Q)\,dS + \int_V \hat{\mathbf{E}}(P_a,\bar{Q})\dot{\boldsymbol{\sigma}}_0(\bar{Q})\,dV + \mathbf{F}\,\dot{\boldsymbol{\sigma}}_0(P_a) \tag{15.43}$$

which is derived in the same way as Eq. (13.59).

The coefficients of $\hat{\mathbf{E}}$ are given by

$$\hat{E}_{ijkl} = \frac{-C_2}{r^{n+1}}\Big[C_3\left(\delta_{ik}\delta_{jl} + \delta_{jk}\delta_{il} - \delta_{ij}\delta_{kl} + C_4\delta_{ij}r_{,k}r_{,l}\right) + C_4\nu\left(\delta_{il}r_{,j}r_{,k} + \delta_{jk}r_{,i}r_{,l} + \delta_{ik}r_{,j}r_{,l} + \delta_{jl}r_{,i}r_{,k}\right) + C_4\left(\delta_{kl}r_{,i}r_{,j} - C_6 r_{,i}r_{,j}r_{,k}r_{,l}\right)\Big] \tag{15.44}$$

and those of **F** by

$$F_{ijkl} = C_{18}\left[C_{12}\left(\delta_{jk}\delta_{il} + \delta_{ik}\delta_{jl}\right) + C_{13}\delta_{ij}\delta_{kl}\right] \tag{15.45}$$

where the constants are given in Table 15.3.

Table 15.3 Constants for fundamental solution $\hat{\mathbf{E}}$

	Plane strain	Plane stress	3-D
n	1	1	2
C_2	-1/4π (1−ν)	-(1+ν)/4π	-1/8π (1−ν)
C_3	1-2 ν	(1- ν)/(1+ ν)	1-2 ν
C_4	2	2	3
C_{12}	1	1	7-5 ν
C_{13}	1-4 ν	(1-3 ν)/(1+ ν)	2-10 ν
C_{18}	-1/8(1- ν)	-(1+ ν)/8	-1/30(1- ν)

The discretized form of (15.43) is

$$\dot{\boldsymbol{\sigma}}(P_a) + \sum_{e=1}^{E}\sum_{n=1}^{N}\Delta\mathbf{R}_n^e\dot{\mathbf{u}}_n^e = \sum_{e=1}^{E}\sum_{n=1}^{N}\Delta\mathbf{S}_n^e\dot{\mathbf{t}}_n^e + \sum_{c=1}^{C}\sum_{n=1}^{N}\Delta\hat{\mathbf{E}}_n^c\boldsymbol{\sigma}_{0n}^c + \mathbf{F}\,\boldsymbol{\sigma}_0(P_a) \tag{15.46}$$

The integrals $\Delta\mathbf{R}_n^e$ and $\Delta\mathbf{S}_n^e$ are evaluated as explained in Chapter 9. The integrals $\Delta\hat{\mathbf{E}}_n^c$ can be evaluated using Gauss Quadrature as explained previously if the point P_a is not one of the cell nodes. If P_a coincides with the nodes of cells, then the integrand tends to infinity with $o(r^2)$ for 2-D and $o(r^3)$ for 3-D problems and special attention has to be given to the evaluation of $\Delta\hat{\mathbf{E}}_n^c$. As explained in Chapter 13 for the case with initial stresses a small zone of exclusion is assumed around P_a and this results in the "free term" $\mathbf{F}\,\dot{\boldsymbol{\sigma}}_0\left(P_a\right)$. Now, however we need to evaluate the strongly singular integral $\Delta\hat{\mathbf{E}}_n^c$ over the cells excluding the spherical region with radius ε.

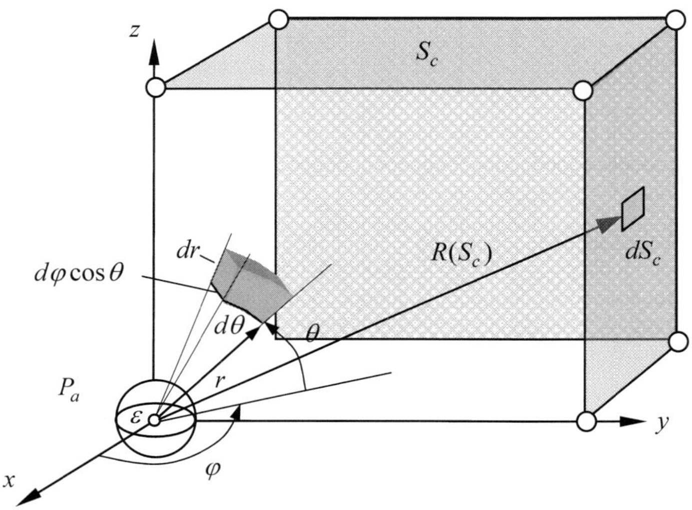

Figure 15.9 Polar coordinates for integration with a spherical region of exclusion

For this the singularity isolation method[9] is used. The singularity is isolated be rewriting the strongly singular domain integral in the form

$$\begin{aligned}\int_V \hat{\mathbf{E}}(P_a,\bar{Q})\cdot\dot{\boldsymbol{\sigma}}_0(\bar{Q})\cdot dV &= \int_V \hat{\mathbf{E}}(P_a,\bar{Q})\cdot\left[\dot{\boldsymbol{\sigma}}_0(\bar{Q})-\dot{\boldsymbol{\sigma}}_0(P_a)\right]\cdot dV \\ &\quad + \dot{\boldsymbol{\sigma}}_0(P_a)\int_V \hat{\mathbf{E}}(P_a,\bar{Q})dV\end{aligned} \tag{15.47}$$

The first integral on the right had side of Eq. (15.47) is weakly singular and can be integrated numerically using the cell subdivision technique. The strong singularity has been moved to the second integral and can be treated semi-analytically. For this the domain is divided into a singular and a regular domain. The singular domain is bounded by the faces of the cells that contain point P_a surrounded by the region of exclusion as shown in Figure 15.9. Using a polar coordinate system with the origin at P_a, the volume integral over the domain is rewritten as

$$\int_{V_c} \hat{\mathbf{E}} dV = \int_0^{\pi}\int_0^{2\pi}\left[\lim_{\varepsilon\to 0}\int_{\varepsilon}^{R(S_c)} \frac{1}{r} dr\right] \hat{\hat{\mathbf{E}}} \sin\theta d\phi d\theta \tag{15.48}$$

where $\hat{\hat{\mathbf{E}}}$ is the part of the kernel which is not a function of r (i.e., term within square brackets in equation (15.43)). Therefore the singularity is isolated and after some transformation[9] the volume integral can be replaced by a surface integral

$$\int_{V_c} \hat{\mathbf{E}} dV = \int_{S_c} \hat{\hat{\mathbf{E}}}(\mathbf{r}\bullet\mathbf{n}) \ln(r) dS \tag{15.49}$$

where $\mathbf{n}$ is the vector normal to S_c, the boundary of the cell surrounding the point P_a. The strong singularity in the kernel $\hat{\mathbf{E}}$ is now isolated and the numerical evaluation with Gauss Quadrature can be performed by

$$\begin{aligned}\Delta\hat{\mathbf{E}}_n^c \approx & \sum_{m=1}^{sc}\sum_{j=1}^{J}\sum_{k=1}^{K}\sum_{l=1}^{L} \hat{\mathbf{E}}(P_a, \bar{Q}(\xi_j,\xi_k,\zeta_l)) \cdot (N_n - \delta(n,P_a)) \cdot J \cdot \bar{J} \cdot W_j W_k W_l \\ & + \delta(n,P_a)\sum_{m=1}^{sc}\sum_{j=1}^{J}\sum_{k=1}^{K} \hat{\mathbf{E}}(P_a, \bar{Q}(\xi_j,\eta_k))\mathbf{r}\bullet\mathbf{n} \cdot \ln(r(P_a,\bar{Q}(\bar{\xi}_j,\bar{\eta}_k))J_s(\bar{\xi}_j,\bar{\eta}_k) W_j W_k\end{aligned} \tag{15.50}$$

where

$$\begin{aligned}\delta(n,P_a) &= 1 \quad \text{if} \quad n \in P_a \\ \delta(n,P_a) &= 0 \quad \text{if} \quad n \notin P_a\end{aligned} \tag{15.51}$$

sc is the number of cell surfaces as indicated in Fig. 15.10 and J_S is the Jacobian of the transformation of coordinates over the cell boundaries (see section 3.9).
The implementation in 2-D follows the same procedure.

15.3.6 Computation of Boundary Stress

The method presented in the previous section for determining the stresses at internal points can not be used for points exactly on the boundary due to the higher singularity of the integral. We have already presented an alternative method for computing the stress tensor on the boundary itself using the variations of the displacements and tractions over boundary elements in Chapter 9. All that is required here is to modify this procedure by taking into consideration the effect of the initial stresses.

The stresses in the local directions $\bar{x}, \bar{y}$ (tangential and normal to the boundary) are given by

$$\sigma_{\bar{x}} = \frac{1}{1-\nu}\left(\frac{E}{1+\nu}\varepsilon_{\bar{x}} + \nu\left(t_{\bar{y}} + \sigma_{0\bar{y}}\right)\right) - \sigma_{0\bar{x}}$$

$$\sigma_{\bar{y}} = t_{\bar{y}} \qquad \tau_{\overline{xy}} = t_{\bar{x}} \quad \sigma_z = \nu(\sigma_{\bar{x}} + \sigma_{0\bar{x}} + \sigma_{\bar{y}} + \sigma_{0\bar{y}}) - \sigma_{0z} \tag{15.52}$$

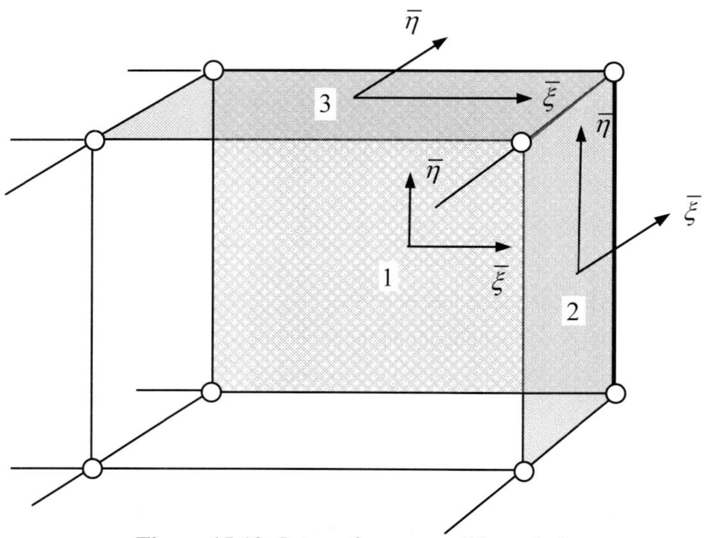

Figure 15.10 Integration over cell boundaries

for plane strain where $\sigma_{0\bar{x}}$ and $\sigma_{0\bar{y}}$ are the initial stresses in local directions. For plane stress we simply substitute ν with $\bar{\nu}$ and E with $\bar{E}$ where

$$\bar{\nu} = \frac{\nu}{1+\nu} \quad \text{and} \quad \bar{E} = \frac{E(1-2\nu)}{(1-\nu^2)} \tag{15.53}$$

and set $\sigma_{\bar{z}} = 0$. In the 3D case, we have

$$\begin{aligned}
\sigma_{\bar{x}} &= C_1(\varepsilon_{\bar{x}} + \nu\varepsilon_{\bar{y}}) + C_2\,(t_{\bar{z}} + \sigma_{0\bar{z}}) - \sigma_{0\bar{x}} \\
\sigma_{\bar{y}} &= C_1(\varepsilon_{\bar{y}} + \nu\varepsilon_{\bar{x}}) + C_2\,(t_{\bar{z}} + \sigma_{0\bar{z}}) - \sigma_{0\bar{y}} \\
\sigma_{\bar{z}} &= t_{\bar{z}} \\
\tau_{\overline{xy}} &= G\gamma_{\overline{xy}} - \tau_{0\overline{xy}} \\
\tau_{\overline{xz}} &= t_{\bar{x}} \quad \tau_{\overline{yz}} = t_{\bar{y}}
\end{aligned} \tag{15.54}$$

where

$$C_1 = \frac{E}{1-\nu^2} \quad ; \quad C_2 = \frac{\nu}{1-\nu} \tag{15.55}$$

The stresses can be transformed to the global coordinate system using (4.37) or (4.39). Indeed the same method may be also used to compute results inside cells. This is much simpler than the method proposed above but also less accurate. The advantage is that the Kernels involved have a much lower singularity and no elaborate schemes are required. Using this method first the displacements are computed at the cell nodes. The variation over a cell is then approximated by the shape functions

$$\mathbf{u} = \sum_{n=1}^{N} N_n \mathbf{u}_n \tag{15.56}$$

The strains inside the cell are computed by

$$\varepsilon_x = \frac{\partial u_x}{\partial x} = \sum \frac{\partial N_n}{\partial x} u_{xn} \quad ; \quad \varepsilon_y = \frac{\partial u_y}{\partial y} = \sum \frac{\partial N_n}{\partial y} u_{yn} \quad \text{etc.} \tag{15.57}$$

The stresses are computed according to Eq. (15.52) and (15.54) but there is no need to use local coordinates and a transformation.

15.3.7 Example

To illustrate the application of the method we present the analysis of a thick-walled cylinder subjected to internal pressure under plane strain conditions. The Von Mises yield function is used and a linear elastic-ideal plastic material behaviour is considered. The material parameters used are:

Youngs Modulus E= 12000 MPa
Poisson's ratio v= 0.3
yield stress Y= 24 MPa

The problem dimensions and loading are shown in Figure 15.11. Due to the symmetry of the problem, only a quarter of the cylinder is analyzed. The mesh used consists of 36 quadratic boundary elements and 72 quadratic volume cells. The solution proceeds in increments where, after each iteration, the yield condition is checked at all the nodes of the cells. If the residual is sufficiently small, convergence is achieved and the analysis is stopped. After applying the full load we can notice that all the points belonging to the first row of the cell are plastic, however the plastic zone goes slightly bit into the second row of the cells.

Some results of the nonlinear analysis are presented. Figure 15.12 shows a plot of the tangential stress distribution. Comparing the boundary element results with the results from the analytical solution[10] we can observe very good agreement.

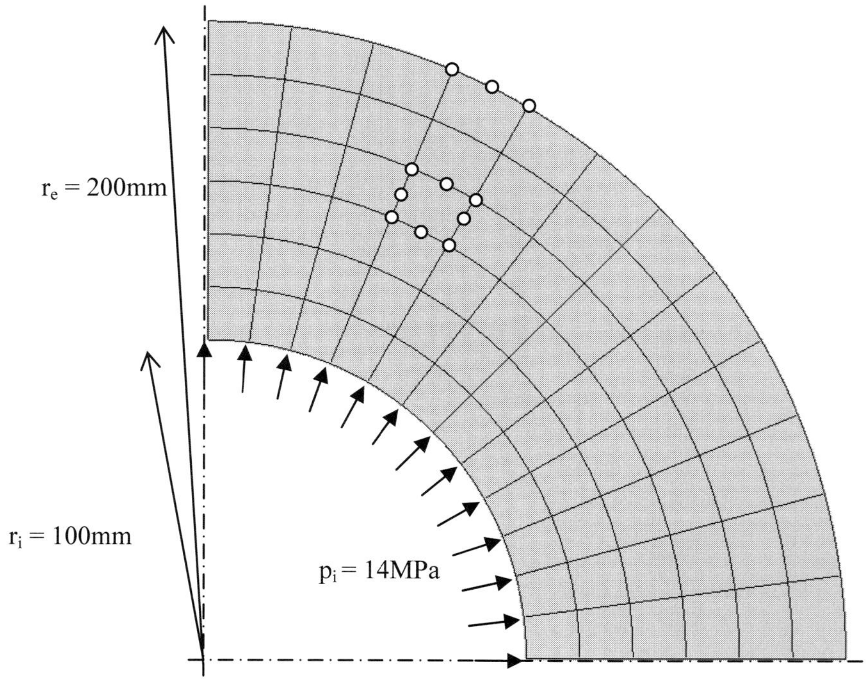

Figure 15.11 Example problem and discretisation used

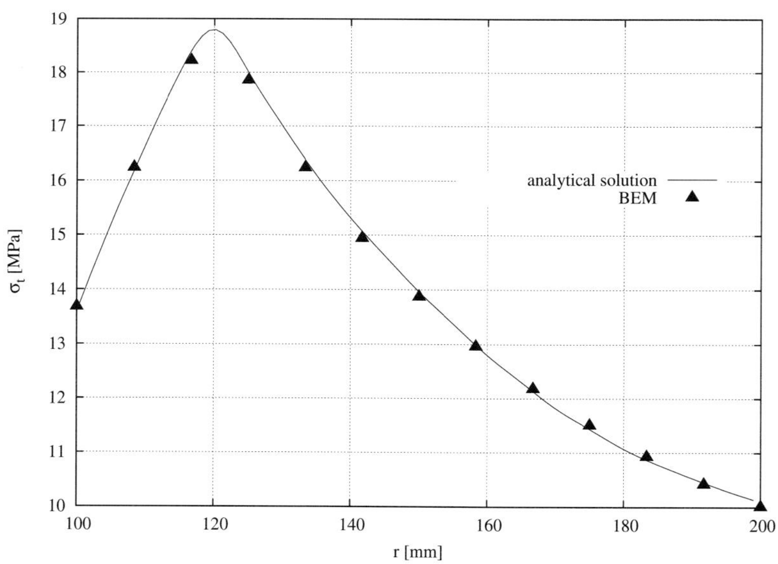

Figure 15.12 Tangential stress distribution in a thick-walled cylinder

15.4 CONTACT PROBLEMS

The second type of nonlinear problem we will discuss here is where boundary conditions change during loading. An example of a contact problem is where interface conditions between regions change. A practical application of this is delamination/slip and crack propagation. For these type of problems we have a condition, similar to the yield condition discussed previously, which determines when the continuity conditions for an interface no longer apply. In the case of crack propagation, for example, we may have a condition based on tensile strength of the material which determines when nodes separate. For problems with joints we may have a criterion based on the angle of friction and cohesion which determines when slip occurs.

In our discussion here we will concentrate on relatively simple problems: ones where contact initially exists and where it is lost due to some conditions being violated. We will see that the theory we will develop[11] can be applied to delamination and joint problems.

15.4.1 Method of analysis

We start with the multi-region method developed in Chapter 10. Consider the beam in Figure 15.13 consisting of two regions.

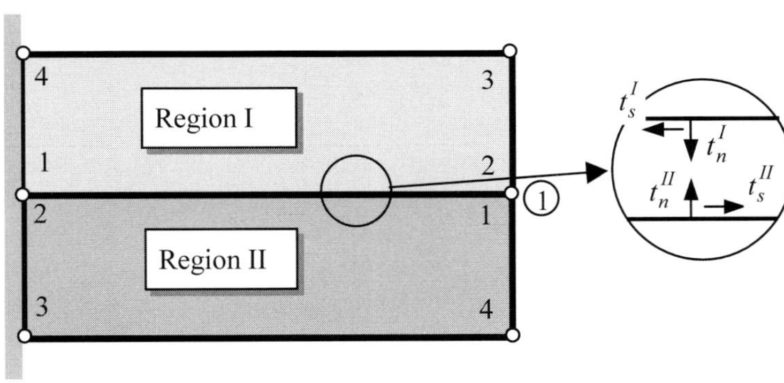

Figure 15.13 Cantilever beam with interface

We recall Equation 11.27 that can be used to compute the tractions at the interface $\{\mathbf{t}\}_c$. For regions I and II we have

$$\{\mathbf{t}\}_c^I = \{\mathbf{t}\}_{c0}^I + \mathbf{K}^I \{\mathbf{u}\}_c^I \quad ; \quad \{\mathbf{t}\}_c^{II} = \{\mathbf{t}\}_{c0}^{II} + \mathbf{K}^{II} \{\mathbf{u}\}_c^{II} \tag{15.58}$$

where $\{\mathbf{t}\}_{co}$ is a vector of tractions assuming fixed interface displacements, $\mathbf{K}$ is the stiffness matrix of the interface nodes and $\{\mathbf{u}\}_c$ is a vector containing displacements at the interface nodes. Since the beam is fixed on the left our problem has only 2 interface unknowns.

For contact problems it is convenient to work with components in direction normal to interface t_n and in a direction tangential to interface t_s instead of global components t_x and t_y. Also, to separate delamination and slip it is required that the local components be used for the displacements as well. The relationship between global and local components is given by:

$$\bar{\mathbf{t}} = \mathbf{T}_g^T \mathbf{t} \;\; ; \;\; \mathbf{u} = \mathbf{T}_g \bar{\mathbf{u}} \tag{15.59}$$

where $\mathbf{T}_g$ is the transformation matrix, as discussed in Chapter 3 and

$$\bar{\mathbf{t}} = \begin{Bmatrix} t_n \\ t_s \end{Bmatrix} \;;\; \mathbf{t} = \begin{Bmatrix} t_x \\ t_y \end{Bmatrix} \;;\; \bar{\mathbf{u}} = \begin{Bmatrix} u_n \\ u_s \end{Bmatrix} \;;\; \mathbf{u} = \begin{Bmatrix} u_x \\ u_y \end{Bmatrix} \tag{15.60}$$

In terms of local components equations (13.58) are rewritten as

$$\bar{\mathbf{t}}_c^I = \mathbf{T}_g \mathbf{t}_{c0}^I + \bar{\mathbf{K}}^I \mathbf{u}_c^I \;\; ; \;\; \bar{\mathbf{t}}_c^{II} = \mathbf{T}_g \mathbf{t}_{c0}^{II} + \bar{\mathbf{K}}^{II} \mathbf{u}_c^{II} \tag{15.61}$$

where $\bar{\mathbf{K}}^N$ is the transformed stiffness matrix , i.e.,

$$\bar{\mathbf{K}}^N = \mathbf{T}_g^T \mathbf{K}^N \mathbf{T} \tag{15.62}$$

The conditions at the interface normally stipulate that the equations of equilibrium and compatibility have to be satisfied , i.e.,

$$\bar{\mathbf{t}}^I = \bar{\mathbf{t}}^{II} \;\; ; \;\; \bar{\mathbf{u}}^I = \bar{\mathbf{u}}^{II} \tag{15.63}$$

We may now define conditions for compatibility. For example the condition

$$t_n \le T \tag{15.64}$$

stipulates that the traction normal to the interface has to be smaller than or, at most, equal to the tensile strength, *T*, of the material. If t_n has reached *T* then delamination occurs, that is, the compatibility condition is no longer applied to that point in the direction normal to the interface.

Analogous to plasticity the yield function can be written as

$$F_1(t_n) = t_n - T \tag{15.65}$$

Another condition may be that the shear traction is limited by

$$|t_s| \le c + t_n \tan\varphi \tag{15.66}$$

where c is the cohesion and φ the angle of friction. If $|t_s| = c + t_n \tan\phi$ slip occurs, that is, the compatibility condition is no longer applied to that point in the direction tangential to the interface.

The corresponding yield function is written as:

$$F_2(t_s, t_n) = |t_s| - c - t_n \tan\varphi \tag{15.67}$$

The consequence is that when either F_1 or F_2 is zero the assembly is changed: Instead of adding all stiffness coefficients we assemble the corresponding stiffness coefficients for region *I* and *II* into different locations in **K**.

Consider, for example, the problem in Figure 15.13. The equations for compatibility at node 1 are (since only one node is involved we have left out the subscript denoting the local (region) node number):

$$u_{n1} = u_n^I = u_n^{II} \quad \text{and} \quad u_{s1} = u_s^I = u_s^{II} \tag{15.68}$$

With the vector of interface unknown only involving the ones at node 1

$$\mathbf{u}_c = \begin{Bmatrix} u_{n1} \\ u_{s1} \end{Bmatrix} \tag{15.69}$$

For the example with only one interface node we may write for the region stiffness matrices (see 11.2.2.)

$$\mathbf{K}^I = \begin{bmatrix} t_{nn}^I & t_{ns}^I \\ t_{sn}^I & t_{ss}^I \end{bmatrix} \quad \text{and} \quad \mathbf{K}^{II} = \begin{bmatrix} t_{nn}^{II} & t_{ns}^{II} \\ t_{sn}^{II} & t_{ss}^{II} \end{bmatrix} \tag{15.70}$$

and the following assembled interface stiffness matrix is obtained

$$\mathbf{K} = \begin{bmatrix} t_{nn}^I + t_{nn}^{II} & t_{ns}^I + t_{ns}^{II} \\ t_{sn}^I + t_{sn}^{II} & t_{ss22}^I + t_{ss}^{II} \end{bmatrix} \tag{15.71}$$

If $F_1 = 0$ then the normal displacement and - as a consequence – also the shear displacement of region *I* are independent of region *II*. The vector of interface unknown is expanded to

$$\mathbf{u}_c = \begin{Bmatrix} u_n^I \\ u_s^I \\ u_n^{II} \\ u_s^{II} \end{Bmatrix} \tag{15.72}$$

And the stiffness matrix is defined as

$$\mathbf{K} = \begin{bmatrix} t_{nn}^{I} & t_{ns}^{I} & 0 & 0 \\ t_{sn}^{I} & t_{ss}^{I} & 0 & 0 \\ 0 & 0 & t_{nn}^{II} & t_{ns}^{II} \\ 0 & 0 & t_{sn}^{II} & t_{ss}^{II} \end{bmatrix} \tag{15.73}$$

If F_2=0 then slip occurs and compatibility does not apply to the shear displacement. The vector of interface unknown is given by

$$\mathbf{u}_c = \begin{Bmatrix} u_{n1} \\ u_s^{I} \\ u_s^{II} \end{Bmatrix} \tag{15.74}$$

In the stiffness matrix only the terms associated with the normal components are added

$$\mathbf{K} = \begin{bmatrix} t_{nn}^{I} + t_{nn}^{II} & t_{ns}^{I} & t_{ns}^{II} \\ t_{sn}^{I} & t_{ss}^{I} & 0 \\ t_{sn}^{II} & 0 & t_{ss}^{II} \end{bmatrix} \tag{15.75}$$

15.4.2 Solution procedure

Only in exceptional cases will it occur that a point is reached when the yield functions are exactly 0. As with plasticity we will have the condition that if the yield function is checked after the application of a load increment with traction $\bar{\mathbf{t}}$ we find that either $F_1(\bar{\mathbf{t}})>0$ or $F_2(\bar{\mathbf{t}})>0$. In the first case this means that the material has been stressed beyond the tensile strength, in the second that the friction law has been violated.

In the first instance the excessive stress, i.e., the one which caused the yield condition to be violated is computed by

$$\Delta t_p = \bar{t}_n - T \tag{15.76}$$

whereas in the second case

$$\Delta t_p = |\bar{t}_s| - c - \bar{t}_n \tan\varphi \qquad (15.77)$$

We now propose the following solution procedure:

1. The system is solved in the normal way using the interface compatibility and equilibrium conditions.
2. The yield conditions F_1 and F_2 are computed at each interface node. If both are zero then the analysis is finished.
3. If one of the yield conditions is greater than zero residual tractions are computed according to equations (15.76) or (15.77).
4. The interface matrix **K** is re-assembled taking into consideration the relaxed continuity conditions for interface points which are separating or slipping.
5. The system is solved with the residual tractions applied as loading in the opposite direction.
6. Points 2 to 5 are repeated until the yield conditions are satisfied at all interface nodes.

The extension of the method to three dimensions is straightforward. In 3-D we have two instead of one shear traction ($\bar{t}_{s1}$ and $\bar{t}_{s1}$) and when we check the yield condition we have to work with a resultant shear traction. This is given by

$$|\bar{t}_s| = \sqrt{(\bar{t}_{s1})^2 + (\bar{t}_{s2})^2} \qquad (15.78)$$

15.4.3 Example of application

As an example of application we present an analysis of the delamination of a cantilever beam. The beam consists of two finite regions described by quadratic boundary elements, as shown in Figure 15.14. At the interface the tensile strength of the material was assumed to be zero. Shear loading is applied to the bottom half of the beam as shown. Figure 15.15 shows the distribution of normal stress at the interface after the linear analysis. It can be clearly seen that the yield condition for tension is violated. Figure 15.16 shows that after iteration step 1, delamination starts as a consequence. Further examples of the application of the method to the modelling of faulted rock can be found in [12,13]. The method can also be applied to the simulation of dynamic crack propagation[14].

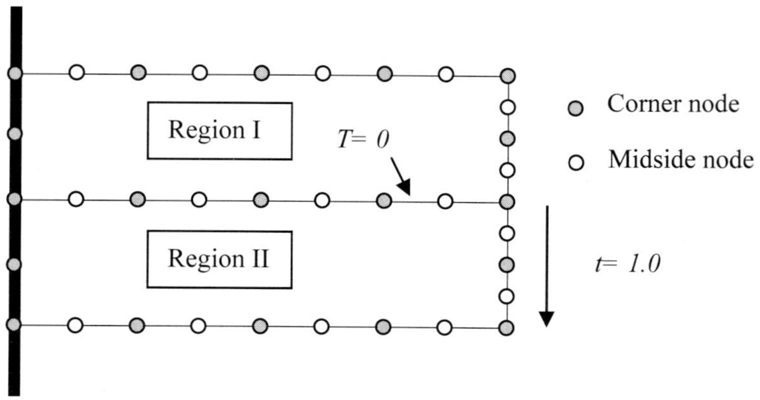

Figure 15.14 Mesh used for cantilever analysis

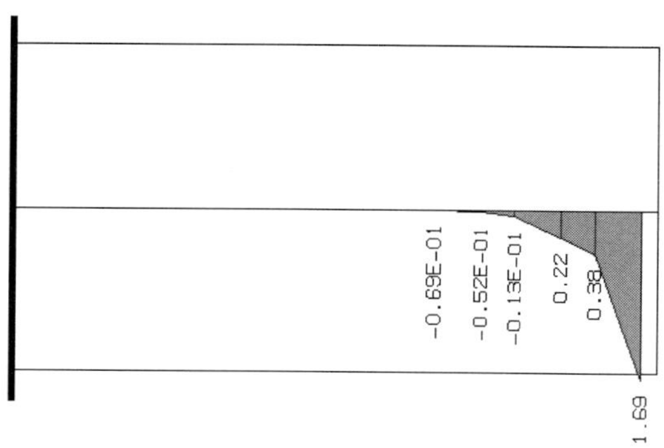

Figure 15.15 Distribution of normal stress at the interface

Figure 15.16 Displaced shape of beam after first iteration

15.5 CONCLUSIONS

In this chapter we attempted to show that the treatment of nonlinear problems is almost as straightforward as with the FEM. Only two types of nonlinear problems have been discussed: plasticity and contact problems. In the first type, additional volume discretisation is needed and the BEM looses a bit of its attraction. However, it was pointed out that the internal cell discretisation does not add to the number of unknown and that all the advantages of the BEM are still retained. It must be admitted, however, that the effort in programming, especially dealing with hyper-singular integrations, is rather involved.

We have also found that for contact problems the BEM is better suited than the FEM because the interface stresses required for checking the yield conditions are directly obtained from the solution. In the FEM these would have to be determined by differentiation of the computed displacement field. The purpose of this chapter has been to demonstrate that any type of nonlinear problem can be solved with the BEM. However, the computer implementation has not been discussed in any detail here because it would be beyond the scope of an introductory text.

15.6 REFERENCES

1. Hill R.(1950) The Mathematical Theory of Plasticity. Oxford University Press.
2. Pande G., Beer G. and Williams J. (1990) Numerical methods in rock mechanics. J.Wiley, Chichester.
3. Simo J.C. and Hughes T.J.R. (1998) Computational Inelasticity. Volume 7 of Interdisciplinary Applied Mathematics. Springer-Verlag New York.
4. Ortiz M. and Popov E. (1985) Accuracy and stability of integration algorithms for elastoplastic constitutive relations. *Int. J. Numer. Methods Eng.* **21:** 1561-1576.
5. Cormeau I.C. (1975) Numerical stability in quasi-static elasto-viscoplasticity. *Int. J. Numer. Methods Eng.* **9** (1).
6. Ribeiro S.A.T. (2006) Elastoplastic boundary element method with adaptive cell generation. Monographic Series TU Graz, Austria.
7. Venturini W.S. (1983) Boundary Element Method in Geomechanics. Springer, Berlin.
8. Telles J.C.F. (1983) The Boundary Element Method Applied to Inelastic Problems. Springer, Berlin.
9. Gao X.-W. and Davies T.G. (2002) Boundary Element Programming in Mechanics. Cambridge University Press, Cambridge.
10. Prager W. and Hodge P.G. (1951) Theory of Perfectly Plastic Solids. John Wiley & Sons, New York.
11. Beer G. (1993) An efficient numerical method for modeling initiation and propagation of cracks along material interfaces. *Int. J. Numer. Methods Eng.* **36** (21), 3579-3594.

12. Beer G. and Poulsen B.A. (1994) Efficient numerical modelling of faulted rock using the boundary element method. *Int. J. Rock. Mech. Min. Sci. & Geomech. Abstr.* **31** (5): 485-506.
13. Zaman, Booker and Gioda (eds) Modeling and Applications in Geomechanics. J.Wiley, Chichester.
14. Tabatabai-Stocker B. and Beer G. (1998) A boundary element method for modeling cracks along material interfaces in transient dynamics. *Comun. Numer. Meth. Engng.* **14**:355-365.

16

Coupled Boundary Element/ Finite Element Analysis

Marriage à la mode
O. C. Zienkiewicz

16.1 INTRODUCTION

In the introduction we compared the boundary element method (BEM) with its main "competitor" the finite element method (FEM). Although in the specific example the impression was given that a BEM analysis would be superior to the FEM this was not meant to imply that this is always the case. In a famous paper written more than two decades ago[1], O.C. Zienkiewicz pointed out that benefits could be gained by combining the two methods of analysis, thereby gaining the "best of both worlds". This was at a time when BEM protagonists claimed that the BE could do everything better and there was almost no collaboration between the two groups. Zienkiewicz, in his inimitable style, chose the title "Marriage a la mode" which shows a double meaning: marriage a la mode means a marriage of convenience, not love, but also there is a double meaning with the word mode (displacement mode =shape function).

There are several reasons why one would want to consider the combination of the two methods:

- Some problems, for example those involving highly heterogeneous material, require additional effort to solve with the BEM.
- For some problems no fundamental solutions of the governing differential equations can be found or in certain cases the solutions are extremely complex.

- Users familiar and happy with a FEM package would want to upgrade the program capabilities by including, for example, an efficient modelling of an infinite domain. Many well known commercial packages have capabilities for the specification of a user defined element stiffness, so the implementation of the Subroutine Stiffness_BEM introduced in Chapter 11 would be fairly straight forward.

As will be seen here, the coupling of the FEM and BEM is not very difficult. Indeed, we have already set the foundation for this in Chapter 11, where we explained how the "stiffness matrix" of a region can be computed. In essence, for coupling we have to find a way of harmonising the differences between the two methods. The main difference is that nodal tractions are used as primary unknown in the BEM, whereas nodal point forces are used in the FEM. The "stiffness matrix" obtained for the BE region turned out to be unsymmetrical and this may cause some problems because symmetric solvers are usually employed in the FEM. Therefore we will also show how the BEM stiffness matrix can be "symmetrised".

16.2 COUPLING THEORY

There are basically two approaches to coupling the boundary and finite element methods. In the first approach, the BE region is treated as a large finite element and its stiffness is computed and assembled into the global stiffness matrix. In the second approach, finite elements are treated as equivalent BE regions and their "stiffness matrix" is determined and assembled, as explained in Chapter 10 for multiple regions. The choice of coupling method depends mainly on the software available for the implementation, i.e., if boundary element capabilities are to be added to a finite element program, or finite element capabilities to a boundary element one. In the following we will discuss the theoretical basis and implementation of each approach. The coupling theory is discussed using problems in elasticity. However, as demonstrated throughout this book, potential problems can be considered in an analogous way.

16.2.1 Coupling to finite elements[2]

The FEM leads to a system of simultaneous equations which relate displacements at all the nodes to *nodal forces*. In the BEM, on the other hand, a relationship between nodal displacements and *nodal tractions* is established.

Consider the cantilever beam in Figure 16.1 consisting of one BE region connected to two finite elements. We refer to the assembly of two finite elements as *Finite Element Region*. Following the procedures in Chapter 11, we can obtain for the BE region a relationship between tractions $\{\mathbf{t}\}_c$ and displacements $\{\mathbf{u}\}_c$ at the interface (equation 11.27)

$$\{\mathbf{t}\}_c = \{\mathbf{t}\}_{c0} + \mathbf{K}_{BE}\{\mathbf{u}\}_c \tag{16.1}$$

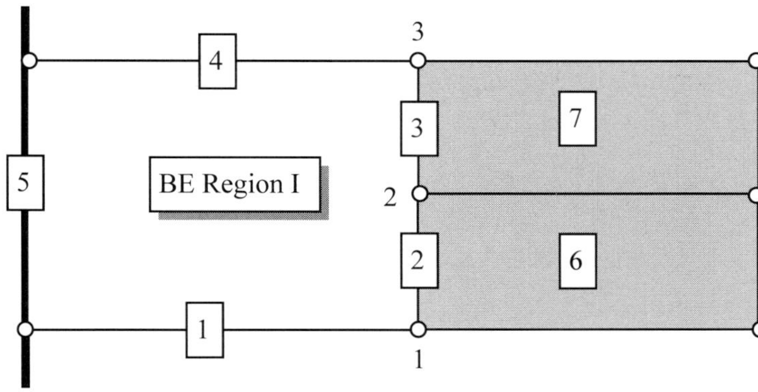

Figure 16.1 Cantilever beam: discretisation into finite and boundary elements

In the above, $\{\mathbf{t}\}_{c0}$ is a vector containing tractions, if all interface displacements are zero, and $\mathbf{K}_{\mathrm{BE}}$ is the pseudo "stiffness matrix" of the BE region.

For the example problem we have

$$\{\mathbf{t}\}_c = \begin{Bmatrix} t_{x1} \\ t_{y1} \\ t_{x2} \\ t_{y2} \\ t_{x3} \\ t_{y3} \end{Bmatrix} \; ; \; \{\mathbf{u}\}_c = \begin{Bmatrix} u_{x1} \\ u_{y1} \\ u_{x2} \\ u_{y2} \\ u_{x3} \\ u_{y3} \end{Bmatrix} \tag{16.2}$$

For the finite element region we can write a relationship between interface displacements and interface nodal forces as

$$\{\mathbf{F}\}_c = \{\mathbf{F}\}_{co} + \mathbf{K}_{FE}\{\mathbf{u}\}_c \tag{16.3}$$

where $\{\mathbf{F}\}_{c0}$ is the force vector at the interface when all interface displacements are zero and $\mathbf{K}_{FE}$ the condensed stiffness matrix of the finite element region which involves only the interface nodes. In equations (16.1) and (16.3) we have already implicitly assumed that compatibility conditions are satisfied (i.e., displacements of the BE and FE regions are the same at nodes 1-3). Figure 16.2 shows the forces that exist at the interface. For the BE region these are boundary stresses, whereas for the FE region these are nodal point forces.

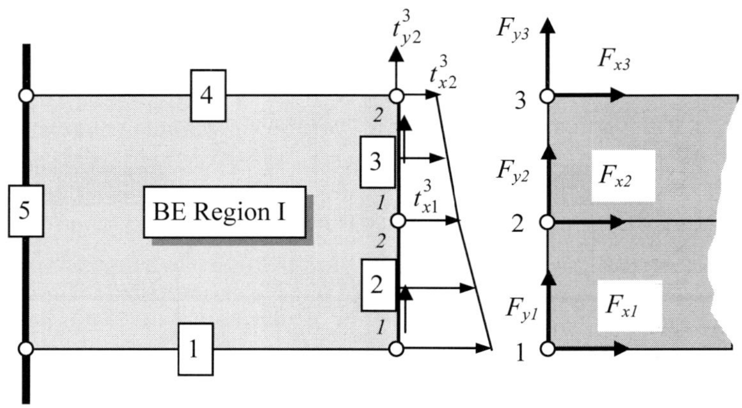

Figure 16.2 Interface between finite element and boundary regions showing interface forces

In the first method of coupling we propose that the boundary tractions be converted into equivalent nodal point forces.

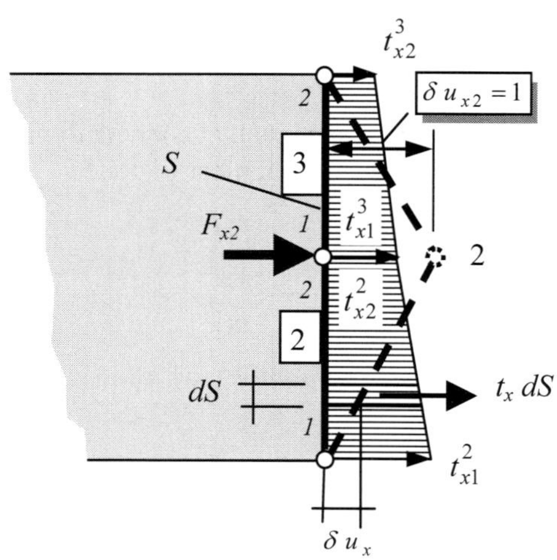

Figure 16.3 Calculation of F_{x2} by principle of virtual work

To compute the x-component of the equivalent nodal point force at node 2, for example, we apply a unit virtual displacement in the x-direction at that point (Figure 16.3). For equilibrium to be satisfied, the work done by the tractions must be equal to that done by the equivalent nodal forces at node 2.

This gives

$$F_{x2} \cdot 1 = \int_S t_x \delta u_x dS \tag{16.4}$$

Substituting the interpolation for tractions and displacements

$$t_x = \sum_{n=1}^{2} N_n t_{xn}^e \quad ; \quad \delta u_x = N_j \, \delta u_{xj}^e = N_j \cdot 1 \tag{16.5}$$

(where j is 2 for element 2 and 1 for element 3) we obtain

$$F_{x2} \cdot 1 = \int_{S_2} (N_1 t_{x1}^2 + N_2 t_{x2}^2) N_2 dS_2 + \int_{S_3} (N_1 t_{x1}^3 + N_2 t_{x2}^3) N_1 dS_3 \tag{16.6}$$

A second equation can be obtained by applying a virtual displacement in y direction. Based on this approach a general equation can be derived for computing the equivalent nodal point force at a point k

$$\mathbf{F}_k = \sum_{\substack{Elements \\ with\ k}} \sum_{n=1}^{N} \mathbf{N}_{jn}^e \mathbf{t}_n^e \tag{16.7}$$

where the outer sum is over all boundary elements that connect to point k, the inner sum is over all nodes of the element and j is the local number of node k.

For 2-D problems we have

$$\mathbf{N}_{jn}^e = N_{jn}^e \mathbf{I} \tag{16.8}$$

where $\mathbf{I}$ is the unit matrix and

$$N_{jn}^e = \int_{S_e} N_j N_n dS_e \tag{16.9}$$

The integration over elements can be conveniently carried out using numerical integration (Gauss Quadrature) with two points. For 2-D problems this gives

$$N_{jn} = \int_{\xi=-1}^{1} N_j N_n J d\xi \approx \sum_{m=1}^{2} N_j N_n J W_m \tag{16.10}$$

whereas for 3-D problems

$$N_{jn} = \int_{\eta=-1}^{1}\int_{\xi=-1}^{1} N_j N_n J d\xi d\eta \approx \sum_{k=1}^{2}\sum_{m=1}^{2} N_j N_n J W_m W_k \qquad (16.11)$$

Equation (16.1) can now be expressed in terms of equivalent nodal point forces by pre-multiplying with **N**

$$\{\mathbf{F}\}_c = \mathbf{N}\{\mathbf{t}\}_c = \mathbf{N}\{\mathbf{t}\}_{c0} + \mathbf{N}\mathbf{K}_{BE}\{\mathbf{u}\}_c \qquad (16.12)$$

where **N** is assembled from element contributions $\mathbf{N}^e_{jn}$.

For the example in Figure 16.1 this matrix is given by

$$\mathbf{N} = \begin{bmatrix} \mathbf{N}^2_{11} & \mathbf{N}^2_{12} & 0 \\ \mathbf{N}^2_{21} & \mathbf{N}^2_{22} + \mathbf{N}^3_{11} & \mathbf{N}^3_{12} \\ 0 & \mathbf{N}^3_{21} & \mathbf{N}^3_{22} \end{bmatrix} \qquad (16.13)$$

Matrix $\mathbf{NK}_{BE}$ is now a "true" stiffness matrix in the finite element sense, i.e., one that relates nodal point displacements to nodal point forces. However, since $\mathbf{K}_{BE}$ is not symmetric, this matrix is also unsymmetrical.

Although there is in principle no problem in dealing with unsymmetrical matrices, and they actually occasionally do occur sometimes in nonlinear FEM analysis, some solvers used for finite elements are specialised in dealing with symmetric system of equations and, in some cases, it may be convenient if all stiffness matrices are symmetric. One way of getting a symmetric matrix is to use the principle of minimum potential energy to derive the equilibrium equations at the interface[3].

Considering for simplicity only the forces/tractions due to interface displacements we can compute the total potential energy at the interface as

$$\Pi = \{\mathbf{u}\}_c^T \mathbf{K}_{FE}\{\mathbf{u}\}_c + \{\mathbf{u}\}_c^T \mathbf{N}\mathbf{K}_{BE}\{\mathbf{u}\}_c \qquad (16.14)$$

where the first expression on the left hand side is the work done by the FE region and the second is the one done by the BE region. Taking the minimum of potential energy we obtain

$$\frac{\partial \Pi}{\partial\{\mathbf{u}\}_c} = \frac{1}{2}\left(\mathbf{K}^T_{FE} + \mathbf{K}_{FE}\right)\{\mathbf{u}\}_c + \frac{1}{2}\left[\left(\mathbf{N}\mathbf{K}_{BE}\right)^T + \mathbf{N}\mathbf{K}_{BE}\right]\{\mathbf{u}\}_c = 0 \qquad (16.15)$$

The operation in the square parentheses means that a symmetric stiffness matrix for the BE region can be obtained by adding the transpose and by halving the result. This way of stating the equilibrium condition is commonly used in the FEM. However, its application here is not quite correct since in the derivation of $\mathbf{K}_{BE}$ an interpolation of both displacement and traction has been assumed at the interface. It has been shown however that acceptable results can be obtained for most applications.

Having obtained a "true" stiffness matrix for the BE region, the further steps in the computation of coupled problems are fairly straightforward. The boundary element region is treated as a super (finite) element and its stiffness is assembled in the usual way to obtain the system equations.

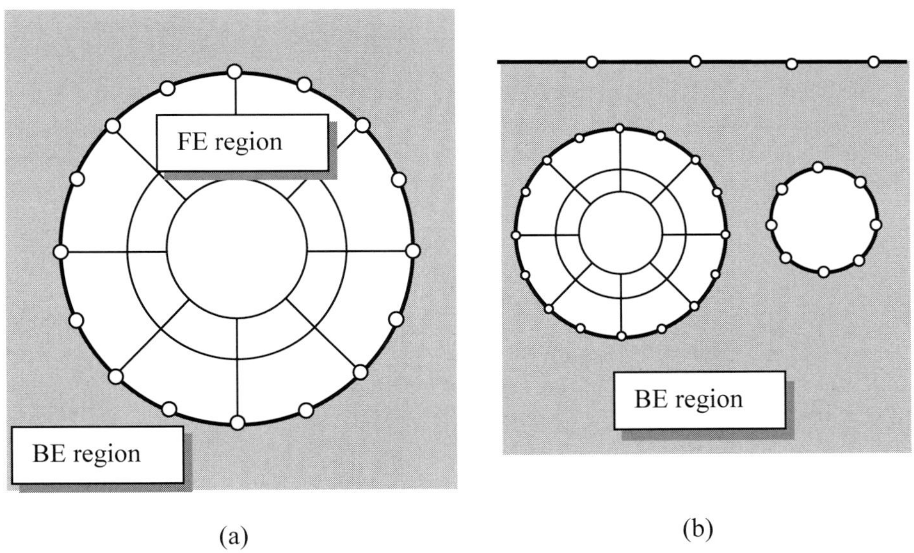

Figure 16.4 Fully and partially coupled discretisations

In the implementation we distinguish between fully coupled and partially coupled analyses. In a fully coupled analysis all nodes of the boundary element region are connected to the finite element region and no loading is assumed at the interface. An example of this type of analysis is the problem of an excavation in an infinite domain solved by a coupled discretisation, as shown in Figure 16.4 (a). In this case, the infinite boundary element region can be considered as a large finite element which accurately represents the effect of the infinite domain. This is a good example of gaining the "best of both worlds" as the alternative to the coupled mesh shown would be either to extend the finite element region a large distance away and apply artificial boundary conditions there or to use infinite finite elements. Both methods require more mesh generation and computational effort and result in loss of accuracy. The reason for a coupled analysis may be that a thin lining is to be modelled, which is done more efficiently with shell elements. In a fully coupled analysis only the stiffness matrix of the BE region needs to be determined and pre-multiplied with **N** in order to change it to a true stiffness matrix that can be assembled.

An example of a partially coupled analysis is shown in Figure 16.4 (b). Here we consider the additional effect of a ground surface and another (existing) excavation. In a partially coupled analysis we first solve the problem with the interface nodes fixed and obtain an interface traction vector $\{\mathbf{t}\}_{c0}$. Then we compute the pseudo stiffness matrix of the region. Before we assemble our finite element system, both $\{\mathbf{t}\}_{c0}$ and $\mathbf{K}_{BE}$ have to be pre-multiplied with **N,** yielding a nodal point force vector as well as a stiffness matrix.

The only additional programming required for the implementation of a coupled analysis capability is the assembly of transformation matrix **N** and the pre-multiplication of the stiffness matrix $\mathbf{K}_{BE}$ and, in the case of a partially coupled analysis, the traction vector $\mathbf{t}_{BE}$ with this matrix. If required, a "symmetrisation" procedure may be applied as explained above.

We develop a function **Mtrans** which returns the transformation matrix **N**, an array of dimension Ndofsc x Ndofsc, where Ndofsc is the number of interface degrees of freedom. The input parameters of this function are number of interface elements, number of interface nodes, incidence vector for each element and coordinates of interface nodes.

```
FUNCTION MTrans(Nelc,Ndofsc,xPc,Incic)
!-----------------------------------------
!   Function returns the assembled matrix N
!   for the conversion of a pseudo stiffnes matrix
!   into a true stiffness matrix
!-----------------------------------------
INTEGER, INTENT (IN):: Nelc          ! No. of interface elements
INTEGER, INTENT (IN):: Ndofsc        ! No. of interface nodes
REAL, INTENT (IN)   :: xPc(:,:)      ! Coords of interface nodes
INTEGER, INTENT (IN):: Incic(:,:)    ! Incidences of interface elem
REAL      ::  Mtrans(Ndofsc,Ndofsc)  ! Function returns array
REAL      ::  MMjn(Ndof,Ndof)
REAL      ::  Glcor(2),Wi(2),Wie(2),Ni(Nodel),Elcor(Cdim,Nodel)
REAL      ::  xsi,eta,Jac,Weit,Mjn
INTEGER   :: Inci(nodel)
Mtrans= 0.
ldim= Cdim - 1
Mi= 2 ; Ki= 1 ; Wie=1.0
CALL Gauss_coor(Glcor,Wi,2)              !  2x2 integration
IF (Cdim == 3) THEN
 Ki=2
 Wie= Wi
END IF
Interface_elements: &
DO Nel= 1,Nelc
  Inci(:)= Incic(nel,:)
  Elcor(:,:)= xPc(:,Inci(:))
 Nodes_of_elem1: &
 DO j=1,nodel
   Nodes_of_elem2: &
   DO n=1,nodel
     Mjn= 0.
```

```
      Gauss_points_xsi: &
      DO m=1,Mi
        xsi= Glcor(m)
        Gauss_points_eta: &
        DO k=1,Ki
          eta= Glcor(k)
          Weit= Wi(m)*Wie(k)
          CALL Serendip_func(Ni,xsi,eta,ldim,nodel,Inci)
          Jac= Jacobean(xsi,eta,zeta,ldim,nodel,Inci,Elcor
          Mjn= Mjn + Ni(j)*Ni(n)*Jac*Weit
        END DO &
        Gauss_points_eta
      END DO &
      Gauss_points_xsi
        MMjn= 0.
      DO nd=1,ndof
        MMjn(nd,nd)= Mjn
      END DO
      nrow= (Inci(j)-1)*Ndof+1
      ncol= (Inci(n)-1)*Ndof+1
      Mtrans(nrow:,ncol:)= MMjn
    END DO &
    Nodes_of_elem2
  END DO &
Nodes_of_elem1
END DO &
Interface_elements
RETURN
END FUNCTION MTrans
```

16.2.2 Coupling to boundary elements

The coupling of finite elements to boundary elements follows the same steps as for the multi-region method discussed in Chapter 11. We may consider an assembly of finite elements as a boundary element region. Using standard FEM procedures we obtain the following system of equations for the finite element region

$$\{\mathbf{F}\}_c = \{\mathbf{F}\}_{co} + \mathbf{K}_{FE}\{\mathbf{u}\}_c \tag{16.16}$$

where the notation has been defined at the beginning of this chapter. The equations which we get for each region in the BEM are

$$\{\mathbf{t}\}_c^I = \{\mathbf{t}\}_{c0}^I + \mathbf{K}_{BE}^I\{\mathbf{u}\}_c^I \tag{16.17}$$

where the roman superscript denotes the region number.

For coupling the finite element region all that is required is to convert (16.16) into a form such as (16.17). This is simply the inverse relationship to (16.12), i.e.

$$\{\mathbf{t}\}_c = \mathbf{N}^{-1}\{\mathbf{F}\}_c = \mathbf{N}^{-1}\{\mathbf{F}\}_{c0} + \mathbf{N}^{-1}\mathbf{K}_{FE}\{\mathbf{u}\}_c \tag{16.18}$$

where the inverse of **N** has to be determined. Since **N** is a sparsely populated and diagonally dominant matrix this does not pose any problems.

After having obtained the pseudo stiffness matrix of the finite element region $\mathbf{N}^{-1}\mathbf{K}_{FE}$ and, for partially coupled problems, the equivalent traction vector $\mathbf{N}^{-1}\{\mathbf{F}\}_{c0}$ we proceed in the same way as for multi-region problems.

16.3 EXAMPLE

The example presented here is that of a circular excavation in an infinite domain. The problem geometry, material properties and initial stress field assumed are shown in figure 16.5 (a). The discretisation into quadratic finite and boundary elements is shown in Figure 16.5 (b). One plane of symmetry is assumed.

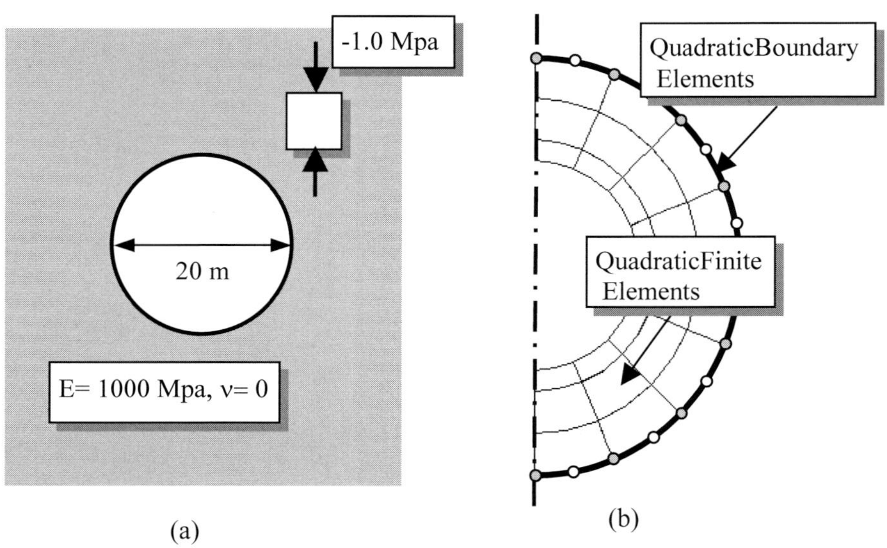

Figure 16.5 Example problem specification and coupled mesh used for analysis

Some results of the analysis are shown here. Figure 16.6 shows the displaced shape and 16.7 the distribution of the maximum compressive stress in the finite element region. The distribution of maximum compressive stress along a nearly horizontal line through the Gauss points of finite elements and inside the boundary element region shows good agreement with the theoretical results.

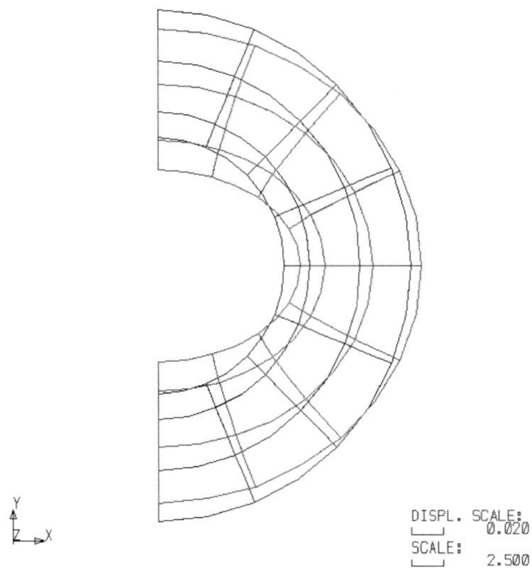

Figure 16.6 Displaced shape after excavation

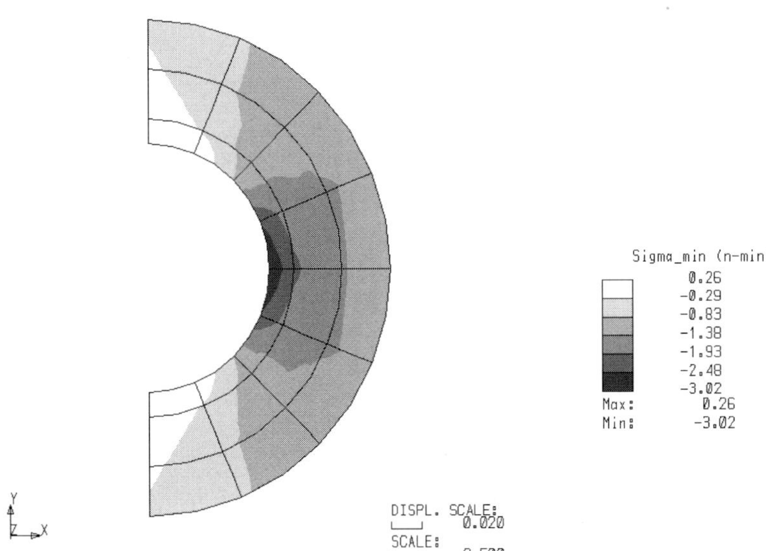

Figure 16.7 Contours of maximum compressive stress

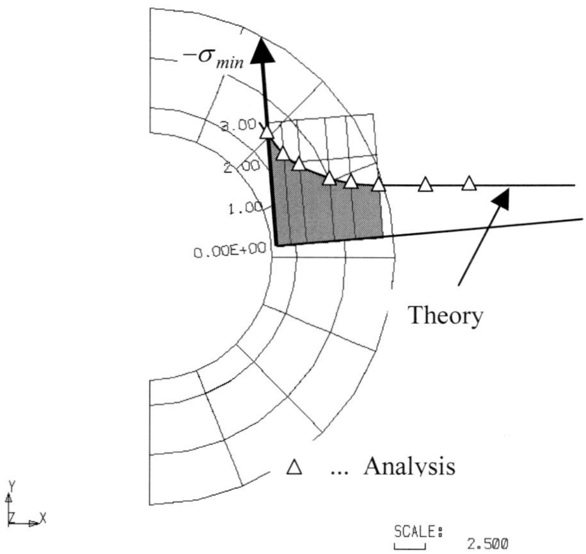

Figure 16.8 Distribution of maximum compressive stress, comparison with theory

16.4 DYNAMICS

Here we extend the coupling method to dynamics. The dynamic equilibrium equations which arise from finite element discretisation (see Bathe[4]) can be written as

$$[\mathbf{M}]\{\ddot{\mathbf{u}}\}+[\mathbf{C}]\{\dot{\mathbf{u}}\}+[\mathbf{K}]\{\mathbf{u}\}=\{\mathbf{F}\} \tag{16.19}$$

where $[\mathbf{M}]$, $[\mathbf{C}]$, $[\mathbf{K}]$ are the assembled mass, damping and stiffness matrices and $\{\ddot{\mathbf{u}}\},\{\dot{\mathbf{u}}\},\{\mathbf{u}\}$ are the acceleration, velocity and displacement vectors. The time may be discretised into n time steps of size Δt. Assuming an average acceleration within the time step the system of differential equations can be transformed into a system of algebraic equations (Newmark method[4])

$$[\bar{\mathbf{K}}]\{\mathbf{u}(t)\}=\{\bar{\mathbf{F}}\} \tag{16.20}$$

where $t = n\Delta t$ and $t_{-} = (n-1)\Delta t$

The “dynamic stiffness matrix” is given by:

$$\left[\overline{\mathbf{K}}\right]=\frac{4}{\Delta t^2}[\mathbf{M}]+\frac{2}{\Delta t}[\mathbf{C}]+[\mathbf{K}] \tag{16.21}$$

and

$$\begin{aligned}\left\{\overline{\mathbf{F}}\right\}=[\mathbf{M}]\left(\frac{4}{\Delta t^2}\{\mathbf{u}(t_-)\}+\frac{4}{\Delta t}\{\dot{\mathbf{u}}(t_-)\}+\{\ddot{\mathbf{u}}(t_-)\}\right)\\+[\mathbf{C}]\left(\frac{2}{\Delta t}\{\mathbf{u}(t_-)\}+\{\dot{\mathbf{u}}(t_-)\}\right)+\{\mathbf{F}(t_-)\}\end{aligned} \tag{16.22}$$

Since we have already worked out a “dynamic stiffness matrix” of the boundary element region in Chapter 14 the coupling procedure is now straightforward. For a fully coupled problem the system of equations is given by

$$\left(\left[\mathbf{N}\overline{\mathbf{K}}\right]_{BE}+\left[\overline{\mathbf{K}}\right]_{FE}\right)\{\mathbf{u}(t)\}=\left\{\overline{\mathbf{F}}\right\}_{BE}+\left\{\overline{\mathbf{F}}\right\}_{FE} \tag{16.23}$$

16.4.1 Example

The example is that of a concrete column embedded in a semi-infinite soil mass. The description of the problem can be seen in Figure 16.9. The top of the column is subjected to a suddenly applied load *p(t)* of 1 MN/m^2. The material properties for the column are: spec. weight= 2500 kg/m^3, E=30 000 MN/m^2, ν=0.2. For the soil we have: spec. weight= 2000 kg/m^3, E=100 MN/m^2, ν=0.2.

Figure 16.9 Description of example

Figure 16.10 shows the mesh used for the analysis it consists of a finite element region that describes the column and a boundary element region that describes the semi-infinite ground. The mesh has 1500 degrees of freedom. Figure 16.11 shows the time-dependent displacements at the top of the column obtained from the analysis. The results compare well with a reference solution with the FEM that used 1 Million elements.

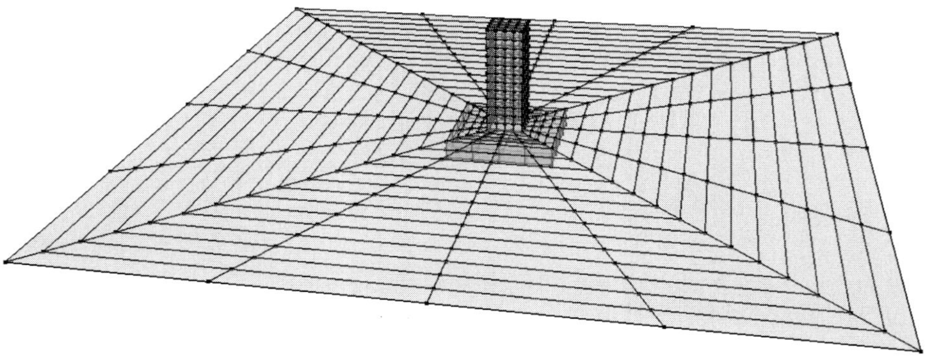

Figure 16.10 Coupled mesh

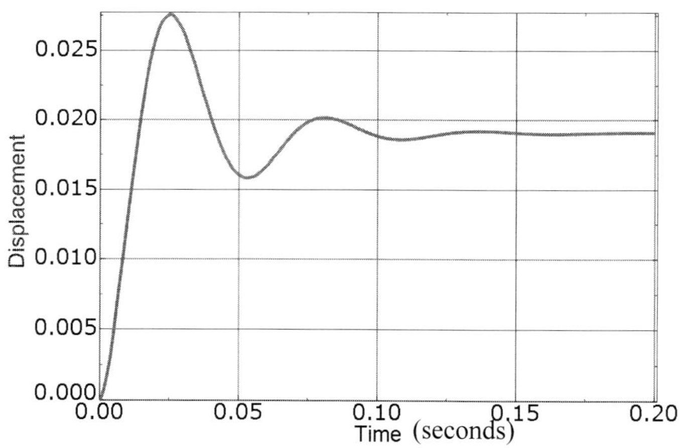

Figure 16.11 Displacement at the top of the column

16.5 CONCLUSION

In this chapter we have shown how the capability of a finite or boundary element program can be easily extended so that the advantages of both methods can be combined giving the user "the best of both worlds". We have shown one example where the capability of the BEM in dealing with infinite domains was exploited. Many other such examples exist and we will show in the next chapter one industrial application that could not have been analysed with either method given the restrictions regarding time and computing resources.

Although it is true that both methods can deal with almost any problem that arises in engineering (and comprehensive text books on the FEM and BEM assert this), it is also clear that they are more appropriate for some applications and less so for others. It should have become clear to the reader, for example, that the BEM is well suited for problems involving a small ratio of boundary surface to volume. Extreme cases of this are problems which can be considered as involving an infinite volume. Such problems exist, for example, in geomechanics[5], where the earth's crust has no lateral boundaries. Another extreme where the ratio boundary surface to volume is very large is the application to thin shell structures.

Another aspect is the importance that is given to surface stresses. As we have seen in Chapter 9, stresses at the surface are computed more accurately with the BEM than with the FEM. We have shown that problems where "body forces" occur in the domain, as for example plasticity problems, etc., can be handled with the BEM but it has to be admitted that implementation is much more involved than with the FEM. A final aspect which is also gaining more importance, is the suitability of the methods for implementation with regards to computer hardware. The future seems to lie in massive parallel processing and we have seen in Chapter 8 that the BEM seems to lend itself to parallel programming.

16.6 REFERENCES

1. Zienkiewicz O.C. ,Kelly D.W. and Bettess P. (1979) Marriage a la mode- the best of both worlds (finite elements and boundary integrals) Chapter 5 of Energy Methods in Finite Element Analysis (ed. R.Glowinski, E.Y. Rodin and O.C.Zienkiewicz), pp. 82-107, Wiley, London.
2. Beer G. (1977) Finite element, boundary element and coupled analysis of unbounded problems in elastostatics. *Int. J. Numer. Methods Eng.*, **11,** 355-376
3. Beer G. (1998) Marriage a la mode (finite and boundary elements) revisited. In Computational Mechanics New Trends and Applications (E.Onate and S.R.Idelsohn (eds).
4. Bathe K.J. (1982) Finite Element procedures in engineering analysis. Prentice Hall.
5. Beer G., Golser H., Jedlitschka G. and Zacher P. Coupled finite element/boundary element analysis in rock mechanics - industrial applications. *Rock Mechanics for Industry*, Amadei, Kranz, Scott & Smeallie(eds). Balkema,Rotterdam. 133-140.

17

Industrial Applications

Grau ist alle Theorie ...
(Grey is all theory ...)
J.W. Goethe

17.1 INTRODUCTION

So far in this book we have developed software which can be applied to compute test examples. The purpose of this was to enable the reader to become familiar with the method, ascertain its accuracy and get a feel for the range of problems that can be solved. The emphasis in software development has been on an implementation that was concise and clear and could be well understood. As pointed out in the introduction to programming, this is not necessarily the most efficient code in terms of storage and computer resources.

If one wants to tackle real engineering problems one is inevitably faced with the need to develop efficient code. The programs developed here would be unsuitable for such a task. Aspects of the software that need to be improved are:

- Greater efficiency in the computation of coefficient matrices by rearranging DO loops, so that calculations that are independent of the DO loop variable are taken outside the loop.
- Greater efficiency in data and memory management so that data are only stored in RAM when they are needed, use of hard disk storage to achieve this (see for example [1]).

It has been shown in Chapter 8 that a significant gain in efficiency can be achieved by using element by element techniques and parallel programming. Indeed, to solve problems at an industrial scale in a short time, special hardware, such as parallel computers may have to be used.

In this chapter we attempt to show applications of the boundary element and coupled methods which have been compiled from a number of tasks that have been carried out over more than two decades using BEFE^2, a combined finite element/boundary element program. The purpose of the chapter is twofold. Firstly, an attempt is made to demonstrate the applications for which the BEM may have a particular advantage over the FEM. These applications include:

- Problems involving stress concentrations at the boundary, such as they occur in mechanical engineering
- Problems consisting of infinite or semi-infinite domains, such as those occurring in geotechnical engineering
- Problems involving slip and separation at material interfaces, such as they appear in mechanical and geotechnical engineering
- Contact and crack propagation problems

The second purpose of this chapter is to show how the very complex problems that invariably arise in industrial applications can be simplified, so that the analysis can be performed in a reasonable short time.

It is very rarely the case that a problem can be modelled exactly as it is. In most cases we have to decrease its complexity. The process of *modelling* a given complex structure requires a lot of engineering ingenuity and experience. When we simplify a complex problem we must ensure that the important influences are retained neglecting other less important ones. For example, in a structural problem some parts of the structure may not contribute significantly to its load carrying capacity but are there because of design considerations.

One very significant modelling decision is if a 3-D analysis needs to be carried out. Obviously this would result in much greater analysis effort. As an example in geotechnical engineering consider a tunnel which is very long compared to its diameter. If we are only interested in the displacements and stresses at a cross section far away from the tunnel face, then a plane strain analysis would obviously suffice. Another way of simplifying a problem is the introduction of planes of symmetry. As we have seen in some of the examples in Chapter 10, this results in considerable savings. Obviously if the prototype to be analysed is symmetric there is no loss in modelling accuracy. In some cases, however, symmetry planes can be assumed without significant loss in accuracy even if the prototype itself is not exactly symmetric.

In the following we will present background information on each application, in some cases together with a story associated with it. We will start with the description of the problem and how it was simplified. We show the boundary element mesh generated and the results obtained. Comments are made on the quality of the results. The problem areas are divided into mechanical, geotechnical, geotechnical civil engineering and reservoir engineering.

17.2 MECHANICAL ENGINEERING

17.2.1 A cracked extrusion press causes concern

A small company in Austria manufactures rolled thin tubes by extrusion. The extrusion press in use was 35 years old and made of cast iron (see Figure 17.1). During a routine inspection cracks were detected on the surface of the cast iron casing, as indicated. The company was in the process of ordering a new press, however delivery was expected to take more than six months. There was some concern that something dramatic might happen during the extrusion process with the press suddenly breaking, meaning not only a danger to lives but also the possibility of losing the press. With full order books the latter was a very serious economic threat.

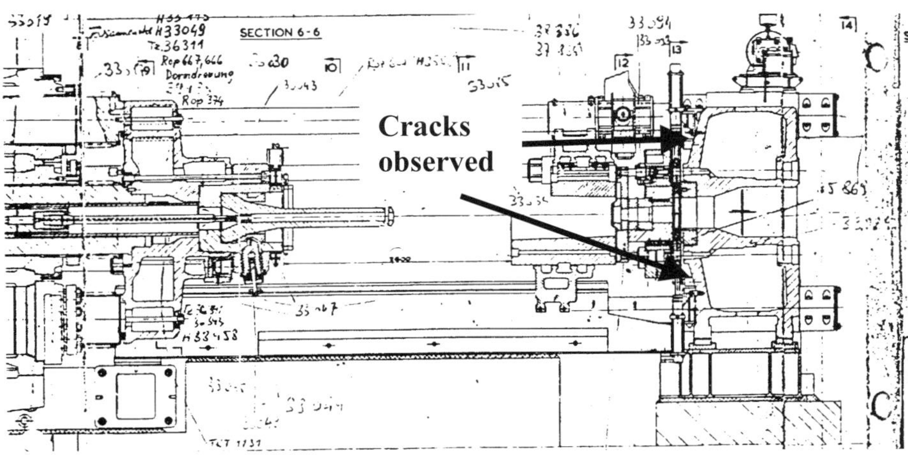

Figure 17.1 35 year old drawing of extrusion press with location of cracks indicated

The aim of the analysis was therefore to determine:

- If the existing cracks would propagate
- If this propagation would lead to a sudden collapse of the structure

The geometry of the part to be analysed was fairly complicated and had to be reconstructed from the original plans. For the purpose of the analysis it was assumed that there were two planes of symmetry, as shown in Figure 17.2, although this was not strictly true.

The cylindrical bar restraining the casing was not explicitly modelled but instead appropriate *Dirichlet* boundary conditions were applied. Each time a tube is extruded the casing is loaded with a force of 3700 tons (37 MN), as shown by the arrows. Although this load is actually applied dynamically it was assumed to be static for the purpose of the analysis.

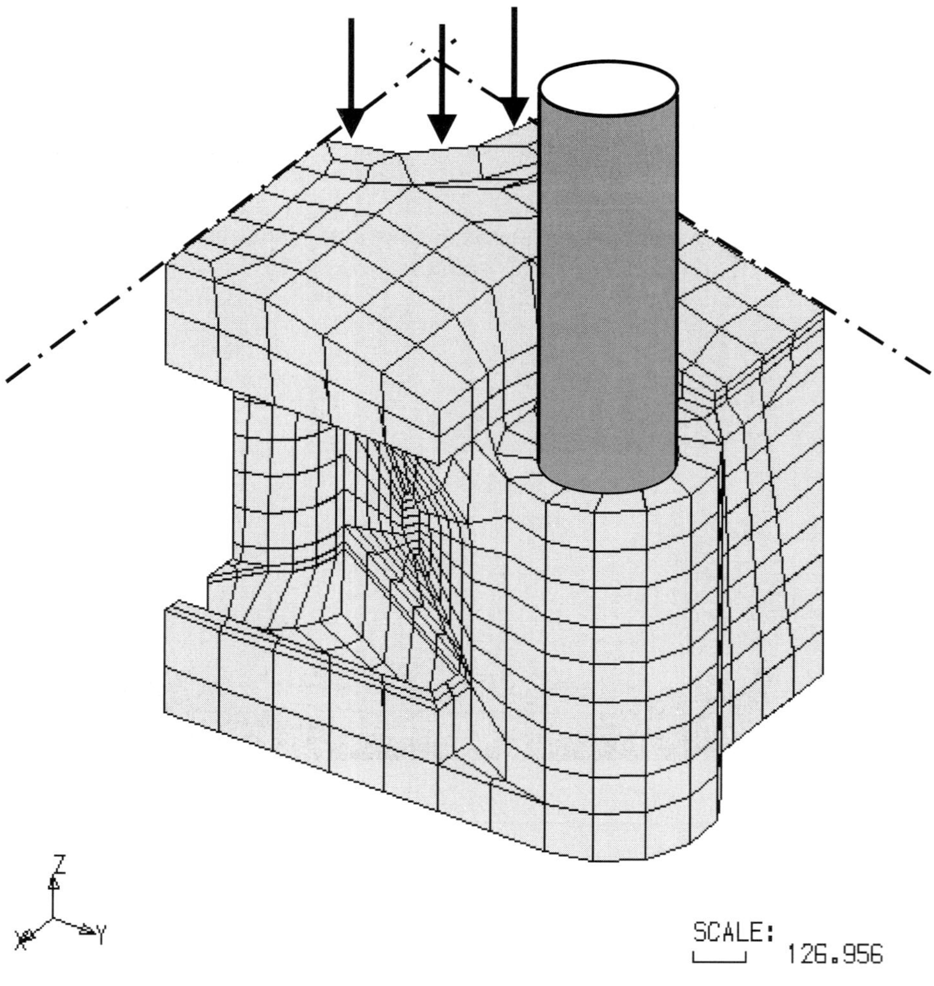

Figure 17.2 Boundary element model showing axes of symmetry and holding bar

The drawing in Figure 17.2 actually looks like a finite element mesh but if viewed from the symmetry planes (Fig. 17.3) one can notice that, in contrast to a FEM discretisation, there are no elements inside the material. The mesh consists of a total of 1437 linear boundary elements and has 4520 degrees of freedom.

There were two reasons why a boundary element analysis was chosen for this problem. Firstly, the generation of the mesh was found to be easier, since no internal elements and connection between surfaces had to be considered. Secondly, the task was to determine surface stresses and then to investigate crack propagation. As outlined previously, the BEM is well suited for this type of analysis.

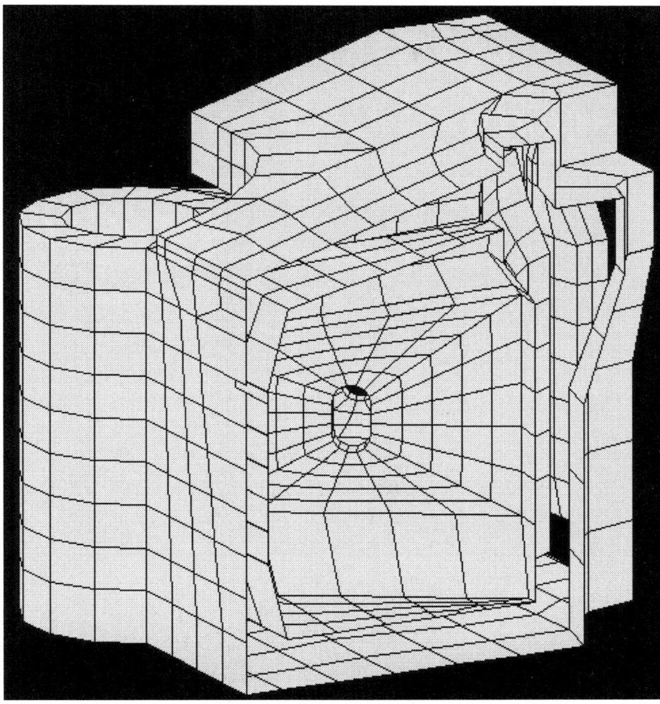

Figure 17.3 Boundary element mesh viewed from one of the symmetry planes

Initially, an analysis with only one region was carried out without considering the presence of cracks. This was done in order to check that the analysis was able to predict crack initiation. The criteria chosen for this was the maximum tensile strength of the material, taking into consideration the dynamic nature of the loading and the number of cycles that the press had so far sustained (approx. 2 million cycles). This analysis was also carried out to see if the model was adequate and to enable the client to get confidence in the BEM analysis proposed. The contours of maximum stress obtained from the single region analysis, shown in Figure 17.4, clearly indicate a stress concentration at the locations where cracks were observed, of a magnitude which would cause crack initiation there after a number of cycles.

After this verification of the model, a multi-region analysis was carried out. For this each of the flanges where the crack was observed was divided into two regions. For simplicity it was assumed that the crack path was known a priori and is in the diagonal direction, as observed. Along this assumed crack path an interface was assumed between regions and the interface was allowed to slip and separate.

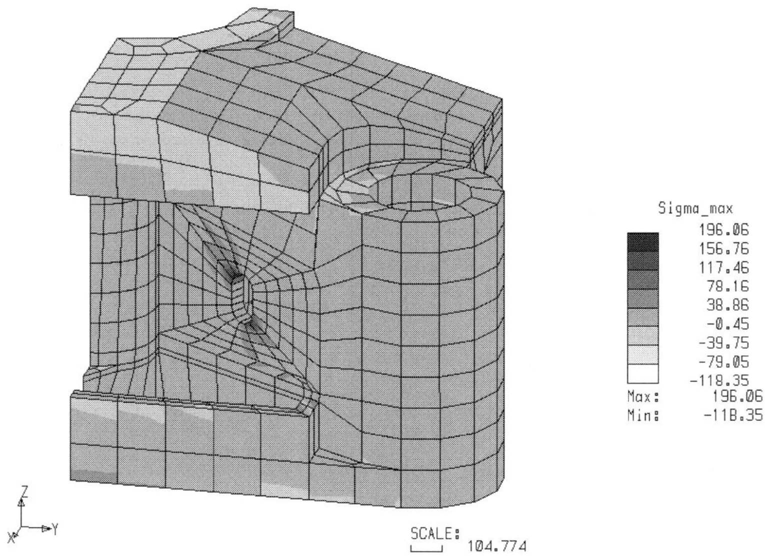

Figure 17.4 Contours of maximum principal stress

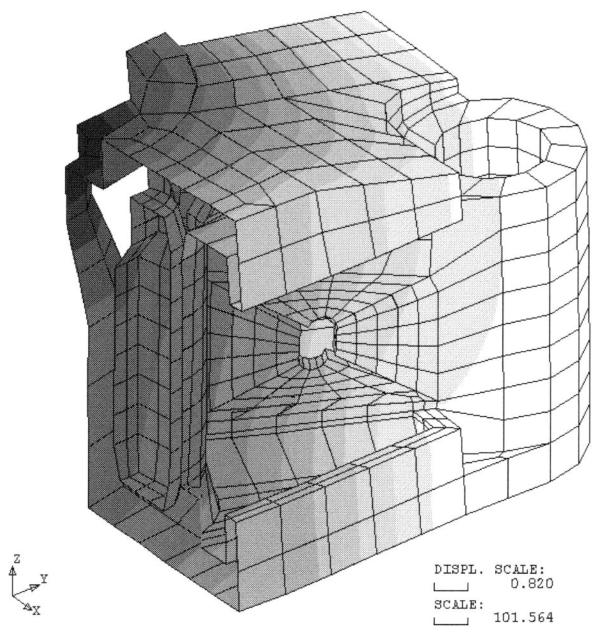

Figure 17.5 Displaced shape showing crack opening

It was found that in the worst case (lowest parameters assumed for the material) the crack would tend to propagate to the corners of the flange (Figure 17.5). However, even with the crack propagated that far the model predicted that there would be no dramatic failure of the casing. Instead, the deformations would become so large that the press would become inoperable.

After half a year the new press arrived and was installed. The old press gave service without any major problems prior to replacement.

The advantages of the BEM over a FEM model may be summarised as:

- The fact that there are no elements inside and no connections were required between elements on opposing boundaries the mesh generation was simplified. The number of unknowns and elements was also reduced.
- The stress concentrations were computed more accurately because they are not obtained using an extrapolation from inside the domain but from boundary results.
- The method was well suited to model crack propagation.

17.3 GEOTECHNICAL ENGINEERING

17.3.1 CERN Caverns

The European Laboratory for Particle Physics (CERN) is the world's largest research laboratory for subatomic particle physics. The laboratory occupies 602 hectars across the Franco-Swiss border and includes a series of linear and circular particle accelerators. The main Large Electron Positron (LEP) accelerator has a circumference of 26.7 km and a series of underground structures situated at eight access and detector points (Fig. 17.6). The LEP accelerator has been operating since 1989 but in 2000 it has been shut down and replaced by the Large Hadron Collider (LHC) in 2005. This will use all existing LEP structures but will also require new surface and underground works. Two new detectors will be installed in two separated cavern systems, called Point 1 and 5.

Here we will present the three-dimensional analysis of the new caverns of Point 5[3,4] (Fig 17.7). This is an interesting application because point 1 of the LHC was analysed using the finite element method and a picture of the results appear in the cover of the book *Programming the Finite Element Method*[5]. According to a report published on this study the mesh had approx 300 000 degrees of freedom and a supercomputer was required to solve the problem.

Initially, an elastic analysis was carried out with the single region BE mesh shown in Figure 17.8. The aim of the analysis was to ascertain the range of validity of 2-D analyses carried out with a distinct element code.

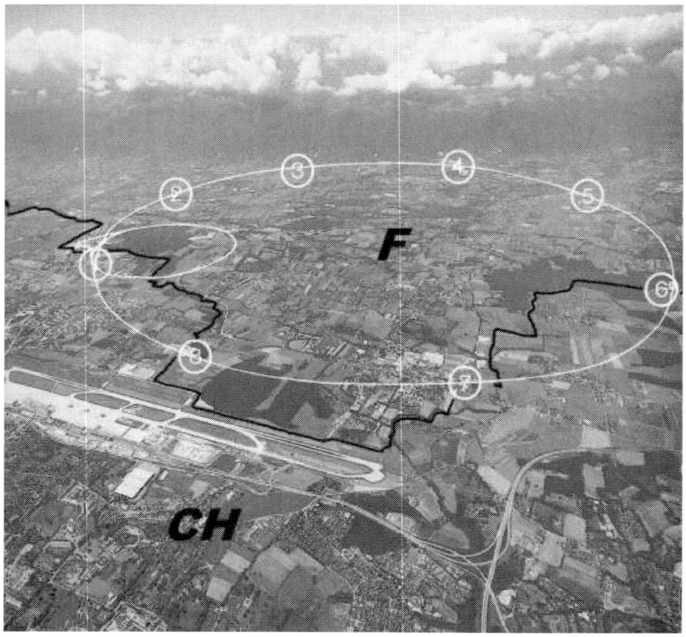

Figure 17.6 Photo showing location of the CERN particle accelerator

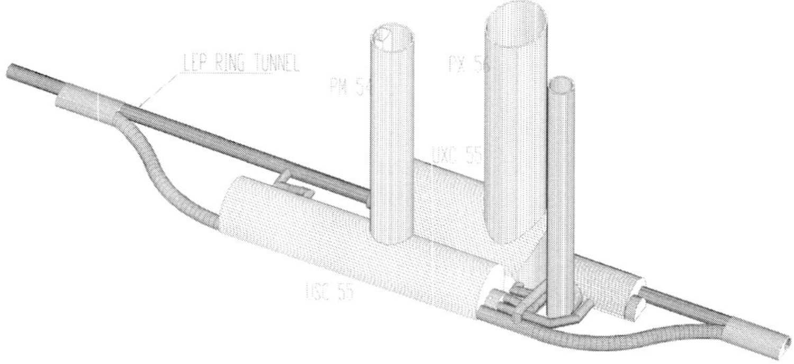

Figure 17.7 Cavern system at Point 5, showing existing and new structures

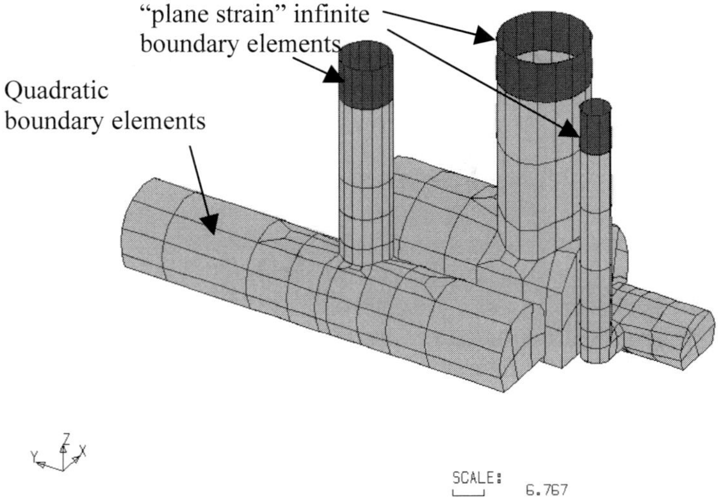

Figure 17.8 Boundary element mesh, single region analysis

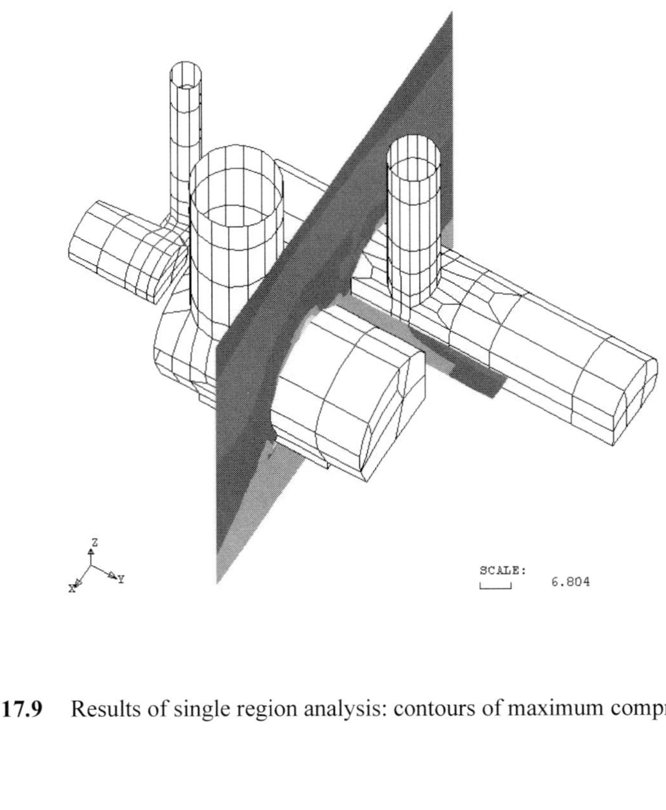

Figure 17.9 Results of single region analysis: contours of maximum compressive stress

The overburden above crown is about 75 m. In the analysis therefore the ground surface was assumed to be sufficiently far away so that its influence on the cavern was neglected. In order to reduce the number of unknowns "plane strain" infinite elements were used, as introduced in section 3.7.2. and as indicated in Figure 17.8. The mesh has a total of 4278 unknowns and the calculation took 10 minutes on a PC. The results of the analysis are shown in Figure 17.9. Here the maximum compressive stress is plotted on two planes inside the rock mass. Looking at the horizontal result plane it can be seen that at a cross-section between the vertical shafts, nearly plane strain conditions are obtained, warranting a 2-D analysis there.

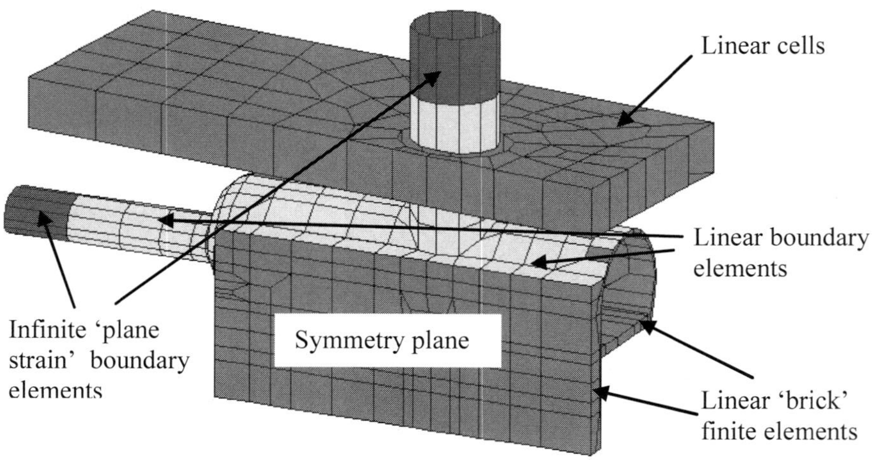

Figure 17.10 Coupled boundary element / finite element mesh of USC55 cavern

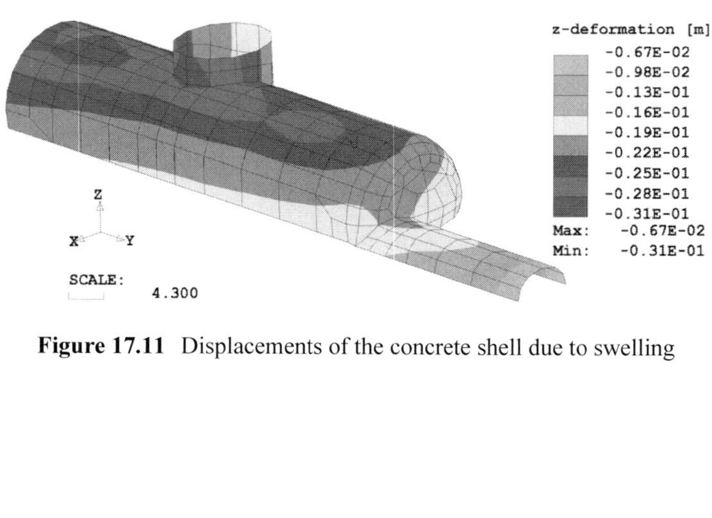

Figure 17.11 Displacements of the concrete shell due to swelling

Geologists found that a portion of the soil above the cavern could swell significantly if subjected to moisture. Therefore, an analysis had to be carried out to determine the effect of swelling on the final concrete lining. Obviously, this cannot be simplified as a 2-D problem because the concrete lining acts as a 3-D shell structure. For this analysis a coupled finite element/boundary element analysis was performed with the thin concrete shell modelled by finite elements. The swelling zone was modelled by linear cells as explained in Chapter 13. In addition a symmetry plane was assumed between the large and the small cavern. Even though in reality no symmetry exists this was thought to be acceptable since the assumption that the second cavern is the same size as the first one would give results that are on the safe side. The main reason for the choice of this mesh was that due to time limitations the job had to be completed quickly and only standard PCs were available for performing the analysis. The coupled mesh of cavern USC 55 is shown in Figure 17.10. The mesh has a total of 7575 degrees of freedom and the run took 45 minutes on a standard PC. Most of the computing time was for computation of the stiffness matrix of the boundary element region

The displacements of the concrete lining due to swelling were determined from the analysis. These are shown in figure 17.11. From these displacements the internal forces in the shell (bending moment and normal force) could be determined and used for designing the reinforcement. The analysis shown here demonstrates that with limited resources available (time and computer), boundary element and coupled analysis offer an efficient alternative to the FEM.

17.4 GEOLOGICAL ENGINEERING

17.4.1 How to find gold with boundary elements

The analysis was performed to test a theory of geologists that gold dust was originally suspended in water and was deposited in the ground in locations that had a significantly smaller amount of compressive stress than the surrounding rock[6]. This seems to make sense, since deposits would naturally occur in voids, i.e., areas where the compressive stress is zero.

Since Australia is one of the richer countries in terms of gold resources the story takes place there. In particular, the analysis concentrates on what is presumed to have occurred in a region of Western Australia (where a deposit was found) during the Precambrian period (about 800 million years ago). The geologists assume that the region was shortened in an approximate east/west direction and that the deposit was formed at approximately 2.5 km of depth below the surface. On this basis it was suggested that a volume of rock of about 2000x2000x1000 m dimension with the geological structure as observed in that area should be analysed. The geological structures are shown in Figures 17.12 and 17.13. Figure 17.12 shows contours of the contact between different rock types, whereas Fig.17.13 shows contours of two faults (termed Lucky and Golden faults).

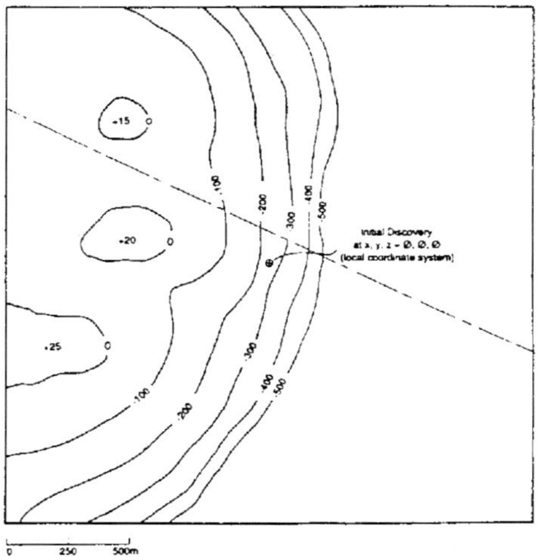

Figure 17.12 Contours of contact between different rock types

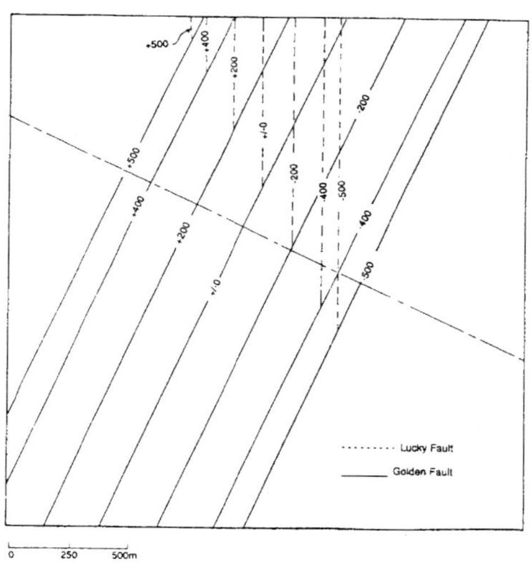

Figure 17.13 Contours of Lucky and Golden faults

It was assumed that the block to be analysed was subjected to 2000 m of overburden (which was subsequently eroded) and to tectonic stresses which were estimated from the presumed shortening of the region.

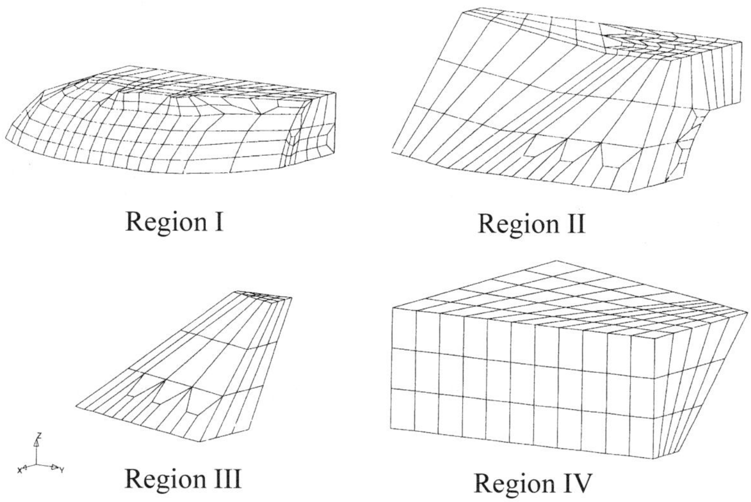

Figure 17.14 Definition of boundary element regions

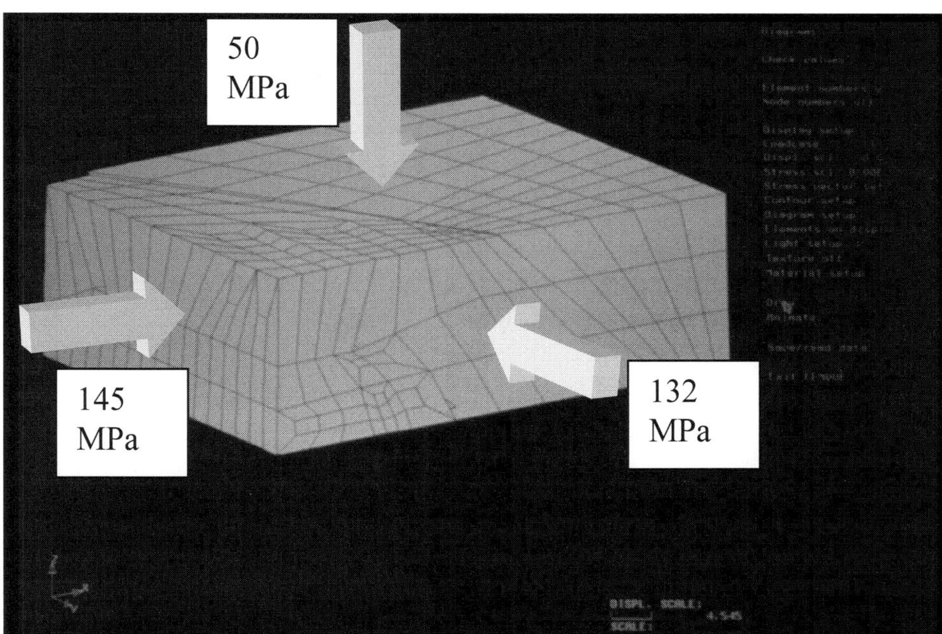

Figure 17.15 Block analysed showing stress boundary conditions applied

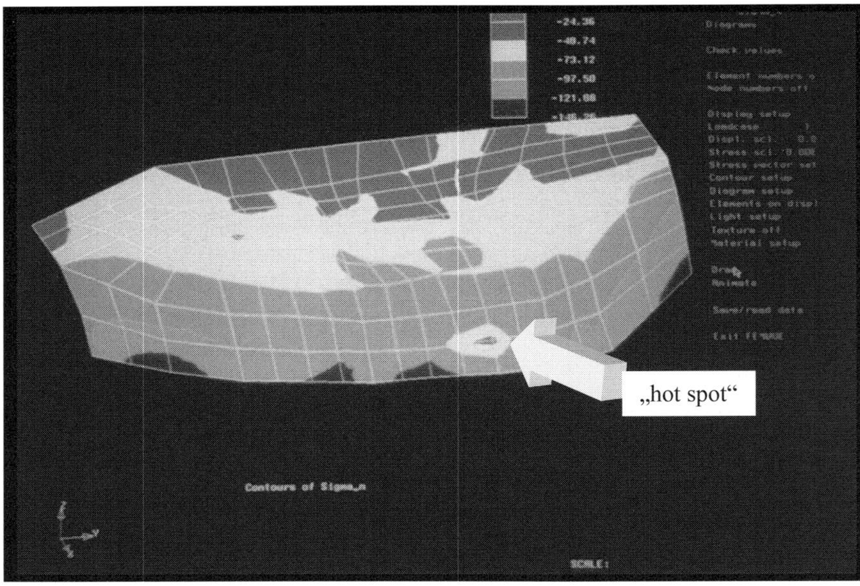

Figure 17.16 Contours of maximum compressive principal stress

For the analysis a multi-region boundary element method was used with special contact/joint algorithms implemented on the interfaces between regions. Figure 17.14 shows a view of the four regions considered. Figure 17.15 shows the block analysed with stress boundary conditions applied. In this figure the deformation of the blocks and the movements on the Golden and Lucky faults can be seen. The results of the analysis can be seen in Figure 17.16 as contours of the maximum (compressive) principal stress on the contact between regions I and II. One can clearly see an anomaly of the compressive stress ("hot spot") and this is near the location where the gold deposit was assumed to be. So the boundary element method was successfully applied to find gold deposits. Note that an analysis with a domain type method would be feasible. However, the mesh generation would be more complicated because of the presence of elements inside the regions and the necessity to assure proper connectivity.

17.5 CIVIL ENGINEERING

17.5.1 Masjed-o-Soleiman underground power house

The Masjed-o-Soleiman hydroelectric scheme is situated in the south of Iran. The powerhouse is situated underground. In 2002 the last of 4 turbines were installed in the existing powerhouse and an extension of the facility to house another 4 turbines was underway. During the excavation of the extension, cracks were observed in the concrete walls of the existing powerhouse, which caused some concern. In addition measurements from pressure cells installed behind the concrete walls recorded seasonally dependent pressure increases that showed an increasing tendency. Following a visit by the panel of experts it was decided to carry out a numerical analysis. The aim of the analysis was to determine the cause of the cracks and to predict if the cracking would get worse because of continuing excavation activity on the extension.

Figure 17.17 View of hydroelectric plant, the powerhouse cavern is inside the mountain on the left of the dam

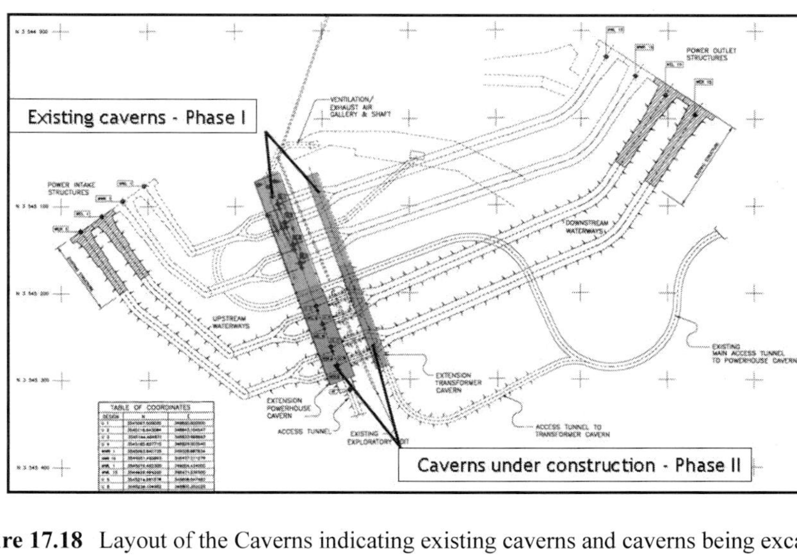

Figure 17.18 Layout of the Caverns indicating existing caverns and caverns being excavated

Figure 17.17 shows a view of the hydroelectric facility and Figure 17.18 a plan of the layout showing the existing powerhouse cavern and the extension under construction. The areas where cracking was observed are shown in Figure 17.19. Special consideration was given to the circled area near the construction of the extension.

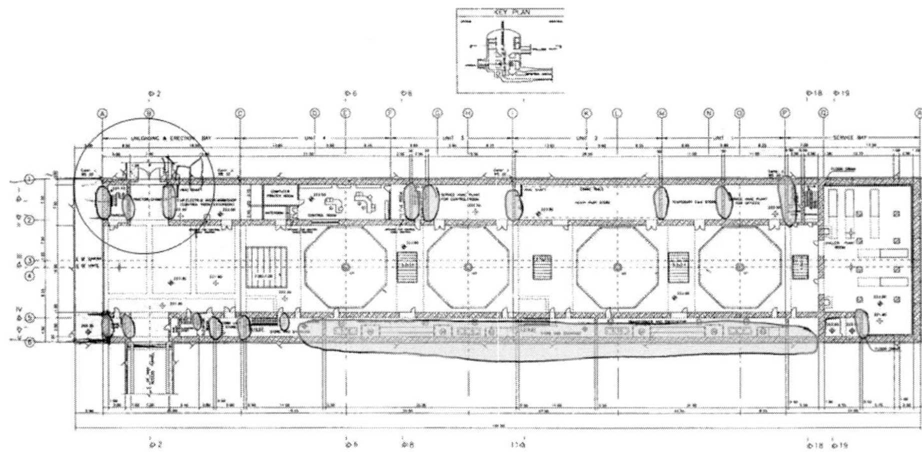

Figure 17.19 Plan of powerhouse depicting areas where cracks were observed

The ground conditions in the vicinity of the caverns, as shown in Figure 17.20, are dominated by layers of very weak mudstone and sandstone.

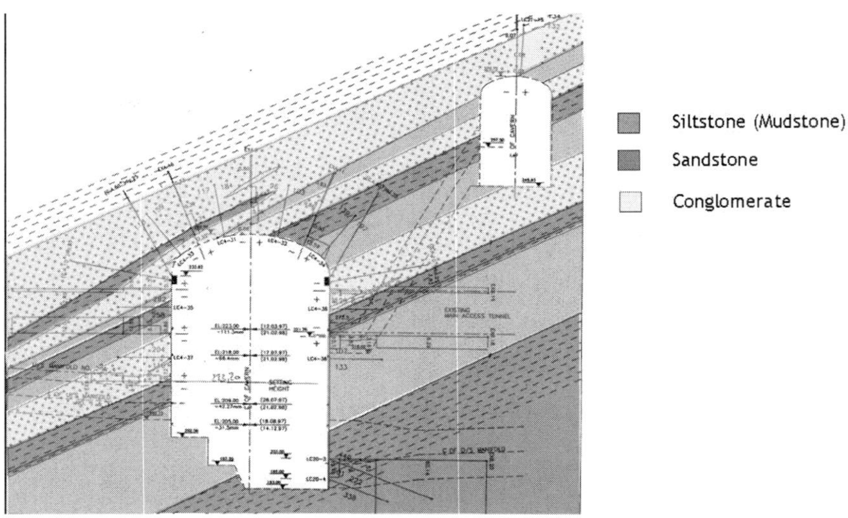

Figure 17.20 Geological conditions near the caverns

To ascertain the fineness of the mesh required for the analysis and the displacement patterns, a 3-D Boundary Element analysis was first carried out.

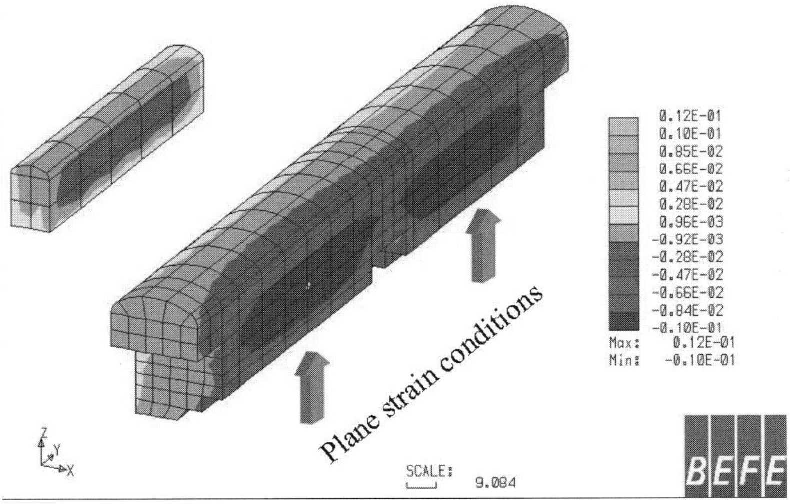

Figure 17.21 Boundary element mesh of caverns and computed deformations

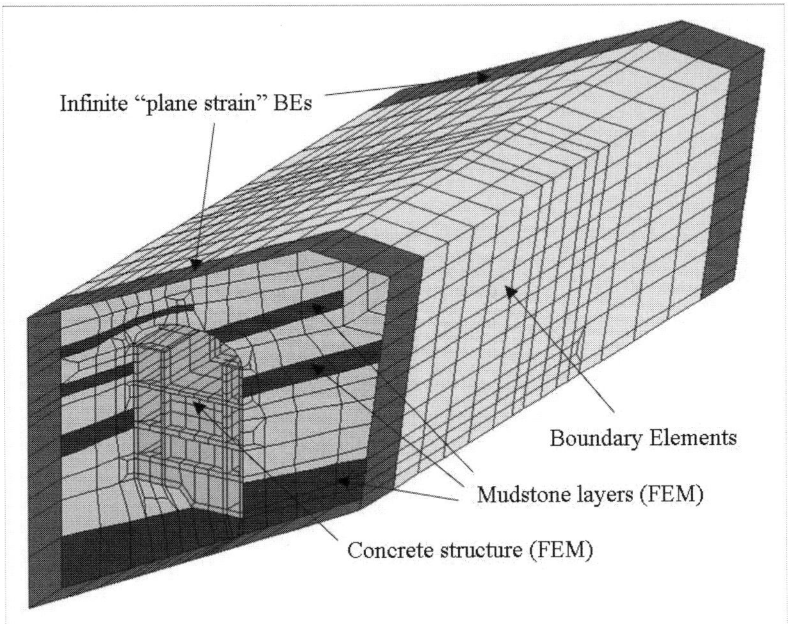

Figure 17.22 Coupled mesh for the analysis of powerhouse cavern and concrete powerhouse structure

However, this analysis does not consider the presence of geological features and non-linear effects, which are important. Figure 17.21 shows the mesh with quadratic (8-node) boundary elements and the result for the case where both caverns are excavated, plotted as displacement contours on the excavation surface. It can be seen that for a large portion of the cavern plane strain conditions can be observed. It was therefore decided that the mesh could be reduced by the use of infinite plane strain boundary elements as they have been introduced in Chapter 3.

For a meaningful analysis, however, the effect of the geological features as well as the non-linear behaviour of the ground had to be considered. For this purpose a coupled finite element/boundary element mesh was constructed as shown in Figure 17.22. Here the rock mass in the vicinity of the cavern, as well as the concrete structure of the powerhouse is discretised into finite elements. Plane strain boundary elements were used to shorten the mesh in the direction along the cavern, taking into consideration the displacement conditions, as depicted in Figure 17.21. This analysis allows to consider the geological features as well as the nonlinear behaviour of the ground (in particular the mudstone layers).

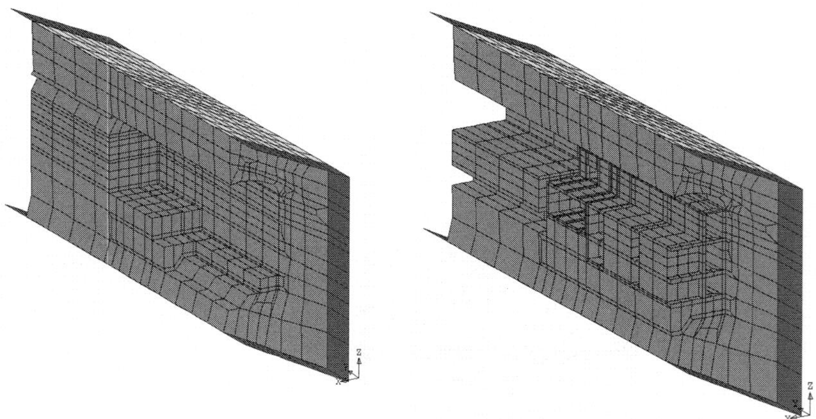

Existing powerhouse cavern excavated

Powerhouse structure installed, excavation of extension, status Nov. 2003

Figure 17.23 Two of the stages considered in the analysis

Several excavation stages were considered and two of these are shown in Figure 17.23. The mesh on the left models the complete excavation of cavern 1, the one on the right the construction of the powerhouse structure and the excavation stage of the extension as existed during the visit of the panel of experts. Figure 17.24 shows the displaced shape on a section through the end of the existing cavern near the extension excavation. The deformation of the FEM-BEM interface can be seen especially on top of the cavern, so an analysis without coupling to BEM would not have yielded meaningful results. A view of the finite element mesh of the powerhouse structure, indicating the location of the cracks near the extension is shown in Figure 17.25.

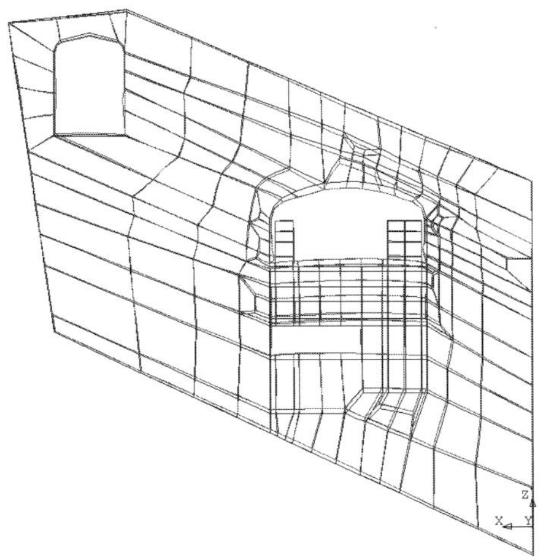

Figure 17.24 Displaced shape in a cross-section through the end of the powerhouse.

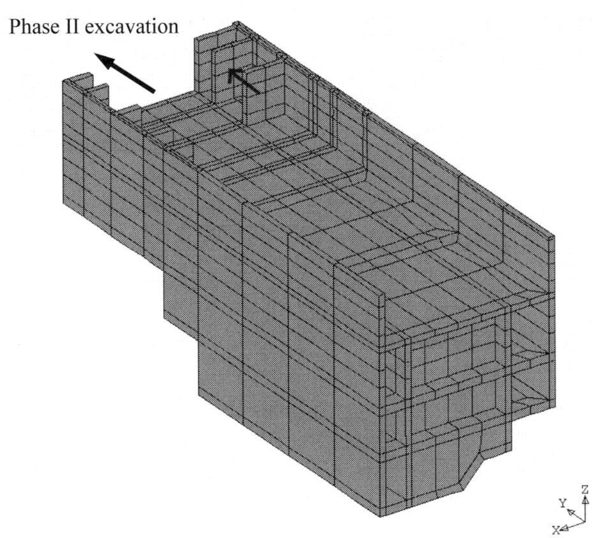

Figure 17.25 View showing the concrete powerhouse and the location of the cracks

Figure 17.26 shows one result of the analysis namely the stress distribution in the concrete wall plotted as principal stress vectors. It can be seen that the observed crack pattern on the right is perpendicular to the maximum computed principal stress.

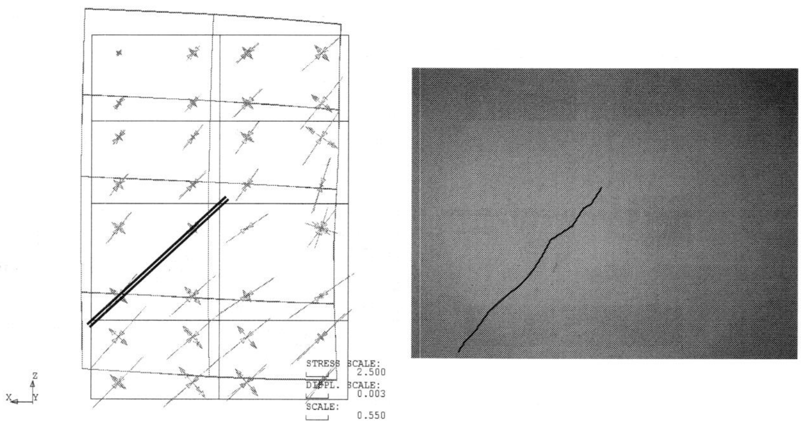

Figure 17.26 Predicted stress pattern in wall and crack pattern observed.

This is a nice example of the use of a coupled analysis because it substantially reduces the effort. Without coupling to BEM the mesh would have to be made much larger, to reduce the effect of artificial boundary conditions to an acceptable level. It should be noted here that with the methods for non-linear analysis described in Chapter 15 and for dealing with heterogeneous ground conditions outlined in Chapter 18 it would have been possible to completely avoid the discretisation into finite elements of the ground surrounding the cavern. All that would be required is to subdivide the ground into cells. However, at the time of the analysis these capabilities were not available.

17.6 RESERVOIR ENGINEERING

17.6.1 Borehole stability

The example relates to some work performed in cooperation with the University of Kuwait. For oil recovery vertical boreholes are drilled to a depth of several thousand meters. In order recover as much oil as possible from one bore hole, deviated boreholes are drilled as shown in Figure 17.27. The angle of deviation of the lateral bore varies from 30° to 60° and the direction of the deviation with respect to the virgin stress field also varies. Since the boreholes are drilled in a highly pre-stressed rock mass, mud pressure has to be applied in order to stabilize the borehole. The questions to be answered by the simulation were with respect to the stability of the rock near the junction of the vertical and the deviated borehole.

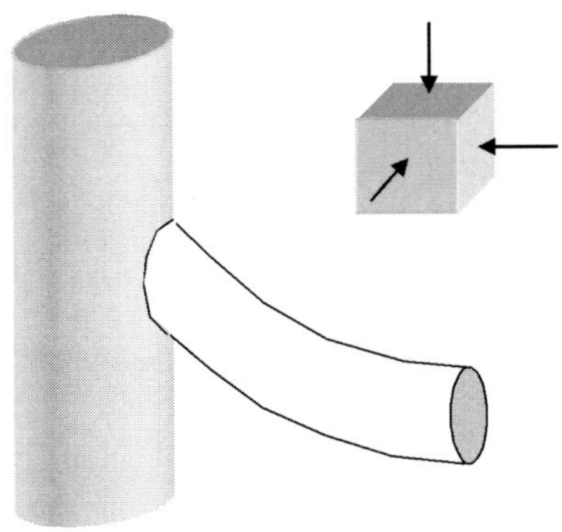

Figure 17.27 Sketch of borehole junction

To determine the areas in the rock mass that are likely to break an elastic analysis was performed. The result of the analysis was a contour plot of a yield function, $F(\sigma)$. The yield functions used were the Mohr-Coulomb and Hoek and Brown models. The mesh used for the analysis for a 45° deviation is plotted in Figure 17.28 on the left.

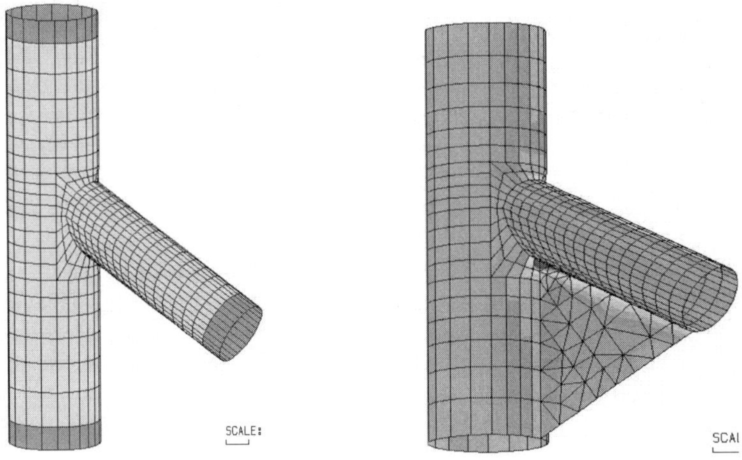

Figure 17.28 Boundary element mesh (left) and results of the analysis (right) plotted on surface and dummy plane

The mesh consists of about 1000 linear boundary elements. Plane strain elements are used at the ends of the mesh, where the boreholes are truncated, to simulate an infinite extent of the boreholes, which was assumed to be realistic since the depth of the junction was 2000 m underground. The analysis took about 5 minutes to run and therefore a great number of runs could be carried out with various virgin stress fields and mud pressures. The contours of the yield function showed for which orientation to the virgin stress filed relative to the deviated borehole and for which mud pressures the failure zone had the minimum extent. An example of the results obtained can be seen in Figure 17.28 on the right. The advantages of a boundary element approach can be summarised as follows:

- The mesh generation was much simpler than with finite elements
- The accuracy of the results was probably significant higher compared with a finite element analysis since no approximation of displacements is assumed in the rock mass. An automatic mesh generation would probably have computed distorted elements in the vicinity of the junction degrading the accuracy at this location.
- The effort would have been significantly larger with the FEM

The analysis presented here, however, is only elastic and therefore does not consider the non-linear material behaviour. In addition it is assumed that the two boreholes are excavated instantly and simultaneously which is not very realistic. A non-linear analysis with internal cells and the sequential excavation algorithm described in Chapter 12 was proposed but results are not available at the time of writing of the book.

17.7 CONCLUSIONS

In this chapter we have attempted to show, on some practical applications, that the method is not only of academic interest but can be used to solve real life problems. We have purposely concentrated on applications where the BEM has been shown to have a distinct advantage over the FEM in terms of effort to generate the mesh and computing resources.

However, we do not make the claim that the BEM is always superior to the FEM and to be fair have included two applications where a combination of the BEM and the FEM leads to best results. Indeed, we believe that the analyst should be given a choice used and make a case for more commercial software which allows the use of either method independently or in combination.

17.8 REFERENCES

1. Beer G. and Watson J.O. (1991) Introduction to Finite and Boundary Element Methods for Engineers, Wiley, Chichester.
2. Beer G. BEFE users manual, CSS, Geidorfgürtel 46, Graz, Austria.
3. Beer G.,Sigl O & Brandl J (1997) Recent developments and application of the boundary element method. *Numerical Models in Geomechanics*. Pietruszczak & Pande (eds), Balkema,Rotterdam, 461-467.
4. Beer G., Golser H., Jedlitschka G. and Zacher P. Coupled finite element/boundary element analysis in rock mechanics - industrial applications. *Rock Mechanics for Industry*. (Amadei,Kranz,Scott&Smeallie(eds). Balkema, Rotterdam. 133-140.
5. Smith I.M. and Griffiths D.V. (1998) Programming the Finite Element Method. J.Wiley, Chichester.
6. Beer G. and Poulsen B.A. (1994) Efficient numerical modelling of faulted rock using the boundary element method. *Int. J. Rock. Mech. Min. Sci. & Geomech. Abstr.* **31** (5):485-506

18

Advanced topics

Sometimes one pays most for the things
one gets for nothing.
A. Einstein

18.1 INTRODUCTION

In this Chapter we shall discuss topics which are advanced in the sense that they cover topics which were still subject to investigation at the writing of the book or that are non-standard engineering applications.

One particular topic is overcoming the difficulty the BEM has to deal with heterogeneous material. As we have seen in Chapter 11 only the consideration of piecewise heterogeneous domain is possible via a multi-region concept. If the heterogeneity is pronounced then the simulation effort can become considerable. Also, for some problems a continuous heterogeneity may have to be assumed. Another topic is the inclusion of linear elements such as reinforcement in concrete technology or rock bolts in tunnelling. The inclusion of these elements in the FEM is fairly straightforward because nodes exist inside the domain to which these elements can be connected. However, in the BEM no such nodes exist. Here, we will show some results which at the writing of the book were fairly new.

The availability of a code, that could be downloaded free of charge for readers of the book "Programming the BEM" has led to applications that are beyond the usual engineering topics. One such application is shown here. It relates to the simulation of piezo-electric materials i.e. materials which show some reaction (deformation) if subjected to an electric current. This application has been chosen because it shows that once the framework of the program has been established even complicated new applications can be implemented relatively quickly by supplying the appropriate fundamental solutions. It also shows that the concept of the program includes flexibility in dealing with any number of degrees of freedom and that the routines for computing Kernel- shape-function products remain unchanged.

18.2 HETEROGENEOUS DOMAINS

18.2.1 Theory

The approach is first explained on a problem with 2 different properties but it will become obvious that the method will also work for a general heterogeneous domain. The example in Figure 18.1 shows a problem of a tunnel being excavated in a domain with 2 different materials (this is a pure Neumann problem). We could solve this with a multi-region approach but here we choose a different method.

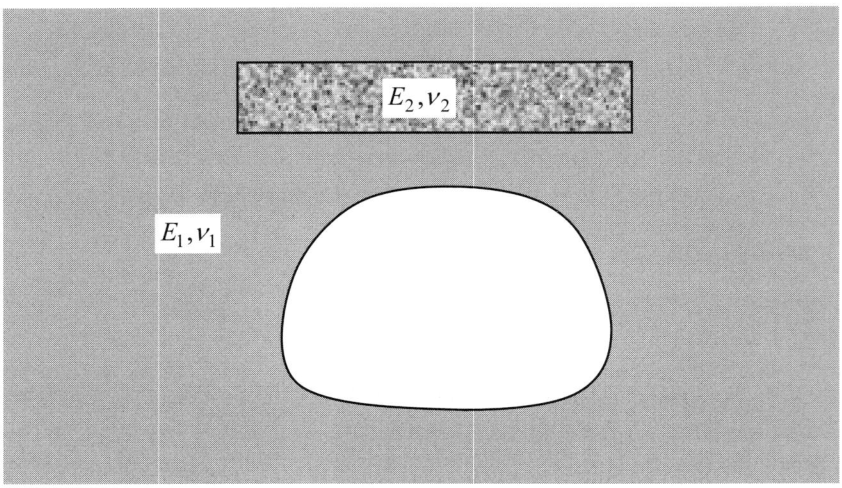

Figure 18.1 Example with heterogeneous domain

The idea is to start with an analysis that assumes that the whole domain has the same properties (E_1, ν_1) which are represented by the constitutive matrix $\mathbf{D}_1$. Hence we solve the following system of equations (see also 7.2)

$$[\mathbf{T}] \cdot \{\mathbf{u}\} = \{\mathbf{F}\} \tag{18.1}$$

Next we compute displacements at the boundary nodes and via post-processing the strains, $\boldsymbol{\varepsilon}$, at the cell nodes P_a

$$\boldsymbol{\varepsilon}(P_a) = \int_S \bar{\mathbf{S}}(P_a, Q)\mathbf{t}(Q)\,dS - \int_S \bar{\mathbf{R}}(P_a, Q)\mathbf{u}(Q)\,dS \tag{18.2}$$

We find that when we compute stresses these should be computed according to:

$$\boldsymbol{\sigma} = \mathbf{D}_1 \cdot \boldsymbol{\varepsilon} \tag{18.3}$$

because this was the assumption made for the analysis. However, this is not correct if the point is inside the inclusion. Here the stresses should be computed according to

$$\boldsymbol{\sigma} = \mathbf{D}_2 \cdot \boldsymbol{\varepsilon} \tag{18.4}$$

where $\mathbf{D}_2$ is the constitutive matrix for material 2. A correction of the stresses has to be made therefore to the results. The proposal is to follow a similar approach as presented for plasticity in Chapter 15. A residual "initial stress" is computed inside the inclusion for the first iteration by

$$\dot{\boldsymbol{\sigma}}_0 = \left(\mathbf{D}_1 - \mathbf{D}_2\right) \cdot \boldsymbol{\varepsilon} \tag{18.5}$$

and this is applied as body force to the system. The computation of the residual $\{\mathbf{R}\}$ is the same as outlined in Chapter 15. For the pure Neumann problem of Fig. 18.1 increments of the displacements due to the body force are computed by

$$[\mathbf{T}] \cdot \{\dot{\mathbf{u}}\} = \{\mathbf{R}\} \tag{18.6}$$

The next increments of $\dot{\boldsymbol{\varepsilon}}$ are computed by

$$\dot{\boldsymbol{\varepsilon}}(P_a) = -\int_S \bar{\mathbf{R}}(P_a, Q)\dot{\mathbf{u}}(Q)\,dS + \int_V \bar{\mathbf{E}}(P_a, \bar{Q})\dot{\boldsymbol{\sigma}}_0(\bar{Q})\,dV + \bar{\mathbf{F}}\dot{\boldsymbol{\sigma}}_0(P_a) \tag{18.7}$$

With this increment $\dot{\boldsymbol{\varepsilon}}$ a new initial stress $\dot{\boldsymbol{\sigma}}_0$ and residual $\{\mathbf{R}\}$ is computed by

$$\dot{\boldsymbol{\sigma}}_0 = \left(\mathbf{D}_1 - \mathbf{D}_2\right) \cdot \dot{\boldsymbol{\varepsilon}} \tag{18.8}$$

The iteration proceeds until the norm of the residual is below a specified value.

18.2.2 Example

This example, shown in Figure 18.2, is a block (2m x 2m) with an inclusion in centre (1m x 1m) under plane strain conditions. The block is fixed at the bottom and loaded on the top surface with a constant pressure of p = 1N/m². The inclusion is assumed to be 10 times softer than the block. The block is discretised with 4x12 quadratic boundary elements and the inclusion is discretised with 6x6 quadratic cells. Figure 18.3 shows the deformed shape of the mesh. The analysis took 39 iterations to converge. The results are compared with a finite element reference solution. Figure 18.4 shows the vertical displacements along the left half of the upper block surface. A good agreement with the reference solution is obtained.

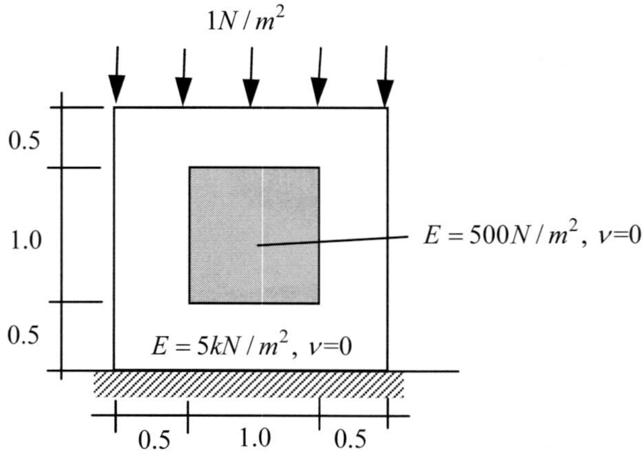

Figure 18.2 Description of example problem

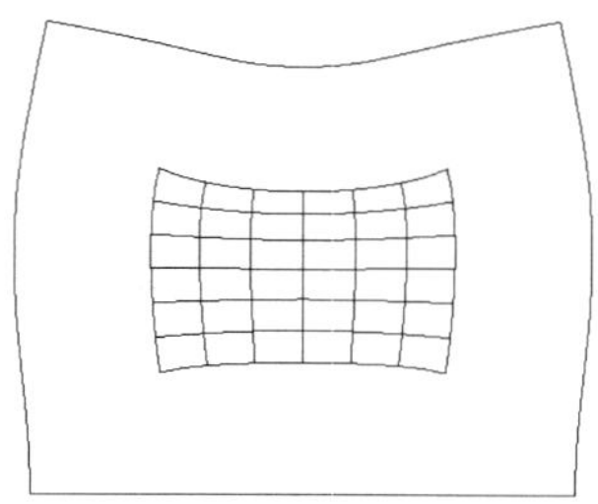

Figure 18.3 Deformed mesh

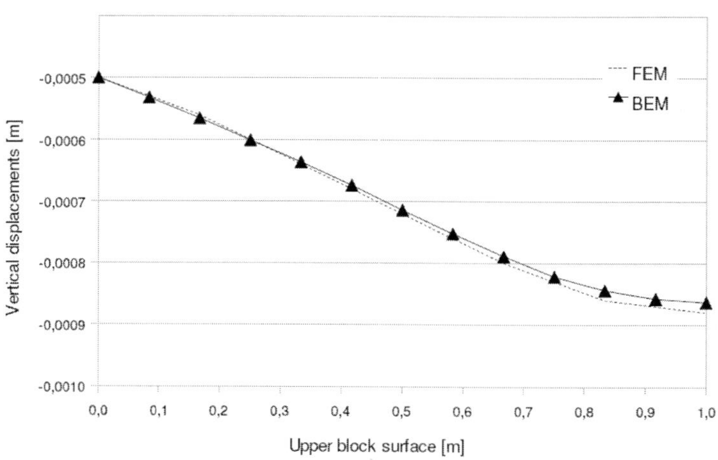

Figure 18.4 Variation of computed displacement on to of the block

18.3 LINEAR INCLUSIONS

18.3.1 Theory

The treatment of linear inclusions follows a similar approach as with heterogeneities[1]. Figure 18.5 shows an example of a tunnel with a rock bolt which will be treated as an inclusion with different material properties. The rock bolt is assumed to be fully grouted i.e. in contact to the rock mass along it's length.

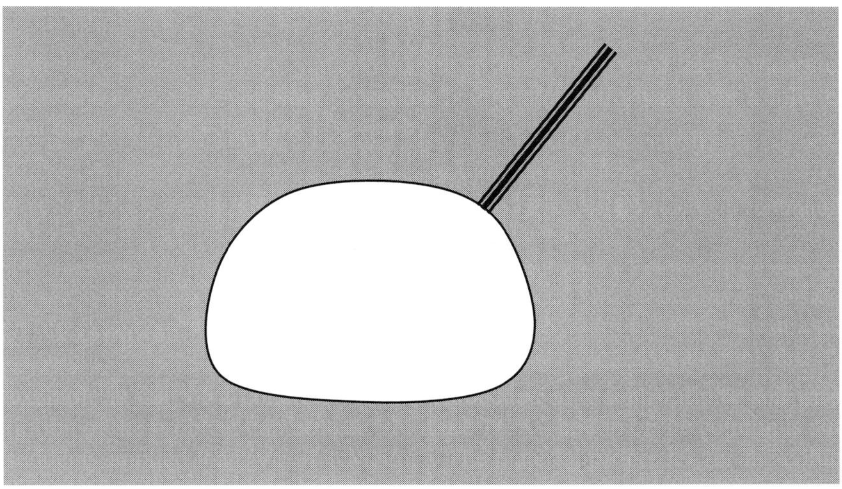

Figure 18.5 Example of a tunnel with a rock bolt

The cross-section of the rock bolt is assumed to be small compared to its length and therefore the variation of the stress across the section can be assumed to be constant. The approach is very similar to the previous one, i.e. first an analysis is carried out without the rock bolt and then a correction made due to the presence of the rock bolt. It is explained on a plane problem but the extension to 3-D is straightforward.
The first system of equations to be solved is for the *Neumann* example in Fig. 18.5

$$[\mathrm{T}]\{\mathbf{u}\}=\{\mathbf{F}\} \tag{18.9}$$

After the first analysis the strain in the direction of the rock bolt is computed. Because of the difference in moduli between the rock and the bolt the stress is different at a point depending if the point lies in the bolt or in the rock (Fig. 18.6). The difference in the stress (in the direction of the bolt $\bar{x}$) is computed by:

$$\sigma_{\bar{x}} = \sigma_{\bar{x}Rock} - \sigma_{\bar{x}Bolt} = (E_{Rock} - E_{Bolt}) \cdot \varepsilon_{\bar{x}} \tag{18.10}$$

and this will be applied as initial stress.

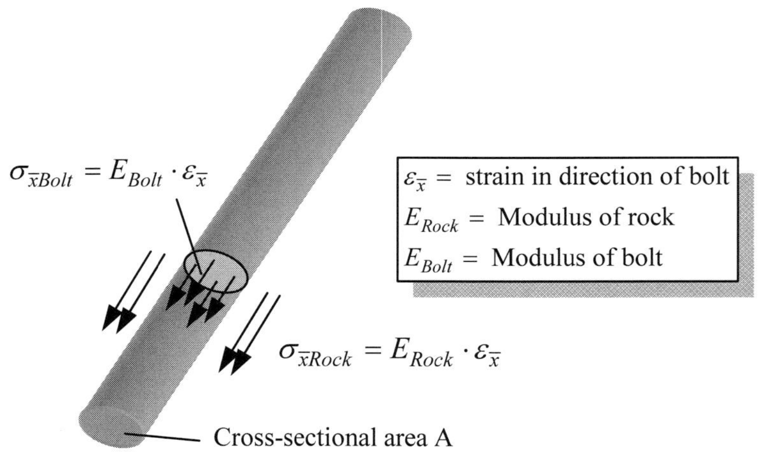

Figure 18.6 Explanation of difference in stress between rock bolt and rock mass

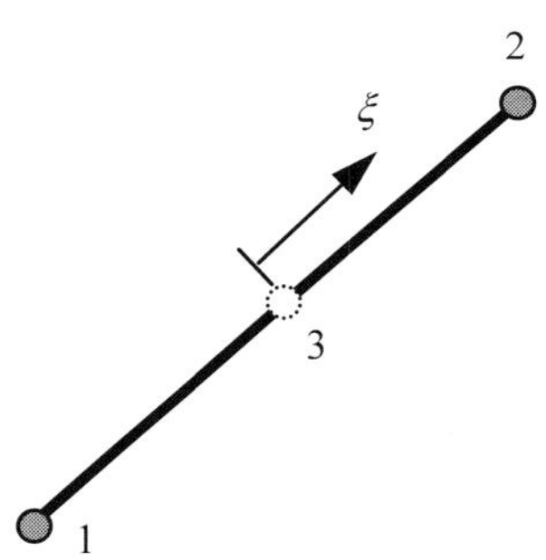

Figure 18.7 Linear cell element

To compute the strain along the rock bolt we use a line cell as shown in Fig. 18.7. First we compute the displacements at the cell nodes using Eq. (9.48)

$$\mathbf{u}(P_a) = \sum_{e=1}^{E}\sum_{n=1}^{N} \Delta\mathbf{U}_n^e \, \mathbf{t}_n^e - \sum_{e=1}^{E}\sum_{n=1}^{N} \Delta\mathbf{T}_n^e \, \mathbf{u}_n^e \tag{18.11}$$

where P_a is a cell node. We introduce a local coordinate system specified by vectors $\mathbf{s}_1$ along the rock bolt and $\mathbf{s}_2$ perpendicular to it (Figure 18.8). The values of the displacement in direction of the bolt $u_{\bar{x}}$ are computed at nodal point n of the cell using the transformation:

$$u_{\bar{x}n} = s_{1x} \cdot u_{xn} + s_{1y} \cdot u_{yn} \tag{18.12}$$

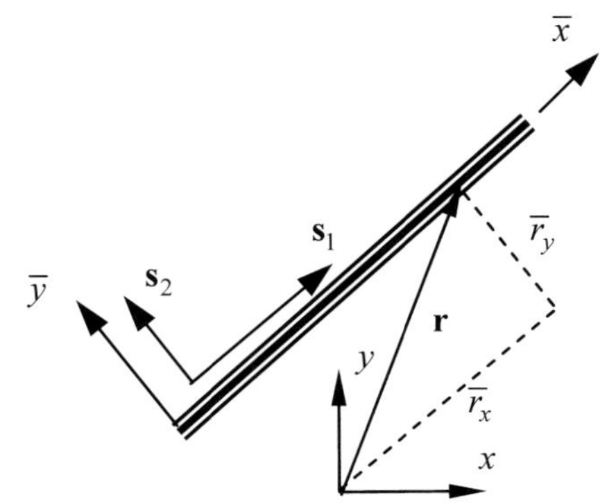

Figure 18.8 Local coordinate system

Then the strain in the bolt at nodal point n can be computed by:

$$\varepsilon_{\bar{x}n} = \frac{\partial u_{\bar{x}}(\xi_n)}{\partial \bar{x}} = \frac{\partial u_{\bar{x}}(\xi_n)}{\partial \xi} \cdot \frac{\partial \xi}{\partial \bar{x}} \tag{18.13}$$

where

$$u_{\bar{x}}(\xi_n) = \sum_{i=1}^{2(3)} N_i(\xi_n) \cdot u_{\bar{x}i} \quad \text{and} \quad \frac{\partial \xi}{\partial \bar{x}} = \frac{1}{J} \tag{18.14}$$

where $u_{\bar{x}i}$ is the displacement in bolt direction at node i of the cell and J is the Jacobian of the transformation from $\bar{x},\bar{y}$ to ξ. The initial stress at node n of a cell c is computed by

$$\dot{\sigma}^c_{0\bar{x}n} = (E_{Rock} - E_{Bolt}) \cdot \varepsilon_{\bar{x}n} \tag{18.15}$$

In general the component $\mathbf{R}_i$ of the residual vector $\{\mathbf{R}\}$ due to the "initial stress" is computed by

$$\mathbf{R}_i = \int_V \mathbf{E}\left(P_i, \bar{Q}\right) \boldsymbol{\sigma}_0\left(\bar{Q}\right) dV \tag{18.16}$$

Where $\mathbf{E}$ is the fundamental solution for strain. $\mathbf{E}$ and $\boldsymbol{\sigma}_0$ are based on the global coordinate system. As the initial stress $\dot{\sigma}_{0\bar{x}n}^{c}$ in (18.15) is based on a local coordinate system the fundamental solution is needed in the local system, too. $\mathbf{E}$ is calculated with the local value $\bar{\mathbf{r}}$ which is computed by

$$\bar{\mathbf{r}} = \mathbf{T}_g \mathbf{r} \tag{18.17}$$

where the geometrical transformation matrix is given by

$$\mathbf{T}_g = \left(\mathbf{s}_1 \ , \mathbf{s}_2\right) \tag{18.18}$$

The fundamental solution for displacement is now given in local coordinates $(\bar{x}, \bar{y})$. Since only the axial stresses are taken into account, the stress vector $\boldsymbol{\sigma}_0$ in (18.16) reduces to a scalar $\sigma_{0\bar{x}}$ and the matrix $\mathbf{E}$ reduces to a vector $\hat{\mathbf{E}}$ in local coordinates, for plane problems we have:

$$\hat{\mathbf{E}} = \begin{bmatrix} E_{\overline{xxx}} \\ E_{\overline{yxx}} \end{bmatrix} \tag{18.19}$$

The residual (18.16) in the local coordinate system is

$$\bar{\mathbf{R}}_i = \int_V \hat{\mathbf{E}}\left(P_i, \bar{Q}\right) \sigma_{0\bar{x}} \left(\bar{Q}\right) dV \tag{18.20}$$

The global residual vector $\{\mathbf{R}\}$ can be obtained with a transformation

$$\mathbf{R}_i = \mathbf{T}_g \bar{\mathbf{R}}_i = \int_V \mathbf{T}_g \hat{\mathbf{E}}\left(P_i, \bar{Q}\right) \sigma_{0\bar{x}} \left(\bar{Q}\right) dV \tag{18.21}$$

The transformed fundamental solution is defined by

$$\tilde{\mathbf{E}} = \mathbf{T}_g \hat{\mathbf{E}} \tag{18.22}$$

The discretised form of (18.21) is computed by

$$\mathbf{R}_i = \sum_{c=1}^{C} \int_{V_c} \tilde{\mathbf{E}}\left(P_i, \bar{Q}\right) \dot{\sigma}_{0\bar{x}} \left(\bar{Q}\right) dV_c \tag{18.23}$$

Assuming that the cross-section of the bolt is small and that the initial stress does not vary across it, the initial stress at point $\bar{Q}$ is only a function of ξ i.e.

$$\dot{\sigma}_{0\bar{x}}(\bar{Q}(\xi)) = \sum_{n=1}^{2(3)} N_i(\xi) \cdot \dot{\sigma}_{0\bar{x}n}^{c} \tag{18.24}$$

Substitution into (18.23) gives

$$\mathbf{R}_i = \sum_{c=1}^{C} \int_{V_c} \tilde{\mathbf{E}}\left(P_i, \bar{Q}\right) \sum_{n=1}^{2(3)} N_n \dot{\sigma}_{0\bar{x}n}^{c} dV_c \tag{18.25}$$

or

$$\mathbf{R}_i = \sum_{c=1}^{C} \sum_{n=1}^{N} \Delta\tilde{\mathbf{E}}_{ni}^{c} \dot{\sigma}_{0\bar{x}n}^{c} \tag{18.26}$$

where

$$\Delta\tilde{\mathbf{E}}_{ni}^{c} = \int_{V_c} \tilde{\mathbf{E}}(P_i, \bar{Q}) N_n dV_c \tag{18.27}$$

If n is not node i then (18.27) may be evaluated using Gauss Quadrature

$$\Delta\mathbf{E}_{ni}^{c} = \mathrm{A} \sum_{k=1}^{K} \tilde{\mathbf{E}}\left(P_i, \bar{Q}(\xi_j)\right) \cdot N_n\left(\xi_j\right) \cdot J \cdot W_j \tag{18.28}$$

where A is the cross-sectional area of the inclusion, which is assumed constant along the cell here and K is the number of Gauss points required.
If n is node i then the Kernel becomes singular. In this case an analytical integration can be carried out[2]

$$\Delta\tilde{\mathbf{E}}_{ni} = \int_{-1}^{1} \left[\int_{A} \tilde{\mathbf{E}}(P_i, \bar{Q}) \cdot dA \right] N_n(\xi) J d\xi \tag{18.29}$$

The increment in displacement due to the body force $\{\mathbf{R}\}$ is computed by

$$[\mathrm{T}]\{\dot{\mathbf{u}}\} = \{\mathbf{R}\} \tag{18.30}$$

The increment in displacement $\dot{\mathbf{u}}$ at a cell node P_a due to an increment in displacement (computed by equation 18.30) is computed for the subsequent iteration steps as

$$\dot{\mathbf{u}}\left(P_a\right) = -\sum_{e=1}^{E} \sum_{n=1}^{N} \Delta\mathbf{T}_n^e \, \dot{\mathbf{u}}_n^e + \sum_{c=1}^{C} \sum_{n=1}^{N} \Delta\tilde{\mathbf{E}}_n^c \, \dot{\sigma}_{0\bar{x}n}^{c} \tag{18.31}$$

where

$$\Delta\tilde{\mathbf{E}}_n^c = \int_{V_c} \tilde{\mathbf{E}}(P_a, \bar{Q}) N_n dV_c \tag{18.32}$$

The iterations continue until the residual vanishes.

18.3.2 Example

The example is that of a circular excavation of radius 10 m in an infinite domain with linear inclusions. In practice this would correspond to a tunnel with fully grouted rock bolts. For the example the assumption is of an internal pressure of 15 MN/m^2. The mesh consists of 40 linear boundary elements for the boundary of the hole and of 24 rock bolts which are discretised into 2 quadratic cells each. A finer mesh with 120 linear boundary elements and 4 cells per rock bolt was also used.

Figure 18.9 Mesh used for the analysis

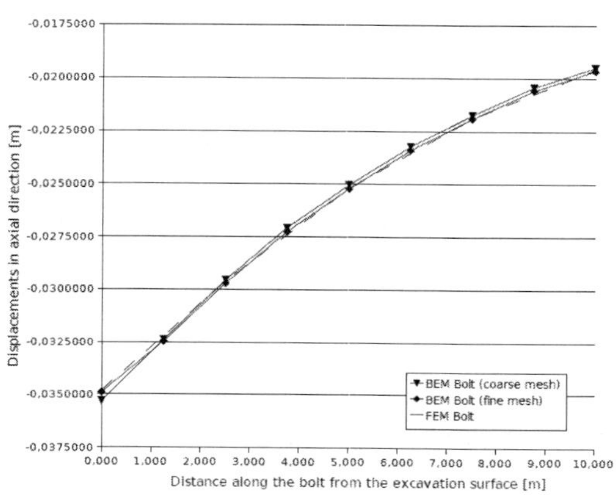

Figure 18.10 Variation of displacement along a bolt axis

For the analysis the following material properties were assumed: E_{Rock}=5 GN/m², ν=0.3, E_{Bolt}=400 GN/m². The bolts have a cross section of A_{Bolt}=0.007854m² and a length of 10m. The analysis took 22 iterations to converge and the results were compared with a finite element analysis with a very fine mesh in Fig. 18.10. In this figure the displacement along the axis of the rock bolt is shown. It can be seen that the solutions compare well with the reference solution.

18.4 PIEZO-ELECTRICITY

This is a good example of how easy it is to implement new capabilities into the general purpose program 7.1 by changing only a few lines and adding subroutines for computing a fundamental solution. This application has been worked out by a PhD student from Venezuela and the application is in piezo-electricity. Some materials react to an applied current by deforming and the deformation also causes an electric potential, so there is an interaction between electricity and deformation. These are known as piezo-electric materials. Apparently some bio-materials exhibit this property and hence the application of this work is actually bio-engineering. For a piezo-electric material we have a coupling of the stresses $\boldsymbol{\sigma}$ and the *electric displacement* **d**[3]:

$$\begin{aligned}\boldsymbol{\sigma} &= \mathbf{D}\boldsymbol{\varepsilon} + \mathbf{e}^T\mathbf{E} \\ \mathbf{d} &= \mathbf{e}\boldsymbol{\varepsilon} + \in \mathbf{E}\end{aligned} \tag{18.33}$$

We define the electric displacement **d** and the electric field vector **E** by

$$\mathbf{d} = \begin{pmatrix} d_x \\ d_y \\ d_z \end{pmatrix} \quad ; \quad \mathbf{E} = \begin{pmatrix} E_x \\ E_y \\ E_z \end{pmatrix} \tag{18.34}$$

A characteristic of piezo-electric materials is that they behave in an an-isotropic way therefore the matrix **D** is an anisotropic constitutive matrix. However, here we restrict ourselves to transversely isotropic materials and the matrix **D** presented in section 4.3.1. is used. For this case the matrix linking the stresses and the electric field vector (piezoelectric matrix) is

$$\mathbf{e} = \begin{pmatrix} 0 & 0 & 0 & 0 & e_{15} & 0 \\ 0 & 0 & 0 & e_{15} & 0 & 0 \\ e_{31} & e_{31} & e_{33} & 0 & 0 & 0 \end{pmatrix} \tag{18.35}$$

Where e_{ij} are material parameters and the matrix relating electric displacement to the field vector is given by

$$\in=\begin{pmatrix} \in_{11} & 0 & 0 \\ 0 & \in_{11} & 0 \\ 0 & 0 & \in_{33} \end{pmatrix} \tag{18.36}$$

Equations (18.32) can be combined into one by

$$\mathbf{\Sigma} = \hat{\mathbf{D}}\hat{\boldsymbol{\varepsilon}} \tag{18.37}$$

where

$$\mathbf{\Sigma} = \begin{pmatrix} \boldsymbol{\sigma} \\ \mathbf{d} \end{pmatrix} \;;\; \hat{\mathbf{D}} = \begin{pmatrix} \mathbf{D} & \mathbf{e}^T \\ \mathbf{e} & \in \end{pmatrix} \;;\; \hat{\boldsymbol{\varepsilon}} = \begin{pmatrix} \boldsymbol{\varepsilon} \\ \mathbf{E} \end{pmatrix} \tag{18.38}$$

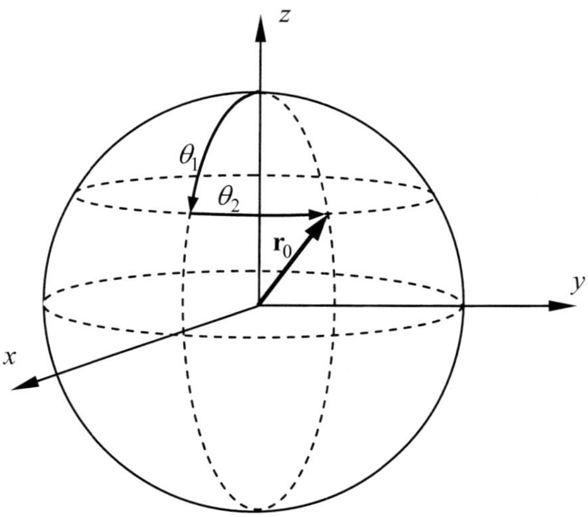

Figure 18.11 Calculation of angles θ_1, θ_2

The governing differential equation can be written as

$$\mathbf{L}\hat{\mathbf{u}} = 0 \tag{18.39}$$

where **L** is the piezo-electric differential operator[3] and

$$\hat{\mathbf{u}} = \begin{pmatrix} \mathbf{u} \\ \varphi \end{pmatrix} \tag{18.40}$$

u is the displacement vector and φ is the *electric potential* that is related to the electric field vector by

$$E_x = \varphi_{,x} \ ; \ E_y = \varphi_{,y} \ ; \ E_z = \varphi_{,z} \tag{18.41}$$

The plane problem has 2+1=3 and the 3-D problem 3+1=4 degrees of freedom. The fundamental solution of the differential equation for unit values of load and electric field can be derived using a Radon transformation [4,5]. The fundamental solution in 3-D for $\hat{\mathbf{U}}$ which combines the displacements and the electric potential is given by

$$\hat{\mathbf{U}}(P,Q) = \frac{1}{r}\mathbf{M}_U(\theta_1,\theta_2) \tag{18.42}$$

r is the distance between P and Q (length of vector **r**) and θ_1, θ_2 are determined according to Figure 18.11, where $\mathbf{r}_0$ is a unit vector in the direction $\mathbf{r}$. The complementary fundamental solution for the tractions and the electric field in direction **n** is given by

$$\hat{\mathbf{T}}(P,Q) = \frac{1}{r^2}\mathbf{M}_T(\theta_1,\theta_2) \tag{18.43}$$

Because of the complexity of the fundamental solution which would take very long for the computation of values at Gauss points of elements a scheme is adopted where a table of values of **M** as a function of θ_1, θ_2 is computed. The required particular values are then obtained by interpolation. We will see that the most difficult part of the implementation is the calculation of the fundamental solution; the other changes in the program are minimal.

18.4.1 Changes required in General_Purpose_BEM

The first change is in the input. Here we allow an additional Analysis Type (=4) in **Toa** and allow 4 degrees of freedom for the combined vector $\hat{\mathbf{u}}$. We have to allow for reading in a larger number of material constants. In addition we may provide some additional information about the size of the table to hold the pre-computed values of **M**. These are computed in the main program and stored in the arrays of rank 4 in **m_GridU** and **m_GridT**. The size of the increments of θ_1, θ_2 in the table are determined by the variables **m_step1** and **m_step2**. The only other change is in the subroutine INTEG3 where additional IF statements are included to call the subroutine that computes the fundamental solutions using the tables.

```
SUBROUTINE Integ3(Elcor,Inci,Nodel,Ncol,xPi,Ndof,E,ny,ko,dUe&
        ,dTe,Ndest,Isym,Toa,m_step1,m_step2,m_GridU,m_GridT)
!---------------------------------------------------
!    Computes  [ dT] e and [ dU] e for 3-D problems
!    by numerical integration
```

```
!-------------------------------------------------------
IMPLICIT NONE
REAL, INTENT(IN)    :: Elcor(:,:)     !   Element coordinates
INTEGER, INTENT(INOUT) :: Toa       !   Type of analysis
!  Increments for table and arrays for storing table
REAL,INTENT(INOUT) :: m_step1,m_step2
REAL,INTENT(INOUT) :: m_GridU(:,:,:,:),m_GridT(:,:,:,:,:)
......
!---------------------------------------------------------------
!     Part 1 : Pi is not one of the element nodes
!---------------------------------------------------------------
Colloc_points:DO i=1,Ncol
....
      IF(Ndof .EQ. 1) THEN
        UP= U(r,ko,Cdim) ; TP= T(r,dxr,Vnorm,Cdim)
      ELSE
        IF(Toa.EQ.3)THEN
          UP= UK(dxr,r,E,ny,Cdim) ; TP= TK(dxr,r,Vnorm,ny,Cdim)
        END IF
        IF(Toa > 3)THEN
  !  piezoelectric solution
          UP=UKa(GCcor,xPi(:,I),NDOF,1,r,m_step1,m_step2,m_GridU)
   TP=TKa(GCcor,xPi(:,I),NDOF,1,Vnorm,r,m_step1,m_step2,m_GridT)
         END IF
      END IF
......
END DO Colloc_points
!---------------------------------------------------------
!     Part 1 : Pi is one of the element nodes
!---------------------------------------------------------
Colloc_points1: DO i=1,Ncol
......
        IF(Ndof .EQ. 1) THEN
          UP= U(r,ko,Cdim) ; TP= T(r,dxr,Vnorm,Cdim)
        ELSE
         IF(Toa.EQ.3)THEN
           UP= UK(dxr,r,E,ny,Cdim) ; TP= TK(dxr,r,Vnorm,ny,Cdim)
          END IF
          IF(Toa > 3)THEN
        UP=UKa(GCcor,xPi(:,I),NDOF,1,r,m_step1,m_step2,m_GridU)
   TP=TKa(GCcor,xPi(:,I),NDOF,1,Vnorm,r,m_step1,m_step2,m_GridT)
          END IF
        END IF
.....
END DO Colloc_points1
RETURN
END SUBROUTINE Integ3
```

Further details and examples are available in [4].

18.5 CONCLUSIONS

In this chapter we have dealt with applications where a solution with the BEM was not previously possible and with applications that go beyond usual engineering problems. The proposed efficient treatment of heterogeneous material and of reinforcement is still a subject of research at the writing of the book, but preliminary results look very promising. Indeed, with a concentrated research effort sponsored by national and European funds it seems possible that the lag in the development of the BEM as compared with the FEM can be shortened.

The availability of a numerical toolbox through the book "Programming the boundary element method" by G. Beer seems to have had a positive effect on the development of the BEM and several applications in "exotic" areas emanated. One such area is piezo-electricity and here the "beauty" of the BEM is revealed: only a new fundamental solution is required to implement a completely new application. Try this with finite elements. Unfortunately the work of the PhD student was just starting at the writing of the paper and we hope to present some interesting applications of this in the area of bio-mechanics (increased healing potential of human bones when subjected to low voltage electricity) in a second edition.

18.6 REFERENCES

1. Riederer K, Prazeres P.G. and Beer G. (2007) Numerical modeling of ground support with the boundary element method. *ECCOMAS Thematic Conference on Computational Methods in Tunneling*, EURO:TUN 2007.
2. Riederer K. (2009) PhD Dissertation, Graz University of Technology, Austria.
3. Gaul L., Kögl M. and Wagner M. (2003) Boundary Element Method for Engineers and Scientists. Springer, Berlin.
4. Duarte V. (2009) PhD Thesis, Universidad Central de Venezuela, Caracas, Venezuela.
5. Thoeni K. Effiziente Berechnung anisotroper Fundametallösungen für die Methode der Randelemente. Diploma Thesis, Graz University of Technology, Austria.

Appendix
Fundamental Solutions

Supplied by Tatiana Ribeiro

The fundamental solutions presented for static elasticity are in indicial notation. Please refer to Chapter 1 for the correlation between indicial and vector notation.

A.1. DISPLACEMENT SOLUTION

The displacement u_i at an internal point P is computed by

$$u_i(P) = \int_S U_{ij}(P,Q)t_j(Q)dS - \int_S T_{ij}(P,Q)u_j(Q)dS + \int_V E_{ijk}(P,\bar{Q})\sigma_{0jk}(\bar{Q})dV + \int_V \Sigma_{ijk}(P,\bar{Q})\varepsilon_{0jk}(\bar{Q})dV \tag{A.1}$$

Where u_i, t_i are the displacements and tractions and $\sigma_{0jk}, \varepsilon_{0jk}$ are initial stresses and strains. The fundamental solutions are given for plane problems by

$$U_{ij}(P,Q) = C\left(C_1 \ln\frac{1}{r}\delta_{ij} + r_{,i}r_{,j}\right) \tag{A.2}$$

and in 3D by

$$U_{ij}(P,Q) = \frac{C}{r}\left(C_1\delta_{ij} + r_{,i}r_{,j}\right) \tag{A.3}$$

For plane as well as for 3-D problems we have

$$T_{ij}(P,Q) = \frac{-C_2}{r^n}\left[\left(C_3\delta_{ij} + (n+1)r_{,i}r_{,j}\right)\cos\theta - C_3\left(n_j r_{,i} - n_i r_{,j}\right)\right] \tag{A.4}$$

$$E_{ijk}\left(P,\bar{Q}\right)=\frac{-C}{r^{n}}\left[C_{3}\left(r_{,k}\delta_{ij}+r_{,j}\delta_{ik}\right)-r_{,i}\delta_{jk}+C_{4}r_{,i}r_{,j}r_{,k}\right] \tag{A.5}$$

$$\Sigma_{ijk}\left(P,\bar{Q}\right)=\frac{C_{2}}{r^{n}}\left[C_{3}\left(r_{,k}\delta_{ij}+r_{,j}\delta_{ik}\right)-C_{5}r_{,i}\delta_{jk}+C_{4}r_{,i}r_{,j}r_{,k}\right] \tag{A.6}$$

where r is the distance between the source point P and the field point Q and n_i the outward normal. The derivative of r with respect to the Cartesian axis j is given by $r_{,j}$. The term $\cos\theta$ is computed by

$$\cos\theta=\frac{1}{r}\mathbf{r}\bullet\mathbf{n} \tag{A.7}$$

and the values for the constants are given in Table A.1

Table A.1 Constants for fundamental solutions

	Plane strain	Plane stress	3-D
n	1	1	2
C	1/8πG(1−ν)	(1+ν)/8πG	1/16πG(1−ν)
C_1	3-4 ν	(3- ν)/(1+ ν)	3-4 ν
C_2	1/4π (1−ν)	(1+ν)/4π	1/8π (1−ν)
C_3	1-2 ν	(1- ν)/(1+ ν)	1-2 ν
C_4	2	2	3
C_5	1	(1- ν)/(1+ ν)	1-2 ν
C_6	4	4	5
C_7	1-4 ν	(1-3 ν)/(1+ ν)	1-4 ν
C_8	-1/8(1−ν)	-(1+ν)/8	-1/15(1−ν)
C_9	3-4 ν	(3- ν)/(1+ ν)	4-5 ν
C_{10}	1	(1-3 ν)/(1+ ν)	1-5 ν
C_{11}	-1/16G(1−ν)	-(1+ν)/16G	-1/30G(1−ν)
C_{12}	1	1	7-5 ν
C_{13}	1-4 ν	(1-3 ν)/(1+ ν)	2-10 ν
C_{14}	1-2 ν	(1-3 ν)/(1+ ν)	1-4 ν
C_{15}	1	(1- ν)/(1+ ν)	1-2 ν
C_{16}	G/4(1- ν)	G(1+ ν)/4	G/15(1- ν)
C_{17}	1-4 ν	1	2+10 ν
C_{18}	-1/8(1- ν)	-(1+ ν)/8	-1/30(1- ν)
C_{19}	G/2π (1−ν)	(1+ ν)G/2π	G/4π (1−ν)

A.2. STRAIN SOLUTION

The strain tensor at an internal point is computed by

$$\begin{aligned}\varepsilon_{ij}(P) &= \int_S \bar{S}_{ijk}(P,Q)t_k(Q)dS - \int_S \bar{R}_{ijk}(P,Q)u_k(Q)dS \\ &+ \int_V \bar{\Sigma}_{ijkl}\left(P,\bar{Q}\right)\varepsilon_{0kl}\left(\bar{Q}\right)dV + \bar{H}_{ijkl}\varepsilon_{0kl}(P) + \int_V \bar{E}_{ijkl}\left(P,\bar{Q}\right)\sigma_{0kl}\left(\bar{Q}\right)dV + \bar{F}_{ijkl}\sigma_{0kl}(P)\end{aligned} \tag{A.8}$$

where

$$\bar{S}_{ijk}(P,Q) = \frac{C}{r^n}\left[C_3\left(r_{,j}\delta_{ik} + r_{,i}\delta_{jk}\right) - r_{,k}\delta_{ij} + C_4 r_{,i}r_{,j}r_{,k}\right] \tag{A.9}$$

$$\bar{R}_{ijk}(P,Q) = \frac{C_2}{r^{n+1}}\left\{\begin{aligned}&C_4\cos\theta\left[\nu\left(r_{,j}\delta_{ik} + r_{,i}\delta_{jk}\right) + r_{,k}\delta_{ij} - C_6 r_{,i}r_{,j}r_{,k}\right] \\ &+C_3\left(n_j\delta_{ik} - n_k\delta_{ij} + n_i\delta_{jk} + C_4 r_{,i}r_{,j}n_k\right) + C_4\nu\left(n_j r_{,i}r_{,k} + n_i r_{,j}r_{,k}\right)\end{aligned}\right\} \tag{A.10}$$

$$\begin{aligned}\bar{\Sigma}_{ijkl}\left(P,\bar{Q}\right) = \frac{C_2}{r^{n+1}}\Big[&C_3\left(\delta_{ik}\delta_{jl} + \delta_{jk}\delta_{il} - \delta_{ij}\delta_{kl}\right) \\ &+ C_4\nu\left(\delta_{il}r_{,j}r_{,k} + \delta_{jk}r_{,i}r_{,l} + \delta_{ik}r_{,j}r_{,l} + \delta_{jl}r_{,i}r_{,k}\right) \\ &+C_4\left(C_3\delta_{kl}r_{,i}r_{,j} + \delta_{ij}r_{,k}r_{,l} - C_6 r_{,i}r_{,j}r_{,k}r_{,l}\right)\Big]\end{aligned} \tag{A.11}$$

$$\begin{aligned}\bar{E}_{ijk}\left(P,\bar{Q}\right) = \frac{C}{r^{n+1}}\Big[&C_4\nu\left(\delta_{il}r_{,j}r_{,k} + \delta_{jk}r_{,i}r_{,l} + \delta_{ik}r_{,j}r_{,l} + \delta_{jl}r_{,i}r_{,k}\right) \\ &+ C_4\left(\delta_{kl}r_{,i}r_{,j} + \delta_{ij}r_{,k}r_{,l} - C_6 r_{,i}r_{,j}r_{,k}r_{,l}\right) \\ &+ C_3\left(\delta_{ik}\delta_{jl} + \delta_{jk}\delta_{il}\right) - \delta_{ij}\delta_{kl}\Big]\end{aligned} \tag{A.12}$$

$$\bar{F}_{ijkl} = C_{11}\left[\delta_{jk}\delta_{il} - C_9\left(\delta_{ik}\delta_{jl} + \delta_{ij}\delta_{kl}\right)\right] \tag{A.13}$$

$$\bar{H}_{ijkl} = C_8\left[C_9\left(\delta_{jk}\delta_{il} + \delta_{ik}\delta_{jl}\right) - C_{10}\delta_{ij}\delta_{kl}\right] \tag{A.14}$$

where the constants are given in Table A.1 .

A.3. STRESS SOLUTION

The stress tensor for an internal point is computed by

$$\sigma_{ij}(P) = \int_S S_{ijk}(P,Q)t_k(Q)\,dS - \int_S R_{ijk}(P,Q)u_k(Q)\,dS$$
$$+ \int_V \hat{E}_{ijkl}(P,\bar{Q})\sigma_{0kl}(\bar{Q})\,dV + F_{ijkl}\sigma_{0kl}(P) + \int_V \hat{\Sigma}_{ijkl}(P,\bar{Q})\varepsilon_{0kl}(\bar{Q})\,dV + H_{ijkl}\varepsilon_{0kl}(P) \tag{A.15}$$

where

$$S_{ijk}(P,Q) = \frac{C_2}{r^n}\left[C_3\left(\delta_{ik}r_{,j} + \delta_{jk}r_{,i} - \delta_{ij}r_{,k}\right) + C_4 r_{,i}r_{,j}r_{,k}\right] \tag{A.16}$$

$$R_{ijk}(P,Q) = \frac{C_{19}}{r^{n+1}}\left\{C_4\cos\theta\left[C_3 r_{,k}\delta_{ij} + \nu\left(r_{,j}\delta_{ik} + r_{,i}\delta_{jk}\right) - C_6 r_{,i}r_{,j}r_{,k}\right]\right.$$
$$\left. + C_3\left(n_j\delta_{ik} + n_i\delta_{jk} + C_4 r_{,i}r_{,j}n_k\right) - C_7 n_k\delta_{ij} + C_4\nu\left(n_j r_{,i}r_{,k} + n_i r_{,j}r_{,k}\right)\right\} \tag{A.17}$$

$$\hat{E}_{ijkl}(P,\bar{Q}) = \frac{C_2}{r^{n+1}}\left[C_3\left(\delta_{ik}\delta_{jl} + \delta_{jk}\delta_{il} - \delta_{ij}\delta_{kl} + C_4\delta_{ij}r_{,k}r_{,l}\right)\right.$$
$$\left. + C_4\nu\left(\delta_{il}r_{,j}r_{,k} + \delta_{jk}r_{,i}r_{,l} + \delta_{ik}r_{,j}r_{,l} + \delta_{jl}r_{,i}r_{,k}\right) + C_4\left(\delta_{kl}r_{,i}r_{,j} - C_6 r_{,i}r_{,j}r_{,k}r_{,l}\right)\right] \tag{A.18}$$

$$\hat{\Sigma}_{ijkl}(P,\bar{Q}) = \frac{C_{19}}{r^{n+1}}\left\{C_3\left(\delta_{ik}\delta_{jl} + \delta_{jk}\delta_{il}\right) - C_{14}\delta_{ij}\delta_{kl}\right.$$
$$+ C_4\nu\left(\delta_{il}r_{,j}r_{,k} + \delta_{jk}r_{,i}r_{,l} + \delta_{ik}r_{,j}r_{,l} + \delta_{jl}r_{,i}r_{,k}\right)$$
$$\left. + C_4\left[C_3\delta_{kl}r_{,i}r_{,j} + C_{15}\delta_{ij}r_{,k}r_{,l} - C_6 r_{,i}r_{,j}r_{,k}r_{,l}\right]\right\} \tag{A.19}$$

$$F_{ijkl} = C_{18}\left[C_{12}\left(\delta_{jk}\delta_{il} + \delta_{ik}\delta_{jl}\right) + C_{13}\delta_{ij}\delta_{kl}\right] \tag{A.20}$$

$$H_{ijkl} = C_{16}\left[C_{12}\left(\delta_{ik}\delta_{jl} + \delta_{jk}\delta_{il}\right) + C_{17}\delta_{ij}\delta_{kl}\right] \tag{A.21}$$

where the constants are shown in Table A1.